THE RISE OF SCIENTIFIC EUROPE
1500 - 1800

THE RISE OF SCIENTIFIC EUROPE
1500 -1800

Edited by David Goodman and Colin A. Russell

Hodder & Stoughton  The Open University

First published in 1991 by

Hodder & Stoughton Limited

Mill Road, Dunton Green

Sevenoaks, Kent

British Library Cataloguing-in-Publication Data

The rise of scientific Europe 1500–1800,

      I. Goodman, David C.   II. Russell, Colin A.

      500.9402

      ISBN 0-340-55861-X

Front cover illustration: The Armillary Sphere 1543, Caspar Medebach. (Nationalmuseet, Copenhagen. Photo: Kit Weiss.)

Back cover illustration: From Johann von Staden, Wahrhattige Historia und Beschreibung Eyner Landtschafft der Wilden, *1557. (Reproduced by permission of the British Library Board.)*

Edited, designed and typeset by The Open University.

This book forms part of an Open University course AS283 *The Rise of Scientific Europe, 1500–1800.* For information about this course please write to the *Central Enquiry Service,* The Open University, PO Box 200, Walton Hall, Milton Keynes, UK, MK7 6YZ.

Printed and bound in the United Kingdom for the educational publishing division of Hodder & Stoughton Ltd, by Thomson Litho Ltd, East Kilbride, Scotland.

Contents

Preface

One of the most remarkable features of modern European history is the emergence in the last 400 years or so of that body of experimental practice and theoretical reasoning that today we call science. That in itself poses a fascinating question: why did science as we understand it arise in Europe and not, say, in China or the Middle East? And there is the further question why science prospered more in some countries of Europe rather than in others – England more than Spain, Sweden more than Russia – in this period?

Such questions amongst others are addressed in this book. Rather than being a history of European science it seeks to be a history of scientific Europe, of its institutions as well as its achievements, of public attitudes as well as private attainment. It is not so much a history of scientific discoveries as of scientific attitudes, though all the major advances are of course included. So, for the critical years 1500–1800, the growth of scientific culture is examined in countries as far apart as Italy and Sweden, Russia and England.

This book has been produced for Open University students taking the second-level undergraduate course entitled *AS283: The Rise of Scientific Europe 1500–1800*; but we believe that it will also be of interest to anyone interested in the development of modern science.

David Goodman

Colin A. Russell

Europe's Awakening *Chapter 1*

by David Goodman

1.1 Problems and approaches

The success of Europe continues to fascinate and challenge professional historians. There has been a spate of books with such titles as *The Rise of the West*, *The Triumph of the West* and *Europe and the Rise of Capitalism*. And two recent studies, *The Military Revolution: Military Innovation and the Rise of the West, 1500–1800* and *The Tools of Empire*, investigate how Europe was able to achieve military supremacy over the rest of the world.[1]

None of these works is the product of European jingoism or of that old-fashioned and discredited Whig historiography which used the past to glorify the present. Instead they are serious, professional and entirely legitimate attempts to analyse and explain some fundamental and distinctive aspects of Europe's history. Europe *was* unique in generating that large-scale economic development which we call capitalism. Europe *alone* experienced the establishment of representative institutions of government. And no other part of the world engaged in an enterprise like the European exploration which led to the discovery of new continents and eventually to that colonization of the world which persisted into the twentieth century.

Europe was also the nursery of modern science, a revolution regarded by a former professor of modern history at Cambridge as of such importance that 'it outshines everything since the rise of Christianity and reduces the Renaissance and Reformation to the rank of mere episodes' (Butterfield, 1957, p.vii). The idea of a European Scientific Revolution seems now to be generally accepted, but opinions differ on what that revolution amounted to. Was it a new view of nature overall, a new set of superior theories, or the perfection of scientific method? There is also disagreement on when the revolution began: a few try to trace it back to the Middle Ages; more look to the early modern period. But all seem to accept that something special occurred in the seventeenth century through the work of such famous scientists as Galileo and Isaac Newton. For one historian their achievement clearly meets his idea of a Scientific Revolution: 'the overthrow of an entrenched orthodoxy; challenge, resistance, struggle and conquest ... grandeur of scale and urgency of tempo. Also some consciousness of a revolution afoot' (Porter, 1986, p.300).

A principal purpose of this book is to investigate another fundamental and largely neglected question relating to the European Scientific Revolution: Why did it occur where and when it did? Why Europe and not Asia? Remarkable signs of promise had occurred, for example, in India, where in the ninth century mathematicians were 'solving isolated problems that in the West had to await the time of Newton' (Hall, 1983, p.30). Too little is yet known of Indian science to attempt an extensive comparison. Much more

---

[1] The authors are respectively W. McNeill (1963); J.M. Roberts (1985); J. Baechler, J. Hall and M. Mann (eds) (1988); G. Parker (1988); D. Headrick (1981).

work has been done on the history of Chinese and Islamic science to justify some exploratory comparisons. But the question also arises within Europe itself: why were the scientific developments concentrated in certain parts of Europe, above all in the west? There are today signs of a growing recognition amongst historians of the importance of a comparative approach. In this book our aim is to throw light on the emergence of modern science in Europe, above all through comparisons between different countries within Europe; some general comparisons between Europe, China and the lands of Islam will also be considered. Comparisons alone can indicate what is common to two different civilizations or regions, and reveal what is distinctive.

Should we define what we mean here by 'science'? There is something to be said for leaving that aside, for the reasons given at the beginning of one general work on the history of science:

> Science throughout is taken in a very broad sense and nowhere do I attempt to cramp it into a definition. Indeed science has so changed its nature over the whole range of human history that no definition could be made to fit. *(Bernal, 1954, vol.1, p.3)*

But because there is no agreed definition of science some indication is needed of the subject-matter of this book. It certainly includes the definition of science given by a distinguished historian of science in terms of knowledge of the external world:

> Science comprises, first, the orderly and systematic comprehension, description, and/or explanation of natural phenomena, and secondly, the tools necessary for that undertaking. *(Clagett, 1963, p.15)*

That definition leads naturally to a concentration on 'the activities and writings of learned men' (Hall, 1983, p.1). Our scope is broader than this because we are concerned also to consider the wider social and political context of scientific development. Therefore all contributors to this book have had to reflect at some stage on very different views of science, ranging from that of the philosopher of science who believes that the 'fundamental feature of science is its ideal of objectivity, an ideal that subjects all scientific statements to the test of impartial criteria, recognizing no authority of persons in the realm of cognition' (Scheffler, 1967, p.1), to the totally contrasting belief of the sociologist of science who supports the position that science is a 'sub-system of society' and that 'not only error or illusion or unauthenticated belief but also the discovery of truth [is] socially conditioned' (Merton, 1973, p.11). The second of these carries the focus away from great individual scientists and on to social groups, social and scientific institutions, and climates of opinion, all of which are important to the aims of this book.

Behind these various definitions of science lie contending models of the origin and development of the sciences. Some of these have been schematized in Figure 1.1.

Model 1 goes all the way back to classical Greece; Aristotle believed that no science was possible until society secured an agricultural surplus, freeing a section of the population from manual labour and permitting the leisure essential for intellectual activity. Model 2 has perhaps been frequently assumed in modern philosophy. Popper regards science as a refined form of common sense, so focussing on a purely internal development of the history of thought. Model 3 is rather more complicated and could raise the question of the origins of religion; in this book some chapters will consider the way religious beliefs may have stimulated particular views of nature or

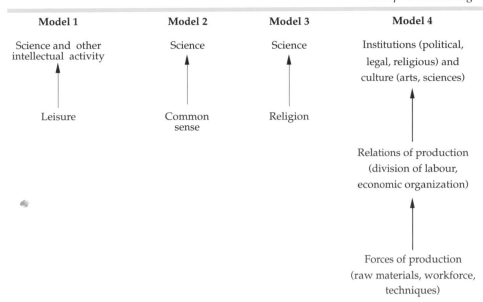

Model 1	**Model 2**	**Model 3**	**Model 4**

Figure 1.1 *Four models of the origins and development of science*

encouraged the investigation of the cosmos through natural science. The fourth model (this list is not of course exhaustive) is the most complex and represents the materialistic theory of Karl Marx, though his views on the subject are not entirely clear. It is easier to describe the crude form of the model as present in the less subtle writings of Marx's disciples. In these the sciences, like all culture, are deterministically produced from purely material causes. Marx himself cannot be dismissed so simply, because he seems to have accepted the independent role of non-material human creativity in social development, and included existing scientific knowledge within his base category of forces of production. Nor is it entirely clear how the arrows between the levels of the model are to be interpreted; some say they mean no more than 'conditions', not 'causes'. While the Marxist model is open to crude misinterpretation it has the attraction of spinning the sciences into the web of society as a whole, and on occasion this type of model may assist our discussion.

This book does not pretend to be another history of science, setting out to provide a comprehensive history of scientific ideas. But the chief features of the development of European science are described within a comparative framework. Scientific developments in different countries and at different times are compared and contrasted to discern the effects of varying social, political and intellectual conditions. Various explanations of scientific growth will be examined, and tentative conclusions proposed for the emergence of modern science in Europe, especially in Western Europe, in the period 1500–1800. Large historical problems do not receive immediate, definitive solutions – only gradual illumination.

1.2 The Greek miracle

When historians speak of the Scientific Revolution they are never referring to that first great burst of European scientific activity which occurred in ancient Greece in 600 BC–200 AD. Yet that remarkable development was a revolution of sorts. Compared to what had been achieved before in the neighbouring civilizations of the Near East, the autocratic monarchies of Mesopotamia and

Egypt, the ancient Greeks brought the sciences to an entirely new level of development. In fact Greek science already contains practically all of the ingredients of modern science.

At the centre of the Greek achievement was the belief that the universe was based on order, and that regularities, the laws of nature, were discoverable by the intellectual faculty which most sharply distinguished humans from animals: reason. The very name which the Greeks gave to the world, *cosmos*, means 'order'. The importance given to the rational explanation of natural phenomena, a new and quite explicit programme of research, is evident throughout the work of Aristotle (384–322 BC), where it had matured into an elaborate doctrine of cause and effect. And it was Aristotle who confidently distinguished two very different styles of understanding the world: the old inferior resort to myths and the more recent explanation by natural causes.

What was the source of this Greek emphasis on the rational investigation of nature? Several classical scholars currently believe that it may have originated in the new political environment of the Greek city-state, 'especially the experience of radical political debate and confrontation in small-scale, face-to-face societies' (Lloyd, 1979, p.266). This is how the argument runs. In Athens and other Greek city-states where democracy was developed, the entire adult male citizenry participated in political debate, discussing the most important matters of public policy and even the constitution itself. To persuade voters or to defeat an opponent's political argument the Greeks developed logical thinking. They then applied this to investigate the most fundamental questions of the world of nature. The argument may be too strong; the political and scientific discussions may be simply two facets of the same intellectual development. It is conceded that a weakness of this explanation is that the earliest Greek rationalistic cosmologies pre-date the rise of the city-state. But some interconnections of Greek political life and scientific mentality are clear. For example the Greek physician Alcmaeon of Croton (fifth century BC) believed that health depended on an equal balance of opposing forces in the body, which he called 'the cold and the hot', 'the moist and the dry' and 'the bitter and the sweet'. Disease arose whenever equality was destroyed and one force predominated over its opposite. The political imagery may be already apparent. Alcmaeon made it explicit, describing the healthy condition as an equality of rights and the imbalance of disease as despotic monarchy. The contrast was natural for a citizen of a democratic city-state, for whom despotism and tyranny were the greatest threats. At any rate the environment of the small Greek city-state, so different from the vast centralized monarchies of the Near East, provides us with a focus for approaching the problem of distinctive scientific developments which can be taken up again in considering early modern Europe.

The uncompromising Greek belief in discoverable regularities in the natural world led to scientific ideas which would continue for centuries to direct mediaeval and early modern research. It was Plato (427–347 BC) who invited astronomers to discover the perfect circular motions which he was convinced lay beneath the apparent irregularities in the perceptible movements of celestial bodies. And it was the Pythagoreans, perhaps Pythagoras himself (sixth century BC), who generated the fascinating image of the universe structured on 'harmony', a theory founded on musical discoveries, probably with a one-stringed instrument, which demonstrated that pleasing music was based on the simplest of arithmetical proportions. When a string of a certain length was plucked, and then one of half that length, the consonance of an octave was produced. By extension the Pythagoreans argued that harmony existed throughout the world, expressible in terms of these simple numerical ratios. They also developed this arithmetical outlook into an atomic theory, presenting the world as made up of invisible unit particles of matter.

Reason, rigorous logic and the whole idea of proof, especially the use of deduction from assumed first principles, or axioms, was one of the principal scientific legacies of the ancient Greeks. This powerful tool was used for the demonstrations and proofs, the connected chains of theorems in Euclid's (c. 300 BC) geometry, and for the *reductio ad absurdum* (the falsification of propositions by deducing absurdities consequent on their temporary acceptance), used to such effect by Archimedes (287–212 BC). Deductive logic was also the tool for physical sciences, notably in optics, where by treating rays of light as straight lines, Euclid and others reduced the science of light to a department of geometry. And Greek astronomy, a science in which the Greeks achieved some of their greatest success, became predominantly a matter of geometry.

But the Greek scientific legacy was also empirical. Aristotle's meticulous observational work in the life sciences is still admired today. And there are impressive examples of the Greek use of experimentation joined to mathematics, so characteristic of modern science. Consider the investigation of the development of the chick embryo performed by physicians associated with Hippocrates (late fifth century BC). They put twenty eggs under hens, and every day up to the day of hatching broke open an egg to observe the changing forms within. Some of the finest experiments were performed during the last phase of Greek science, in the second century AD. Claudius Ptolemy (*fl.* 127–141), most famous for his enormously influential geocentric system of astronomy, also carried out an experiment in optics which clearly involved a successful attempt to correlate variables (see Figure 1.2). Equally impressive physiological experiments were performed by Galen of Pergamum (129–c. 200). Working with animals he investigated the effects of cutting the spinal cord between different vertebrae, observing the varying degrees of paralysis which resulted. His many other experiments include the use of ligatures to establish that the function of the ureters was to conduct urine from the kidneys.

Ptolemy lived and worked, probably throughout his life, in Alexandria; Galen travelled widely throughout the Graeco-Roman world, and visited Alexandria, which had long replaced Athens as the chief centre of Greek culture. Since the third century BC this capital of Greek Egypt had been notable as a place of patronage for scholars, and conspicuously for men of science. This was no city-state, but the centre of one of the autocratic monarchies into which the empire of Alexander the Great had split. Greek Egypt was ruled by the dynasty of the Ptolemies, who were anxious to recruit scholars from all over the Greek world to bring fame and prestige to their city and above all to themselves.

The advantages which this brought to scientific research are illuminating. Through crown patronage the great anatomist Herophilus of Chalcedon (third century BC) was provided with funds and permission to dissect the bodies of criminals from the royal prisons. This resulted in notable advances in the knowledge of the brain: Herophilus was able to identify the principal chambers of the brain, describing the cerebrum and cerebellum. He also distinguished the motor and sensory nerves and, in sharp contrast to Aristotle, made the brain instead of the heart the centre of intellectual activity. This and other medical research in Alexandria was associated with an institution of the Crown, the Museum. There had been earlier scholarly

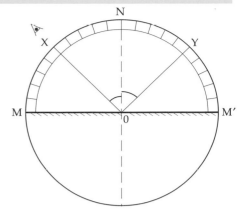

Figure 1.2 Ptolemy's apparatus consisted of a copper disc graduated in a scale of degrees, a thin bar of iron polished into a reflecting mirror MM', and a type of sighting tube, placed along a position such as XO. With his eye looking along XO he moved a coloured marker over the surface of the copper disc until at the position Y its reflected image became visible through the sighting tube. His measurements showed that the angle NOX was always equal to the angle NOY, an experimental demonstration of the law of reflection.

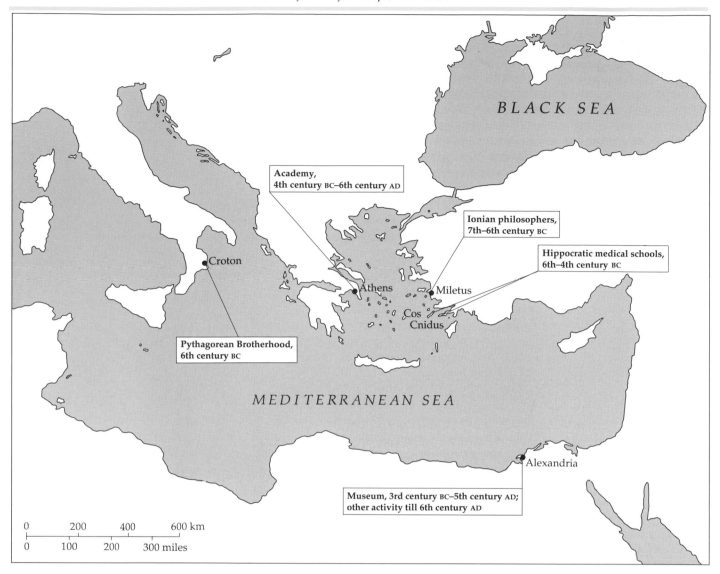

Academy,
4th century BC–6th century AD

Ionian philosophers,
7th–6th century BC

Hippocratic medical schools,
6th–4th century BC

Croton

Athens

Miletus

Cos

Cnidus

Pythagorean Brotherhood,
6th century BC

BLACK SEA

MEDITERRANEAN SEA

Alexandria

Museum, 3rd century BC–5th century AD;
other activity till 6th century AD

| 0 | 200 | 400 | 600 km |
| 0 | 100 | 200 | 300 miles |

Map 1.1 Chief centres of Greek science.

institutions in Athens – Plato's Academy and Aristotle's Lyceum – where communities of scholars came together, chiefly for teaching purposes. Plato's Academy emphasized mathematics; the Lyceum, because of the founder's influence, was strong in biological sciences. Both the Academy and the Lyceum also had religious functions as centres for the worship of the muses. Aristotle's Lyceum seems to have been influential in the formation of Alexandria's Museum. The idea of collaborative scientific research which was practised at the Academy became pronounced at the Museum, and it was funded by the Crown along with its teaching by public lectures. Apart from the medical research the details of exactly what was taught are not known. The stimulus given by royal patronage to the sciences at Alexandria was also evident in the creation of the great Library, which according to one contemporary housed half a million papyri. Eratosthenes (third century BC), famous for his measurement of the circumference of the earth, was appointed by Ptolemy Euergetes I to be the librarian of this institution.

Greek scientific activity did not last. Why did it come to an end? The answer is not easily given, but the most likely reason seems to be the paucity of scientists and their isolation. The Athenian institutions and the Alexandrian Museum were rarities. Over the Greek centuries most scientific activity was uncoordinated, and poor communications often resulted in duplication of effort or ignorance of what had been achieved. Education in Greek schools concentrated on music, poetry and gymnastics, not on science; with the exception of Alexandria there was no strong government or social encouragement of the sciences.

For Europe to have developed the sciences further from these Greek foundations, knowledge of Greek, close contact with Greek scientific texts and sustained interest in what they might teach were all necessary. But in the centuries after the fall of the Roman Empire in the West, none of these conditions was satisfied. For centuries the part of the world which held most promise of scientific advance was not Greece, nor anywhere else in the West, but in the Far East – China.

1.3 The stimulus of Chinese science

1.3.1 Contacts with the West

At the other end of the world lay the vast and populous land mass of China. Remote though it was from the West, there had been occasional contacts even with the ancient Greeks: in the second century AD a Greek merchant called Alexander sailed along the coast of Indo-China. And later in the same century the Roman Emperor Marcus Aurelius dispatched a trading mission by land which communicated directly with Huan-ti, the emperor of China. Archaelogical discoveries of Roman coins and a bronze lamp in the Gulf of Tongking and the Mekong Delta confirm the existence of trading links between China and the classical Mediterranean world. The opportunity for Chinese contacts with Western culture increased in the Middle Ages through the intermediary of the expanding world of Islam. The Chinese Empire had reached its furthest extension in the sixth century AD, incorporating the region of Fergana, close to Samarkand. It was in this region at this time that the Muslim Empire, expanding north-east from Persia, encountered the Chinese and defeated them. And later, direct contact between China and the West was achieved by the overland travels of the thirteenth-century Venetian Marco Polo.

All of these contacts provided the Chinese with opportunities to become acquainted with the culture of classical antiquity; yet the evidence indicates a lack of interest by the Chinese in Greek science. Not till the seventeenth century, when Jesuit missionaries became active and influential in China, did the Chinese become familiar with Euclidean geometry and Ptolemaic astronomy. Before that occurred Chinese science had developed in its own way, apparently unassisted by any Greek influence, with quite remarkable results.

1.3.2 Achievements

The environment of Chinese scientific activity was very different from the world of ancient Greece. China had never experienced the rise of city-states. Instead there had been a millenium of warring feudal dynasties (from *c.* 1500 BC) which came to an end with the first unification of all China under a single emperor in 221 BC. After that the pattern was one of alternating disintegration into a few huge regional states and imperial reunification, a pattern which

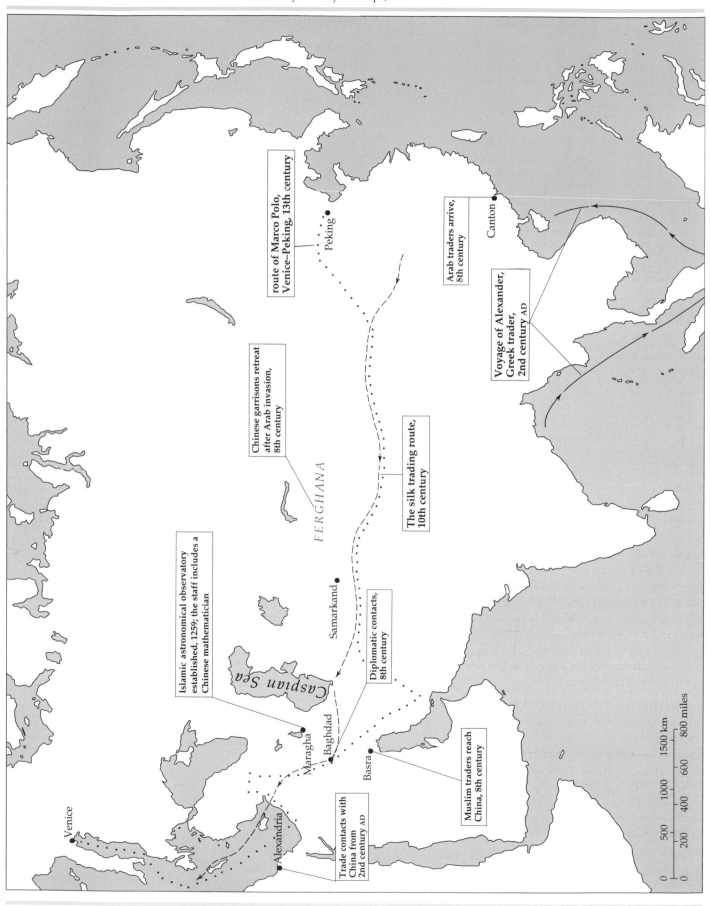

route of Marco Polo, Venice–Peking, 13th century

Arab traders arrive, 8th century

Voyage of Alexander, Greek trader, 2nd century AD

Chinese garrisons retreat after Arab invasion, 8th century

The silk trading route, 10th century

Islamic astronomical observatory established, 1259; the staff includes a Chinese mathematician

Diplomatic contacts, 8th century

Muslim traders reach China, 8th century

Trade contacts with China from 2nd century AD

Peking

Canton

FERGHANA

Samarkand

Caspian Sea

Maragha

Baghdad

Basra

Venice

Alexandria

1500 km					
800 miles					
0	200	400	600	1000	
0		500			

persisted into the twentieth century. One of the recurring characteristics of Chinese science over this very long period is the repeated patronage given by emperors to men of science.

Only now are we beginning to appreciate the magnitude of the Chinese achievement in science and technology – that vast field of engineering and invention which is sometimes entirely independent of theoretical science and in other cases so intimately connected with it that the two cannot be separated. It is above all Joseph Needham's pioneering, mammoth multi-volume *Science and Civilisation in China* (1954–) which has given the Western world a much fuller knowledge of this neglected area.

In its long and continuous cultural history – China seems to have had no 'Dark Ages' – two periods stand out: a philosophical period, of importance for theories about nature, dating roughly from 600 to 200 BC and therefore coinciding with the centuries of Greek philosophical and scientific thought; and a later period of peak scientific achievement under the Sung emperors in the eleventh and twelfth centuries.

The overall picture is one of precocious scientific and technological activity, carrying China far ahead of any knowledge existing in contemporary Europe. Paper, printing (the earliest trials date from the sixth century AD), gunpowder (ninth century AD) and the magnetic compass (eleventh century AD) have long been accepted as world-transforming inventions originating in China. To these Needham adds the following: the production of cast-iron tools (fourth century BC), 'eighteen centuries before Europe'; the stern-post rudder in ships (first century AD), well over a thousand years before its introduction in Europe; the use of wrought-iron chains for suspension bridges (*c*. 600 AD); and the mechanical clock (eighth century AD), again centuries before its use in the West.

Of these, the magnet has the closest connections with physical science. It seems that the Chinese interest in magnetism developed from a type of magic. As early as the first century AD Chinese diviners were operating with a spoon-shaped magnet which was thrown onto a bronze board marked with representations of the Earth and celestial bodies; and they knew that the spoon always orientated in a north–south axis. In the eleventh century this familiar orientating property of the magnet was exhibited apparently for the first time by suspending, pivoting or floating an iron needle in the form of a fish. One of the greatest of Chinese scientists Shen Kua (1031–95) not only used these pivoted magnets but possessed a sense of accuracy in observation which led him to discover the phenomenon of magnetic declination – the fact that a magnetic needle aligns itself a few degrees east or west of the geographical north–south line.

By the early twelfth century these magnets were being used by the Chinese for navigation, and later in the same century by Italian ships in the Mediterranean. Needham argues plausibly that this Chinese success in magnetism may be connected with the marked preference in their scientific thinking for a type of field theory: the action of one body on another at a distance through intervening space. By contrast the alternative type of explanation of natural phenomena, postulating underlying physical contact or collisions between atoms or other particles of matter, is far less evident in Chinese than in Western science.

Observational astronomy was one of the strongest developments in the Chinese scientific tradition. This received patronage from the emperors because of their interest in the astrological fortunes of their dynasty, and for

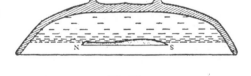

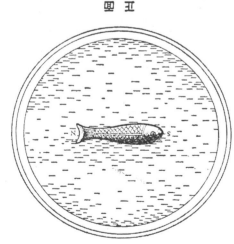

Figure 1.3 Floating iron magnet. (From R. Dawson, The Legacy of China, *Oxford, Clarendon Press, 1964, p.253.)*

Map 1.2 Contacts with China. (facing page)

Figure 1.4 Reconstruction of the great astronomical clock erected by Su Sung, Han Kung-lien and their collaborators at the Imperial Palace, Kaifeng, Homan province, in AD 1088–92. The water-wheel rotated a celestial globe and an armillary sphere, and also an elaborate series of jack figures which announced the time without any dial clock-face. As the sphere was used for observational purposes, this was the first of all clock-drives, such as are used for modern telescopes. (From R. Dawson, The Legacy of China, *Oxford, Clarendon Press,1964, p.262.)*

its importance in the regulation of the official calendar. Chinese astronomy was a function of the imperial civil service. The result was not just the production of accurate star catalogues, but that indispensable feature of observational astronomy, continuity of observation over centuries, to a degree probably unequalled by any other civilization.

The earliest known records of sunspots came from China in the first century BC, fifteen hundred years before Europe, and by 600 AD Chinese astronomers had discovered that the tails of comets always pointed away from the sun. The astronomers made full use of rich facilities: government-endowed observatories with elaborate instruments, some of them by the eleventh century incorporating clockwork mechanisms to drive a rotating sighting instrument. When Shen Kua planned an extended series of observations on the planets, measuring their positions 'three times a night for five years', he was undertaking a project the like of which was not to be seen in Europe 'until the time of Tycho Brahe five centuries later' (Sivin, 1975, p.378). The impressively bold time-scales which Chinese astronomers could envisage is shown by the calculation of I-Hsing in 724 AD that the last general conjunction of the planets – that is when they were all situated in the same straight line – 'occurred 97 million years before' (Needham, 1969, p.45).

Large-scale observational projects combined with considerable organization were evident in the eighth-century geodetic survey 'covering 2500 kilometres of territory from Indo-China to the borders of Mongolia' (Needham, 1969, p.31). This survey was related to the determination of the shape of the earth,

Figure 1.5 Apparatus for alchemical distillation, Han dynasty, around first century AD. (From R. Taton, La Science antique et médiévale, Presses Universitaires de France, Paris, 1957, plate 16. Photo: Sir J. Needham.)

and Needham comments that 'it is doubtful if anywhere else in the world at that time such projects could have been successfully completed' (p.31). Maps of the Chinese Empire survive from the twelfth century, but imperial officials were already employed in the third century AD to produce maps with a rectangular grid system.

Accurate observation is also evident in the fine Chinese illustrations of medicinal plants of the twelfth century, including *Ephedra sinica*, whose extract, ephedrine, checks excessive mucous production in the nasal membranes (as in the common cold). Impressive Chinese medical therapy includes a fourteenth-century description of the cure of the deficiency disease beri-beri by suitable diet. There was a long tradition of examinations for medical students in China going right back to the late fifth century AD. The Chinese, earlier than anyone else, established these systems of official examinations – entry to the powerful civil service was achieved by success in examinations, mostly requiring knowledge of literature.

The experimental side of Chinese science is most evident in its ancient alchemical tradition. The earliest evidence of alchemical experiments comes from China in 133 BC. It has been suggested that the word 'alchemy' comes from a corruption of *lien chin shu*, pronounced 'lin kem shut', Chinese for 'the art of transmuting gold'. But that is not established and others derive the word from ancient Egyptian or Greek. What does seem clear is that Chinese alchemy from the start was dedicated not only to making gold from less prized metals but also to the preparation of medical concoctions to prolong life, even seeking to achieve immortality on earth. The consequence of this was the accumulation of expertise in basic chemical techniques such as distillation and sublimation, involving the construction of special chemical apparatus, and a growing familiarity with the properties of a variety of chemical substances like mercury and its compounds, and salts of tin.

The principal practitioners of Chinese alchemy, and probably the most important source of Chinese scientific thought, were the Taoists, followers of a philosophical, religious and magical system of belief dedicated to the contemplation of the 'Tao', or order of nature. But the Taoists were not mere armchair meditators. Their texts, dating from about 300 BC, show that they also believed that wisdom could be achieved from the manual operations of craftsmen. And it was to defeat the ageing process that Taoists practised gymnastics, breathing exercises and the experimental preparation of alchemical medicines, elixirs of life which they were convinced would bring immortality to humans. According to Needham this goal of immortality persisted up to the Sung dynasty (960–1126) when it became less prominent and was replaced by the more modest aim of curing diseases through alchemical preparations. The Taoist texts also reveal that their alchemical experiments were performed within an overall view of nature as uncreated and eternal.

1.3.3 Limitations

Some of the limitations of Chinese science are apparent from the same Taoist texts. Unlike the contemporary ancient Greeks, Taoists tended to present nature as inscrutable. The universe had a structure, but this was not discoverable by rational inquiry. Instead the world was like a mysterious organism whose component parts spontaneously fitted together to form a harmonious whole. This style of thinking is clear in the *Chuang Tzu*, a Taoist text of around 300 BC:

> Life springs into existence without a visible source and disappears into infinity ... Heaven cannot help being high, the earth cannot help being wide, the sun and moon cannot help going round, and all things of the creation cannot help but live and multiply. Such is the operation of the Tao. The most extensive 'knowledge' does not necessarily know it; 'reasoning' will not make men wise in it. *(Needham, 1954– , vol.2, p.39)*

Although the word 'creation' appears in this translated extract, Needham assures us that Taoists were 'generally averse' to the idea of created nature, and that they failed to develop the idea of laws of nature, and further that throughout the whole of Chinese philosophical and scientific thought there is hardly a trace of that idea (Needham, 1969, pp.311, 320 and 322). Can this marked difference from the ancient Greek foundation of European science be explained by the existence of very different legal systems: the constitution-framing society of the Greek city-state, and the customary law of ancient Chinese society with its abhorrence of imposed rationally devised codes of law?

A related shortcoming of Chinese science is the absence of the idea of proof, so important in Western science since the time of the ancient Greeks. Chinese mathematicians were able to introduce perhaps the earliest of all decimal numerical systems, representing any number in terms of nine different characters and leaving a blank space for zero; they could give accurate approximations for the value of π; and even solve individual summation of series of the type $1^2 + 2^2 + 3^2 + 4^2 + ... + n^2$. But rigorous proofs and demonstrations are not to be found in Chinese mathematics; instead there was the solving of particular problems and manipulation of numbers in a magical or game-like way. And with no developed deductive geometry, Chinese astronomy never took the form, so characteristic of the Western tradition, of elaborate geometrical schemes designed to describe the complicated movements of the planets. As for the absence in China of physical explanations of the planetary motions, which would become so important in European science, it is well worth reflecting on the absence in Chinese belief of a divine Creator constructing the cosmos on a rational plan.

I think it is fair to say that from what is so far known about Chinese science there is a predominant image of conservatism of thought. There seem to be no scientific revolutions, no hard-fought challenges to established theories or views of nature. One of the fundamental theories of Chinese science, the five-element theory, probably dating from around 300 BC, portraying the material world in terms of a cyclical movement of the 'forces' of water, fire, earth, wood and metal, was still current in the eighteenth century. Side by side there existed another widely held theory of Yin and Yang, in which the world is seen in terms of these opposing principles signifying contrasting pairs like darkness and light, female and male. Yin and Yang alternately rise, dominate the world, then fall, a process which Needham compares to a wave-like succession, even to a primitive scientific wave theory. Yin and Yang theory probably originated in the fourth century BC and has persisted into the twentieth century. And the distinctive form of Chinese medicine, acupuncture, in which needles are inserted into numerous selected points on the body surface, has an uninterrupted tradition of 2,500 years.

1.3.4 Influence on the West

What was the effect of all this Chinese scientific activity on the West? Despite the undoubted contacts between China and Europe there is little hard evidence of a Chinese stimulus to European science. Needham is nonetheless persuaded that the whole of the West's early development of magnetism (Petrus Peregrinus in the thirteenth century; Gilbert and Kepler in the sixteenth century) had been built from Chinese foundations; that the West's greatest observational astronomer in the sixteenth century, Tycho Brahe, was plotting the positions of celestial bodies by means of a Chinese reference system of coordinates; that after the time of Galileo, the ancient Chinese doctrine of infinite, empty space directly influenced Western cosmologies; that mediaeval and later Western alchemy with its theories of polar opposites may well have been inspired by the Yin and the Yang; and even that the 'whole science of immunology' in the West originated from Chinese variolation, the technique of immunizing against smallpox by innoculating a minute quantity of a smallpox pustule into the nostril. But all of this is surmise. Confirmation of any of these could come one day from the discovery of the reception and translation of some Chinese scientific text in mediaeval or Renaissance Europe; not a single example is yet known.

Less original here, but with greater plausibility, Needham alleges that Chinese alchemy reached the mediaeval West through the intermediary of the Arabs. The idea of alchemically prepared medicines as the key to health and longevity did become conspicuous in the West, and historians of alchemy have long been inclined to accept a Chinese influence. But again there is little in the way of conclusive evidence. One of the basic texts of Arabic alchemy, the Jabir Corpus, describes an unspecified alloy as *khar sini* meaning the 'Chinese barb' and therefore strongly suggesting the influence of Chinese alchemy.

1.4 The spread of Arabic science

If the Chinese influence on Western science was tenuous, the stimulus given by another Eastern culture much closer to Europe's borders, Islam, was clear, strong and of the greatest importance for the scientific development of the West. In the seventh century the armies of the prophet Mohammed, fired by the conviction that Allah was on their side, marched out of Arabia and achieved a spectacular succession of military victories which brought much of the East Roman (Byzantine) and Persian empires under their rule. Arab forces overran Syria, Mesopotamia and Persia, and simultaneously fanned west along the coast of North Africa, conquering Egypt, Libya and the Maghreb. From there, in 709, they crossed the Straits of Gibraltar and occupied almost the whole of the Iberian peninsula. And in the east these enormous conquests for Islam continued through Kabul, Samarkand and up to the borders of China, and across the Indus into the Indian sub-continent, securing the whole province of Sind.

The vast territory which for centuries would remain under the banners of Islam was situated between Europe and India. The numerals we use today are still called 'Arabic', but in fact they originated in India and were transmitted by Arab mathematicians to Europe across the intervening lands of Islam. That is just the most familiar example of Arabic influence on European science achieved through this unified land bridge.

1.4.1 Motives for Muslim interest in science

The conquerors, Arabian clansmen, nomads and merchants, ruled subject populations who practised different religions and spoke a variety of languages. In accordance with the principles of the Koran other religions were generally tolerated, and forced conversions were rare. So long as the subject people paid their taxes to the conquerors there was little disturbance to everyday life. It was the conquerors themselves who were changed as they settled as administrators of subject towns and came into contact with long-established traditions of foreign cultures.

The holy books of Islam urged the faithful to seek knowledge wherever they might find it; one stirring holy dictum informed Muslims that 'the ink of scholars is worth more than the blood of martyrs'. But this referred principally to religious study. For the devout Muslim nothing mattered more than the study of the Koran with its central idea of a single God, Allah, and the duty of every human being to obey his will as revealed by his prophet, Mohammed. Secular studies for the most devout took second place and were even viewed with suspicion if, as in the case of the sciences, they had an infidel, non-Islamic origin. The religious schools of Islam, the *madrasahs*, consequently taught almost no science. Yet the faithful found that the sciences could also be of service to Islam. Knowledge of astronomy was essential to establish the true direction of Mecca, indicated in every mosque by the mihrab, a niche towards which all Muslims turned in prayer. Astronomy also settled the times for the five daily prayers and the exact beginning and end of the fast-month of Ramadan.

1.4.2 Contacts with Greek culture

The Arabs were in close contact with Persian and Indian science, but the most important influence on them by far was the Greek heritage. In 642 the Arab armies had taken Alexandria, the strongest and most recent of centres of

Greek science. Constantinople, the centre of living Greek culture, had withstood Arab sieges and would not fall to Islam until the Ottoman conquest of 1453; but long before this the capital of the Byzantine Empire was an important source of Islam's knowledge of Greek science.

The most immediately fruitful encounter with the Greek scientific tradition seems to have come from Arab contacts with their Christian subjects in Persia, the Nestorians. The Nestorians' homeland had been Syria, where in the recent past Christian missionaries had brought not only the religion of Jesus but Greek culture as well. In fifth-century Syria doctrines of questionable orthodoxy had been taught by Nestorius, bishop of Constantinople. He seemed to deny that Christ was at once God and man, a belief which led to his condemnation by the pope and Council of Ephesus, and banishment to Egypt. His Syrian followers fled persecution and found refuge in neighbouring Persia in the city of Gundishapur. When Persia was overrun by the Arab armies, especially when the caliphs (successors of Mohammed) moved their capital to the new Arab city of Baghdad (762), the Arab rulers found a neighbouring group of Nestorian scholars, theologians and physicians immersed in Aristotelian philosophy and the Greek cultural tradition. The Nestorians no longer spoke Greek, but Syriac, a version of Aramaic. While Syriac was not Arabic, it was a semitic language, and would have assisted preliminary communication with Arabs until these Christians mastered Arabic. The Nestorians would soon transmit a large corpus of Greek scientific knowledge to their Arab overlords, translating mostly from Syriac into Arabic, and also directly from Greek to Arabic.

1.4.3 Translating Greek science into Arabic

Patronage was important in this process of cultural transmission. From his palace at Baghdad, the Abassid Caliph al-Mansur (*r.* 754–75) brought over Nestorian physicians from Gundishapur, some 200 miles away (see Map 1.3) to serve as court physicians. The translation of ancient Greek medical and scientific texts also began, continued under Caliph Harun al-Rashid (*r.* 786–809) and reached a climax in the reign of al-Ma'mun (*r.* 813–33), when a whole academy dedicated to translation was established called Bayt al-Hikma, the 'House of Wisdom'. And it was here that one of the most prolific translators, Hunayn ibn-Ishaq (*c.* 808–73), and his son and nephew for the first time made available in Arabic dozens of Galen's medical treatises, and in general the most important scientific works of the ancient Greeks. Hunayn, a Nestorian Christian and physician, knew Greek and Syriac, and learnt the Arabic he needed in Basra.

From the translating centre of Baghdad came Arabic versions of Euclid's *Elements* (using manuscripts obtained in Constantinople); mathematical works of Archimedes and Apollonius; the astronomical treatise by Ptolemy which they called *Almagest* ('the greatest'); various logical and scientific works of Aristotle; pseudo-Aristotelian and neo-Platonic philosophical texts; and magical texts associated with the shadowy figure of Hermes Trismegistus.

From Baghdad scientific communication radiated west to Córdoba in Muslim Spain, partly as a result of travelling medical students training at the Adudi hospital, which was established in Baghdad in the tenth century by the caliph's officials and dedicated to the teaching of Greek medicine and philosophy. And in Córdoba itself, under the patronage of the Arab emirs, a search began for Greek manuscripts, and a large library was eventually built

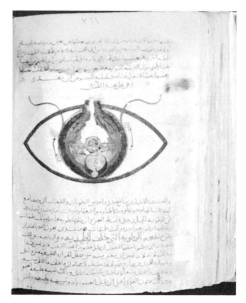

Figure 1.6 A page from Hunayn's Galenic optical treatise, showing an illustration accompanying an accurate anatomical description of the eye. Diseases of the eye, common in the Near East, stimulated Arabic studies of ophthalmology, but optical science was also pursued out of intellectual curiosity. (Middle East Archive, London)

Figure 1.7 The caraway plant, from an Arabic manuscript of Dioscorides' Materia medica. *(Bibliothèque Nationale, Paris)*

up. From Constantinople the Byzantine emperor presented the emir of Córdoba with a copy of Dioscorides' *Materia medica*, the principal classical source of pharmacology. The translation into an exquisitely illustrated Arabic manuscript was achieved by the collaboration of a monk from Constantinople, who could read Greek, with the emir's Jewish physician Hasday ben-Shaprut. The Arabic language possessed the requisite precision and flexibility to deal with a whole range of technical terms from Greek science. Loans from the Greek were common; for example *dysenteria* was Arabicized as *ad-dusantariya*.

The long-term significance of these translations went beyond transmission across cultures. In some cases the Greek original was later lost and the only surviving text was in Arabic. This is true of Apollonius' *Conics*; Galen's principal anatomical work, *On Anatomical Procedures*; and his commentary on the Hippocratic text *Airs, Waters and Places*, which was believed lost until the complete Arabic version was rediscovered in Cairo in 1971.

The translations had provided a solid basis of Greek knowledge from which Arabic science could develop. The phrase 'Arabic science' is likely to mislead, because several of its leading practitioners were neither Arab nor Muslim, but Jews, Christians or Persians living under the rule of Islam. But since their output was written in Arabic, the term does have a real meaning.

What interest did the recovery of the Greek scientific and philosophical corpus have for the world of Islam? For some Muslims the logic and reasoning of Aristotle was seen as a valuable tool to defend the basic tenets of Islam and even to elucidate some of the more puzzling passages in the Koran. In twelfth-century Córdoba ibn-Rushd (1126–98) (known in the West as Averroes) wrote dozens of commentaries on Aristotle's works, and sought to show there was a harmony between faith and reason in Islam. Some of the devout were hostile to this rationalizing tendency in philosophy, and for a while Averroes was banished under suspicion of heresy. On the other hand aspects of Aristotle and Galen, especially the emphasis on the evidence of purpose in the world of nature, harmonized well with the Muslim belief in a universe created by God according to a divine plan.

1.4.4 Going beyond the Greeks

The stimulus of Arabic science came not only from transmitting a lost ancient wisdom, but in independent thought in the form of commentaries on the Greek texts, and in scientific advances beyond the Greeks. The authors of Arabic scientific works produced not merely encyclopaedic compilations of existing knowledge, nor only slavish copies of Aristotle and Ptolemy. Some criticized their Greek masters. The Persian physician and alchemist al-Razi (865–925) wrote a book entitled *Doubts Concerning Galen*, in which he recorded how his experience of diseases at the hospital of Baghdad, where he was chief physician, simply did not agree with what Galen had written. This supported his conviction that the Greeks had not perfected the sciences and that there was much still to be done. He contributed towards this by demonstrating for the first time that smallpox and measles were two distinct diseases. More criticism of Galen later came from Maimonides of Córdoba (1135–1204), rabbi and court physician in Cairo, who indicated contradictions in Galen's various treatises. And the Iraqi physician Abd-al-Latif corrected Galen's description of the lower jaw. His observations of the skeletons of victims of the great famine of 1200 showed that the lower jaw was a single bone and not, as Galen alleged, a jointed two-part structure. More exciting than this, ibn-al-Nafis of Damascus (1210–88) rejected Galen's assertion that the body's blood flowed through the septum, the fleshy membrane separating the two ventricles of the heart. Instead he postulated an alternative route for the blood, through the lungs.

Details of Aristotle's physics were questioned by the Persian scholar al-Biruni (973–1050). And Ptolemy, another of the Greek scientific authorities most revered in the world of Islam, was criticized and corrected by the greatest of Islamic astronomers, al-Battani. Through his observations over the period 877–918 at the observatory of al-Raqqah on the Euphrates he was able to give more accurate descriptions than Ptolemy's on the orbits of the Sun and Moon, and more accurate values of the apparent diameters of those two celestial bodies. And contrary to Ptolemy he argued correctly that annular solar eclipses were possible.

Other notable advances were in Arabic mathematics, where Omar al-Khayyami (*c.* 1050–1123) solved cubic equations; and in connection with geometrical astronomy, trigonometry was introduced for the first time in about 1000 AD. Impressive experimental work was performed in alchemy and optics. The alchemists discovered powerful new solvents through distillation processes, and achieved a more elaborate classification of chemical substances. And in experimental optics some of the most outstanding Arabic scientific research was due to ibn al-Haytham (*c.* 965–*c.* 1039), who worked in Cairo and was known in the West as Alhazen. This included the earliest use of the *camera obscura* ('dark room'), in which solar light is studied by focussing it on a wall by means of a slit in a screen. His investigations of the behaviour of light and his improved theory of visual perception would for centuries be admired in the West. In the thirteenth century the Persian Kamalad-Din explained the sequence of colours in the rainbow by experimenting with glass spheres filled with water, models for raindrops.

In another respect Arabic science may have developed beyond the Greeks by encouraging a less prejudiced view of manual work; for scientific experiments can be seen as a type of manual work. But this is debatable because it is doubtful whether the ancient Greeks really were hostile to experiments and the practical application of scientific knowledge, or scornful of the crafts. The issue is important because controversial interpretations have been proposed suggesting that scientific revolutions have depended on the close association of the manual skills of the craftsman and the intellectual theorizing of the scholar. Merchants were certainly accorded honour and esteem in the Islamic world. And abu-al-Qasim al-Zahrawi (Abulcasis) (*c.* 936–*c.* 1013), court physician in Córdoba, suffered no disgrace from the manual practice of surgery. In Ottoman Turkey, the sultans even felt obliged to learn a craft.

Yet none of this added up to a scientific revolution in the Islamic world. The criticism of the Greeks did not lead to any dramatic developments. Aristotle, Ptolemy and Galen remained the revered teachers of scientists in Córdoba, Cairo and Baghdad. Scientific knowledge certainly advanced but reached a peak, without the overturning of ancient systems, in the twelfth and thirteenth centuries. In the eastern Islamic world, court patronage at Baghdad came to an end with the collapse of the Abbasid dynasty and the eventual fall of Baghdad to the invading Mongols. At Islam's western extremity, in Andalucía, the Ummayad caliphate of Córdoba disintegrated in the eleventh century into twenty emirates. It is interesting that this new environment of competing petty states here produced some of Muslim Spain's greatest scientists and philosophers, like ibn-Rushd (Averroes), physician and the most prolific commentator of Aristotle's works. In the end the most notable effect of Arabic science was the stimulus it gave to the West through the conservation and transmission of the Greek scientific heritage.

1.5 The revival of the Latin West

1.5.1 Conditions influencing intellectual life

In the year 1000, while Chinese and Arabic science were blossoming, Europe remained scientifically stagnant. But contrary to what was once assumed, there had never been a 'Dark Age' in the West in which the lamps of culture went out all over Europe.

The Romans had acquired a huge empire which united Europe, the Levant and Africa. They had absorbed Greek culture, though without much notable addition to the sciences. In the fourth century the emperor Constantine's conversion made Christianity the official religion of the empire. His transference of the capital from Rome to the Greek city Byzantium, renamed Constantinople, moved the centre of gravity of the empire eastwards. A growing division between the Latin-speaking West and the Greek-speaking East Roman Empire would be intensified by the increasing differences in doctrine between the Greek and Roman churches, eventually leading in 1054 to that schism which has never been healed (see Chapter 13). But long before that the Latin West's links with the Greek world had been weakened by the fall of the Roman Empire in the West. In the fifth century Goths, Vandals, Lombards and other invading Germanic tribes had conquered the whole of Western Europe from Britain to the Adriatic. The political unification of Europe was over. Instead there were a variety of barbarian kingdoms. Knowledge of Greek became rare in the Latin West while in the surviving East Roman Empire contact with ancient sources and the Greek language was a living tradition. This was of fundamental importance for the distinct intellectual developments of different parts of Europe. In Western Europe there was a strong sense of a lost, superior classical culture which for centuries would drive scholars to search for relics of ancient learning, to recover what had been lost and even to advance beyond the ancients. In the Byzantine Empire there was not the same mentality, simply because the Greek sources had never been lost; nor was there the same ambition to improve on the ancients. From at least the twelfth century the image of scholars often evoked in the West was of dwarfs standing on the shoulders of giants, meaning the geniuses of ancient Greece. It is hardly conceivable that a scholar from mediaeval Constantinople would ever have imagined that metaphor.

Another peculiarity of the intellectual environment of the Latin West relates to the position of the Church. In the Eastern Roman Empire the head of the Greek Church, the patriarch of Constantinople, was dominated by the emperor. But in the Latin West there was no longer an emperor, nor did the barbarian kings attempt to interfere with the Church. Consequently a much more powerful ecclesiastial authority, the pope, emerged who constructed an elaborate international Church centred on Rome. The popes also sought secular power, eventually becoming territorial princes of Italy. And when after the coronation of Charlemagne in 800, emperors reappeared in the West, they had far less power than before and had to contend with a powerful pope. The struggle in the West between secular and ecclesiastical authority, between the power of the pope and the emperor, was fierce and partly fought on the field of reasoned argument, one of the causes of the Latin West's intense cultivation of logic, a basic intellectual tool.

1.5.2 Greek science regained: translation

From Greek into Latin

When the Roman Empire fell knowledge of Greek science in the West amounted to nothing more than the cosmology of Plato's *Timaeus*, and even that in an incomplete text. But knowledge of the Greek language was still common, especially in Italy. Boethius (*c.* 480–524), a Roman aristocrat and government official of the barbarian Ostrogothic kingdom of Italy, planned an ambitious programme to recover the treasures of Greek thought by translating the entire works of Plato and Aristotle into Latin. But his arrest and execution for alleged treason put an end to the project. By that time he had translated a part of Aristotle's logical works. Also in Italy, in Ravenna, at the same time, a few medical texts by Galen and the Hippocratic physicians were translated into Latin.

After that there were no further translations by Westerners from Greek originals for over five centuries, by which time knowledge of Greek was very scarce, and the only real prospects of translation came from contacts with the Greek-speaking world. In the twelfth century the rising naval power of Venice had established close trading connections and military alliance with the Byzantine Empire. Probably a reflection of these close contacts, James of Venice (*fl.* 1136–48) became acquainted with Byzantine philosophers and learnt enough Greek to provide the West with the first Latin translation of Aristotle's *Physics*. Also in the mid-twelfth century another fundamental Greek scientific treatise was translated into Latin, Ptolemy's *Almagest*. A manuscript copy had been sent by the Byzantine emperor to the king of Sicily and there was still sufficient knowledge of Greek in that island for translation to be possible.

The conquest of Constantinople by the Turks in 1453 with the consequent fleeing of Greeks into Italy is still sometimes presented as the West's first opportunity for many centuries to recover Greek manuscripts and learning. In fact Constantinople fell before this – in 1204 to the Venetians and crusaders, who ruled the region until 1261. Why this did not result in more translations is yet to be explained. During this period of Latin rule William of Moerbeke, a Flemish Dominican monk, visited Greece, learned the language and was later appointed archbishop of Corinth. The most prolific translator from the Greek, his vast output gave the West reliable Latin translations of Archimedes' mathematical treatises and almost the whole of Aristotle's works.

From Arabic to Latin

Before these direct translations from the Greek, the Latin West rediscovered the treasury of Greek science through the intermediary of Arabic sources. And it was principally via the roundabout means of translating Arabic texts that scientific interest in the Latin West rose sharply. The lost science of the Greeks was recovered for the first time along with stimulating Arabic commentaries and discoveries.

Spain, Italy and Sicily were the outstanding centres of what soon became a hectic translating industry. Conditions in Spain were highly favourable for

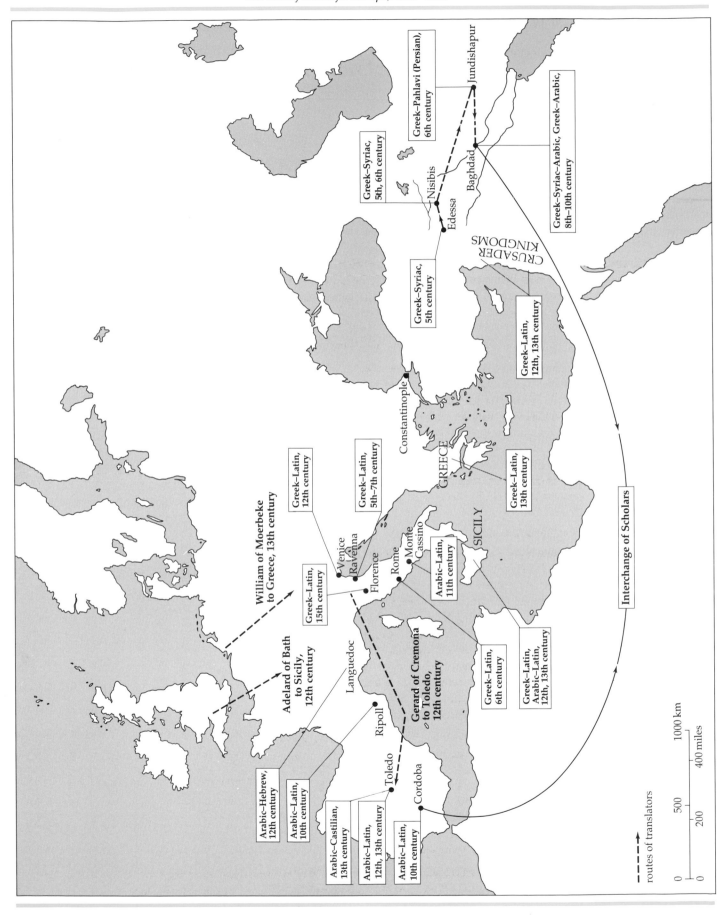

Greek–Pahlavi (Persian), 6th century

Jundishapur

Greek–Syriac, 5th, 6th century

Greek–Syriac–Arabic, Greek–Arabic, 8th–10th century

Nisibis

Baghdad

Edessa

Greek–Syriac, 5th century

CRUSADER KINGDOMS

Greek–Latin, 12th, 13th century

Constantinople

Greek–Latin, 12th century

Greek–Latin, 5th–7th century

GREECE

Greek–Latin, 13th century

William of Moerbeke to Greece, 13th century

Venice

Ravenna

Florence

Rome

Monte Cassino

SICILY

Interchange of Scholars

Greek–Latin, 15th century

Arabic–Latin, 11th century

Adelard of Bath to Sicily, 12th century

Languedoc

Gerard of Cremona to Toledo, 12th century

Greek–Latin, 6th century

Greek–Latin, Arabic–Latin, 12th, 13th century

Ripoll

Toledo

Cordoba

Arabic–Hebrew, 12th century

Arabic–Latin, 10th century

Arabic–Castilian, 13th century

Arabic–Latin, 12th, 13th century

Arabic–Latin, 10th century

1000 km

400 miles

500

200

0

0

routes of translators

this cultural transmission. The peninsula's population was the most ethnically mixed in Europe: Christians living in close proximity with large semitic populations of Jews and Moors. Within Muslim Spain there lived Christians fluent in Arabic, and Jews bilingual in Hebrew and Arabic, languages which have some very similar vocabulary as the transliterations in Table 1.1 show.

Table 1.1

Arabic	Hebrew	English
Madrasa	Midrash	Study; also a religious school
Bayt al-Hikma	Beit Ha-chochma	House of wisdom
Manhaj	Minhag	Custom
Nar	Noor	Fire
Shams	Shemesh	Sun

The earliest known Western translation of an Arabic scientific text was made in the tenth century at the Benedictine monastery of Santa Maria de Ripoll in the Christian county of Barcelona, which bordered a Muslim state. This was a translation of technical treatises on the use of two astronomical instruments: the astrolabe and quadrant. Through the network of European monasteries, copies of this translation were soon diffused to the distant German monastery of Reichenau.

In the early phases of the *reconquista*, that gradual reconquest by the Christians of Muslim Spain, Toledo was recaptured (1085) and quickly became the most important Spanish translating centre. Under the patronage of its archbishops, Moors, Jews and Christians collaborated to produce the desired Latin translations. A productive team of two, Domingo Gundisalvo, a Christian priest, and John of Seville, who may have been a Jew or Arabized Christian, provided many translations. John is thought to have translated from Arabic into the vernacular Castilan; the priest subsequently put this into Latin.

Spain acted as a magnet for Westerners, so eager for the recovery of knowledge they knew to exist there, that some devoted their entire lives to the enterprise. They include Robert of Chester, the Italian Plato of Tivoli, and Hermann of Carinthia. The most active was Gerard of Cremona (*c.* 1114–87), who had gone to Toledo in search of Ptolemy's *Almagest.* There he learned Arabic and with assistants translated a huge body of Arabic texts, astronomical, mathematical, medical, alchemical and magical: Euclid's *Elements*; Ptolemy's *Almagest*; Aristotle's *Physics* and *On the Heavens*; Hippocratic medical treatises; several works by Galen; as well as original Arabic and medical works, notably Avicenna's encyclopaedic survey of medicine and Abulcasis's surgical treatise. All these now became available to the Latin West, the greatest of boosts for its scientific development.

In central Italy in the eleventh century an Arabic-speaking African became a Christian monk and was soon translating Arabic medical treatises, which became influential in the Latin West. Constantine the African had left his native Carthage with his medical books and entered the principal Benedictine monastery of Monte Cassino. His translations of Galen's *Art of Medicine*, Hippocratic treatises, an Arab medical encyclopaedia by Haly Abbas, and works by the Jewish physician Isaac Israeli all influenced teaching at nearby Salerno, where the West's first medical school was rising.

Impressive translating activity was also evident in twelfth-century Sicily. As in Spain the ethnically mixed population facilitated cultural transmission of

Map 1.3 Centres of translating Greek science, 500–1500. (facing page)

Figure 1.8 Top right: the prince of Muslim Tunis presents an Arabic manuscript of the encyclopaedic medical treatise by the Persian physician al-Razi to envoys of Charles of Anjou, ruler of Sicily. Top left: the envoys deliver the manuscript to Charles. Below: the upper section shows Charles giving the manuscript to the Jewish translator Furaj; the lower section shows Faraj at work on the Latin translation, completed in 1279. Known under the title Liber continens *it became very influential in Western medicine. (Bibliothèque Nationale, Paris)*

Arabic science. The island had been conquered by the Arabs in the ninth century and reconquered for Christendom by the Normans in the eleventh century. The indigenous Greeks, Jews, Arabs, Berbers, and now Normans, brought an astonishing variety of languages to the island. Ptolemy's *Optics* was one of many important Latin translations from the Arabic. An Englishman, Adelard of Bath, learned Arabic in Sicily or Syria and translated the first complete Latin version of Euclid's *Elements*, and al-Khwarizmi's treatise, which taught the Latin West how to use astronomical tables.

When Norman rule in Sicily was succeeded in the thirteenth century by the Hohenstaufens and the Angevins, patronage of translation continued. Frederick II gathered Jews, Arabs and Christians together for this purpose. And under Charles of Anjou, Faraj ben Salim, a Jew, translated several Arabic medical treatises.

Consequences

One result of the translations was the introduction of strange-sounding new words, adopted from the Arabic to enrich the scientific vocabulary of the Latin West. The technical terms entered Western astronomy (*azimuth*, *zenith*, *nadir*), mathematics (*algebra*, *cipher*, *zero*) and alchemy (*alcohol*, *alkali*, *alembic*, *realgar*). The name of al-Khwarizmi, the Muslim mathematician who

ᴴHVMANI CORPORIS OSSA PARTE ANTERIO⸗
RIEXPRESSA.

Foramina quæ in harum triũ chartarum delineatione conspici possunt, sunt in temporum osse auditorius meatus: post mamillarem processum vnum, per quod interna iugularis in cerebrum mergit:in facie circa oculorum sedem quatuor, primum ad frontem, secundum ad nares, tertium ad maxillam superiorem, quartum ad temporalem musculũ:duo quoq, in maxilla inferiori. Et per hæc singula ramulus tertij paris neruorum excidit.

Ód͂͂νes, dentes, שִׁנַּיִם scinaim, plurimum triginta duo. τομάς, incisorij, מחתכות mecbathchim, octo: κυνόδοντες, canini, בלב calbym quatuor. μύλιται, molares, maxillares, מחננות thochnim viginti. omnes disparibus radicibus suos alueolos subeunt.

B Clauiculæ, κλῶδες, claues, iugula, תרקוה tharkuha, Furculæ: vtrumq, os literam.ſ.refert, figura inæquabili.

C Ακρώμιον, summus humerus, processus superior scapulæ, à Galeno in lib.de vſu par.κεφακευ διer ad rostri coruini similitudinem nominatus, אל זגם charton, huius appendix cuius principio claues per arthrodian dearticulantur, propriè κατακλῶδὲν quasi od clauiculas dicitur, Rostrum porcinum.

D Processus scapulæ interior inferiórque ab anchoræ similitudine ἀγκυροειδὴς dictus, ey hunc sæpe κορακοειδὲν ey sigmoeidè Gale. vocauit. qnos עין bachateph, Oculus scapulæ.

E Pectoris os, σχῖρος, פֵּתֵן bechaseh, Cassos, septem constat ossibus, sicuti costæ quæ illi alligantur, per vnionem potius, quàm per coarticulationem, parte inferiori iunctis:id ab vtroq, latere lunatum est.

F Cartilago ξιφοειδὴς, ensiformis, quo nomine totum os quoque dicitur, ע̇ך̇ך̇ך̇ alchangri, Ensifoidis, Malum granatum, Epiglottalis cartilago.

G Βραχίων, brachium, humerus Celso ey Cisari, זרוֹעַ Zeroach, Adiutorium brachy, Asetb: hoc tilιæ osse minus est.

H Sinus, humeri caput veluti in duo tubercula diuidens.

I Humeri orbita trochleis similis.

K Cubitus, πῆχυς, קנה h kaneh, Asaid, quibus nominibus etiam tota hæc pars dicitur, vlna. Focile maius, זנד עליון zenad elion, huius acutus processus ad brachiale ὠλεκράνον nominatur.

L Radius, κερκὶς זנד תחתון zenad tbachthon, Focile minus brachy.

N Brachiale, καρπὸς, רב reseg, Raseta, Raseba, ossibus disparibus octo ey duplici ordine distinctis constat, in superiori tribus, in inferiori quatuor: hec simul figuram intrinsecus cauam, ey extrinsecus gibbam constituunt:istorum cum Celso non incertus numerus est.

O Μετακάρπιον, palma, pecten, מסרק masrek, Postbrachiale ossibus quatuor Galeno, non quinque, vt alys cunplurimis, conformatum est.

P Δάκτυλοι, digiti, אצבעות esbaoth singuli ex ternis ossibus conformantur, priori semper interno dio in subsequentis sinum subeunte.

Q Μύλη, ἐπιγονατὶς, patella, rotula genu, מגן הארכובה magen harcubach, scutum genu, Aristatur: os rotundum breuis scuti instar.

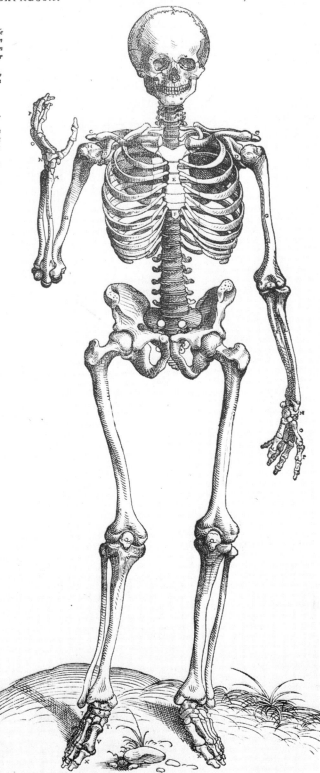

R Ασφχαλος, talus, קרסל karsul, Balistæ os, Cauilla, Chabab, Alscebi:aliqui malleolum hodie malè vertunt.

S Nauiforme, σκαφοειδὲς, nauiculare, זורקי zorki.

T τάρσος, רשג reseg, Raseta pedis, quatuor ossibus constat, quorum maximum extrinsecus situm à cubi figura dicitur κυβοειδὲς, tesseræ os, תרדי thardy, Exagonon, Grandinosum, Nerdi. Religua tria nominibus carent, sed καλκανεῦ nonnullis nominantur. Bii vidimus dextrum pedem vno abundare.

V Plantæ, planum, πεδίον, pecten, מסרק masrek, ossibus quinque construstum est, cui succedunt pedis digiti, X. qui omnes ex ternis internodys côstant, magno tamũ excepto, qui inter alios ex du plici osse construstus est.

Ossiculum illud quod ad primum pollicis articulum apparet, vnum ex sesaminis ossibus est: ey in illo duntaxat loco duo in vtroq, pede obseruauimus.

HVMANI CORPORIS OSSA NONNVLLI IN DVCENTA QVADRAGINTA OCTO, ALIQVI VERO,
in alium numerum redigunt, ego excepto hyoide quod integrum fere ex sex ossiculis per syncondrosim vnitis conformatur, ey sesaminis ducenta ey quadraginta sex putauerim sequentis tabellæ disticho comprehensa.

Figure 1.9 Anatomy's linguistic heritage. (Published by permission of the British Library Board.)

introduced Hindu numerals to the West, became *alguarismo* in Castilian and later *guarismo*, both meaning 'arithmetic'. In English the same surname produced *algorism* with the same meaning.

The lasting importance of the Arabic and Hebrew heritage in Western medicine is clear from the plates illustrating the *Tabulae sex* (1538) of the great anatomist Vesalius of Brussels (see Figure 1.9). Their accompanying texts include numerous Arabic and Hebrew words for bones and blood vessels. The breast bone, 'E', is named not only by the Latin term *pectoris os*, but also *Hechaseh* (from the biblical Hebrew *He-Chazeh*) and *Cassos* from the Arabic *Al-qass*.

Not just the Greeks but the authors of Arabic scientific works would for centuries be regarded as authorities by Western men of science. Al-Battani would be frequently quoted by Copernicus on the subject of the Sun's motion, and Tycho Brahe, Kepler and Galileo were keenly interested in al-Battani's observations.

Another consequence of the translations was to set up a tension in the Latin West between secular Greek thought and Christian doctrine, between faith and reason. Was the new secular learning a help or a threat to Christian orthodoxy? A bigoted reply came from the Italian cardinal Peter Damian (1007–62) who opposed the use of logic in theology and advised monks to avoid secular studies. The statutes of the Dominican order in 1228 prohibited monks studying sciences without special permission and ordered them to confine their reading to theology. On the other hand Berengar, eleventh-century master of a church school in Tours (France), saw Aristotelian logic as a means to clarify mysteries like the transformation of bread and wine into the body and blood of Jesus Christ. His analysis brought a demand for retraction from the pope. The debate between faith and reason, religion and science would echo through the lecture halls of the West's new educational institutions.

1.5.3 The universities

The centres of learning in the mediaeval West had been the monasteries, rural and often intentionally remote. From the eleventh century education moved to the towns as the monasteries were superseded by the new urban cathedral schools (Chartres, Rheims, Paris, Canterbury, Salamanca), where already in the twelfth century pupils were being taught some of the newly translated Aristotelian and Arabic doctrines.

But it was the universities which above all would become the home of the recently recovered Greek and Arabic learning, formally transmitting this through precise curricula of study which became so firmly rooted over the centuries that change would require a struggle. The universities had sprung up spontaneously from the twelfth century, in some cases developing from the cathedral schools (Paris, Salamanca). Paris and Bologna became models for other universities, and by 1350 there were already some thirty of them in Europe: fourteen in Italy, eight in France, seven in the Iberian peninsula, two in England and only one (Prague) in Central and Eastern Europe. By 1500 the number had increased to around seventy; the chief areas of expansion had been France, Italy and Central Europe.

Patronized by princes and especially by popes and prelates, the traditional clerical control of education was maintained, notably at Paris and Oxford, where Dominicans and Franciscans were prominent in university teaching,

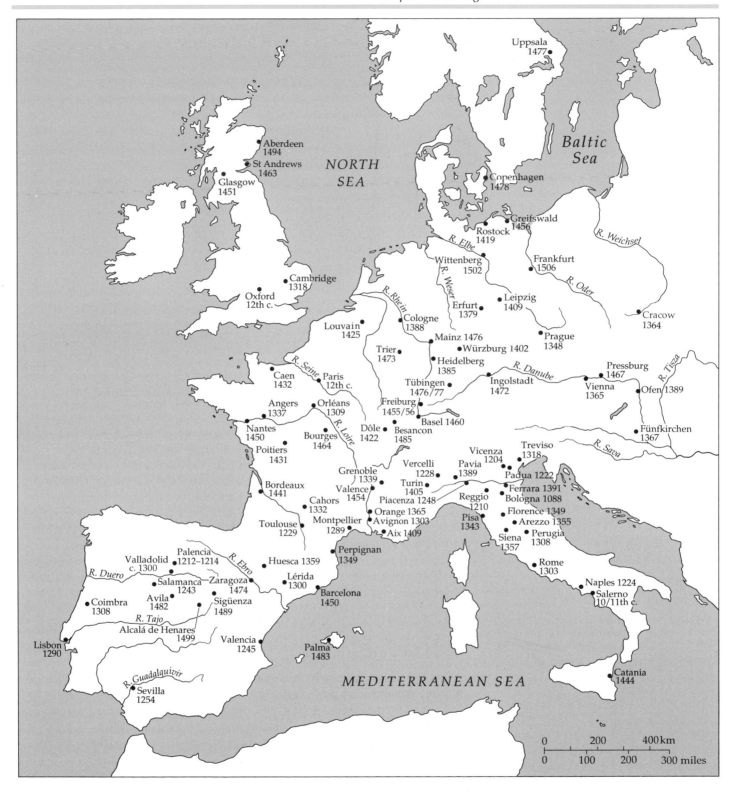

Map 1.4 *The spread of European universities.*
(Taken from "Putzger, Historischer Weltatlas",
101st ed., page 54. Copyright 1990, Cornelsen
Verlag, Berlin.)

and theology was the main field of study. Elsewhere the friars were less conspicuous, and civil and canon law (Bologna, Salamanca) or medicine (Padua, Montpellier) were dominant.

It was in the arts faculty that mediaeval students learnt their science. The quadrivium – the term goes right back to the Roman Boethius – was a course of study consisting of arithmetic, geometry, astronomy and music; the

Figure 1.10 Teaching at the university of Montpellier in the fourteenth century. Bernard de Gourdon, master of medicine, is seated before a class. His teaching texts on the book-rests are the now-standard authorities: Galen, Avicenna and Hippocrates, here represented on the right as three persons in Western mediaeval dress carrying their famous works. (Bibliothèque Nationale, Paris)

trivium comprised logic, rhetoric and grammar. Together these were 'the seven liberal arts' which served as a four-year foundation course for the higher studies of theology, law and medicine.

Although the emphasis of study and the organization of university life varied greatly from place to place, everywhere the same sources became the authoritative texts: Galen, Avicenna and Hippocrates for medicine; Ptolemy and John of Sacrobosco (perhaps English, his thirteenth-century *Sphera* was a simplified version of Ptolemy's *Almagest* and some Arabic astronomical works); and especially Aristotle for logic and practically anything concerning the natural world. Teaching took the form of masters reading texts from the authoritative authors, commenting and elucidating difficult passages. Masters also practised 'disputation', posing questions and then using logical argument to establish or refute propositions. The same technique, known as 'scholasticism', was reflected in the books written by such masters as Thomas Aquinas. It was another sign of the importance attached to reason in the West, and there were several in the university world who were convinced that logic was the most important of the seven liberal arts, basic for the sciences and all other knowledge.

Aristotle, whose logic, profundity and comprehensive surviving works had made him 'the Philosopher' for the Islamic world, also became the prince of wisdom in the Latin West. But his introduction in the university world had been far from smooth. In 1210 a strongly worded decree from the bishop of Paris and other bishops at the synod of Sens prohibited the arts faculty of the university of Paris from reading any books by Aristotle on natural philosophy, or any commentary on them; transgressions would be punished by excommunication. The university's statutes of 1215 incorporated this prohibition. The reason for the ban was the teaching of pantheism, which contrary to Christian doctrine omitted the idea of a transcendent God, making no separation between God and the universe. That doctrine did not belong to Aristotle but rather to neo-Platonic developments of his thought; the two were not yet clearly separated and so Aristotle was blamed. The prohibition was confined to Paris; at Oxford Aristotle could be freely read, and similarly at the university of Toulouse, where the authorities sought to boost student recruitment by advertising the unrestricted teaching there of books forbidden in Paris.

But Aristotle was too valuable to be given up and others believed he might even be of service to the faith. In 1231 Pope Gregory IX instructed a group of Parisian masters to expurgate Aristotle's works, eliminating what was erroneous or offensive to Christianity, leaving the remainder for useful study, just as 'a comely woman found in the number of captives is not permitted to be brought into the house unless shorn of superfluous hair and trimmed of sharp nails' (Thorndike, 1944, pp.39–40; this alludes to Deuteronomy 21:10–12). For reasons that are not clear the expurgation was never made, but it is evident that the less hostile attitude to Aristotle had by the 1250s resulted in the official acceptance of Aristotle's scientific works within the Paris arts faculty. Still this was not the end of the affair. Siger of Brabant, master of arts at Paris, was teaching students that the world was eternal, and running deterministically, again views derived more from neo-Platonic commentaries than from Aristotle himself. And Aquinas, convinced that faith and reason could only be in harmony, drew freely on Aristotle, including a proof of the existence of God: the argument that the cause of motion in the universe must be an ultimate 'unmoved mover', otherwise the question would fall into an infinite regress. These rationalizing approaches intruded into the preserve of the university theologians, who urged Etienne Tempier, bishop of Paris, to put a stop to them. He obliged. In the 1270s he issued two strong condemnations of Aristotelian doctrines. Siger fled to Italy. Many of his doctrines and some of Aquinas's were cited in Tempier's detailed list of 219 offensive propositions. These included the following statements: that God could not move the heavens with rectilinear motion because a vacuum would result; that the world is eternal; that God could not do the impossible; that the only wise men are philosophers; that theological debate was based on fable; that human will is subject to the power of celestial bodies; that nothing happens by chance, only by necessity; that nothing can be made out of nothing (so denying the first Creation); and that God cannot make several worlds.

A few days after Tempier's condemnation, Archbishop Kilwardby, a Dominican who had taught at Paris, visited Oxford and prohibited the teaching of 30 propositions there. The effects of these bans are not clear. One historian of science, Pierre Duhem, went so far as to see the 1270s as the time when modern science began, because the condemnation encouraged scholars to consider alternative ideas like the possibility of a vacuum or the plurality of worlds. Fourteenth-century physicists like Oresme did in fact consider these bold theories. But the fact remains that Aristotelian doctrines retained their powerful hold over the universities. Tempier's condemnation was annulled in 1325 out of reverence for Thomas Aquinas, leaving free discussion. Henceforth Aristotelianism became the established orthodoxy of the Western academic world; and through Aquinas, an intimate synthesis of Aristotle and Christianity was achieved.

The peaks of mediaeval science in the West were reached in the universities of Oxford and Paris in the development of theoretical and experimental optics, and in the progress towards a mathematical treatment of moving bodies.

1.5.4 The Italian Renaissance

The appetite of European scholars for the recovery of ancient wisdom was far from satiated, and in the fifteenth century the search for long-lost texts was pursued with ardour. Remote monasteries were visited in the hope of unearthing some forgotten classical manuscript, and textual analysis developed with the aim of removing the accretions introduced by Arabic intermediaries, thereby displaying the shining purity of the classical originals.

This was the work of the Italian humanists, men whose chief interest was in classical literature, but who at the same time recovered texts of scientific importance. Poggio Bracciolini journeying through Europe in search of classical inscriptions and Latin literature made an exciting discovery in 1417 in the monastery of St Gallen (east of Zurich). This was the full text of the Latin poem *De rerum natura* (On the nature of things) written in the first century BC by Lucretius, the Roman follower of Greek atomic philosophy. For centuries only fragments were known in the West; now its vivid portrayal of the universe as a plurality of worlds, of atoms of matter moving in vast empty spaces and ruled without any concern of the gods, would become influential. Another humanist, Guarino da Verona, whose main interest was to educate the young by instilling the virtues of classical antiquity and to compose speeches in the style of Cicero, also discovered a scientific text which had been lost for centuries: the Roman encyclopaedic medical treatise of Celsus.

The Platonic revival

But the most far-reaching development came from the translations of Platonic works by the Florentine humanist Marsilio Ficino (1433–99). The mediaeval West knew only a small part of Plato's work. Then, from around 1400, Greek manuscripts of Plato's works began to arrive in Italy from Constantinople. Eagerly encouraged by the patronage of Cosimo de' Medici, the ruler of Florence, Ficino completed the first Latin translation of the entire works of Plato, publishing them in 1484. He also translated Plotinus and Proclus, disciples of Plato who developed the master's philosophy in a mystical way into what is known as neo-Platonism. Closely associated with this were the Hermetic magical texts attributed to the mythical Egyptian sage Hermes Trismegistus. Ficino translated these as well, believing them to be inspired by pre-Christian primitive wisdom, though subsequently they were shown to be eclectic productions of the early Christian centuries.

This revival of Platonic and neo-Platonic philosophy was one of the most characteristic aspects of the Italian Renaissance. The centre of it all was the informal Platonic Academy which Ficino founded near Florence. Aquinas had combined Aristotle with Christianity; now Ficino sought a harmonious blend of Christianity with Platonic philosophy.

The influence of all this on contemporary science is clear but difficult to assess. Plato had made not a single scientific discovery, but his philosophy, very different from that of his pupil Aristotle, emphasized the fundamental importance of thinking about the world in a mathematical way, and that message did inspire Renaissance natural philosophers. Then there is the thorny question of the magical side of neo-Platonism with such beliefs as celestial emanations reaching earth, and their capture and control by words, music or ritual to produce marvellous effects.

There are historians like Frances Yates who have gone so far as to see this active, magical and manipulative approach to nature as the principal cause of the Scientific Revolution of the seventeenth century. And with this interpretation others have been led to believe that the scientific thought of Copernicus and Newton was inspired by the magical philosophy translated and developed by Ficino. At the time of writing the pendulum seems to be moving in the opposite direction and there is talk of exaggeration and error (Vickers, 1984), but the question is far from settled and the debate is sure to continue.

Communication through print

While Ficino was a boy receiving his humanistic education in Florence, a technological invention was announced in Germany which would soon revolutionize communications in the learned and non-learned worlds. Johan Gutenberg of Mainz – so far as is known, without any assistance from Chinese techniques – had invented a method of mechanical printing. For the first time in the West hundreds of identical copies of texts could be produced through the use of type; before this copies had been produced by hand with the inevitable introduction of errors. Now a text could be distributed to a far greater number of readers scattered throughout Europe, facilitating study and debate. Eventually books would become cheaper – no longer the priceless manuscript or prohibitively expensive rare copy which only a rich university might afford.

From Mainz, printing-presses quickly spread to other cities, above all in Italy, where there was most wealth, the largest literate population and the European centre for paper production. Venice, through its highly developed commerce, became the printing capital of Europe, followed in importance by Paris.

The output of printed books was at first predominantly religious – Gutenberg's first publication had been the Bible – with classical literature in second place. But scientific texts were also published. In fact science had special gains from the printing revolution. It facilitated the communication of numerical information laborious to copy by hand, like the abundant data of astronomical tables. In 1584 the Danish astronomer Tycho Brahe would set up a printing press in his observatory at Hveen. Printing was also important in anatomy and botany for reproducing accurate, detailed illustrations, essential to those sciences. After Dioscorides' classical work on botany was published in 1544, successive editions appeared, each improved as a result of information communicated by observers in different regions of Europe. Similarly when the first scientific periodicals were published in the seventeenth century, the latest discoveries were rapidly communicated and experimental results and theories supported or questioned by the much expanded audience. When the great astronomer Kepler wrote in 1606 that 'now the world is alive and is indeed in a state of intense excitement' he was thinking of the effects of printing and reflecting that 'the number of authors whose writings are printed is greater than the number of all the authors over the past one thousand years' (in Rosen, 1967, pp.142–3).

In the universities edition after edition of standard Greek or Arabic texts were now published with commentaries, facilitating access to those texts and the comparison of different versions. Two hundred editions of Sacrobosco's *Sphere*, a simplified manual of Ptolemaic astronomy, were published by 1600, and between 1500 and 1674 no less than sixty Latin editions of Avicenna's *Canon*. That encyclopaedic Arabic medical treatise of the eleventh century had become a basic text for medical students in universities since the thirteenth century, containing Galenic physiology, symptoms and causes of disease, lists of drugs, as well as Aristotle's theories of matter and his depiction of the universe. This mediaeval textbook, perpetuated by printing, was still in the curriculum at Bologna as late as 1800 (Siraisi, 1987, p.3). The professors at Padua and Bologna who lectured on Avicenna were led to develop commentaries which were then published. What is interesting about these is that they incorporated recent scientific developments even when they conflicted with Avicenna's text. Giovanni Costeo, professor of medicine at Bologna, published a commentary in 1589 which, referring to Avicenna's statement that the Earth was at rest at the centre of the universe, informed readers of the very different ideas of Copernicus and the support for them. He also conceded that the Galenic anatomical details in the *Canon* were contradicted by recent Vesalian anatomy, and in general recommended personal observation to decide on the truth of anatomical texts (Siraisi, 1987, pp.266 and 330). The cantankerous Swiss physician Paracelsus burned Avicenna's *Canon*, asserting that the human body was the only trustworthy book.

According to Elizabeth Eisenstein, a historian of printing, printed books quickened the downfall of Galenic anatomy, Aristotelian physics and Ptolemaic astronomy. But the example of Avicenna's *Canon* shows this was not a simple process; printing also had the opposite effect of prolonging the life of an established but increasingly contradicted university text-book. Eisenstein tends to argue that printing explains why the Scientific Revolution occurred when it did (Eisenstein, 1980, pp.687–8). While this has brought printing back into the debate, the argument seems to be a simplification.

1.6 Conclusion

The universities and printing-presses were important features of town life, and it was in the towns that European science developed. It is well worth reflecting on this urban environment, because the towns of mediaeval and Renaissance Europe were quite different from those elsewhere in the world. Florence, Venice, Cologne, Mainz, Basel and Paris were very different from mediaeval Baghdad, Peking or ancient Athens. Only in mediaeval and Renaissance Europe did towns develop with autonomous self-government and a strong commercial or manufacturing character, often becoming centres for long-distance trade. It is sometimes argued that the strong commercial life of European towns stimulated science through the need for calculations for commercial transactions. It is true that schools in mediaeval Florence taught boys arithmetic and the use of the abacus, but this was no more than science at the most elementary level. Nevertheless the towns, with their large and more literate populations, engaged in a rich variety of occupations, created a distinctive environment which may well have been important for scientific development.

The sciences, like painting, have their own internal traditions: painters have produced canvases which seem unaffected by conditions in the outside world; similarly natural philosophers have written about internal problems in optics set by previous generations. But paintings and scientific works can also be affected by conditions in the world. And it is a principal task of the chapters which follow to consider scientific developments in different parts of Europe, raising points for comparison and contrast, to see how conditions peculiar to Italy or Russia, Spain or Poland, England or France, might have had an effect; and to consider also why the scientific movement went further in some parts of the West than in others, and far further than the science of China or the lands of Islam.

Sources referred to in the text

Bernal, J. (1954) *Science in History*, vol.1, Harmondsworth, Penguin.

Butterfield, H. (1957) *The Origins of Modern Science 1300–1800*, London, G. Bell.

Clagett, M. (1963) *Greek Science in Antiquity*, 2nd edn, New York, Collier-Macmillan.

Eisenstein, E. (1980) *The Printing Press as an Agent of Change*, Cambridge, Cambridge University Press.

Hall, A.R. (1983) *The Revolution in Science 1500–1750*, London, Longman.

Lloyd, G.E.R. (1979) *Magic, Reason and Experience: Studies in the Origins and Development of Greek Science*, Cambridge, Cambridge University Press.

Merton, R.K. (1973) *The Sociology of Science*, ed. N. Storer, Chicago, University of Chicago.

Needham, J. (1954–) *Science and Civilisation in China*, Cambridge, Cambridge University Press.

Needham, J. (1969) *The Grand Titration: Science and Society in East and West*, London, Allen and Unwin.

Porter, R. (1986) 'The scientific revolution: a spoke in the wheel', in R. Porter and M. Teich (eds), *Revolution in History*, Cambridge, Cambridge University Press.

Rosen, E. (1967) 'In defense of Kepler', in Archibald R. Lewis (ed.), *Aspects of the Renaissance*, Austin, University of Texas Press.

Scheffler, I. (1967) *Science and Subjectivity*, Indianopolis, Ind., Bobbs-Merril.

Siraisi, N. (1987) *Avicenna in Renaissance Italy: The Canon and Medical Teaching in Italian Universities after 1500*, Princeton, NJ, Princeton University Press.

Sivin, N. (1975) 'Shen Kua', in C. Gillisipie (ed.), *Dictionary of Scientific Biography*, vol.12, New York, Scribner's.

Thorndike, L. (1944) *University Records and Life in the Middle Ages*, New York, Columbia University Press.

Vickers, B. (ed.) (1984) *Occult and Scientific Mentalities in the Renaissance*, Cambridge, Cambridge University Press.

Copernicus and his Revolution   *Chapter 2*

by Colin A. Russell

2.1 The scholar from Torun

In the year 1543 European science entered a dramatic new phase. That is not to deny important discoveries and developments in the Middle Ages and in the earlier part of the sixteenth century. The growth of a mathematical, experimental approach to nature goes back many centuries. With the advent of the printing press, readers in early sixteenth-century Europe had ready access to scientific texts of Albertus Magnus, Roger Bacon, Robert Grosseteste, Nicholas of Cusa and other giants of the past. Now, by mid-century, the eccentric and outrageous self-publicist Paracelsus had laid the foundations of medical chemistry; from the genius of Leonardo da Vinci had come scientific instruments, engineering models and anatomical drawings of a wholly new degree of accuracy and realism; observers of the heavens were reinterpreting ancient astronomical data in attempts to reform the Julian calendar; Columbus had discovered America and Magellan had encircled the globe.

However, in 1543 two books appeared that heralded such a dramatically new approach to nature that, together, they signal the beginning of the Scientific Revolution. Of the first little need be said at this point. It was the *De humani corporis fabrica* (Concerning the fabric of the human body) of the Brussels-born anatomist Andreas Vesalius (1514–64) which, it has been argued, rendered all previous works on the subject out of date and elementary. Such were the new standards of accuracy of the work that it displayed for the first time complex details of muscles and bones, the brain and especially the heart. Through the new technology of block-cutting and printing, and employment of Venetian artists, Vesalius transformed mere 'illustrations' into definitive diagrams. Despite his adherence to old ideas of Galen about the heart and blood, he may truly be hailed as the founder of modern anatomy.

Yet it was not a study of the 'sensitive and animate body' that began the profound transformation of humanity's understanding of itself which attended the rise of modern science. That first step was taken one day in the early summer of 1543 in the sickroom of an aging official of the Cathedral Church in Frauenburg.

The tiny town of Frauenburg (Frombork)[1] is now in Poland, but at that time was in the small state of Ermland, once part of the Duchy of Prussia. It lies on the Baltic coast, about midway between the modern ship-building town of Danzig (Gdansk) and the Lithuanian border. On a small hill, overlooking the Vistula Lagoon and the Baltic Sea beyond, stands the fourteenth-century cathedral. This served the whole independent diocese of Ermland (Warmia), the only diocese that had not been taken over by the Order of the Teutonic Knights, by whom (except on the seaward side) Ermland was completely encircled. The Knights, moreover, were extremely hostile to Poland to their

Figure 2.1 Nicholas Copernicus in middle life. (Ronan Picture Library)

---

[1] For modern names of towns see Appendix and map.

south and the whole area was frequently devastated by war – as it has been for all the centuries since. For this reason, and perhaps because the Bishop had links with both warring factions and therefore would sometimes get involved as a mediator, the Cathedral authorities had the bright idea of

Map 2.1 The world of Copernicus.

enclosing their precinct with immense defensive walls, the height of a two-storeyed house, and incorporating within them at strategic points seven tall towers.

It was either in one of these towers (which still stands), or in a more conventional residence nearby, that the final act of one drama was played out and the first act of an other immeasurably greater was ushered in. Two men are in the room. On a bed lies an obviously dying man, victim of some condition causing haemorrhage, paralysis and mental degeneration. The room is his, or rather belongs to the cathedral in whose service he has spent his last twenty years. His companion, George Donner, is a fellow canon of the church, and under special instructions to care for the sick man. These orders came from Tiedemann Giese, who had been a colleague and close friend for all those years and more. Recently appointed Bishop of Kulm (Chelmno) in West Prussia, Giese had written urgently to Donner and may even have made the long journey eastward to bid farewell to his old companion. It is from his own pen that we learn of the events on 24 May 1543.[2]

The story can be quickly told. Some time during the day a messenger arrives, hotfoot with a package from the Bavarian town of Nüremberg, 600 miles to the south west. Addressed to the dying man, it is nothing less momentous than the final pages of a book from his own pen, a book moreover that will be destined to shake the whole Western world. Yet for the Bishop this is to be a bitter-sweet moment of supreme irony. For him the book is the culminating triumph of years of support, encouragement, counsel and persuasion, for his friend had needed all of these before he would eventually agree to publish. Giese has high stakes in the matter, for on all important matters he and the author had been of one mind, together with their younger friend who had taken charge of the practicalities of publication. Now the long-anticipated moment has arrived but, for the author, far too late. Even as the final pages are placed beside him life is all too clearly slipping away; within hours he is dead and the Cathedral authorities will be making arrangements for burial within the church. Recalling the poignancy of that day Giese wrote: 'He had lost his memory and mental vigour many days before; and he saw his work completed only at his last breath upon the day that he died'.

The dying author was, of course, Nicholas Copernicus and his book *De revolutionibus orbium coelestium* (Concerning the revolutions of the heavenly bodies). If any one publication may be said to have ushered in the European Scientific Revolution it was surely the volume that lay in the lifeless hands of Copernicus that early summer day in 1543.

Nicholas Koppernigk or Kopernik (who later Latinized his name to Copernicus) was born on 19 February 1473 in Thorn (Torun), a town on the Vistula founded in the thirteenth century by the Teutonic Knights as a base for their proposed conquest of Poland. Because it was a border city between Germany and Poland, and because Copernicus' ancestors may have come from German stock but owned allegiance to the Polish king, it is pointless to label the astronomer with either nationality, and he is 'claimed' today by both sides. He was the youngest of four children born to a merchant, also Nicholas, who had some years previously moved to Torun from Cracow. Here he could take advantage of the convergence of trade routes from east and west, the market thronged with foreign and domestic traders, and the great highway of the Vistula down which merchandise was shipped to the Baltic seaport of Danzig.

---

[2] Studies of Ermland church records have recently cast some doubt on this exact date, but it is certainly accurate to within a week.

Figure 2.2   Birthplace of Copernicus, Torun. (Wydawnicto 'Arkady', Warsaw)

Figure 2.3   Statue of Copernicus in Torun. (Wydawnictwo 'Arkady', Warsaw)

Figure 2.4   Lucas Waczenrode, 1447–1512: Uncle of Copernicus and (from 1489) Bishop of Ermland. (Biblioteka Narodowa, Warsaw)

When Copernicus was only ten years old his father died, and a little later the young family was 'adopted' by their mother's young brother, Lucas Waczenrode. This man was a scholar. Having studied at Cracow and Bologna he taught for a while in Torun before entering the Catholic priesthood. Thereafter promotion was rapid. When Copernicus was placed under his care in 1483 he had already been a Canon of Frauenburg for four years. In 1489 he became Bishop of Ermland, with his Cathedral at Frauenburg and his palace at Heilsberg (Lidzbark).

Copernicus is said to have continued his schooling at St John's School, Torun, but was then sent to a school at Wloclawek, attached to the cathedral. Here he seems to have learned some astronomy from a teacher with the improbable name of Vodka (and the even more improbable pseudonym of Abstemius!). Master and pupil are said to have constructed a sundial on the cathedral's south wall. The school had strong links with the University of Cracow, to which ancient and prestigious institution Waczenrode sent his nephew in 1491/2.

The city of Cracow – from which Copernicus' ancestors had come – is situated in the heart of Poland. It lay on the main European trade routes, south–north and east–west. Virtually the capital of Poland from the twelfth century,[3] it is still dominated by the royal castle and cathedral on Wavel Hill. One of the greatest of Cracow's many other churches, St Mary's, with its

---

[3] The capital was transferred to Warsaw in the sixteenth-century by King Sigismund III of the Swedish dynasty of Vasa; see Chapter 12.

immense asymmetric west front, overshadows the vast market-place, exceeding even that of Torun in importance. Here commerce was of matters intellectual and spiritual as well as financial and material. Traders from all over Europe and beyond would exchange tales of threatened wars with Turks and Tartars, of Reformation rumblings in Northern Europe, of the artistic marvels of Renaissance Italy and even of strange ideas from the East about the movement of the heavens. And in this city was one of the oldest universities in Europe,[4] the University of Cracow, founded in 1364, and reconstituted in 1400 (as the Jagiellonian Academy) by King Wladyslaus Jagiello. It was as cosmopolitan as the market-place, attracting students from all over Europe, united in their common language of Latin. In a still united 'Christendom', graduates of one university could teach in any other, and Cracow was exceptionally favoured in its teachers from other countries, especially Italy. Perhaps for this reason it became one of the earliest universities in the north to respond to the new 'humanism' of the Renaissance.

Figure 2.5   The Collegium Maius, Cracow, where Copernicus studied. (Wydawnictwo 'Arkady', Warsaw)

Copernicus studied the usual mix of subjects then universal in European universities – classics, mathematics, philosophy, astronomy and so on (see Chapter 1). The faculty of mathematics and astronomy was the second oldest in the world (after Bologna). The great mathematician Albert Brudzewski taught there, and we now know that an astronomical observatory existed in Copernicus' time, in 1494 receiving a gift of instruments that still survive. Amongst the lectures attended by Copernicus was a course on the very latest thing in astronomy, a book of about 1473, *Theoricae novae planetarum* (New theories of the planets), by the German astronomer Georg Peurbach. It included one of the most up-to-date accounts of classical Ptolemaic astronomy, the almost universally accepted scheme that placed the Earth at the centre of the universe and explained the movements of stars and planets by complicated combinations of circular motion.

Copernicus left Cracow without taking an MA (a common situation then). His uncle proposed that he should pursue a career in the church and accordingly despatched him to Bologna to study canon law. For over three

[4] Only Prague predated it in Central Europe, being founded in 1348.

Figure 2.6 Bologna. (The Trustees of the British Museum, London)

years Copernicus attended lectures at this famous law school but soon found there was more to life than law. Amongst his luggage brought from Cracow were two bound volumes of mathematical and astronomical texts, indicating perhaps the direction in which his secret thoughts were turning. In Bologna was the eminent professor of astronomy, Domenico da Novara. Together they observed the heavens, noting a discrepancy between the actual and calculated times for the occultation (eclipse) of the bright star Aldebaran by the Moon (9 March 1497).

The year 1500 being a jubilee year for the church, Copernicus joined the two hundred thousand pilgrims to Rome for the special celebrations. He remained there for a whole year (during which time the murder of the husband of Lucrezia Borgia, daughter of the Pope, created a scandal for the papacy of monumental proportions). The eyes of Copernicus appear to have been directed more to the heavens; he observed a lunar eclipse on 6 November, and may have given informal lectures on astronomy.

In 1501 Copernicus paid a brief visit to Poland in order to take up an ecclesiastical office secured for him through the good offices of his uncle. Almost immediately, however, he returned to Italy to complete his studies. His chosen destination was the law school at Padua, established in the thirteenth century by *emigrés* from Bologna. This was a university recently declared to be the foremost in the Venetian Republic and thus able to attract the best teachers available. He graduated DCL in 1503, though for some reason took his degree at the University of Ferrara. His legal studies over, he turned his attention to another subject that had made Padua world-famous: medicine. Through lectures on the traditional medicine of Galen and the Greeks, and by watching dissections of recently-executed criminals, he may be supposed to have learned enough of the healing arts to stand in good

Figure 2.7 A view of Padua at the close of the fifteenth century from Schedel's World Chronicles. *(Wydawnictwo 'Arkady', Warsaw)*

stead anyone whose vocation was to be the Church, with its pastoral care of the sick. And here Copernicus learned Greek, so he was now able to read in the original the writings of the early astronomers. Almost certainly he took part in the debates and discussions on such matters as the eternity and structure of the universe, for in Padua were circulating many radical ideas of the Renaissance, from sober conjecture to wild speculation.

On returning once more to his homeland Copernicus might have expected at last to occupy his canonical stall at Frauenburg Cathedral. But it was not to be. His uncle, in need of a personal assistant to help administer his territory, transferred Copernicus to his castle at Heilsberg. With the official function of 'physician to the Bishop', he found time not merely to exercise his administrative and healing skills (both of which were well spoken of) but also to translate Greek letters and even to make a few astronomical observations from a high tower in the castle. Most important of all, he put on paper some daring new thoughts about cosmology in a booklet entitled *Commentariolus* (or Little Commentary), intended for the eyes of a few trusted friends only.

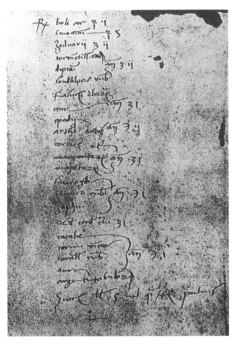

Figure 2.8 Prescription written by Copernicus. (Wydawnictwo 'Arkady', Warsaw)

In 1512 Waczenrode died and Copernicus returned to Frauenburg. But not for long. In 1516 he was appointed by the Cathedral Chapter to take charge of two of their estates to the east, and was relocated to Allenstein Castle. He remained here, on and off, for many years during which war ravaged the countryside. His skill and experience in administration led to considerable social standing, and he wrote a treatise on taxation and coinage reform. But all the time he was brooding over the inadequacies of the old, Earth-centred model of the universe. Both at Allenstein and at Frauenburg (where he occupied one of the seven towers) he watched the heavens, recording eclipses, occultations and other celestial phenomena. And at Frauenburg he met Tiedmann Giese and shared with him the theory that was slowly being formed in his mind.

Figure 2.9 Allenstein: Castle of the Ermland Chapter. (Bibliotheka Narodova, Warsaw)

Figure 2.10 Copernicus' room, Allenstein. (Wydawnictwo 'Arkady', Warsaw)

There, perhaps, matters might have for ever rested had it not been for the visit to Frauenburg in 1539 of a young professor of mathematics at Wittenberg, George Joachim. Otherwise known as Rheticus, Joachim was burning with a desire to see for himself the manuscript about astronomy that Copernicus was rumoured to be writing. He brought with him gifts of astronomical and mathematical books. His visit lasted two and a half years and Rheticus became a devoted disciple. Having seen the manuscript he sought and obtained permission to publish a *Narratio prima* (or First account) of Copernicus' ideas. This appeared in 1540 and soon Copernicus was resolved to publish the full text. Encouraged as ever by Giese, who presumably entrusted the manuscript to Rheticus, all Copernicus now had to do was wait. The arrangements for printing in Nuremburg were left to Rheticus, but as he was moving to Leipzig an alternative agent had to be found. This was a local Lutheran clergyman Andreas Osiander. The upshot was the appearance of the complete *De revolutionibus* at the author's dying bed in May, 1543.

The achievement of Copernicus has indeed become illustrious. When the eighteenth-century philosopher Kant wished to impress upon his readership the importance of his own work he rather immodestly called it a 'Copernican revolution in philosophy'. Perhaps it is equalled in scientific importance only by the 'Darwinian revolution', with which it shares far-reaching implications on the nature and place of man, as well as a timid reluctance to publish on the part of the author.

In a strictly scientific sense Copernicus radically changed the physical model of the universe, the 'world-picture' as it is sometimes called. He replaced a cosmos centred on a fixed, immovable Earth with one in which the Earth was

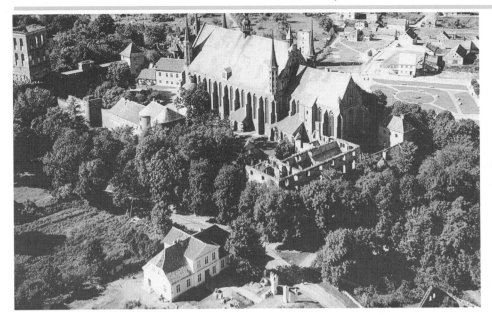

Figure 2.11 *Frauenberg. (Wydawnictwo 'Arkady', Warsaw)*

Figure 2.12 *Copernicus' Tower, Frauenberg. (Wydawnictwo 'Arkady', Warsaw)*

just one other planet circling round the Sun. In fact his achievement was essentially mathematical, involving a transposition of co-ordinates. It was not a *discovery* provoked by accumulation of his own observations, but was rather an almost intuitive choice, made after much thought and in defiance of common sense. Clearly the Copernican Revolution poses several urgent questions. How could a simple mathematical device (for that is what it was) initiate such profound changes in the way humanity thought about itself? How can we explain the conversion of a *world-picture* into a *world-view* (or *Weltanschauung* as the Germans call it) involving questions of value and purpose? We may also enquire why it was able to trigger off a veritable explosion of scientific growth. But there is another question altogether, perhaps the most important of all for the clues it may offer into the causes of the Scientific Revolution in Europe: why was it that Copernicus actually made his choice as and when he did? To begin to answer that question we have to retrace humanity's steps to the cosmological beliefs of the ancient world.

With the new availability of purer and more refined versions of the classics that had not been mediated via the Arabs it now became possible in the fifteenth century to examine much more reliable copies of many ancient writings, including Greek works on science.

2.2 The heritage of ancient astronomy

2.2.1 The empirical basis of ancient and mediaeval astronomy

What facts about the universe were available to ancient and mediaeval observers? In other words, what was the empirical basis to their astronomy? The most obvious thing was the Sun. People were aware of the way in which it moved daily through the sky and also of the fact that the daily movement was combined in some way with an annual movement so that it did not always appear in the same part of the sky at the same time of the day. For example, as the summer progressed, so the Sun became higher in the heavens at mid-day and in winter it became very much lower. Next in importance to the Sun came the Moon; the monthly cycle of the Moon had of course been known since antiquity, as indeed had the fact that it shines by reflected light and not by any energy of its own. After many centuries of observation it became possible to predict with some degree of accuracy eclipses of Sun and Moon, solar eclipses occurring when the Moon was interposed between the Earth and the Sun, and lunar eclipses when the Earth was interposed between the Sun and the Moon. Ptolemy significantly improved that accuracy.

Then there were the so-called fixed stars. They do not change their position relative to each other and they constitute the great back-drop of the sky at night. Of course they too appear to move about the Earth, going right round from east to west once every twenty-four hours. Particularly fascinating was that softly glowing region of the night sky known as the Milky Way. There were also the five planets known since antiquity: Mercury, Venus, Mars, Jupiter and Saturn (although we ought to notice that the Moon and Sun were also sometimes called planets in the Middle Ages). Unlike the fixed stars planets do not 'twinkle', and they have a very odd movement of their own, wandering across the sky in a seemingly erratic fashion. Indeed the word 'planet' is derived from the Greek *planetes*, wanderer. Finally there were those rare visitors known as comets and the spectacular displays of shooting stars called meteors.

Acquaintance with these bodies and with details of their movements led to the only knowledge concerning the universe that we today should regard as important. There was also a cult of astrology, now scientifically disreputable but with many devotees in the Middle Ages. The interesting part of the heavens, where the known planets moved, was divided into twelve parts, 'the signs of the zodiac'. These are thought to have been suggested by fanciful resemblances of objects like crabs and lions to star patterns in those areas, and astrologers made much of these. However the distinction between astronomy (a science) and astrology (a pseudoscience) is in fact quite modern, probably first attributable to de Messa of Seville in 1595.

These then were the main *facts* known about the heavenly bodies. The pre-Copernican world-picture accommodated these facts in ways that derived directly from the Greeks.

2.2.2 The two-sphere universe

We do not know who was the first to suggest that the Earth was a motionless sphere floating in space, but the idea is certainly recognizable in Plato and may have been derived from the Pythagoreans (fifth century BC). Given this assumption, the simplest possible explanation of the apparent movement of the Sun round the Earth is obtainable by taking two kinds of circular motion and combining them together. This is represented in Figure 2.13, where the stationary Earth, E, is situated at the centre of a vast celestial sphere, studded with the fixed stars.

The annual movement of the Sun is represented by the circular path *a b c d*. On the other hand its daily motion is represented by the complete revolution of the sky carrying the Sun circle with it in the direction of *f b e d*.

This scheme was succesful in accounting for the movement of the Sun but it could not be applied to the more erratic behaviour of the Moon and planets. Nevertheless for all its imperfections it had an importance that could hardly be overestimated; the Platonic idea of the circularity of celestial motions lasted two thousand years – well after the time of Copernicus in fact.

2.2.3 *The Aristotelian universe*

Although Aristotle seems to have reacted against the earlier Greek over-emphasis on geometry, he perpetuated the tradition of a universe encircling a motionless, spherical Earth. He also made discoveries in biology of the first importance. His system of physics, by which he tried to codify all the kinds of changes which are part of our ordinary experience, was much less valuable in the long run but of far greater influence until the Renaissance.

Aristotelian physics proceeded from the assumption that there are two kinds of motion, natural and unnatural. On this Earth, and indeed anywhere beneath the Moon's orbit, it is in the nature of things to move in straight lines; stones fall vertically downwards, fire to tends to fly vertically upwards, in each case by a kind of 'homing-instinct'. Everything has its proper place and, unless prevented, would tend to move there. All other motion was unnatural – the trajectory of an arrow, for instance, was not a straight line because forces were impressed upon it from the bow and also, it was believed, by the pressure of the air which the arrow displaced during its flight. This last idea, with the air coming round to the back of the arrow and continuing the effect of the twanging bow-string, was necessary because of Aristotle's belief that for motion to occur there must be a continual application of force. Aristotle's sytem was self-consistent and very hard to disprove in practice. And its adoption by the mediaeval church made opposition all the harder (see Chapter 3).

In this Aristotelian scheme, the spherical Earth again lies motionless at the centre of a finite, spherical universe. Between the Earth and the outer ring of fixed stars lie concentric spheres (crystalline and invisible) upon which are embedded the Sun, Moon and planets. Figure 2.14 shows how this universe was conceived, but it does not try to depict the idea that more than one sphere was usually associated with each heavenly body, as in fact became the case in Graeco-Roman times.

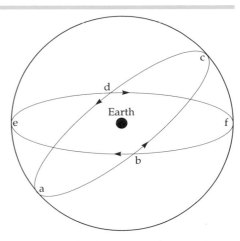

Figure 2.13 The plane f b e d is known as the plane of the celestial equator; the plane a b c d is called the ecliptic because it is in this plane that eclipses take place, for the movement of the Sun is always in this plane and it is crossed at two points by the Moon. For an observer in the northern hemisphere, the point c represents the maximum height the Sun attains in the sky and is called the summer solstice, and similarly a is known as the winter solstice. The points b and d are known as the spring and autumnal equinoxes respectively because when the Sun is at those positions day and night are of equal length.

Figure 2.14 The Ptolemaic system. (Mansell Collection)

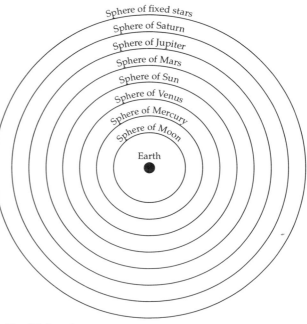

Simplified version

The nearest sphere to the Earth is that of the Moon, and marked a fundamental division in the Aristotelian universe. This applied to the laws of physics, for example. For sub-lunar bodies (which included those on Earth) the *natural* form of motion was in a straight line (rectilinear); but the heavenly bodies above the Moon's sphere moved only with circular motion; indeed in those exalted regions there was only one kind of change possible, and that was the change of position, i.e. motion. The heavens were incorruptible. But on this imperfect and corruptible Earth all kinds of change could and did take place: birth, death, change of qualities and so on. This idea of two different kinds of physics – celestial and terrestrial – persisted right up to the seventeenth century when Newton showed that the same physical laws we know on Earth are applicable throughout the whole of the universe.

Another long-lived error was Aristotle's belief that above the Moon was only one element (the aether) while below were the four terrestrial elements: earth, water, air and fire, each striving to attain its proper place in the universe. Thus fire tended naturally to move (in straight lines of course) to the outside of the Earth's region and earth tended to go towards the centre; hence flames flew upwards and stones dropped downwards. The illusion that the composition of the Earth is different from that of the rest of the universe was not finally destroyed until as late as the nineteenth century when studies in spectroscopy showed that the elements present in the stars and Sun are also present here on Earth.

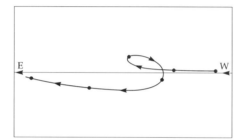

Figure 2.15 A loop of retrogression

2.2.4 The spheres of Eudoxus

In some respects it could be said that the key to the basic cosmological problem in the Middle Ages lay in the behaviour of the five planets then known: Mercury, Venus, Mars, Jupiter and Saturn. Their erratic behaviour had baffled and infuriated generations of Greek thinkers, up to Plato himself. It seemed impossible to reconcile their celestial meanderings with either the supposed divinity of heavenly bodies or with any simple concept of circular motion.

The diagram in Figure 2.15 is derived from recent observations of the planet Mars. Its location against the background of fixed stars is noted at the same time on many successive nights and the points are then joined up to give a curve which represents the planet's path across the sky (a 'loop of retrogression'). The straight line in the diagram is the ecliptic. The early Greek model, with its combination of two different circular motions, can adequately account for the path followed by the Sun, but is useless when applied to the more complex behaviour of the planets. Nevertheless it was the starting point for one of the most remarkable intellectual accomplishments of ancient science. Eudoxus of Cnidos, a younger contemporary of Plato, took the two original spheres and added two more of his own inside them, giving an array represented in Figure 2.16. The planet P is attached to a sphere (4) whose axis is fixed to the one next to it (3). This in turn revolves on another axis, and so on for (2) and (1). Thus there are four spheres, one inside the other (as Aristotle supposed), each sphere rotating on a different axis and each having its own period of rotation (a new concept).

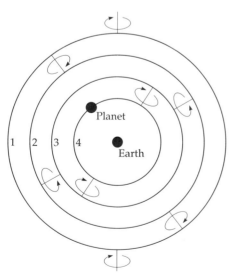

Figure 2.16 The spheres of Eudoxus

Sphere	Axis	Direction
1	Polar axis	East to West
2	90° to ecliptic	West to East
3	In ecliptic	West to East
4	Different for each planet	East to West

Nevertheless, for all his success, Eudoxus failed to provide a completely satisfactory system. His scheme could not explain the motion of Mars, and it had one other flaw, failing to account for two further related facts:

- the Moon, Mars, and some other planets vary appreciably in brightness (as can easily be seen with the naked eye for Mars and the Moon);
- the Moon varies in apparent size (and therefore distance).

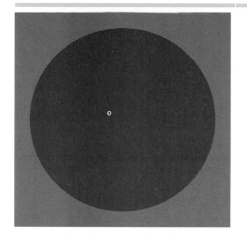

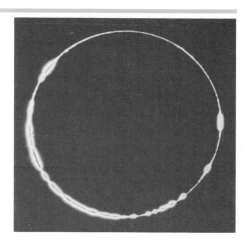

The variation in brightness can be due to a number of causes but no explanation, even in antiquity, could account for the variation in brightness of the planet Mars unless its distance from the Earth were not constant. This is quite incompatible with the scheme devised by Eudoxus, however.

We must now try to take stock of the position the planetary theory had reached in about the second century AD.

Figure 2.17 In a total eclipse of the Sun (left) the whole of its disc is covered by the Moon; but in an annular eclipse (centre) there is a small annulus (or ring) round the edge, which the Moon does not obscure. At the right is a photograph of an annular eclipse.

2.2.5 Ptolemy's solutions

It is clear that all these attempts were, in varying degrees, unsuccessful. Yet rather than jettison the concept of circular motion about a stationary Earth, astronomers tried yet further ways out of the impasse. Various early solutions were incorporated into the sytem of astronomy codified by Claudius Ptolemy in the second century AD. He used various combinations of circular motion, particularly the following:

The eccentric: the planet P moves uniformly round a circle, but the Earth E is displaced from the centre C. A terrestrial observer would therefore see the planet at different distances throughout its revolution (Figure 2.18).

The epicycle: the planet circles round a point which in turn encircles the Earth.

The equant: imagine the planet moving on an epicycle as before, with the Earth displaced from the centre (Figure 2.18). Ptolemy now drew attention to another point which he called the equant point (Eq). This was the same

Figure 2.18 Three Ptolemaic devices

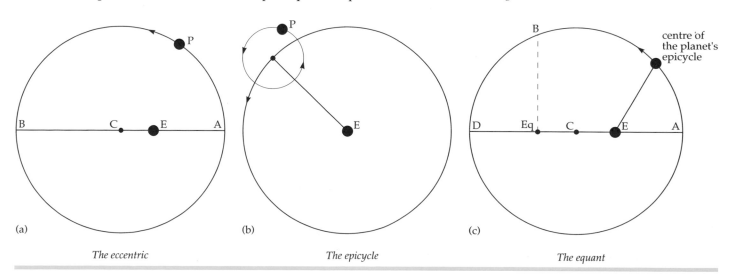

(a)   (b)   (c)

The eccentric   *The epicycle*   *The equant*

distance from the centre as the Earth but on the opposite side. He now supposed that the planet's motion would be uniform *only if viewed from the equant point*. Thus an imaginary observer there would see the planet always take the same time to go through any given angle. For instance it would move through the right angle from A to B in exactly the same time as through the right angle from B to D and so on. But for an observer on Earth of course the planet moves a lot faster when in his vicinity than when it is further away. In mathematical language, the planet has uniform angular velocity about the equant point.

In this kind of way Ptolemy was able to treat each of the planets separately and obtain very good agreement with the results of observation. He became the first astronomer to provide a satisfactory account of nearly all the planetary phenomena then known. It thus became the task for all astronomers for the next thousand years to introduce such modifications as were necessary for the Ptolemaic system to be given an increased accuracy. But the basic ideas continued throughout that long period, though in the fifteenth century good technical advances were made in the Islamic world.

2.3 The achievement of Copernicus

The monumental achievement of Nicholas Copernicus lies disclosed in the pages of his *De revolutionibus*. The nearest thing to a posthumous publication, it remains his most impressive memorial. Written in Latin, it was translated into German by C.L. Menzzer in 1879, and into English by C. G. Wallis in 1939 as part of the *Great Books of the Western World* series. Other English translations have since appeared by Edward Rosen and Alistair Duncan. Excerpts have appeared in many languages.

The volume called *De revolutionibus* consists of six books, together with a dedication and two prefaces (to which we return later). The contents can be briefly summarized as follows:

Book I: a general statement of the book's central thesis, i.e. the centrality of the Sun and the planetary character of the Earth, and concluding with an exposition of trigonometrical methods for planetary calculations.

Book II: applications of these trigonometrical methods to movements of the heavenly bodies, concluding with a star catalogue (basically a corrected and updated version of Ptolemy's catalogue).

Book III: a more refined theory giving expression to his revolutionary proposal concerning the movements of the Earth, again with geometrical constructions and tables.

Book IV: the motion of the Moon, still treated as a satellite of the Earth but with greatly improved accuracy.

Books V and VI: the movements of the planets (respectively in longitude and latitude).

The true nature of Copernicus' achievement can be gauged by comparing Figure 2.14, the Ptolemaic Earth-centred universe, with Figure 2.20, showing the Copernican universe with the Sun at its centre. The transposition is one of staggering simplicity (though for clarity epicycles are omitted from both).

Copernicus thought that the Earth, far from being immobile and fixed, was subject to at least three kinds of motion:

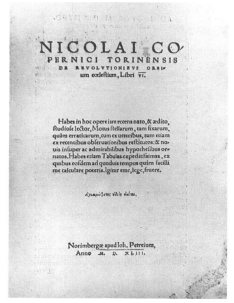

Figure 2.19 Title page of the first edition of De revolutionibus. *(Wydawnictwo 'Arkady', Warsaw)*

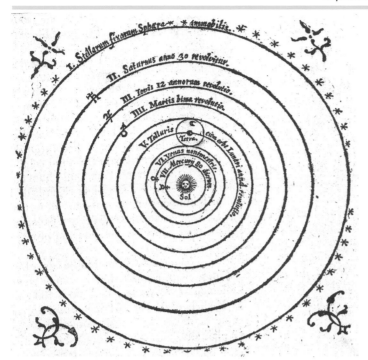

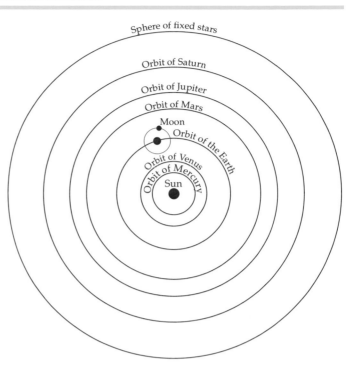

Figure 2.20 The Copernican universe

Annual motion: by journeying once a year round the Sun an observer on Earth will receive a succession of slightly different pictures of the universe. If a planet is similarly encompassing the Sun, but at a different rate, its appearance against the background of fixed stars will be readily explainable. A 'loop of retrogression' is thus a consequence of the movement of planet and Earth (see Figure 2.15).

Daily rotation: the rotation of the Earth accounts not only for day and night but also for the duration and character of the seasons. This supposes also that the axis of rotation is not vertical to the plane of the Earth's orbit. The tilting axis of a rotating Earth enables more sunlight to shine on a given area of the northern hemisphere at summer solstice than at any other time, and so on.

Precession: a rather complicated effect arises when the pattern of fixed stars is viewed over a very long period of time. Ptolemy knew of the effect but failed to offer a convincing explanation. Copernicus explained it in terms of a very slow 'wobble' of the Earth's axis as it spins, rather like that of a spinning-top. As we now know, the axis itself moves round slowly in a circle, meaning that any one part of the Earth faces slightly different directions over the whole cycle, which actually takes 26,000 years to complete (Figure 2.21).

Even this does not exhaust the achievement of Copernicus, for his scheme had an additional merit, of a rather different kind. It seemed *simpler* than the one it replaced and therefore *more elegant*. True, he still used Ptolemy's epicycles and other devices, and it is not true (as sometimes alleged) that there were fewer of them. But he did describe a universe with a new unity and simplicity. During his lifetime, and for long after, no single observation was made that decisively confirmed the doctrine of a moving Earth, but its instant appeal to those prepared to make the considerable mental leap necessary was its austere grandeur, its apparent simplicity and its intellectual integrity.

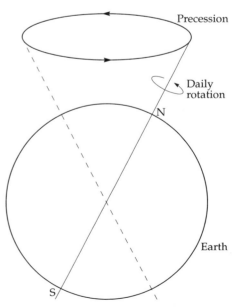

Figure 2.21 The rotation of the Earth's axis

For all his achievement we do well to remember that in just over a century the Copernican mechanisms had all been swept away by the work of Kepler and (later) Newton. Circles were replaced by ellipses, and the *Copernican system* evaporated in the light of universal gravitation. But the *Copernican theory* (a Sun-centred universe) lasted much longer, while the denial of a central motionless Earth is as acceptable to us as it was to Copernicus and his friends at Frauenburg.

And so we come to the most crucial question of them all: why did he do it? What was there in Renaissance Europe that prompted such a cosmic revolution? Did his theories 'just happen', or were they connected causally to the whole nexus of social and intellectual forces in which he lived? These are the questions that occupy the rest of this chapter.

2.4 The roots of the Copernican Revolution

Copernicus, like many another Renaissance man, faced a deep and complex dilemma. In essence, perhaps, it was simply a variation on the general theme of innovation v. tradition, an attempt to unloose the shackles of the past. But however grandly his work is described in terms of a liberation of the human spirit, a release over mediaeval bondage, or even a triumph of reason over superstition, it has to be admitted that he was a practical man of the world confronted with a very difficult *technical* problem. With the apparent lack of solutions to that problem he was profoundly dissatisfied. In eventually emerging with his own answer he had discovered in the new Renaissance culture subtle hints and promptings that could suggest and then underpin the most daring technical strategy for solving his problem. Four things seem to have spurred him on.

2.4.1 Criticism

In a word Copernicus was displeased with Ptolemy. The great astronomical work of this second-century Greek astronomer, *The Almagest*, had recently been paraphrased into Latin direct from the original Greek by the Köningsberg astronomer Johannes Müller, otherwise known as Regiomontanus.[5] The new printed paraphrase, though far from ideal, was greatly to be preferred to the corrupt mediaeval versions then available.[6] No longer was the problem chiefly a literary one (how can we be sure this is what Ptolemy actually said?), but rather a scientific one (how could he possibly be so wrong?).

There were actually several problems. Let Copernicus speak for himself. The 'mathematicians' were Ptolemy and his followers:

> Nothing except my knowledge that mathematicians have not agreed with one another in their researches moved me to think out a different scheme of drawing up the movements of the spheres of the world. For in the first place

Figure 2.22 Johannes Müller (more commonly known as Regiomontanus = 'the man from Königsberg'), 1436–76: German astronomer and writer. (Bildarchiv Preussischer Kulturbesitz)

[5] Regiomontanus (1436–76) was a student of Peurbach (Section 2.1).

[6] The first printed Greek edition of *The Almagest* was brought to Copernicus by Rheticus in 1539, but the astronomer had encountered it in Latin translation long before. A *printed* Latin translation (from the Arabic) appeared in 1515, and another (from the Greek) in 1528. But Copernicus may have seen earlier translations in manuscript, as well as the *Epitome* of it by Regiomontanus and Peurbach (1496).

mathematicians are so uncertain about the movements of the Sun and Moon that they can neither demonstrate nor observe the unchanging magnitude of the revolving year. *(Wallis, 1939, p.507)*

In other words their theory lacks both internal consistency and the capacity to measure the astronomical year. For a churchman who was much concerned with accurate predictions of festivals like Easter this was a serious flaw. Nor was Ptolemy deficient merely in astronomy; his geography was even worse:

> Ptolemy in his *Cosmography* extends the inhabitable lands as far as the median circle, and he leaves that part of the Earth as unknown, where the moderns have added Cathay and other vast regions as far as 60° longitude, so that inhabited land extends in longitude further than the rest of the ocean does. And if you add to these the islands discovered in our time under the princes of Spain and Portugal and especially America – named after the ship's captain who discovered her – which they consider a second *orbis terrarum* on account of her so far unmeasured magnitude – besides many other islands heretofore unknown, we would not be greatly surprised if there were antipodes... *(Wallis, 1939, p.513)*

Figure 2.23 Nicholas Copernicus in later life. (Mansell Collection)

This setback to the high authority of Ptolemy as a geographer inevitably helped to undermine his reputation in astronomy. Here is a further indictment:

> In setting up the solar and lunar movements and those of the other five wandering stars, they [Ptolemy's interpreters] do not employ the same principles, assumptions, or demonstrations for the revolutions and apparent movements. For some make use of homocentric circles only [i.e. with the same centre], others of eccentric circles and epicycles, by means of which, however, they do not fully attain what they seek. *(Wallis, 1939, p.507)*

It seems that the confusion alluded to was that between the Aristotelian model (with homocentric spheres) and the complex apparatus of the true Ptolemaic system (which even then was unsatisfactory). This differentiation between two much-used models might not have mattered were it not for one further consideration. What is really needed, according to Copernicus, is one 'sure scheme for the movements of the machinery of the world'. As it is, followers of Ptolemy can only have, at best, a partial picture of the universe:

> They are in exactly the same fix as someone taking from different places hands, feet, head, and the other limbs – shaped very beautifully but not with reference to one body and without correspondence to one another – so that such parts made up a monster rather than a man. *(Wallis, 1939, p.507)*

Even this does not complete the fusillade of accusations levelled at the unfortunate Ptolemy and his disciples. It is enough, however, to emphasize the great antipathy towards his scheme that grew in Copernicus's mind over several decades. The challenge now was to develop a viable alternative.

2.4.2 Observation

The classic picture of how a scientific revolution begins is usually something like this. For many years, even centuries, science works within a theoretical framework that serves it well. As new facts come to light they are incorporated in the structure, possibly causing slight modifications as time goes by. At length, however, the weight of newly discovered facts becomes so intolerable that minor adjustments are no longer feasible and the time-honoured framework eventually collapses. There is then a major reconstruction ahead and the new theoretical structure emerges from the wreckage of the old. The revolution has arrived.

Such, it has been suggested, was the situation when the old Ptolemaic framework was replaced by the new Copernican one. If so, there should be an impressive list of astronomical observations that were simply incapable of being accommodated within the traditional scheme. In the case of Copernicus it is plain why he was dissatisfied with Ptolemy, but not at all clear which astronomical *facts*, which observations, led him to replace the geocentric by a heliocentric model.

He was not, it must be said, a great observer. To be sure, he had an observatory in the tower at Frauenburg and at least occasionally used it. But in all the pages of *De revolutionibus* he needed to employ only 27 of his own observations. He lived before the invention of the telescope and his main concern was to measure angles between heavenly bodies at specific times, but he admitted that his best accuracy was only about 10 minutes of arc.

However there is evidence that he attached much significance to a few observations. In Bologna in 1497 he noticed that the disappearance of the bright star Aldebaran behind the Moon (Section 2.1) did not occur at the time predicted on the Ptolemy model. Conceivably this observation stirred the first doubts in his mind. Again, we know that the observations of a slow change in the background of fixed stars preoccupied him for years, and may indeed have induced him to construct the observatory at Frauenburg. His eventual resolution of the problem in terms of precession (Section 2.3) was possible only by abandoning the geocentric system.

Having said that, we are bound to admit that the observational evidence on which to base a revolution seems surprisingly thin. We are compelled to look elsewhere for the intellectual origins of Copernicanism.

2.4.3 The classics

Rediscovery of the ancient authors usually known to us as 'the classics' was one of the most conspicuous marks of Renaissance culture. At Wloclawek the young Copernicus encountered such Latin authors as Virgil and Cicero, an acquaintance that would be deepened at Cracow. Here studies in the philosophy and geometry of the Greeks introduced him to Euclid, Plato and Aristotle, though always in Latin translation. But it was under Italian skies that classical learning made an indelible mark upon him. From his own writing it becomes clear that the cosmology of Copernicus was deeply in debt to ideas once circulating in the classical world and half-forgotten almost until his own day.

A first, and almost trivial, example is remarkable only for its reminder of the great gulf between Copernican science and that of our own day. Few modern scientific treatises would go to Virgil for an illustrative quotation. In *De revolutionibus* Copernicus answers objections to his notion of a moving Earth: why does it *feel* stationary, why don't things fly off? and so on. His answer lies in the relativity of motion:

> Things are as when Aeneas said in Virgil; 'We sail out of the harbour, and the land and the cities move away.' As a matter of fact, when a ship floats on over a tranquil sea, all the things outside seem to the voyagers to be moving in a movement which is the image of their own, and they think on the contrary that they themselves and all the things with them are at rest. So it can easily happen in the case of the movement of the Earth that the whole world should be believed to be moving in a circle. *(Wallis, 1939, p.519)*

In similar vein a marginal note in one of the books he read at Padua refers to a passage from Cicero. This quotes an opinion of one Nicetas of Syracuse that if the Earth rotated everything would look just the same as if the Earth were to be motionless and the sky were to move.

However the classics had more to offer than illustrations of relativity. There were clear signals from antiquity that some philosophers long before Christ had entertained quite specific ideas of a moving Earth. Copernicus attributes these words to Plutarch:

> Some think that the Earth is at rest; but Philolaus the Pythagorean says that it moves round the fire with an obliquely circular motion, like the Sun and Moon. Heraclides of Pontus and Ecphantus the Pythagorean do not give the Earth any movement of locomotion, but rather a limited movement of rising and setting around its centre, like a wheel. (*Wallis, 1939, p.508*)

The speculations of these and other Greeks were unquestionably in circulation in the Italy visited by Copernicus in the years around 1500. They are worth a brief inspection.

Philolaus was a Pythagorean of the fifth century BC. Accepting the general idea of many spheres he supposed that these revolved not about the Earth but rather a huge Central Fire, as in Figure 2.24.

Unfortunately this Central Fire was never seen because it was always faced by the uninhabited part of the Earth. It is important to note that this was not the Sun (for this also rotated about the Fire); but unlike the schemes of Ptolemy, the ancient system of Philolaus did suppose that the Earth was on the move, going round the Fire once a day. This was rejected by Plato and his followers. The mysterious 'Counter-earth' may have been added to preserve the symmetry of the cosmos. Pythagoreans were always very interested in number and symmetry. But the location and orbit of this strange body must always be conjectural.

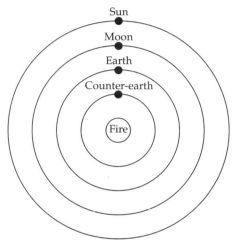

Figure 2.24 The universe according to Philolaus

Heraclides (fourth century BC) supposed Venus and Mercury only to revolve around the Sun while the latter encircles the Earth (Figure 2.25). This might have well turned out to be the start of a new outlook: if two of the planets revolve around the Sun, why not the rest? In the figure, E represents the Earth, S the Sun, V Venus and M Mercury.

However, this step was not generally taken. Instead Apollonius of Perga produced his own epicyclic system for the superior planets and the chance of moving to a completely heliocentric universe was lost.

Heraclides, however, had gone further than this for he also suggested that the Earth rotated daily on its own axis, so accounting for the rising and setting of the Sun and stars in a manner analogous to that of Philolaus. But this again was largely a lost cause and Heraclides remained a voice in the wilderness.

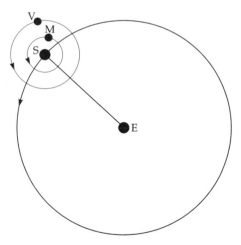

Figure 2.25 The universe according to Heraclides

Aristarchus was an important figure, and has been called 'the Copernicus of antiquity'. Aristarchus (third century BC) was an extremely capable astronomer and in his work on the *Sizes and Distances of the Sun and Moon* gave one of the earliest attempts to determine astronomical distances by applying geometry to observations with the naked eye. His measurements of the relative distance of Sun and Moon were wrong by a factor of about twenty, but this was due to inaccuracies in observation, not to any inherent defect in his methods. In the MS of *De revolutionibus* there was a passing reference to Aristarchus' belief in the mobility of the Earth, but this was removed before publication. No doubt in the fifteenth century there was a *general* awareness of Aristarchus' suggestion, but the fullest statement (and that was brief) was in Archimedes' *Sand-reckoner*, which appeared in 1544 after Copernicus' death.

Apollonius and Aristarchus moved in different directions from the starting point provided by Heraclides. It so happens (as we now know) that Apollonius was wrong and Aristarchus was right. Yet his ideas bore no fruit for 1,700 years and the conception of a heliocentric universe was therefore stillborn.

Were these wild ideas of any significance in suggesting to Copernicus a heliocentric alternative to Ptolemy? It is hard to believe that they had no impact, particularly in view of his references to them. Even they, however, seem less important than the collection of ideas and ideals enshrined in a cultic movement that burgeoned in Renaissance Italy during the time of Copernicus: neo-Platonism.

In 1509 the artist Raphael decorated the walls and ceiling of the Stanza della Segnatura of the Vatican Palace in Rome. There is a fresco depicting 'Dispute concerning the holy sacrament', whose rich theological symbolism was a constant reminder of the power and triumph of the church. Opposite is another Raphaelite fresco: 'The school of Athens', portraying Plato, Socrates, Aristotle and five other Greek philosophers. At this meeting place of Jerusalem and Athens there could be no doubt of the exalted position held by the classical authors. Although the fresco was executed nine years after Copernicus's first visit to Rome it splendidly embodies the penetration of Christian culture by the rejuvenated ideas of Plato. That Copernicus was aware of them there can be no shadow of doubt.

A leading role in the attempted union of Platonic and Biblical ideas was played by the Florentine scholar Marsilio Ficino (1433–99). He translated Plato's *Works* into Latin (1485) and a copy owned (and annotated) by Copernicus has been reported. It is also known that other books by Ficino were circulating in Cracow at the time that Copernicus was a student there. Another supporter of neo-Platonism was Domenico da Novara with whom Copernicus worked (and possibly lodged) in Bologna.

Neo-Platonism was an other-worldly philosophy in which mathematics was the key to understanding God and the universe. Of all mathematical figures the circle was deemed the most perfect. Plato, like Aristotle after him, reduced all movements in the skies to circular motion, either simply or in combination. And this became for Copernicus what Butterfield described as 'almost an obsession for circularity and sphericity'. He described 'that fantasia of circles and spheres' as 'the trade-mark of Copernicus, the very essence of undiluted Copernican thought' (Butterfield, 1957, p.31).

However, Ptolemy was also committed to circularity so how may we explain his overthrow? Copernicus wrote of those who, while employing epicycles and the other devices, nevertheless 'have in the meanwhile admitted a great deal which seem to contradict the first principles of regularity of movement [*or* uniformity of motion]'. In his alternative scheme the one device formally omitted is the equant. This, while permitting truly circular motion, admits *uniform* circular motion only about an imaginary point and not about the Earth. It thus had a fictional quality unworthy of neo-Platonism. It is a measure of the conservatism of Copernicus that he finds fault with Ptolemy because he has departed from true Platonic philosophy.

There was one final feature of classical thought that resurfaced in the Renaissance, became one of the most distinctive marks of neo-Platonism and burned itself into the consciousness of Copernicus. This was the central theme of many of Ficino's writings, including *De sole et lumine* (On sun and light) which we know was available in Poland, and probably familiar to Copernicus.

Writers as diverse as Christopher Columbus and Leonardo da Vinci expressed great admiration for the Sun. Sun worship was to become a durable tradition, being expounded by many later authors, such as Campanella in his *City of the Sun* (1620). Much earlier Michelangelo's 'Last Judgment' in the Sistine Chapel presents Christ as the Sun; this fresco has been described (with some exaggeration) as 'a pictorial vision of heliocentricism'. More remarkably still the chalice of the Communion Service was often replaced by a vessel ('monstrance') surmounted by a golden symbol of the rising Sun. The change may be seen by comparing Raphael's preliminary drawing and the final execution of his fresco 'Dispute concerning the Holy Sacrament' of 1509.

It is surely this tradition that is visible behind that most famous passage of *De revolutionibus*, Copernicus's own hymn to the Sun:

> In the centre of all rests the Sun. For who would place this lamp of a very beautiful temple in another or better place than this from which it can illuminate everything at the same time? As a matter of fact, not unhappily do some call it the lantern; others, the mind, and still others, the pilot of the world. Trismegistus calls it a 'visible God'; Sophocles' Electra, 'that which gazes upon all things.' And so the Sun, as if resting on a kingly throne, governs the family of stars which wheel around. *(Wallis, 1939, p.526)*

In these ways we can begin to see a little of the impact upon Copernicus of the revival of classical learning in Renaissance Europe. This particular scientific revolution manifestly owed at least as much to cultural influences remote from traditional science as to specific astronomical discoveries. It therefore raises much more general questions about the dependence of

Figure 2.26 Raphael: Dispute Concerning the Holy Sacrament. *(Scala Foto, Florence)*

science on a wider culture and lays open the possibility that the rise of scientific Europe will require explanations at several levels, and not merely those in the conventional scientific terms of today.

2.4.4 Aesthetic economy

Even the three sets of explanations advanced so far do not fully account for the radical proposal of a heliocentric universe. The unreliability of existing authorities (Ptolemy), the pressure of observational data and the mystical notions embedded in his culture (the classics) were obviously important to Copernicus, just as similar considerations are at the root of many another scientific revolution. Clear parallels may be suggested for the intellectual transformations associated with (say) Newton and Darwin. Yet there is one other common feature of many scientific advances, great and small. It is the human desire to *simplify*, to reduce the number of relevant parameters needed to explain a phenomenon. The famous 'Ockham's Razor' of the fourteenth century can be rendered 'entities are not to be multiplied beyond necessity'. This simplifying principle has moulded the shape of science from the days of Copernicus to those of sub-atomic particle physics. The scientist has an aesthetic longing for *economy*.

Elegance and simplicity were the true hall-marks of reality, and especially mathematical elegance. Novara became one of the first to criticize Ptolemy's system on the grounds of its complexity, with its multiplication of 'devices' like the equant to 'save the phenomena', or accord with observations. In similar vein Copernicus complained about the failure of Ptolemaic astronomers to deduce the true shape of the universe, with 'the unchangeable symmetry of its parts'. He at least made do with *one* scheme for the whole planetary system, whereas Ptolemy needed a separate one for each planet.

Consider the Copernican universe. To be sure prediction of the position of each planet still needed an epicyclic combination of two separate motions, but one of these was now *the motion of the Earth* and was therefore common to all five major planets. So Copernicus was able to delight in a new-found *symmetria*, and (as Rheticus said) everything is now linked together as if by a golden chain. The scheme immediately included the retrogressions of the planets in one unified scheme, explaining (as Ptolemy could not) why the retrogression of any planet occurred at opposition (i.e. when planet-Earth-Sun are in a straight line). And when Copernicus celebrated 'a sure bond of harmony for the movement and magnitude of the orbital circles' he was simply rejoicing in the elegance of a scheme whereby the major planet remotest from the Sun (Saturn) takes longest to complete one circuit (30 years), whereas the nearest (Mercury) has the shortest period (88 days). All the others fit neatly in order in between. Elegance, simplicity and economy, these were the hallmarks of his revolutionary proposal, and one can hardly doubt that they played an important part in leading him to it.

2.5 Copernicus and the Reformation

There are certain obvious similarities between those two great discontinuities in European thought: the Protestant Reformation in the west and the Copernican Revolution in the east. Both involved trenchant criticism of contemporary practice and theory; the Roman Catholic Church and the Ptolemaic astronomers were each indicted for their errors. Secondly, both movements involved reversion to ideals of a distant past. Luther and his followers called men back to the doctrines and practice of the New Testament Church, uncluttered by the accretions and modifications of a dozen or more centuries. Copernicus recalled astronomers to the truly Platonic ideals of circularity, rejecting Ptolemy's compromises and 'fudges' as unworthy. Yet, in the third place, both reformers were seen to share a reluctance to be radical enough. Luther's 'compromise' over the doctrine of the mass ('consubstantiation') was followed by the more thorough-going ideas of Calvin and Zwingli. Similarly, it has been said of Copernicus that 'he did not see how rich he was', and it required a Kepler to sweep away the surviving Ptolemaic devices in his scheme and replace his circles with ellipses.

Such apparent parallels can be perilous for the historian who is required not merely to call attention to analogous trends but, if possible, to establish some kind of causal connection between them. The questions that need to be pressed are therefore these:

- In what ways (if any) did the changing climate of European religious thought affect the conception and publication of *De revolutionibus*?

- In what ways (if any) was the spread of Copernicanism in Europe promoted or impeded by the Reformation?

The second question is a major concern of the next two chapters; examination of the first question occupies the remainder of the present chapter. There unfolds a tale of complex cross-currents, contrived ambiguities and delicately shifting personal relationships. It is most easily told by considering five central figures: Copernicus and four of his most influential friends. Between them they cover a spectrum of attitudes and disclose something of the theological and social forces at work.

2.5.1 John Dantiscus

The Prussian humanist John Flachsbinder (1485–1548) took the Latinized name Dantiscus (from his native city of Danzig). As Polish ambassador to the Emperor Charles V in the Netherlands he came into close contact with the teachings of the reformer Desiderius Erasmus (1469–1536). This ardent advocate of Christian humanism sought a return to early Church practice and teaching and promoted fresh translations of the New Testament from the original Greek. Yet Erasmus remained within the old Church, advocating reform rather than schism and tolerance instead of dogmatic inflexibility. He thus maintained good relationships with progressive members of his own Church as well as leaders of the emerging Protestant bodies. Under his influence many efforts were made before the Council of Trent (from 1545) to effect reform from within, and these ideals of a peaceful synthesis made a profound impression on Dantiscus.

In true Erasmian spirit Dantiscus paid for a new Latin translation of the Psalms (1532) by Johannes van Campen, otherwise known as Campensis, professor of Hebrew at Louvain. This was printed at Nuremburg by J. Petreius who also printed *De revolutionibus* eleven years later. Both Dantiscus and Campensis deplored the reactionary stand by certain prelates, notably the ultra-conservative Cardinal Aleander. And each maintained friendly relations with such leading Protestants as Philip Melanchthon, whom Dantiscus first met at Wittenberg in 1523.

Partly through the influence of Campensis (who taught at Cracow in the 1530s) and Dantiscus himself, Erasmianism was strongly felt in early

Figure 2.27 Desiderius Erasmus, c. 1469–1536: Dutch theologian and classicist (the first man to teach Greek at Cambridge), translator of the Greek New Testament and fervent advocate of reform within the Catholic Church. (Courtesy of the Galleria Barberini, Rome. Photo: National Portrait Gallery, London)

Figure 2.28 Philip Melanchthon, 1497–1560: Professor of Greek at Wittenberg, a leading Renaissance humanist, Protestant strategist and educational reformer. (Niedersächisches Landesmuseum, Landesgalerie, Hanover)

Figure 2.29 Johann Flachsbinder (more commonly known as Dantiscus = 'the man from Danzig'), 1485–1548: Prussian diplomat and eventually Bishop of Kulm (1530) and of Ermland (1538). (Published by permission of the British Library Board.)

sixteenth-century Poland. Copernicus came into close contact with Dantiscus' ideas when the latter became Bishop of Ermland, and thus his immediate superior, in 1538.[7] Whether through old age, corruption of high office or for other reasons Dantiscus thereafter began to adopt a harder line to dissident opinions, denouncing Lutheranism as heresy. This made it rather difficult for Copernicus who, at that very time, was entertaining Rheticus, the Lutheran visitor who was to play a crucial role in publishing his theory. Copernicus was peremptorily ordered to discontinue relations with a fellow canon, Alexander Sculteti, a known enemy of reaction. And there was also the affair of Anna Schilling. This highly talented daughter of a Torun goldsmith had served Copernicus for many years as housekeeper, but for some reason the arrangement incurred episcopal wrath and repeated instructions to terminate the relationship (whatever it really was). Eventually, in 1538, the astronomer capitulated to Dantiscus' demands.

Yet although 'the prelate and the humanist had opposite interests' (Hooykaas, 1984, p.22), humanist values reasserted themselves after these incidents and in 1541 Dantiscus wrote cordially to Copernicus with an epigram for the title-page of *De revolutionibus*. However ambivalent he may have been with respect to Protestant theology he appears to have had no difficulties in accepting a heliocentric universe. With the later eclipse of Erasmianism in Catholic Europe matters were to take a very different turn for the Copernicans.

2.5.2 Tiedemann Giese

Tiedemann Giese, friend, companion and admirer of Copernicus, was the bishop who succeeded Dantiscus at Kulm and Ermland. Two things about him are certain: he was a liberal follower of Erasmus and a man of great influence on Copernicus.

In a work of 1525, *Antilogikon*, Giese sought to engage with a Lutheran bishop, George von Polentz. Yet admitting 'I dislike any kind of strife', he often agrees with his opponent, condemning only empty rituals and superstitions and deploring the lack of constructive dialogue between the two parties in the Church. As for the Catholic practice of burning Protestants, as in Danzig, 'the wild animals behave more gently to their kind than do Christians to theirs'. Even more Erasmian in tone was his unpublished *De regno Christi* (On the reign of Christ) of 1536, making concessions to the reformers while maintaining other aspects of Catholic doctrine. Manuscript copies were sent to Erasmus and Melanchthon, seeking their opinions. The friendship between Copernicus and Tiedemann Giese is well documented. As colleagues at Frauenburg they represented the Cathedral chapter in ecclesiastical discussions, cooperated in post-war reconstruction in Ermland, even spent summer holidays together. Copernicus became personal physician to Giese, travelling over to Kulm to treat a malarial infection. When it came to publishing, Copernicus encouraged Tiedemann Giese with his *Antilogikon*, clearly identifying himself with its general stance, while *De revolutionibus* was published at the strong urging of Giese. In the difficult matter of Anna Schilling it was Giese who eventually persuaded Copernicus to obey orders from his bishop.

Figure 2.30 Tiedmann Giese, 1480–1550: Bishop successively of Kulm and Ermland, advocate of Church reform and close friend of Copernicus. (Published by permission of the British Library Board.)

Thus there is substantial likelihood that Copernicus endorsed the Erasmian spirit of toleration pervading parts of the Catholic Church in sixteenth-

[7] He had previously (1530–8) been Bishop of Kulm, being succeeded there by Tiedemann Giese (who later succeeded him at Ermland).

century Poland. The isolation at the end of his life may partly reflect the changing situation at Frauenburg where attitudes towards Protestantism were noticeably hardening. But questions have long remained as to how far his relaxed approach to the Reformation was able to accommodate the new system of cosmology. It now appears that some of them may be given a more than tentative answer. The clues lie in the work and writings of a third member of the Copernicus circle.

2.5.3 Joachim Rheticus

Rheticus, or Georg Joachim Iselin (1514–76), was one of Copernicus' most ardent disciples. He was a professor of mathematics at Wittenberg who, after learning of the then unpublished Copernican theory, resolved to seek out the author in his Frauenburg canonry. During his stay with Copernicus of two and a half years, he learned much technical astronomy and imbibed the spirit of his master. They even joined Tiedemann Giese on a long visit to his bishop's residence at Löbau. Having published his *Narratio prima* in 1540 Rheticus eventually became instrumental in starting through the press the *De revolutionibus* itself. Copernicus owed Rheticus a great debt of gratitude.

Unlike Dantiscus and Giese, Rheticus was a Lutheran, and a layman at that. That his friendship with Catholics like Giese and Copernicus was so firm would have been truly remarkable had it not been for the Erasmian outlook shared by them both. Clearly they must all have faced up to the theological implications of Copernicanism (see Chapter 3), but if so left behind few traces of their conclusions. Tiedemann Giese was known to have written a theological defence of Copernicus, *Hyperaspisticon* (Shieldbearer), but it does not seem to have survived. And remarks by Giese about a little treatise by Rheticus on the same subject served only to tantalize scholars for the next four centuries. No such treatise was discovered. However in 1973 it was at last located by the Dutch historian of science Professor R. Hooykaas, anachronistically included in a collection of seventeenth-century tracts. Although it is anonymous the internal evidence leaves practically no room for doubt that this is indeed the lost document.

Rheticus' approach is one that became commonplace in the years after Galileo. He just happens to be the first to enunciate it. Taking a long-established patristic view of the Bible he argues that its purpose was never to teach 'scientific' truth but rather spiritual and moral realities. Similarly, as Copernicus himself wrote, 'astronomical matters are written for the astronomers', and not, by implication, for the theologians. So Biblical references to a stationary Earth are concessions to the 'vulgar' people, who would not be able to understand anything else (nor would they need to). Yet Rheticus is not consistent, and having established the irrelevance of Scripture to a cosmological description he then uses it to dispose of his conservative opponents (by arguing that some passages *do* refer to a moving Earth, for example). This is in character with what is known of Rheticus himself. Impetuous enough to forsake his post in search of Copernicus, he was also considered by his teacher Melanchthon to be 'Enthusiasmus' – i.e. temperamentally unstable. Yet this cannot undermine the contention that Copernicus and his immediate circle felt no threat to their views from Scripture and that they believed in the reality of the new heliocentric system. And underpinning all their understanding lay a very Erasmian view of the church which, while committing them to a fully Christian philosophy, allowed them much liberty in interpretation. It was this very flexibility of mind that enabled them to initiate the Copernican Revolution.

2.5.4 Andreas Osiander

It is well known that the first edition of *De revolutionibus* contained an anonymous Preface by the Lutheran clergyman Andreas Osiander (1498–1552). Having studied at the university of Wittenberg, Osiander became, in 1522, the leading Protestant minister in Nuremburg. A participant in the Marburg Colloquy of 1529 (a vain effort to unite Protestants in Europe), and then at the more famous Diet of Augsburg[8] (1530), Osiander was later to propose a modified version of Luther's doctrine of justification by faith. No mean theologian, he had given Copernicus advice about the possible reception of a book declaring the Earth in motion. Would theologians and scholastic philosophers be up in arms? He replied to the effect that, since astronomical theories are not articles of faith, it is immaterial whether they are true in reality or false. Present them as mere calculating devices and theologians will be silent. Writing to Rheticus on the same day he gave the same advice: by stressing the hypothetical nature of the various schemes 'peripatetics and theologians will be easily placated'.

It is an ironic fact of history that it was Osiander to whom fell the task of seeing *De revolutionibus* through the final stages of production. True to his conviction as to the desirability of heading off opposition even before it developed, he added his own Preface 'To the reader: concerning the hypotheses of this work'. There was no intention to deceive, pretending the Preface was the work of Copernicus, for Copernicus is lauded as another individual. But it roused the wrath of Giese, for it was entirely unauthorized and moreover in contradiction to the views of Copernicus and himself. Yet it must be admitted that, in the circumstances then prevailing, Osiander was technically correct, since no empirical facts were then known that unambiguously distinguished between Copernican and Ptolemaic systems. Either could be correct, therefore each had to be a hypothesis.[9] Yet Osiander's argument had another significance.

Figure 2.31 Andreas Osiander, 1498–1552: Lutheran clergyman and theologian from Nuremburg, and author of the anonymous Preface to the first edition of Copernicus De Revolutionibus. *(Bildarchiv Preussischer Kulturbesitz)*

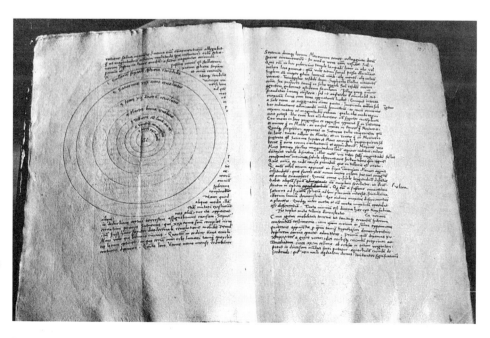

Figure 2.32 Copernicus' own copy of his book De revolutionibus. *The hole left by his compass in the left-hand page is visible in the original. (from Erich Lessing,* Discoverers of Space, *courtesy of Erich Lessing and Search Press Ltd)*

---

[8] The Diet of Augsburg was the church assembly at which the Lutheran confession of faith was presented to the Emperor Charles V.

[9] Only the discovery of stellar parallax by F.W. Bessel in 1838 gave the first *unambiguous*, direct, observational proof of a moving Earth.

The Osiander episode suggests that at that early stage (in 1541) some opposition was to be expected from the Protestant camp. Had Osiander but known it Melanchthon was even then considering the case for censure against proponents of a moving Earth. Yet Melanchthon retained a good relation with Rheticus (for all the latter's 'enthusiasm'), gave him letters of recommendation and invited him to his daughter's wedding. It seems that any categorization of individuals, let alone movements, into one side or the other is far too simple. Like many another after him Osiander was *anticipating* possible reaction, certainly not *predicting* it.

2.5.5 *Copernicus*

Our final glimpse must be of Copernicus himself: reading, calculating, writing, hesitating, recalculating and rewriting and only 'going public' after his ideas have been 'leaked' all over Europe. Even in faraway Ermland he is aware of the impending cataclysm about to befall the Catholic Church, and in his own country 'martyrdom' is acquiring a sinister new meaning. Before the storm ultimately breaks, unleashing pent-up forces of unpredictable and devastating power, he must decide what to do.

There are powerful arguments for doing nothing at all. For one thing, Copernicus is a loyal member of the Catholic Church, and, though not a priest, holds the high office of a canonry within it. Openly to challenge the received wisdom about the cosmos (or anything else) might be seen by an institution with enough troubles on its hands already as the last straw, an act of defiance if not treachery. Does it really matter how one performs astronomical calculations if disaster is just around the corner? Does it even matter if the Earth is in motion if the foundations of civil society are shaking? Then again, he has many other matters to attend to, both as a church administrator and as a physician. Only two years previously (1541) he was asked by Duke Albert of Prussia to give medical help to a trusted servant who was seriously ill in Königsberg. The visit lasted many weeks but his mission was successfully accomplished. And, at the age of 70, one is surely entitled to a period of calm and tranquillity in one's personal affairs?

Yet there are voices that will not be silenced, not least those of Giese and Rheticus. And they are articulating a conviction deep in his own consciousness. The argument is not really about calculating devices, despite all that Osiander had written (and was about to write). It is about truth, for by whatever route he had come to his momentous conclusion Copernicus appears to have no doubt that his system represents reality. This conviction is shared with friends (which is why the usually gentle and peaceable Giese was later goaded to impotent fury by Osiander's anonymous Preface). For all the raging storms outside Copernicus is aware of the still small voice within, to which men have given many names: conscience, inner conviction, the voice of rational dissent, and so on. And *that*, together with a passionate search for truth, was the essence of Reformation theology.

And so Copernicus is reminded of the quest of Erasmus and his friends. If they appealed to the New Testament he must appeal to what others were to call the 'book of nature'; if they were to emphasize the rights of individual believers he must do the same for himself; if they sought for truth through dialogue and enquiry then he dare not remain silent. To this task he must now set himself with all the skill at his command.

In this way we can begin to understand the strategy that he chose to use. He had to publish, and by every means he had to fend off opposition within his own Church. So *De revolutionibus* was dedicated to the Pope, and contained commendatory references to Cardinal Schönberg and Bishop Giese. Possible theological difficulties are hardly touched upon. The confidant and medical adviser to a whole succession of bishops knew enough of the ecclesiastical mind to avoid needless controversy. In that spirit he gladly acknowledged help from the Catholic Giese but made no mention of the Lutheran Rheticus. But it was not a case of persuasion 'at all costs'. Copernicus refused to accept reduction of his system to a mere hypothesis; for him it was *true*, and he makes no attempt at a cover-up.

So he writes to the Pope:

> In order that the unlearned as well as the learned might see that I was not seeking to flee from the judgment of any man, I preferred to dedicate these results of my nocturnal study to Your Holiness rather than to anyone else; because, even in this remote corner of the Earth where I live, you are held to be most eminent both in the dignity of your order and in your love of letters and even of mathematics... But what I have accomplished in this matter I leave to the judgment of Your Holiness in particular and to that of all other learned mathematicians. *(Wallis, 1939, p.509)*

The judgment of His Holiness and his immediate successors was suspended for some time. The judgment of the learned mathematicians, on the other hand, was swift in some places, slow in others. In each case the revolutions of the heavenly bodies were accompanied by two of the greatest revolutions in human thought: the Protestant Reformation and the Copernican Revolution.

Sources referred to in the text

Butterfield, H. (1957) *The Origins of Modern Science, 1300–1800*, London, Collins.

Hooykaas, R. (1984) (translation, annotations, commentary and other material), *G.J. Rheticus' Treatise on Holy Scripture and the Motion of the Earth*, Amsterdam, North Holland.

Wallis, C.G. (1939) *Great Books of the World*, vol.16. Chicago, Encyclopaedia Britannica Inc.

Further reading

Armitage, A. (1971) *The World of Copernicus*, Menston, S.R. Publishers.

Bienkowska, B. (1973) (ed.), *The Scientific World of Copernicus*, Dordrecht, Reidel.

Bietkowski, H. and Zonn, W. (1973) *The World of Copernicus*, Warsaw, Arkady.

Boas, M. (1970) *The Scientific Renaissance 1450–1630*, London, Fontana.

Crombie, A.C. (1969) *Augustine to Galileo*, Harmondsworth, Penguin.

Dillenberger, J. (1961) *Protestant Thought and Natural Science*, London, Collins.

Hooykaas, R. (1973) *Religion and the Rise of Modern Science*, Edinburgh, Scottish Academic Press.

Jobert, A. (1974) *De Luther à Mohila: la Pologne dans le crise de la Chrétianité 1517–1648*, Paris, Institut Études Slaves.

Kuhn, T.S. (1966) *The Copernican Revolution*, Harvard University Press.

Nebelsick, H.P. (1985) *Circles of God*: *Theology and Science from the Greeks to Copernicus*, Edinburgh, Scottish Academic Press.

Acknowledgements

A small amount of this chapter appeared in Course Units for an earlier, now defunct, Open University Course, A201 *Renaissance and Reformation*. I am grateful to all the present Course Team and the External Assessor for their comments and suggestions. Most particularly I should like to thank Professor Owen Gingerich for his very detailed and constructive criticism. And I also gladly acknowledge the kindness of the Royal Astronomical Society for granting access to their library, as well as the University of Cambridge for almost continuous use of its University Library.

Appendix

Place names

Old German names	*Modern Polish names*
Allenstein	Olstyn
Danzig	Gdansk
Ermland	Warmia
Frauenburg	Frombork
Heilsberg	Lidzbark
Kulm	Chelmno
Löbau	Lubawa
Thorn	Torun

Personal names

Given names	*Latinized names*
Johannes van Campen	Campensis
Nicholas Koppernigk	Copernicus
(or Niclas Kopernik)	
Johann Flachsbinder	Dantiscus
Johannes Müller	Regiomontanus
Georg Joachim Iselin	Rheticus

The Spread of Copernicanism in Northern Europe

Chapter 3

by Colin A. Russell

3.1 Introduction: the paradox of the Copernican Revolution

Any observer of the Europe of the sixteenth and early seventeenth centuries is immediately confronted by a paradox and a puzzle, well expressed by a distinguished European scholar in these words:

> The Copernican 'Revolution' was no revolution. As we now see, it may have inaugurated a revolution in science, but this took one hundred and fifty years to accomplish itself and before 1600 it went on at a very slow pace. Even then the acceptance of the Copernican system was far from general. *(Hooykaas, 1973, p.58)*

The paradox of the first sentence conveys some of the ambiguities inherent in the vexed concept of a revolution in science (let alone the 'Scientific Revolution'). Is it, or is it not, a quick, dramatic, almost instantaneous phenomenon? The question is largely one of semantics. But there remains a puzzle, or really a whole cluster of puzzles, centred round the problem of an immensely slow rate of acceptance. Is it indeed the case that the Copernican theory had a take-up period of nearly a century and a half? Did it lie fallow in European culture, as it were, germinating so slowly that for all those years people simply forgot about it? Or was there uneven development, with a few enthusiasts here and there to provide inconspicuous nuclei for the later growth of Copernicanism? Answers to all these questions presuppose some agreed means of measuring the extent to which a theory may be adopted, and should not depend on purely subjective impressions. And if, and when, such answers may be found there remains the all-important question: why?

This chapter will examine the situation in Northern Europe. Chapters 4 and 5 will touch on similar problems for the more southerly countries of Italy (4) and Portugal and Spain (5). Here the first problem is to identify, if possible, the actual centres where Copernicanism first took root, and then to report the pattern of its later development in the early seventeenth century. The concluding sections will attempt an analysis of the data and a tentative answer to the questions arising.

The traditional way to ascertain whether, and how far, a new theory is adopted is by means of citations. You simply examine all known cases of published works explicitly referring to it or its author. That is as true for today's scientific ideas as it is for those of four centuries ago (except that modern writers are immortalized in computerized citation indices and the problem is instantly soluble). For the age of Copernicus, however, things are not quite as easy. In fact there are very few incidental references to either him or his theory much before 1600 (though this is certainly not the case after that date). For this reason it has even been suggested that virtually no Copernicans existed in the sixteenth century.

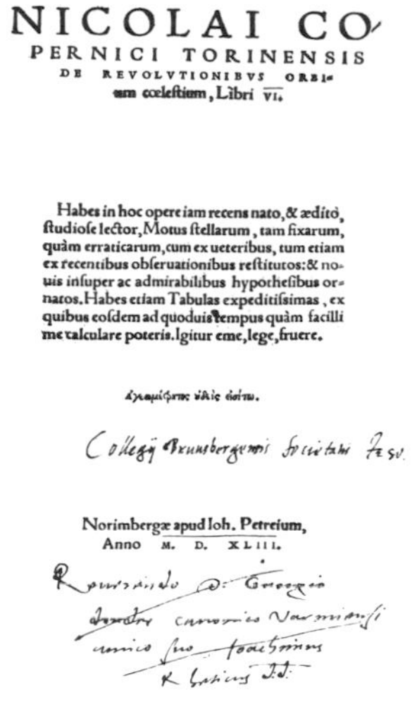

NICOLAI CO
PERNICI TORINENSIS
DE REVOLVTIONIBVS ORBI-
um cœleſtium, Libri VI.

Habes in hoc opere iam recens nato, & ædito,
ſtudioſe lector, Motus ſtellarum, tam fixarum,
quàm erraticarum, cum ex ueteribus, tum etiam
ex recentibus obſeruationibus reſtitutos: & no-
uis inſuper ac admirabilibus hypotheſibus or-
natos. Habes etiam Tabulas expeditiſsimas, ex
quibus eoſdem ad quoduis tempus quàm facili
me calculare poteris. Igitur eme, lege, fruere.

Ἀγεωμέτρητος οὐδεὶς εἰσίτω.

Collegij Braunsbergensis Societatis Jesu.

Norimbergæ apud Ioh. Petreium,
Anno M. D. XLIII.

Reuerendo D. Georgio
Donnero canonico Varmiensi
amico ſuo faciendum
R. Lossius Jt.

Figure 3.1 *Title page of first edition of*
Copernicus' De revolutionibus. *This copy was
for nearly a century lodged in the Jesuits' Library
at Braunsberg before being captured by the Swedes
and brought to Uppsala. The lower inscription
reveals that it was presented by Rheticus to
George Donner, Canon of Warmia at Frauenburg,
a colleague of Copernicus himself. (University of
Uppsala)*

However, there are other strategies available. One can, for instance, examine even the mere handful of references that have been found and assess whether they are simply passing allusions or represent a rather more sustained attempt to disseminate the doctrines concerned. In other words, one has to evaluate their significance. When this is done for Copernicus it will be seen that the wholly negative view just mentioned is not sustainable. Several books do exist which clearly sought to proclaim the Copernican theory. How far they were successful is not presently the point. Their very existence proclaims an importance for the ideas that cannot be gainsaid. No author is likely to have written such polemics without at least some support in his own circle of friends and disciples. There, at least, Copernicanism was known and favoured.

Another technique has recently been applied with remarkable results. Copernicus was of course the author of a book, *De revolutionibus*. It might be possible to track down surviving copies and thus ascertain where they went and to whom they belonged. This method has been used by Professor Owen Gingerich in what he once described as 'the great Copernicus chase'. He started with the published Book Auction Records from 1888, his subsequent trail taking him to the great public libraries and private collections in Europe and America. Starting with the assumption that 'it is entirely possible that more people alive today have read through *De revolutionibus* than in the entire sixteenth century', he was driven to the conclusion that this was in fact quite wrong.

As it was not uncommon in the sixteenth century for owners to add their own annotations in the margins of books (and *De revolutionibus* had very wide margins) it is sometimes possible to identify not only the owner but also his attitudes to many of the questions raised by the book. And of course there were sometimes book-plates as well, giving unambiguous evidence of ownership (though not necessarily of the book's first owner). Amongst copies examined were those associated with Rheticus, Reinhold and Kepler. Altogether, Gingerich had by 1979 been able to locate 245 copies of the 1543 edition, from which he estimated that a print-run of 400–500 copies was likely (Gingerich, 1979).[1] This fact alone makes it highly improbable that Copernicus was largely unheard of in the half-century immediately following his death. This is confirmed by another simple fact. *De revolutionibus* went into a second edition in 1566 (Basle), and yet a third in 1617 (Amsterdam). No one would bother to republish a book without significant contemporary demand, and it seems incredible that Copernicus had sowed entirely on stony ground or that everyone used his book simply as a calculating device. Nevertheless it does not follow that the reader or publisher had to accept all its cosmology; thus N. Mullerius of Groningen, who edited the 1617 edition, denied the revolution of the Earth but found the book indispensable for teaching mathematical astronomy. It is clear that any general lack of European enthusiasm for Copernican cosmology must betoken some fairly profound and widespread objections. Ostensibly, these were of two kinds: scientific and religious.

---

[1] The number now stands at over 260 (private communication).

3.2 *Scientific objections to Copernicanism*

Two things are remarkable in this connection. In the first place, few objections (scientific or otherwise) may be found in literature of the sixteenth century – i.e. during the period when Copernicanism was generally ignored. This could be because it was also unknown (which, as we have seen above, is unlikely) or because the objections were so self-evident that writers declined to state the obvious. The latter seems to be more probable. Secondly, the spokesmen for the opposition were rarely active in astronomy, but were quite often poets or philosophers.

The obvious difficulty with Copernicanism is the common-sense reluctance to accept that the Earth is in motion through space.[2] It just does not feel that way. The French political writer and philosopher Jean Bodin (1530–96) scorned Copernicanism for that reason, in his *Universae naturae theatrum* saying:

> No one in his senses ... will ever think that the Earth, heavy and unwieldy from its own weight and mass, staggers up and down around its own centre and that of the Sun; for at the slightest jar of the Earth, we would see cities and fortresses, towns and mountains thrown down. *(From Stimson, 1917, pp.46–7)*

The Englishman John Donne wrote in similar vein:

> Why may I not beleeve that the whole earth moves in a round motion, though that seeme to mee to stand, when as I seeme to stand to my Company, and yet am carried in a giddy, and circular motion, as I stand? *(Donne,* Devotions upon emergent occasions, *XXI)*

Quite apart from the lack of giddiness there was the vexed question of a missile thrown vertically upwards, yet not landing to the west, in the kind of way to be expected if launched from a chariot or other moving object. The point was made in an anti-Copernican poem of 1578, *The week, or the creation of the world,* by the Gascon poet William du Bartas (1544–90), which was widely circulated in English translation:

> So, never should an arrow, shot upright,
> In the same place upon the shooter light;
> But would do, rather, as, at sea, a stone
> Aboard a ship upward uprightly thrown;
> Which not within-board falls, but in a flood
> Astern the ship, if so the wind be good ...

Armed with these and other reasons he ironically observed that

> 'twere superfluous
> T'assail the reasons of Copernicus
> Who, to save better of the stars th'appearance,
> Unto the earth a three-fold motion warrants.   *(From Kuhn, 1966, pp.190–1)*

In more jocular mood, and rather later (1642), Sir Thomas Browne (1605–82) commented in his *Pseudoxia epidemica*:

> It ... is no small disparagement unto baldness, if it be true what is related by Aelian concerning Aeschilus, whose bald-pate was mistaken for a rock, and so was brained by a Tortoise which an Aegle let fall upon it. Certainly it was a very great mistake in the perspicacity of that Animal. Some men critically disposed, would from hence confute the opinion of Copernicus: never conceiving how the motion of the earth below, should not wave him from a knock perpendicularly directed from a body in the air above. *(From Meadows, 1966, p.72)*

Figure 3.2 Jean Bodin (1530–1596), French Roman Catholic political writer who rejected Copernicanism in 1597 on grounds of commonsense. (Bibliothèque Nationale, Paris)

---

[2] We now know the Earth to have a rotational velocity of about 1038 mph (0.18×10^3 m sec^{-1}) and an orbital velocity of about 648,000 mph (11×10^3 m sec^{-1}); although these precise figures were not known in the sixteenth century their approximate orders of magnitude were available.

All these objections eventually vanished with the new post-Galilean mechanics. But more serious difficulties remained for a longer time. These concerned the Copernican view that the stars were an immensely great distance away. Otherwise their angular distances would be different when viewed from opposite extremes of the Earth's annual orbit. No such discrepancy was observed for several centuries, though the greatest observer of the sixteenth century, Tycho Brahe (see below), attempted to discover one. Many astronomers were uneasy about the vast empty spaces between stars, which was as much an aesthetic difficulty as a scientific one. The economy thought to have been achieved by Copernicus in terms of mechanism was dangerously near being lost in terms of scale. And the vast speeds envisaged in the Ptolemaic system were being replaced by vast distances (and therefore sizes).

Astronomers in all the Northern European countries were aware of these problems, which probably played some part in delaying the general acceptance of a heliocentric universe. However considerations of a quite different kind were also involved in forming attitudes to Copernicanism. These were essentially religious in nature, and owed much to the renewed theological awareness of Reformation Europe.

3.3 Religious attitudes to Copernicanism

It used to be said that initial opposition to the Copernican doctrines came from the Protestant Church, that Roman Catholic hostility did not erupt till the trial of Galileo (1633) and that the critical issue in both cases was the authority of the Bible. Each of these statements is a considerable distortion of the facts as we now have them, though, as is often the case with wild generalizations, each has a small element of truth.

At most, if not all, the European universities in the sixteenth century scholastic Aristotelian philosophy was still entrenched. This was true in the Protestant world (as at Wittenberg) and in that of Roman Catholicism (as at Paris). If Copernicanism were seen as destroying the very foundations of scholastic philosophy the traditionalists would have been unlikely to give in without a fight. To this end all weapons might be serviceable, notably those of common-sense 'scientific' arguments. But if the authority of Church and Bible could be invoked they would be in an even stronger position. The most generally satisfactory explanation for the delayed acceptance of Copernicanism is along these lines, with Scripture viewed through strictly Aristotelian spectacles and made the ostensible basis for an assault on Copernican cosmology. In this context it is actually hard to avoid the cynical view of historical conflicts that they are rarely about principles, but nearly always about power. For this reason, instead of dealing with the detailed theological arguments,[3] we concentrate on the responses from several power-centres in Northern Europe.

3.3.1 Germany: the 'Wittenberg interpretation'

The place from which Rheticus came to become Copernicus' first disciple, and to which he briefly returned later, was the old German town of Wittenberg. This has been described as the birthplace of the Reformation, not

[3.] These have often been discussed elsewhere (e.g. Dillenberger, 1961; Hooykaas, 1973, 1974; Russell, 1985).

chiefly because it was to the door of its Castle Church that Martin Luther nailed his famous 95 Theses, but rather because it was in the tower of Wittenberg's Augustinian monastery that he first conceived the immense task before him of reinstating the authority of the Bible within the Church.[4] In view of Luther's reputation for opposing Copernicanism it is remarkable that this same town was not only home to Rheticus but was also to nurture an influential school of theologians with decisive views on the same subject.

Countless books in the history of science have implied that Luther was responsible for Protestant opposition to Copernicanism, or even that he led a crusade against it. The facts are otherwise. Everything hangs upon a famous remark Luther is alleged to have made about Copernicus: 'the fool would upset the whole art of astronomy'. The phrase appeared in a conversation of 1539 first reported in a late edition of Luther's *Table Talk*; an earlier edition does not have it. Three points may be made: its authenticity is not certain; even if genuine its condemnation is hardly decisive – 'the designation "fool" is comparatively mild when used by Luther for an enemy'! (Dillenberger, 1961, p.37); and it cannot refer to *De revolutionibus* since that was published four years later and after Luther was dead (Norland, 1953, pp.273–6). Luther, it may be added, had a positive view of astronomy though he rejected the pretensions of astrology which he held in clear distinction.

If Luther had no authenticated attitude to Copernicanism the same cannot be said of his colleague Philip Melanchthon (1497–1560), Professor of Greek from 1518, and noted educational reformer. 'A charismatic man, beloved teacher, and talented humanist, Melanchthon was also a brilliant administrator with a gift for finding compromise positions' (Westman, 1986, p.82). He was above all a Renaissance humanist, with a high view of the classics and especially Aristotle. He promoted the study of Aristotelian works, including a thirteenth-century popular treatise which he republished, *On the Sphere* by Sacrobosco (John of Holywood). He held to the authority of Scripture and the unity of all knowledge (as propounded by Aristotle). Indeed, neglect of the sciences within the university would adversely affect everything else, including theology. So he promoted astronomical studies as a means of disciplining the mind and revealing the beauty of creation, thus initiating a strong tradition of mathematical astronomy which took hold in German, Danish and Swedish universities in the sixteenth century.

It is perhaps not surprising that Melanchthon had initial difficulties with Copernicanism, though he subsequently modified his attitude. There were two problems. By rejecting Aristotelian cosmology Copernicus seemed to Melanchthon to be not only undercutting an important part of humanistic thought but opening the gate to a return to dangerous pre-Aristotelian ideas, as in his echo of Aristarchus' views. That could even lead to the atheistic notions of Democritus. On the other hand, Copernicanism could also be held to conflict with a literal interpretation of a few Biblical passages which seemed to say that the Sun, not the Earth, was in motion.[5]

Melanchthon presented his own synthesis in a work on physics, *Doctrinae physicae elementa sive initia* (1545), reiterating the old arguments (physical and theological) for an immobile Earth and casting his astronomical discourse within a Ptolemaic framework. His rejection of heliocentricity is measured and relaxed, and his correspondence has favourable references to Copernicus. The latter was to be praised for correcting errors of the past and

Figure 3.3 The Castle Church at Wittenberg on whose door Martin Luther nailed his 95 Theses (the sixteenth-century equivalent of instant media publicity). (Illustration reproduced from Here I Stand *© 1977 Roland H. Bainton. Used by permission of Abingdon Press.)*

Figure 3.4 Martin Luther (1483–1546). A priest of Saxon origins, who became gradually disenchanted with abuses in his own Catholic Church. Making a decisive break with mediaeval Catholicism in 1517, he nailed to the door of Wittenberg Church his 95 Theses against the sale of indulgences, thus initiating the Reformation which spread like wildfire through Germany and beyond. It transformed the political face of Europe and the world, but for Luther himself was a liberating personal discovery of the original Gospel of Christ. (Bildarchiv Preussischer Kulturbesitz)

[4] As Luther himself was to put it 'the Holy Spirit unveiled the Scriptures for me in this tower'.

[5] These included Psa. 78:69 ('the earth which he had established for ever'), Josh. 10:12 ('Sun, stand thou still!'), Eccles. 1:4 ('the earth abideth for ever').

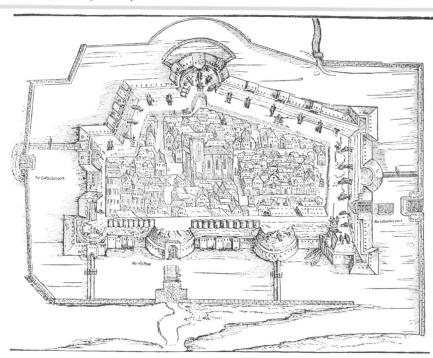

Figure 3.5 View of Wittenberg. (Mansell Collection)

for establishing a celestial mechanism based upon threefold circular motion. Copernicus, like Melanchthon himself, was to be regarded as a moderate reformer, rather than a dangerous revolutionary.

Melanchthon's ideas have been encapsulated in what one scholar has called 'the Wittenberg interpretation' of Copernican astronomy (Westman, 1986). This interpretation saw Copernicus as a moderate reformer, proposing a mathematical hypothesis that returned to the wisdom of the ancients and avoided the compromise solutions of Ptolemy. In that sense Copernicus might be justifiably compared to Melanchthon himself. The question of whether the Copernican idea was *true*, or even probable, was largely set to one side. Melanchthon's influence may be perceived in the small circle of followers who gathered round him in Wittenberg: Reinhold, Rheticus and Caspar Peucer (his future son-in-law), together with their own later disciples. All except Rheticus opposed the Earth's motion, but even he was encouraged and supported by Melanchthon. Others from Northern Europe, like the Polish prelate Dantiscus, had visited Melanchthon at Wittenberg and maintained cordial relationships with him. In the very ambivalence of this influential Wittenberg school towards Copernicanism lies one promising explanation for the slow acceptance of heliocentric ideas. A way had been shown for men of moderation to accept what was good in Copernicus and to reject unnecessary flights of fancy. At this time no compelling reasons existed in astronomy for abandoning a Ptolemaic world-view. Had Copernicanism been an issue at Wittenberg in 1542 it is doubtful if Rheticus would have been appointed Dean of Arts. There was no need to throw away the baby with the bath-water. The 'Wittenberg interpretation' was a perfect expression of the Christian humanism pervading the north-western countries of Europe in the sixteenth century.

3.3.2 *Switzerland: the doctrine of 'accommodation'*

If Wittenberg was in some ways the spiritual home of Lutheranism, that other major movement of the Reformation known as Calvinism was centred on the

Swiss town of Geneva. Here, in the thinking of John Calvin (1509–64), a refugee from France and leading Protestant theologian, ideas were developed and formulated that had a considerable effect on the fortunes of science at that turbulent period of European history.

Switzerland saw a remarkable growth of scientific thought in the sixteenth century, a growth which Ramus (see below) attributed directly to the impact of the Reformation. This was specially true of Basle which, as well as being an industrial centre, had found itself host to several important ecclesiastical councils and reforms. It was an obvious town from which to publish the second edition of *De revolutionibus*. It was to Basle that Calvin first fled after his expulsion from France in 1535, and from here that he issued the first edition of his monumental *Christianae religionis institutio* (better known as Calvin's *Institutes*). However by 1541 he had settled in Geneva, where he became the foremost theologian of early Protestant Europe.

Calvin encouraged the study of science: 'Being placed in this most delightful theatre let us not decline to take a pious delight in the clear and manifest works of God' (*Institutes*, I, Chapter 5). He wrote more specifically of astronomy: 'Astronomy is not only very pleasant, but also very useful to be known; it cannot be denied that this art unfolds the admirable wisdom of God' (*Commentaries on ... Genesis*, 1554, I, pp.86–7).

Yet there remained the problem of apparent contradictions with Scripture. This applied not merely to Copernicanism but to any seeming conflict of description, e.g. as to whether the 'waters above the Earth' were a literal ocean or simply clouds. Drawing on the much earlier (Augustinian) belief that in the Biblical text the Holy Spirit 'accommodated himself' to the usage of everyday life, he wrote of Moses, the assumed scribe of Genesis:

> Moses wrote in a popular style things which, without instruction, all ordinary persons endued with common sense, are able to understand; but astronomers investigate with great labour whatever the sagacity of the human mind can understand. Nevertheless, this study is not to be reprobated, nor this science to be condemned. *(Ibid. p.86)*

Hence, 'Moses adapts his discourse to common usage'. This doctrine of accommodation was highly influential in all the countries where Calvinism was to flourish.

In the light of these values it is surprising to read the oft-repeated statement that Calvin took the lead in opposing Copernicanism by citing Psalm 93:1 ('The world is firmly established, it cannot be moved'), condemning heliocentric notions, and asking 'Who will venture to place the authority of Copernicus above that of the Holy Spirit?' The mythical nature of this alleged quotation has been independently exposed by Hooykaas and Rosen. In the unlikely setting of a sermon on I Corinthians Calvin did deny the rotation of the Earth, but he was talking of its daily rotation, not about its movement about the Sun, and argued not that Scripture did deny it but that it was 'against common sense'. Conceivably he had never even heard of Copernicus. But his open attitude to Scripture certainly facilitated belief among his followers.

However, as if to emphasize the anti-Copernican inertia that had to be overcome in the sixteenth century, and that no man had the final word, it is noteworthy that Calvin's immediate successor at Geneva, Theodore Beza (1519–1605), held highly to Aristotle (as befitted a professor of Greek) and used Aristotelian teaching as a basis for rejecting the doctrine of a moving Earth.

Figure 3.6 John Calvin (1509–1564). French-born theologian who spent most of his life in Geneva. He became a leader of the Reformation in its second phase, writing extensively on the doctrines of the church and its ministry, and emphasizing the authority of Scripture. (Bibliothèque Publique et Universitaire de Genève)

3.4 European Copernicanism to 1600

3.4.1 Early Copernicanism in Germany

Probably the first disciple of Copernicus was Joachim Rheticus (see Chapter 2). As we have already seen he helped to propagate his master's ideas in his *Narratio prima* of 1539, for a long while the best brief introduction to Copernicanism. And it was Rheticus who sent *De revolutionibus* to the press a couple of years later. Just before Copernicus' death in 1543 Rheticus returned to Wittenberg and after many adventures settled, from 1554 to his death in 1574, in Cracow. This town had the same longitude as Frauenburg, from which Copernicus had made his astronomical observations. It was therefore specially suitable for his disciple to continue – even complete – the work of his master. And so we find this restless, turbulent mathematician supporting his laudable enterprise by working as a physician in the tradition of another revolutionary, Paracelsus.

In place of the Greek tradition in astronomy, Rheticus deferred to that of the Egyptians (real or imagined). So he had erected in the Polish town that symbol of ancient Egypt, the obelisk, than which 'no astronomical instrument would be better'. And from Cracow he conducted a correspondence with (among others) the French philosopher Pierre de la Ramée, or Ramus as he is more often known. Ramus admired Copernicus though dissented from his heliocentric view of the universe. In fact, he had an important role to play later in the growth of Copernicanism. Now he was to encourage Rheticus to abandon all artificial hypotheses in astronomy, and received rather strange letters in which his correspondent roundly deprecated the hypotheses of Ptolemy but kept ominously quiet about his own views of the Copernican scheme. Perhaps he just wanted to please his reader, a characteristic trait in Rheticus.

How far Rheticus' recently-discovered *Treatise on Holy Scripture and the Motion of the Earth* (see Chapter 2, Section 2.7.3) played any part in the reception of Copernicanism must be a matter for conjecture. It was not printed until 1651 and whatever happened to its manuscript until then is entirely unknown (Hooykaas, 1984). But it was evidently not through Rheticus, but through a slightly older colleague, that the cause of Copernicus was to be commended in Northern Europe. He was Erasmus Reinhold, senior professor of astronomy and mathematics at Wittenberg at the time that Rheticus held a junior professorial post in the same university.

Reinhold (1511–53) held Copernicus in high esteem. Even before *De revolutionibus* appeared he referred to its author as a 'second Ptolemy' (1542). Eight years after its publication he brought out his own astronomical tables, *Prutenticae tabulae coelestium motuum* (Prussian tables of celestial motion), named after his patron the Duke of Prussia and derived directly from those of Copernicus, enlarged and corrected. This revisionist embodiment of Copernicanism proved an indispensable aid to astronomers (and astrologers) for eighty years or so. Every time it was used it was a silent testimony to the original author. As one historian has argued:

> Since the tables were known to derive from the astronomical theory of the *De revolutionibus*, Copernicus' prestige inevitably gained. Every man who used the Prutenic Tables was at least acquiescing in an implicit Copernicanism. *(Kuhn, 1966, p.188)*

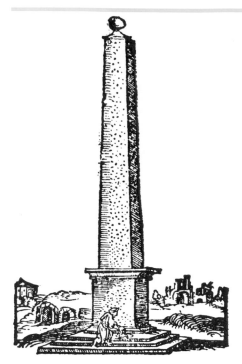

Figure 3.7 Obelisk erected in Cracow by Rheticus (from the frontispiece of his book Ephemerides novae, *1560). This was a symbol of Egyptian science, intended to indicate the demise of the Greek tradition in astronomy (on the doubly questionable assumption that Egyptian science was 'better' than Greek, and that Copernicus was emancipated from the latter). (Basel University Library)*

Figure 3.8 Petrus Ramus (Pierre de la Ramée) (1515–72), French philosopher and mathematician, murdered for his conversion to Protestantism. A professor in Paris and a humanist, he advocated sweeping academic reforms, including the novel use of diagrams to facilitate comprehension. He admired Copernicus, rejecting Aristotelian hypotheses though not fully accepting the Copernican system. (Bibliothèque Nationale, Paris)

However, it must not be supposed that Reinhold was a convert to the Copernican system. Though he must have mentioned it, apparently he did not teach it. This impression is confirmed by an examination of his own copy of *De revolutionibus*;[6] its annotations clearly imply a man preoccupied with technical calculating devices but seemingly indifferent to the possibilities of heliocentricity. Reinhold was the most influential astronomer in Northern Europe.

Another early German convert was Christopher Rothmann, astronomer to Wilhelm, Landgrave (= Count) of Hesse-Cassel, and correspondent of the Danish astronomer Tycho Brahe. By the late 1580s Rothmann's letters were advocating the Copernican theory, unsuccessfully as it happened, though a projected book on the subject was never published. A more influential German astronomer was Michael Maestlin (1550–1631), professor at Tübingen. Though his *Epitome* of astronomy of 1588 was Ptolemaic in conception, later editions added Copernican appendices. It seems he was particularly impressed by his own observations of the new star (see below), and by the end of the century was publicly praising Copernicus, demanding a radical renovation of astronomy and lecturing on both old and new cosmologies. One of his students was Johannes Kepler.

It is interesting that even today the one place in Europe where first editions of *De revolutionibus* outnumber the second is Germany (East and West) (Gingerich, 1979, p.82). Thus this was not only the initial focus of the book, which had been printed in Nuremburg, but also a major centre for Copernican ideas. But there were other places too.

3.4.2 The spread of Copernicanism to England

In English writers may be found some of the earliest evidence for the spread of Copernicanism. One of the first was John Field (or Feild) (*c.* 1525–87), a Yorkshire man who held some office as a public teacher of science in London. In 1556 he published his *Ephemeris anni* 1557, a series of astronomical tables based upon the *Prutenticae tabulae*, and alluding to Reinhold, Rheticus and Copernicus (whose labours were 'Herculean'). The book contained a preface by the famous magician-cum-alchemist John Dee who, though commending the tables, did not consider that to be a proper place to evaluate the Copernican hypothesis. Neither he nor Field apparently committed himself on that matter.

Not so their contemporary Robert Recorde (1510–58). In the same year (1556) he produced *The Castle of Knowledge*, one of a graduated series of books intended to teach mathematics to artisans. Recorde himself was no rustic labourer but physician to Queen Mary. His skills in medicine were matched by those in mathematics and Greek, and he was a very early pioneer in popular science education. This book, like the others, contained a dialogue between master and scholar. They discuss Copernicanism, which the scholar is inclined to ridicule but which the master commends, promising to expound it on another occasion, after which the scholar would be 'as earnest then to credit it as you are now to condemn it'.

Thomas Digges (*c.* 1546–95) was a mathematician who graduated from Queens' College, Cambridge. Impressed by a new star (see Section 3.5.2), he determined its position relative to four others, and demonstrated that it must

[6] At the Royal Observatory, Edinburgh (Gingerich, 1979, p.81).

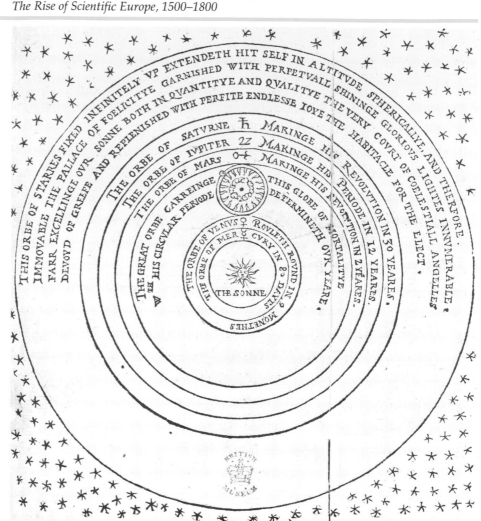

Figure 3.9 The Copernican ('Pythagorean') system according to Thomas Digges, from his A Perfit Description of the Caelestiall Orbes (1576). (Reproduced by permission of the British Library Board.)

Figure 3.10 William Gilbert (1540–1603), 'father of magnetical philosophy'. Born in Colchester, he was a Fellow of St John's College, Cambridge, President of the Royal College of Physicians and physician to Elizabeth I and James I. His book De magnete (1600), declaring the Earth to be a magnet, incorporated seventeen years' experimental work and was the first great scientific book to be published in England. (Colchester Town Hall)

be far out in space.[7] Just because it was new it contravened the old Aristotelian notion of changeless realms beyond the sphere of the moon. This was consistent with, but did not actually prove, the Copernican system. This was reported in his book *Alae seu scalae mathematicae* (Mathematical wings or scales), 1573.

Shortly after this, Digges revised an almanac of his father Leonard Digges (*d.* 1558), originally cast in Ptolemaic form. His filial duty was accompanied by a reluctance to perpetuate what he was now pretty sure was the erroneous 'doctrine of Ptolemy'. Accordingly the new edition of *A Prognostication Everlasting* (1576) included 'a perfect description' of the Copernican system. There was even a diagram, but Thomas Digges now went beyond even Copernican teaching, and distributed the stars through an infinite space. Was he, perhaps, expressing mystical opinions of the mentor once appointed by his father, the ubiquitous John Dee? Or did his blend of the astronomical heaven with the Heaven of Christian theology owe something to the traditions of Queens' College where Erasmus had but recently studied? Or was he simply overwhelmed by the staggering distances implied by a total lack of observable stellar parallax? Whatever the reason, Digges Jr. not merely promoted Copernicanism in England; he profoundly modified it as well.

[7] This followed from an absence of observable diurnal parallax.

One more figure from sixteenth-century England gave his aid to the Copernican cause. Like Thomas Digges he combined mysticism with rational science, and like Recorde he was a royal physician, in his case to Queen Elizabeth I. William Gilbert (1540–1603) published his epochal work in the last year of the century, 1600; *De magnete* (On the magnet). (See also Chapter 7.) Born in Colchester, and an MA of St John's College, Cambridge, he became justly known as 'the father of magnetical philosophy', his great treatise being the culmination of seventeen years' experimental work. It brought together all kinds of magnetic phenomena, described the properties of the lodestone as never before, gave the subject a new quantitative dimension, and sought to establish a comprehensive theory of magnetism (an 'effluvium' penetrating the lodestone). Its one unifying thesis was of the Earth as a huge magnet, suspended in space. Gilbert considered its behaviour as analogous to that of a floating magnet which rotates in a magnetic field. He thought he could in this way explain the precession of stars, and he came near to identifying gravitation with magnetism.

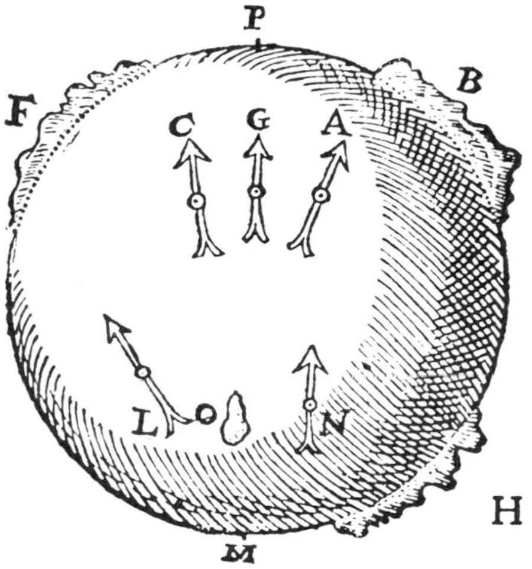

Figure 3.11 Illustration from Gilbert's De magnete, *published in 1600.*

These four Englishmen were thus amongst the first to publish the Copernican doctrine, Recorde and Digges Jr. being amongst the earliest of its vocal supporters. We return later to consider possible reasons for this early aspect of English science. How far they influenced developments on the continent is not wholly clear; one inhibiting factor must be that Recorde and Digges wrote in English, not Latin. It may be significant that one 'foreigner' to visit the London of their day was Giordano Bruno. No doubt there were other crypto-Copernicans in England at this time. Thomas Harriot (1560–1621) may have been one of them; he was certainly an avowed Copernican a little later. He was mathematical tutor to Sir Walter Raleigh, owner of one of the few copies of *De revolutionibus* known to have been then in England.

3.4.3 On counting Copernicans

To summarize the situation in Europe before 1600, we may fairly point to some diversity of opinions. At one extreme there is the old view that hardly anyone embraced Copernicanism in the sixteenth century. Some modern writers have modified that impression, though not too strongly. Thus Westman argues that 'we can identify only ten Copernicans between 1543 and 1600', using the term to mean those 'who actively adopted its radical proposals' (Westman, 1986, p.85). His list is as shown in the box.

Copernicans between 1543 and 1600	
German:	Rheticus, Maestlin, Rothmann, Kepler
Italian:	Galileo, Bruno
English:	Digges, Harriot
Spanish:	de Zúñiga
Dutch:	Stevin

All of these, however, were in some sense at least, 'professional' astronomers or men of science. Gilbert may have been excluded because his *De magnete* did not appear until exactly 1600, but his ideas must have been formulated well before then. It is hard to divine why Recorde should have been excluded, unless it is because he was a 'mere' popularizer. And of course there is always the possibility that more converts remain to be discovered.

At the other extreme it is also right to stress the large numbers of *De revolutionibus* in circulation, with the inference that Copernicanism must have been quite widely known if not actually accepted with all its radical proposals. And in between these two emphases it is likely that many astronomers accepted the book as a calculating device while rejecting its heliocentric world picture as an objective description of how the universe really was. How far this spectrum of attitudes could have survived we can never know, for towards the end of the sixteenth century events in Europe demonstrably accelerated a new swing towards Copernicanism.

3.5 New elements in the climate of astronomical thought

3.5.1 Signs on Earth

Since the late fifteenth century the voyages of discovery, especially by the Portuguese, had given a hard knock to the traditional geography expounded by Ptolemy. This was recognized by Copernicus (Chapter 2, Section 2.6.1), and by his disciple Rheticus (Hooykaas, 1984). Realization that the tropics were habitable, and that all land-masses were not joined together, had already undermined faith in the authority of Ptolemy as a commentator on nature. This trend continued throughout the sixteenth century as explorers from many European countries returned home with fabulous stories, and unheard-of products to corroborate them. It was not just that Ptolemy's geography was wrong in detail, even in important detail. It was rather a question of scale, of the importance of new evidence now multiplying at a

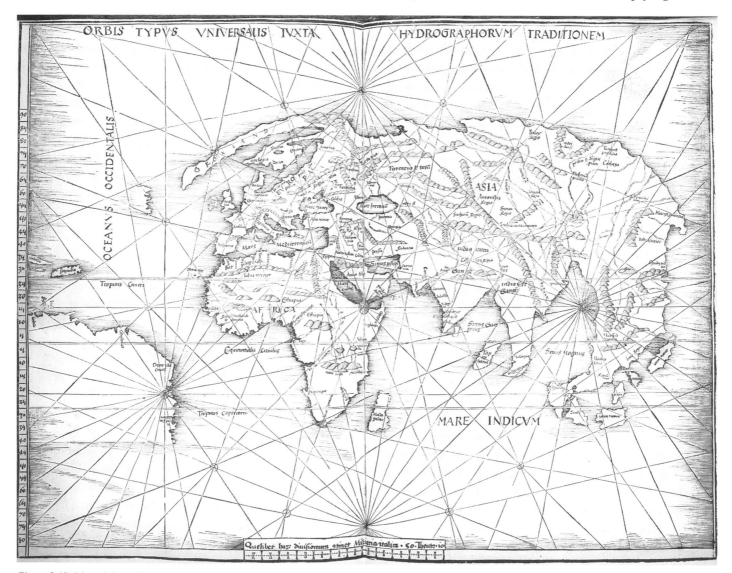

Figure 3.12 Map of the world included in a late (1513) edition of Ptolemy's Geographia. *(Reproduced by permission of the British Library Board.)*

prodigious rate, and of the failure of traditional philosophy. As the Portuguese poet Camões put it:

> I have seen things of which coarse mariners (who have only long experience as their teacher) proclaim as true and real; yet learned men, who penetrate the secrets of the world by ingenuity or science, declare that they are false and illusory. *(From Hooykaas, 1983: our translation)*

Or as an English rector in London, William Watts, wrote in 1633, 'the thoughts of the philosophers have been contradicted by the unexpected observations of the navigators' (Hooykaas, 1973, p.37).

With the almost explosive growth of new knowledge about the earth came, fortunately, new methods of depicting it. Thus modern mathematical geography may be said to have originated with the maps of the Flemish cartographer Gerard Mercator (1512–94) whose famous 'Mercator projections' are still in use. These reinforced public awareness of the new discoveries. This was even more true of the globes which from the early sixteenth century had displayed the new discoveries. Important centres of globe-making included the Low Countries (especially Amsterdam) and Nuremberg. The globes were not devoid of social or political significance, particularly in the case of the imperial powers. On receiving one from the London merchant-adventurer William Sanderson, Queen Elizabeth is said to have graciously replied 'The whole earth, a present for a Prince', and one of her best-known portraits shows her hand, as though divine, overarching a terrestrial globe (Wallis, 1965). In England the 'imperialism of the mind' found ready expression in conquering the whole cosmos in the name of Copernicanism.

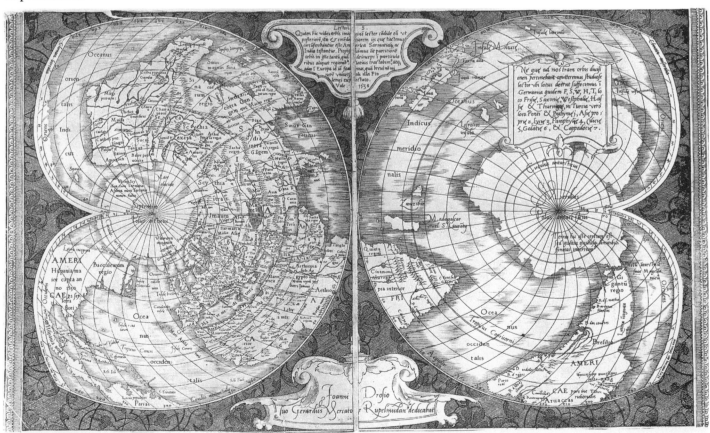

Figure 3.13 Map of the world (1538) by Gerald Mercator, using the famous 'Mercator projection' which, in due course, enabled cartographers to display the whole world on a rectangular map. (Library of Congress)

Figure 3.14 Terrestrial globe of 1541 by the Flemish cartographer Gerard Mercator (1512–1594): 41 cm. diameter, made of papier mâché. (Jagellonian University Museum, Cracow)

Figure 3.15 Armada portrait by George Gower of Queen Elizabeth I (1588). The Queen's fingers lightly touch a terrestrial globe, symbolizing her spiritual authority and imperial ambitions. (By kind permission of the Marquess of Tavistock and the Trustees of the Bedford Estate.)

Remarkably, the pioneers of the new geography, the Portuguese, played little part in disseminating the new astronomy. It has been argued that Jesuit monopoly in education was responsible for this lack of progress (Hooykaas, 1983, p.596). If social conditions played some part in facilitating Copernican ideas in England, it is likely that a different social environment in Portugal had the opposite effect. But if discoveries on Earth evinced an ambiguous response, that cannot be argued for the much more spectacular revelations that were to come from the skies above.

3.5.2 Signs in the heavens

A series of the most remarkable coincidences in the whole of the history of science took place in the 1570s. Nor did they lack drama. At the very time when Copernicanism was beginning to struggle for acceptance, two events far out in space forced into human consciousness the utter untenability of the old Aristotelian cosmology that Copernicus had begun to overthrow.

The first of these was the appearance of a new star in the constellation of Cassiopeia. Put like that it sounds prosaic enough, but this was no ordinary new star. Its brightness was prodigious – greater than that of any heavenly body except the Sun, Moon and (at times) Venus. Sharp-eyed people could see it even before sunset. Today we call it a supernova, the visible result of a gargantuan nuclear explosion in space. Only four or five have been seen in recorded history,[8] and this was the brightest of them all. It must have seemed symbolic. For over two years it blazed down upon a continent emerging from

Figure 3.16 Comet of 1547, as illustrated in an Islamic MS now in Istanbul University. (From Cambridge Illustrated History of Science.*)*

[8] In 1006, 1054, 1572, 1604 and (possibly) 1987.

the darkness of mediaeval thought. Then it disappeared gradually, and was gone by 1574 leaving not a visible trace behind. But it is still there, recognizable to modern astronomers as a strong radio-source.

The new star was more than symbolic, however. Its appearance generated all kinds of apocalyptic fears and hopes. It has been said that it caused a greater shock to European thought than had ever been generated by the Copernican hypothesis. Beza thought it the same star as had appeared at the birth of Christ. It was widely believed (by Catholics) to herald the victory of their church over the reformers. And of course it caught the attention of astronomers. There was a crucial question to solve: was it, or was it not, beyond the sphere of the Moon's orbit? According to Aristotle the only changes that occurred above that sphere were changes of position (i.e. motion). Hence if the old theory were right the new star must be nearer than the Moon. And so, without telescopes or sophisticated instrumentation, men went to work all over Europe.

In Germany the young Michael Maestlin attempted to locate it precisely with reference to four adjacent stars, holding a thread before his eyes so that it cut across one pair of stars with the new star in the middle, the observation then being repeated with the other pair. Other observers included the Landgrave of Hesse-Cassel and two astronomers at Wittenberg (Peucer and Schuler). In the Low Countries Cornelius Gemma of Louvain wrote a long book on the subject, and in other continental countries it received much attention. Across the English Channel astrologers like Dee gazed at the visitor with uncomprehending awe whilst Thomas Digges examined it by a similar method to Maestlin, with a straight ruler six feet long.

Whereas most of these observational programmes were not embarked upon to test Copernicanism, but to test whether this was merely a meteorological phenomenon, it soon became clear that its parallax was small, i.e. that its distance must be great. If so, the implications for the old Aristotelian model were inescapable. Digges and Maestlin were convinced but the most telling observations were made by a Danish astronomer, Tycho Brahe (1546–1601), the greatest observer of the pre-telescopic age. Tycho used a new sextant to locate it in relation to nine of its neighbours. His conclusion placed it far beyond the sphere of the planets, in Aristotle's supposedly incorruptible heaven. He compared observations from all over Europe, and published his *De nova stella* in 1573. The beliefs of ancient astronomy were being shaken to their very foundations. Nor was this all.

By 1577 Tycho Brahe had been established by the King of Denmark at an observatory on the island of Hveen which he called 'Uraniborg'. On the evening of 13 November of that year he looked up from the agreeable task of fishing for his supper, and noticed another new object in the darkening western sky. Realizing that it could not be Venus (the first and most obvious explanation) he waited until after sunset and then recognized it as a comet, with a brilliant tail subtending some 22°. Over the next two months he examined it with the best instruments that had yet been placed at the service of astronomers. The result was that this new visitor also had no parallax and must therefore be in the regions above the Moon. Not only was Aristotle's incorruptible heaven discredited but so also were his crystalline spheres bearing the planets, for how could a comet penetrate them? Later workers confirmed his results and extended them to other, less spectacular, cometary intruders. Although astrologers and other devotees of the occult had a field day, eventually the cosmological significance of these new heavenly objects was apprehended far beyond the astronomical communities. But by 1646 Sir

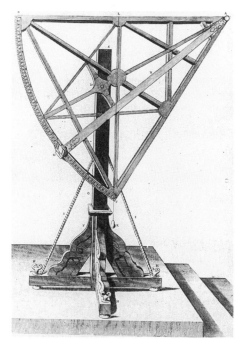

Figure 3.17 Astronomical sextant used by Tycho Brahe. (Mansell Collection)

Figure 3.18 Tycho Brahe's mural quadrant. This device, mounted on a west wall in his observatory of Uraniburg, enabled the elevation of a heavenly body to be accurately measured. Light from a star or planet enters through the hole high up in the wall on the left, and its elevation is determined by observing it through a movable eyepiece (E). Such an arrangement was far more stable and reliable than movable quadrants. Note also the diagonal dotted lines on the scale between vertical markings, a kind of primitive vernier scale enabling measurements of unprecedented accuracy to be made. Also prominent are two duplicate clocks, measurements of time being as important as measurement of angle. The two observers can be identified, with Tycho himself directing operations from a table and recording results. The quadrant wall is adorned with a picture, framed by the scale, of other activities in the observatory (which is why Tycho appears twice in the same illustration!). The mural picture depicts Tycho's dog ('an emblem of sagacity and fidelity') and, behind his master, two small portraits of King Frederick II (Tycho's benefactor) and Queen Sophia. (Ronan Picture Library and Royal Astronomical Society)

Thomas Browne permitted himself, in his *Pseudodoxia epidemica*, to doubt whether one should 'attribute terrible astrological effects to comets and blazing stars, *now that it was known that they were above the moon*' (Meadows, 1969, p.2, our italics). The signs in the heavens could not be denied.

3.5.3 Modifying the programme

The full programme of Copernican cosmology was so ambitious, and raised so many uncomfortable questions, that attempts to modify it are hardly surprising. The first of these was advanced by none other than Tycho Brahe, the so-called 'Tychonic system'. Dating from 1588 it was a compromise between the Ptolemaic and Copernican systems. As with Ptolemy the Earth is

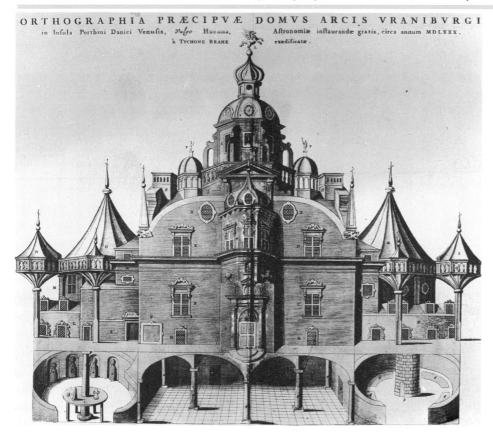

ORTHOGRAPHIA PRÆCIPVÆ DOMVS ARCIS VRANIBVRGI in Infula Porthmi Danici Venufia, *Vulgo* Hueuna, à TYCHONE BRAHE Aſtronomiæ inſtaurandæ gratia, circa annum MDLXXX. exædificatæ.

Figure 3.19 Uraniburg, the observatory created in 1576 for Tycho Brahe on the island of Hveen, Denmark. (Mansell Collection)

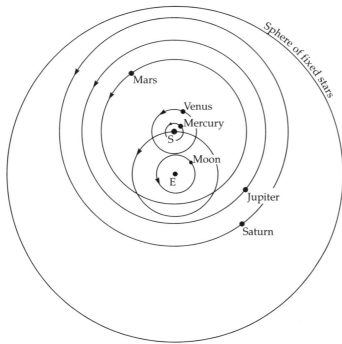

Figure 3.20 The Tychonic system.

stationary, and at the centre of the Sun's orbit, one rotation taking place every twenty-four hours. But, as Copernicus envisaged, the Sun carries with it the planets (all, that is, except the Earth).

It was not the case that Tycho's measurements, better than any before him, had suddenly disclosed some new fact which decisively disproved Copernicanism. Such an arrangement is in fact mathematically equivalent to both Ptolemaic and Copernican systems, and certainly fitted as well as either of them with the observational data. However Tycho was not a great mathematician and did not offer a mathematical basis for his system.

The Tychonic alternative was offered for at least three reasons. The lack of stellar parallax would mean a universe of immense size if the Earth really were moving, as in Copernicanism. And there were the two old arguments for a stationary Earth: the notion that it was too sluggish to move and that Scripture was against it.

Tycho's observations (which were of course independent of his system) were eventually to lead to a far greater and more enduring synthesis. But paradoxically the advocacy of his system seems also to have promoted that to which it was an intended alternative, the Copernican universe. One of its great achievements was to discredit the celestial 'spheres' of Aristotle, thus helping further to break down the old cosmology. Merely undermining the Ptolemaic and Aristotelian edifice was, in the long run, an advantage for the spread of Copernicanism. As Francis Bacon observed, the Tychonic system was a stepping-stone from the Aristotelian to the Copernican model of the universe.

Some of Tycho's pupils, such as the Dutch cartographer Willem Janszoon Blaeu (1571–1638), spread his fame far and wide through a succession of magnificent globes (several of which have survived). But Tycho's most famous disciple was to go much further than that.

Johannes Kepler (1571–1630), came of a Protestant family in Würtemberg and studied at Tübingen under Michael Maestlin, from whom he probably first learned of Copernicanism. Having obtained a post teaching mathematics at a Protestant academy at Graz (in Austria) he wrote his first book *Mysterium cosmographicum* (1596), a mixture of wild speculation and shrewd

Figure 3.21 Terrestrial globe of 1599 by the Dutch cartographer Willem Blaeu (1571–1638): 34 cm. diameter, made of papier mâché. (Jagellonian University Museum, Cracow)

Figure 3.22 Johannes Kepler (1571–1630), Bavarian-born astronomer and mathematician and pupil of Tycho Brahe. His ideas are shot through with mysticism and some of them appear bizarre today. Yet he discovered the three famous laws of planetary motion named after him, the first of which declares that planets move in ellipses, not circles. (Wellcome Institute)

mathematical deduction. He supposed that the planets were impelled by an *anima motrix*, or moving spirit in the Sun, and that a relation existed between inter-planetary distances and the well-known regular solids. One such regular solid could, he thought, be interposed between adjacent planets without any interference.

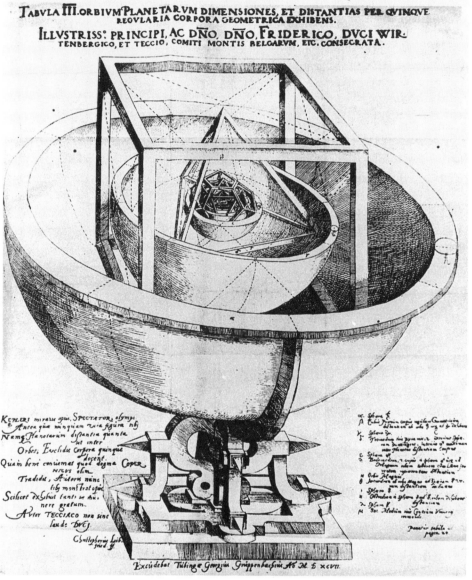

Mercury		
separated by octahedron	8 faces
Venus		
separated by icosahedron	20 faces
Earth		
separated by dodecahedron	12 faces
Mars		
separated by tetrahedron	4 faces
Jupiter		
separated by cube	6 faces
Saturn		

Figure 3.23 Kepler's diagram of the five regular polyhedra: cube, tetrahedron, octahedron, icosahedron and dodecahedron, supposedly related to the orbital sizes of the five (major) planets. (Ronan Picture Library)

Though fanciful and, so far as we know, entirely fortuitous, this notion gave its author a strong incentive to obtain better data in order to establish it beyond doubt. In Kepler's view a heliocentric system gave the only simple explanation for the fact that the Earth and planets are nearest when in a straight line: sun–earth–planet. However he outdid Copernicus in exalting the Sun by referring planetary orbits to the Sun rather than the centre of the Earth's orbit. And in *Mysterium cosmographicum* he praised the example of Maestlin, publicly advocated the Copernican system, and mentioned an 'entire dissertation' he had written in defence of Copernicanism, thus becoming one of the few well-known disciples of Copernicus in the sixteenth century.

There followed one of those apparently chance encounters that were to have immense consequences for the development of science. Kepler sent a copy of his book to Tycho Brahe. Though not himself a Copernican the Danish astronomer was impressed by the younger man's work and extended an invitation to join him at Uraniborg. At first Kepler declined, preferring to live in German lands. As the century ended, however, the expanding Counter-Reformation in Austria was making things increasingly uncomfortable for Protestants like Kepler (nominally a Lutheran but with sympathies towards Calvinism). Moreover he was by this time anxiously searching for even more accurate data. So, in 1600, he decided to join Tycho, by then settled in Prague. It was to be a case of mutual admiration. Kepler rejected any conclusions differing from Tycho's results by as little as 8'. As it happened, Tycho's life was almost over, but into Kepler's hands came his priceless observational data. And out of these results, and his own observations, Kepler created the fabric of a new system that swept away the planetary devices of Copernicus, made redundant the regular solids of his own invention, and paved the way for the Newtonian synthesis at the end of the century. Retaining the Sun as the focal point of our system, he described the planetary orbits no longer in terms of circles but of ellipses. This immense step forward was proclaimed in his *Astronomia nova* of 1609, together with two of his three famous laws of planetary motion (the last was announced in his *Harmonices mundi* of 1619).

Although the Copernican system seems to have been modified almost beyond recognition, its central thesis is retained: a planetary system (including the Earth) rotating about the Sun. In some ways Kepler was more Copernican than Copernicus, for his devotion to heliocentricity was the more complete. Gone are epicycles and equants, and the Sun itself, not some imaginary point, is now at the major focus of each planetary ellipse. As he said, Copernicus simply did not see how rich he was. But in his later disciple he had found the most powerful advocate for an essentially Copernican system.

3.5.4 *Through a glass clearly*

One last but important factor needs to be mentioned in the circumstances that led to the spread of Copernicanism through Europe. This was the first use of the telescope in astronomy.

The discoveries of Galileo in 1610 (see Chapter 4) gave an entirely new kind of boost to Copernicanism, for here was visual evidence that all could understand. It was particularly because his discoveries were subversive of Aristotelian philosophy that they helped to undermine Ptolemy. The satellites of Jupiter demonstrated that the Earth was not the only centre of motion; the phases of Venus suggested it is a dark planet like the Earth and if it rotates around the Sun, then so may the Earth; sunspots and other solar phenomena pointed to a body far from perfect and unchanging, and so on. The first impact of these discoveries was admittedly greatest in southern Europe, but the book in which they were disclosed, *The Starry Messenger* (1610), soon sold out its 500 copies printed in Venice, and a second edition was published at Frankfurt within a few months.

Thus many features of European thought around 1600 conspired to pave the way for a wide acceptance of Copernicanism. It would, however, be wrong to add to them the alleged birth of an experimental method, for this was also part of the Aristotelian legacy. Thus in Galileo's *Dialogues Concerning Two World Systems* the use of experiment is actually advocated by Simplicius, the Aristotelian. The challenge as Galileo saw it came not from experiment but from mathematical speculation.

3.6 Later centres of Copernicanism in Northern Europe

3.6.1 England

As seen previously (Section 3.4.2) Copernicanism gained an early foothold in England. Yet in the first decades of the seventeenth century its victory was far from complete. Several notable writers opposed it resolutely, often on the familiar grounds of common-sense or Scripture. One of their number was the London rector Thomas Tymme (*d.* 1620), translator of Ramus and popular author who opposed the theory in 1612. Another was William Barlowe (*d.* 1625), chaplain to Prince Henry and one of the first to write about the magnetic compass; his *Observations and Experiments Concerning the Nature and Properties of the Lodestone* (1616) denied the Copernican hypothesis. Alexander Ross (1591–1654), a Scot who became chaplain to Charles II, published a strongly anti-Copernican treatise *Commentum de terrae motu circulari* in 1634. Later still the notable Puritan divine and fierce opponent of astrology, John Owen (1616–83), attacked the heliocentric theory on grounds of Biblical exegesis in a tract on *The Day of Sacred Rest* (1671).

There were some Englishmen who held to a 'partial Copernican' view, accepting the diurnal rotation of the Earth but keeping an open mind (or even a sceptical one) about its annual passage round the Sun. Of great influence was the *De magnete* of William Gilbert (1600) who supposed that magnetism could explain the Earth's rotation. The book contained a preface by his friend Edward Wright (*c.* 1558–1615), Fellow of Caius' College, Cambridge, mathematician and hydrographer. Although denying the Earth's annual motion Wright defended its daily rotation against attacks in the name of Scripture; he argued that 'Moses accommodated himself to the understanding and the way of speech of the common people, like nurses to little children'. Another 'partial Copernican' was the Puritan writer, Nathaniel Carpenter (1589–*c.* 1638), Fellow of Exeter College, Oxford, and 'the father of English geography'. He was one of the targets of the attack by Alexander Ross. An avowed anti-Aristotelian, he claimed in his *Philosophia libera* (1622) to be 'bound to nobody's word except to those inspired by God'. His *Geographie Delineated Forth* of 1627 proposes the daily rotation of the Earth in the name of 'philosophical liberty'.

The British opponents of Copernicanism before about 1640 were relatively few, at least according to the published books available to us. Very few of the technically competent rejected the system. The following table lists the principal writers who rejected Copernicanism (C), some of whom also rejected Ptolemy (P) and Tycho (T), relapsing into a kind of cosmological agnosticism, or else devising their own variation (Russell, 1972).

Author	Date	Against	For
Nathanael Torporley	1602	P, T, C	
Thomas Lydiat	1605	P, T, C	
Arthur Hopton	1613	C	
Samuel Purches	1614	P, C	?T
Richard Allestree	1620	C	
Francis Bacon	1623	P, C	
George Hakewill	1627	C	
John Swan	1635	P, C	T

There is little doubt that English Copernicanism owed much to the new uses of the magnetic compass in voyages of discovery. As the century passes this became more obvious, together with another phenomenon, the rise of English Puritanism. Although not all Puritans fully accepted a Copernican universe (as Tymme, Owen, Carpenter and even Wright testify), many did, and especially those associated with the infant Gresham College (see Chapter 9). An early advocate was Mark Ridley (1560–1624), one-time physician to the Tsar of Russia, and later Censor to the Royal College of Physicians in London. In a treatise on the magnet (1613) Ridley came down clearly in favour of Copernicanism. Another physician-turned-astronomer was John

Bainbridge (1582–1643), a supporter of the new astronomy. A published description of the comet of 1618 (another new heavenly body) led to his appointment as Savilian professor of astronomy at Oxford. His colleague, in the Savilian chair of geometry, was Henry Briggs (1550–1630), having previously been professor of geometry at Gresham College. He was author of an early set of Napierian logarithms and was well-known both for his Copernicanism and his writings on navigation. A friend of Briggs, Henry Gellibrand (1597–1636), had been a professor of astronomy at Gresham College. Another devout Puritan he was responsible for a revisionist almanac (with Roman Catholic saints replaced by Protestant ones), and for his pains was unsuccessfully prosecuted by Archbishop Laud. He also wrote mathematical works and discovered the change in magnetic variation with time. After visiting Philip Lansbergen, a Dutch apostle of Copernicus, he remarked that what Lansbergen would style 'a truth' he himself would 'readily receive as an hypothesis' (1635).

After the Civil War it has been said that, in England at least, the ideas of Ptolemy were dead though those of Copernicus and Tycho Brahe still strove for succession (Hill, 1974, p.159). From the 1650s the purge of Royalists from Oxford, the rise of Baconianism and the assembly of kindred Parliamentary spirits were to lead to the Royal Society. As Puritanism triumphed, so did the Copernicanism with which it was to become strongly linked. There were even apparent parallels between politics and cosmology: as the Earth had lost its place in the hierarchy of the universe, so had the king lost his within the state.

Amongst the English proponents of Copernicanism at this period was one of the founders of the Royal Society, the mathematician John Wallis (1616–1703), Savilian professor of geometry at Oxford (whose mathematical achievements included invention of the sign α for infinity). More in the public eye was his colleague John Wilkins (1614–72), who became a powerful advocate for Copernican doctrines. Brother-in-law to Oliver Cromwell, he was successively Warden of Wadham College, Oxford, Master of Trinity College, Cambridge, and (after the Restoration) Bishop of Chester. In his *Discovery of a New World ... in the Moon* (1638) Wilkins argues that other worlds than ours may be inhabited, and in so doing attacks those who use Scripture to deny it, thus undermining some of the case against Copernicanism: 'it is evidently besides the scope of Scripture to discover anything unto us concerning the secrets of science' (Hooykaas, 1974, p.72). Wilkins frequently quoted with approval Calvin's Old Testament commentaries, and in 1640 argued for the planetary nature of Earth in his *Discourse Concerning a New Panet*. He was but one of many Puritans who helped to further the cause of Copernicanism in England. So thoroughly had they done their work that by 1654 Seth Ward (1617–89), Savilian professor of astronomy at Oxford and later Bishop of Exeter and of Salisbury, could write that almost everyone in England with any knowledge of astronomy had accepted Copernicanism as true or at least as the most convenient hypothesis available (Hooykaas, 1974, p.76).

Figure 3.24 John Wallis (1616–1703), mathematician and Savilian professor of geometry at Oxford. Remarkable for his ability to decipher coded messages (and so employed by Parliament and William III), and a founder member of the Royal Society. (National Portrait Gallery, London)

3.6.2 The Netherlands

Despite important differences in their political and commercial situations, England and the Low Countries, or Netherlands, had rather striking similarities in the fortunes of Copernicanism within their borders. Both areas of Europe saw the rise of a strong Calvinistic movement, and with it a remarkable new emphasis on scientific study.

One scholar has concluded 'that in the sixteenth century Copernicus was widely known, but that there were no Copernicans in the Netherlands', perhaps reflecting the Dutch preoccupation with trade and with expulsion of the Spaniards (Hooykaas, 1983, p.640). After 1600 the Netherlands began to catch up with England. The same author has argued that two other trends in science were necessary for the Copernican world picture to be widely acceptable: a mechanistic world view, and a commitment to experiment (or observation). The second of these seems to have to come to England before it reached the Netherlands.

The first committed Copernican in the Netherlands seems to have been Simon Stevin (1540–1620), engineer to Prince Maurice of Orange. His Copernicanism was tinged with a degree of mystical romanticism about a mythical Golden Age in which such ideas would have been known. No such mythology impeded Isaac Beeckman (1588–1637), theologian and physician, of Middleburg in Zeeland. Founder of Europe's first meteorological station, he was so committed to a mechanical philosophy that he established a college in Amsterdam to teach it. He attempted to find a mechanistic explanation for the motion of the Earth as taught by Copernicus. The Copernican cause was well served by two other men from the Low Countries. Nicholas Mulerius (1564–1630) of Groningen published the third edition of *De revolutionibus* in 1617. Willem Blaeu (1571–1638) not only produced globes for teaching purposes (see Section 3.5.3) but also wrote a popular manual of instructions, referring to 'the natural thesis of Copernicus' (1634). He went on to give encouragement to another distinguished Copernican in the Netherlands, Philip van Lansbergen.

Lansbergen (1561–1632) was a Reformed minister in Zeeland. In 1619 he wrote *Progymnasmatum astronomiae*, a learned treatise arguing the reality of the Earth's motion and defending the idea against cosmological readings of the Bible. For so doing he joined Carpenter in the firing-line of the implacable Ross. At the instigation of Blaeu he wrote a second book, *Bedenckingen*, but in the vernacular, making similar points and reiterating that though Scripture was profitable for doctrine etc. 'it is not meet for instruction in geometry and astronomy' (1629). Lansbergen's crusade was continued by a disciple, Martinus Hortensius (1605–39), and his own son Jacob in a book saturated in Calvinistic doctrine (1633). This was a response to a series of attacks by anti-Copernicans in the Netherlands most notably from Libertus Fromondus (see below). Significantly, he noted how people from Holland and Zeeland became Copernicans on sea-voyages, and referred to the 'Calvinistic-Copernican system'. Though not intended as a compliment this term gives unwitting testimony to the strength of an alliance that later historians have often noted.

Not all opposition to Copernicanism came from Roman Catholic sources in the Netherlands, however. The first Rector of the university of Utrecht, Gisbertus Voetius (1588–1676), though a Calvinist, did not accept the doctrine of accommodation, and so opposed Copernicanism on religious grounds. Interestingly, he did not recommend to his students Calvin's commentary on Genesis! A similar position was taken by the Leiden professor Joannes Cocceius (1603–69).

The case of Voetius falsifies the belief attributed to Galileo that all Calvinists were Copernicans. Yet there was sufficient truth in the maxim, for the percentage of Copernicans in the population seems to have been unusually high. By 1657, Christiaan Huygens (1639–96) of The Hague claimed that only those of slow wits and superstitious veneration of human authority could

Figure 3.25 Simon Stevin (c. 1548–c. 1620), Flemish pioneer in mechanics and hydrostatics who brought decimal fractions into common use. His statue is in Place Simon Stevin, Bruges. (A.C.L. Brussels)

Figure 3.26 Philip van Lansbergen (1561–1632), minister of the Reformed Church in Ghent, in Flanders, and author of a spirited defence of Copernicanism (1629). (Reproduced by permission of the British Library Board.)

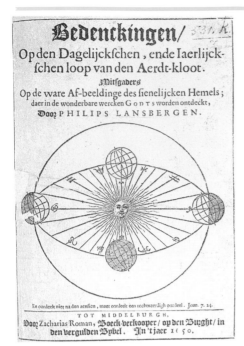

Figure 3.27 Title page of the first (Dutch) edition of the defence of Copernicanism by Philip van Lansbergen, Bedenckingen op den dagelijckschen, ende iaerlijkschen loop van den Aerdt-kloot *(Reflections on the daily and yearly course of the terrestrial globe), 1650. (Reproduced by permission of the British Library Board.)*

remain unconverted to the 'divinely invented system' of Copernicus. He was probably speaking for a majority of his educated fellow-countryman.

3.7 Conclusion: where resistance lingered

3.7.1 Belgian lands

It is instructive to contrast the effects of Calvinism and Catholicism within the Low Countries on cosmological thinking. Still under Spanish rule was the southern Netherlands with its capital at Brussels, now part of the Belgium, where Catholicism was the dominant faith. As in mainland Spain, the decree against Galileo (see Section 4.2.2) was accepted and scientific activity was severely inhibited. One Catholic scholar has written 'There was little astronomic activity in the Spanish Netherlands during the seventeenth century, apart from a few Jesuit mathematicians' (Russell, 1989). As late as 1691 a professor at Louvain, Martin van Velden, on proposing a public defence of Copernicanism as 'indubitable', was opposed by his faculty, threatened with dismissal, and only withdrew 'voluntarily' after being heavily leant on by the Papal Nuncio.

The other part of modern Belgium, the independent bishopric of Liège, was also Catholic territory though less repressive than its neighbour. That Copernicanism was in the air is clear from an anti-Copernican treatise by a theologian from Louvain, Libertus Frodmondus (1631), a Roman Catholic priest originally from Antwerp in the Spanish Netherlands. In his *Anti-Aristarchus* (1631) Fromondus defends the 1616 Decree against Galileo, and in his *Vesta, sive anti-Aristarchi vindex* (1634) the 1633 Decree also. The country's leading astronomer was Godefroid Wendelin, who defended Copernicanism before the Papal Nuncio in about 1626. However, after the 1633 Decree against Galileo he judiciously, if ineffectively, disguised a heliocentric view in his *Teratologia cometica* (1652) by avoiding explicit reference to the planetary Earth, speaking of it merely as 'the third body'. One wonders who was fooled by this cosmological euphemism.

3.7.2 France

As in Belgium so in Catholic France the effects of the 1633 Decree were considerable. Even before that date Copernicanism was indicted by the Paris astronomer J.B. Morin in his *Famosi et antiqui problematis de telluris motu* (1631). News of Galileo's condemnation caused alarm among Catholic scholars, and several with Copernican sympathies went scurrying for cover. Descartes had almost completed written a work in that tradition, *Le Monde*, but even though he was then living in Holland, he abandoned it at once (it was published posthumously in 1664). Descartes' works were placed on the Index of prohibited books. Mersenne was similarly intimidated, having spoken favourably of Galileo in his *Questions theologiques practiques* (1634). He promptly issued a sanitized version of the book, to circulate simultaneously with the original edition!

Gradually a few bolder spirits emerged with public support for Copernicanism. In the early 1630s Ismael Boulliau (1605–94), a Catholic priest, had feared to issue his own pro-Copernican work, *Philolaus*, though he published it anonymously from the relative safety of Amsterdam in 1639. In

1645, throwing caution to the winds, he published in Paris and under his own name a book that openly defied his church's decree and clearly proclaimed a heliocentric universe, *Astronomia philolaica*. Although he incurred the wrath of Morin he appears to have escaped public censure from Rome. For a whole generation French astronomers had to live with the tension generated by the Decree. Some, as Gassendi, took refuge in the ambiguity of its authority; it was not an infallible utterance of the church. Others, as Descartes himself, resorted to the oldest strategem of believers faced with an apparent challenge from a scientific proposition: 'it's only a theory'. Others merely expressed their discomfort and continued to sit on the fence. By the 1660s opposition to Copernicanism continued to be expressed, but rather from the philosophers and theologians than from the astronomers. Within another decade the popularization in France of Cartesianism had ensured the success of the heliocentric hypothesis with which it was closely tied.

In conclusion, it might seem that on one hand Roman Catholicism impeded the progress of Copernican teaching, while Protestantism promoted it. While there is much truth in that proposition as a general statement, it is too simple, for several reasons. Thus it may be pointed out that much Catholic opposition to the teachings of Descartes arose not from his espousal of heliocentricity but because his metaphysics was seen to undermine the theology of the Eucharist. A similar point has been made by Redondi, rather more contentiously, about the condemnation of Galileo (see Chapter 4). The simple proposition may also be refuted from three final examples.

3.7.3 Germany

The first of these is furnished by a country in which Copernicanism made early progress: Germany. Rheticus had done his work well. The early concentration in Germany of copies of *De revolutionibus* testifies to that (see Section 3.4.1). Yet in the seventeenth century Germany was not one nation but an association of several hundred principalities united (or divided) by a common tongue and a degree of common culture. The great division of Europe into Catholic and Protestant was perfectly reflected in microcosm in the German lands. In general the Protestant universities flourished, and with them the Copernican theory survived. However only very few German universities, as Altdorf and Rostock, permitted the Copernican-based philosophy of Cartesianism to be taught. Theological attacks continued to come from a reformed theology that had not yet purged itself of Aristotelianism. Not all theologians held such a view, the most influential Cartesian among them being Christoph Wittich (1625–87) who defended the work of Copernicus in a book of 1659, *Consensus veritatis*. Elsewhere astronomical science was even more moribund.

3.7.4 Scotland

The slow progress of Copernicanism in this northern half of Britain contrasts sharply with events south of the Border, and further stresses the complexity of questions about favourable conditions for progress in science.

At first sight it seems that Scotland had many advantages in the seventeenth century. It had a strongly Calvinist tradition (more so than England), and it had twice as many university towns: Edinburgh, Glasgow, St Andrews and Aberdeen (two colleges, King's and Marischal). But the country was economically poor and had gone through troubled times politically. The

Union with England in 1707 was to polarize the community still further, as the 1715 and 1745 rebellions testify. But in the seventeenth century times were generally hard and many students emigrated to study and teach overseas.

When Duncan Liddell left Aberdeen to study on the Continent he encountered Copernicanism at Breslau. Returning to Scotland in 1607, he died shortly afterwards (1613), bequeathing to Marischal College a chair of mathematics and the books he had acquired in Europe, not least *De revolutionibus* in both first and second editions. There is, regrettably, no evidence that the fortunes of Copernicanism in Scotland were helped by his generosity. Another Scot, Thomas Seggerth, met Galileo and Kepler in his continental travels and appears to have become an enthusiastic advocate for Copernicanism though not, it seems, in his native Scotland.

Within Scotland itself few astronomical books of any complexion appeared until the late seventeenth century. We must, however, except the strongly anti-Copernican work of Alexander Ross (Section 3.6.1). The first published defence of Copernicanism in Scotland appears to have been the *Optica promota* of James Gregory (1663), who later occupied mathematics chairs at St Andrews and Edinburgh. As for the university teaching it has been suggested, on the basis of graduation theses, that by the 1660s Copernicanism was being mentioned at Edinburgh and by 1682 was being actively promoted. Elsewhere it may have been advocated a few years earlier. The predominantly Aristotelian emphasis in all Scottish Universities seems to have been at the heart of the strongly anti-Copernican movement.

3.7.5 Sweden

The last example comes from Sweden, a notionally protestant country with an established Lutheran church. Here Copernicanism, after a promising early start, made no perceptible progress until nearly the end of the seventeenth century. It was in the 1660s that a young astronomer, Nils Celsius, defended Copernicanism in a thesis, and in so doing brought down upon his hapless head the wrath of the theology faculty at his university (Uppsala). In fact the church authorities of Sweden were then engaged in a vendetta against Cartesianism, and its Copernican foundations were similarly unacceptable. Here, then, opposition to Descartes was not because he challenged the Eucharistic doctrines of the church but because he was irreconcilable with the Aristotelian scholasticism to which the Swedish Lutherans were then committed. Only when the King in 1689 gave a limited approval to Cartesian philosophy did the erosion of Aristotelianism really begin in Sweden. The first professor of astronomy publicly to advocate Copernicanism was Per Elvius, and that was not until well into the eighteenth century. Where Protestantism was shackled to Aristotle it was as impotent as Catholicism to give positive help to astronomical science. Precisely for this reason the importance of Calvinistic and Puritan theology for the development of European science stands out in even higher relief.

Sources referred to in the text

Calvin, J. (1554) *Commentaries on the First Book on Moses called Genesis*, tr. J. King, 1948, Grand Rapids, Erdmans.

Calvin, J. (1879) *Institutes of the Christian Religion*, tr. H. Beveridge, Edinburgh, Clark.

Dillenberger, J. (1961) *Protestant Thought and Natural Science*, London, Collins.

Dobrzycki, J. (ed.) (1972) 'Études sur l'audience de la théorie héliocentrique', *Studia Copernica*, 5.

Gingerich, O. (1979–80) 'The great Copernicus chase', *American Scholar*, 49, pp.81–8.

Hill, C. (1974) *The Century of Revolution 1603–1714*, London, Nelson.

Hooykaas, R. (1973) *Religion and the Rise of Modern Science*, Edinburgh, Scottish Academic Press.

Hooykaas, R. (1974) 'The impact of the Copernican transformation', in AMST283 *Science and Belief: From Copernicus to Darwin*, Block I, Units 1–3, Milton Keynes, The Open University Press.

Hooykaas, R. (1983) 'The reception of Copernicanism in England and the Netherlands', in *Selected Studies in the History of Science*, Acta Universitatis Conimbrigensis, Coimbra, pp.635–63.

Hooykaas, R. (1984) *G.J. Rheticus' Treatise on Holy Scripture and the Motion of the Earth*, Amsterdam, North-Holland.

Kuhn, T.S. (1966) *The Copernican Revolution*, Cambridge, Mass., Harvard University Press.

Meadows, A.J. (1969) *The High Firmament*, Leicester University Press.

Norland, W. (1953) 'Copernicus and Luther: a critical study', *Isis*, 44, pp.273–6.

Russell, C.A. (1985) *Cross-currents: Interactions Between Science and Faith*, Leicester, Inter-Varsity Press.

Russell, J.L. (1972) 'The Copernican system in Great Britain', in Dobrzicky, pp.189–239.

Russell, J.L. (1989) 'Catholic astronomers and the Copernican system after the condemnation of Galileo', *Annals of Science*, 46, 365–86.

Stimson, D. (1917) *The Gradual Acceptance of the Copernican Theory of the Universe*, New York, Baker and Taylor.

Wallis, H. (1965) 'The use of terrestrial and celestial globes in England', *Actes du XIᵉ Congrès International d'Histoire des Sciences*, Warsaw and Cracow, iv, 204.

Westman, R.S. (1986) 'The Copernicans and the churches', in D.C. Lindberg and R.L. Numbers, *God and Nature*, pp.76–114, Berkeley, University of California Press.

Further reading

Armitage, A. (1971) *The World of Copernicus*, Wakefield, S.R. Publishers.

Dreyer, J.L.E. (1963) *Tycho Brahe*, New York, Dover.

Gingerich, O. (1973) 'Copernicus and Tycho', *Scientific American*, 229, Dec., 86–101.

Grant, E. (1984) 'In defense of the earth's centrality and immobility: scientific reaction to Copernicanism in the seventeenth century', *Trans. American Phil. Soc.*, 74, part 4.

Johnson, F.R. (1937) *Astronomical Thought in Renaissance England*, Baltimore, Johns Hopkins Press.

Mocgaard, K.P. (1972) 'How Copernicanism took root in Denmark and Norway', in Dobrzycki, pp.117–51.

Russell, J.L. (1989) 'Catholic astronomers and the Copernican system after the condemnation of Galileo', *Annals of Science*, 46, 365–86.

Sandblad, H. (1972) 'The reception of the Copernican system in Sweden', in Dobrzicky, pp.241–70.

Chapter 4

Crisis in Italy

by David Goodman

4.1 The Italian environment

4.1.1 A European focus

Italy in 1500 was in many ways a region of special importance for Europe. One of the most highly urbanized areas in the world, its cities were amongst the most populous of Europe: Milan, Venice and Palermo had around 100,000 inhabitants; Naples, approaching 200,000. And Italian cities were the richest in Europe, the result of late-mediaeval textile manufactures, long-distance trade and banking. Venice continued to function as a great centre for the import of oriental spices; Genoa and Florence were famous throughout Europe for their banking houses. Associated with this urbanization and prosperity, Italy had higher levels of literacy and perhaps numeracy. Double-entry book-keeping was an Italian invention designed to facilitate the recording of accounts of expanding businesses. Elementary arithmetic was taught in schools in Florence and Venice in response to the demand of merchants who wanted to prepare their sons for a commercial career. And the growing recognition in Florence of the importance of literacy in everyday life explains the high enrolment in the city's private schools; around 10,000 boys, some ten per cent of the population, are thought to have attended in the 1340s. In fifteenth-century Italy private schoolmasters taught the elements of grammar, prose writing and mathematics to the sons of the wealthy. In Mantua, Vittorino da Feltre offered the same literary and mathematical education to the rich and, free of charge, to the poor.

High literacy went hand in hand with book collecting. The wealthy accumulated large libraries. And soon after the invention of printing Venice quickly became Europe's greatest centre of book production, because of its large reading public and its extensive commercial links throughout Europe which facilitated distribution. When in the 1570s Philip II of Spain was searching for books to build up a great library for his new palace at the Escorial, he knew that Venice would be the most important place to look; accordingly he directed his ambassador in Venice to find out what books were available and soon large purchases were made on his behalf, many of them scientific works.

The unusually developed opportunities for primary education in Italy led pupils to the region's numerous universities which included some of the oldest and most distinguished in Europe. Bologna and Padua were famous for law; and for medicine, Padua had become a leading centre. Their fame attracted not only Italians. Copernicus was one of the many foreigners who travelled far to study at an Italian university.

Italy had also become the focus of European culture through the Renaissance, a wide-ranging literary, artistic and philosophical revival (see Chapter 1) originating in Florence and then spreading to other parts of Italy and Europe. Princes competed keenly for the services of Italian artist-engineers like

Leonardo da Vinci (1452–1519), the son of a Florentine notary and one of the geniuses of the age. One reason for this princely patronage was to enhance military power in a period of expanding warfare. Leonardo and other Italians offered explosive chemicals, numerous other inventions for attack, and a superior system of defensive fortifications based on regular geometrical forms. And in peacetime rulers used the talents of these Italians to promote the reputation of their dynasties by constructing magnificent palaces; writing laudatory histories or painting flattering allegorical portraits of themselves and their ancestors; or simply reflecting glory through the presence in their courts of a gifted scientist or painter. The greatest efforts of patronage were sometimes exerted by the pettiest of rulers anxious to present a more powerful image to the world. This was true of fifteenth-century Urbino, whose dukes, Federico da Montefeltro and his son Guidobaldo, spent lavishly to secure leading talents. Leonardo da Vinci was employed by Lodovico Sforza, Duke of Milan, and at the end of his life, he was painter and engineer to Francis I of France. Leonardo's passion for painting was closely connected with his investigations in anatomy – his illustrations of muscles were of unprecedented accuracy. Through dissection he was able to provide truer representations of the uterus and of the optic nerves. And by the technique of injecting wax he made casts of the cavities of the brain which showed the true shape and size of the cerebral ventricles. Leonardo's experiments, discoveries and inventions cover a remarkable range: from an appreciation of the organic origin of fossils to the discovery of the annual ring formation in the growth of trees; ideas on the strength of materials and experiments on flight using model birds; proposals for a steam-powered cannon and the idea of ball bearings to reduce friction in machines. But in all of this output what mattered most in the end was that much of Leonardo's brilliance was hidden in numerous notebooks which he never published. At a time when the spread of scientific knowledge was fostered by printing and soon by academies, the effect of Leonardo's scientific research was nullified by his failure to communicate.

Figure 4.1 Portrait of Friar Luca Pacioli (c.1445–1517) with his patron, Duke Guidobaldo da Montefeltro of Urbino. Pacioli, a Franciscan and itinerant teacher of mathematics, once employed by a Venetian merchant, had written an encyclopaedic treatise on mathematics (Summa de arithmetica, geometrica, proportioni et proportionalita, Venice, 1494) which mostly described practical arithmetic such as double-entry book-keeping. (From the painting by Jacopo de Barbari. Archivi Alinari, Florence.)

Renaissance culture, famous universities and prosperity all made Italy a focus of European attention. And so of course did Rome, the home of the papacy, and until the Reformation the spiritual centre of the Western Church. The power of the Popes, temporal and spiritual, was a particularly important influence in Italian cultural developments.

4.1.2 Italy's diversity: varied conditions affecting science

'Italy' in the sixteenth century has real meaning as a geographical area and also as a linguistic entity – Tuscan Italian was becoming the common written language of the peninsula. But until the nineteenth century Italy remained politically fragmented, a collection of states of varying size from the large Kingdom of Naples to the minute Republic of San Marino. And it is important to consider the diverse conditions existing in different parts of Italy for any analysis of the country's scientific development.

Florence

Florence in the early fifteenth century was a city-state with a long tradition of republicanism. Ruled by wealthy merchants, an oligarchy and no democracy, its constitution was intended to prevent the tyranny which had overtaken the Duchy of Milan. Through military conquest Florence had annexed a considerable area, much of Tuscany, including in 1406 the city-state of Pisa, once an ambitious trading republic with possessions in the Levant and now reduced to a subject city with fine buildings and a university.

Florence's expansion coincided with the rise of the Medici, a family whose soaring wealth enabled it to penetrate and finally dominate the government of Florence. That wealth had come from the family banking firm, especially after 1413 when Giovanni di Bicci de' Medici was appointed to handle the enormous revenues of the papacy and was paid a lucrative commission. Giovanni's son, Cosimo de' Medici (1389–1464) inherited the most flourishing banking house in Europe and became the ruler of Florence, the founder of a dynasty which with short interruptions would remain in power into the eighteenth century.

Partly out of scholarly interest but also to bring prestige to his family, Cosimo invested heavily in culture, bestowing patronage on architects to construct impressive public buildings and on philosophers to recover the ancient wisdom of Plato. He had a passion for collecting books and searching for Greek manuscripts. And he supported the philosopher Marsilio Ficino, paying for his education, buying him a house and sponsoring his famous complete edition of Plato's works.

This patronage, motivated by scholarship and politics, would become a family tradition; succeeding Medici rulers would continue to erect magnificent palaces, commission ornate tombs and frescos to glorify the dynasty, and employ men of learning to share in the fruits of wisdom and add lustre to the court. But Florence was not yet a princely court. Not until the stifling of republicanism after 1530 did the Medici become titled princes. That was achieved with the help of the soldiers of the Emperor Charles V who, with the aim of increasing Habsburg power in Italy, agreed to convert Florence into a hereditary duchy of the Medici family, on condition that loyal support was given to the Habsburgs. This plan was supported out of family interest by Giulio de' Medici, now Pope Clement VII. The loss of freedom in Florence was symbolized by the intimidating citadel immediately constructed by Alessandro, the first duke.

Then came another elevation of the Medici status from Dukes of Florence to Grand Dukes of Tuscany. After further territorial expansion, including the city of Siena, Cosimo I de' Medici persuaded Pope Pius V in 1569 to grant him the title of Grand Duke. Under his autocratic rule dissent was firmly suppressed and the employment of spies introduced. Medici patronage continued under him and his son Francesco, whose obsessive interests led

Map 4.1 Natural philosophy in Italy, 1550–1650

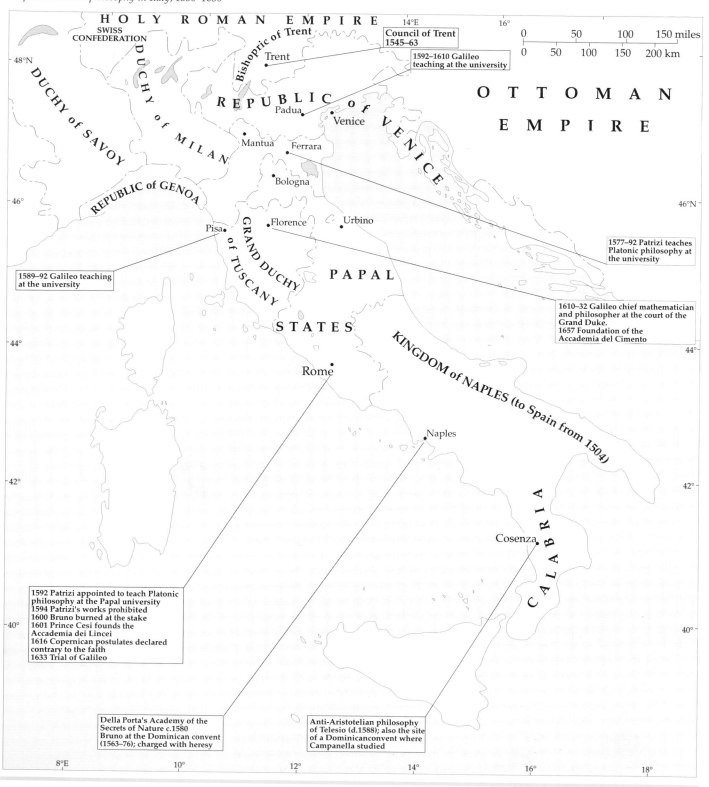

Council of Trent 1545–63

1592–1610 Galileo teaching at the university

1589–92 Galileo teaching at the university

1577–92 Patrizi teaches Platonic philosophy at the university

1610–32 Galileo chief mathematician and philosopher at the court of the Grand Duke.
1657 Foundation of the Accademia del Cimento

1592 Patrizi appointed to teach Platonic philosophy at the Papal university
1594 Patrizi's works prohibited
1600 Bruno burned at the stake
1601 Prince Cesi founds the Accademia dei Lincei
1616 Copernican postulates declared contrary to the faith
1633 Trial of Galileo

Della Porta's Academy of the Secrets of Nature c.1580
Bruno at the Dominican convent (1563–76); charged with heresy

Anti-Aristotelian philosophy of Telesio (d.1588); also the site of a Dominican convent where Campanella studied

him to delegate affairs of state to others, to allow him time for his mineral collections and alchemical experiments. Francesco was succeeded by his brother Ferdinando I, a ruler who returned to the direction of state affairs, patronized scholars and employed Galileo in 1605 as tutor in mathematics to his grandson, the young prince Cosimo. And it was the young Grand Duke Cosimo II, dominated by his mother Christina, who would receive Galileo in 1610 as the court's chief mathematician and philosopher. Future events would confirm the warnings Galileo received of the risks of settling in an autocratic state, which employed spies and was allied to the inflexible champions of Catholicism, the Habsburgs and the Papacy (Hale, 1977).

Rome

Florence was governed by merchants; Rome by priests. The city of Rome was the centre of a large territorial state, the Papal States, occupying much of central Italy. The Pope, the elected successor of St Peter and the supreme spiritual leader of Western Christendom, was also the temporal ruler of a central Italian state which stretched from the borders of Naples to the republic of Venice. The papacy was not occupied by a dynasty but by a succession of elected individuals, usually chosen from powerful Italian families. Like the secular Italian princes, the Popes used patronage of artists and scholars to strengthen their authority. Nicholas V (1446–55), like Cosimo de' Medici, sponsored the translation of Greek manuscripts, was an avid collector of books, in effect establishing the nucleus of the great Vatican library, and began a wholesale re-planning of Rome designed to impress the world with a magnificent capital. The same policy was continued by Giovanni de' Medici when he became Pope Leo X (1513–21). Entirely in the spirit of his family tradition, he collected classical manuscripts, gave patronage to scholars, employed Raphael to decorate the rooms of the Vatican and assisted the rebuilding of St Peter's. All of this converted Rome into the most sumptuous court in Europe.

Not all Popes were friends of learning; the character of the papacy varied with the individuals elected and with prevailing political and religious conditions. There was no guarantee that a liberally minded humanist would retain these values after election to the papacy. For the Popes of the sixteenth century the most pressing problem was the German Reformation which threatened to destroy the unity of the Western Church, and eventually did so. From Germany Lutheran doctrines, and from Geneva similarly anti-papal Calvinist theology spread into Italy. The Popes became increasingly concerned with heresy. The humanist Alessandro Farnese, who became Pope Paul III (1534–49), continued to favour artists and scholars, but also established a central Congregation of the Inquisition or Holy Office (1542) to investigate the most serious cases of heresy with the assistance of anonymous denunciations and the use of torture. Worried by spreading Protestantism, Paul III also promoted the summoning of a General Council of the Church which eventually met at Trent.

The first Index of prohibited books was issued by the Congregation of the Inquisition in 1557 under Paul IV. And in 1571 Pius V introduced a new feature making the Index a continuously revised list. Books could be totally prohibited or put back into circulation after the removal of offensive passages. The stern Sixtus V (1585–90) appealed to scholars in universities throughout Europe to protect the faith by collaborative detection of heretical views in printed books. And in a bull of 1586 which must have made natural philosophers more cautious, he broke with the earlier more liberal papal attitude to astrology, condemning those forms of astrology which clearly denied human free will and in general all magical practices; rigorous punishment was prescribed for those who continued to practise and for those who possessed or read books dealing with these matters. The atmosphere in sixteenth-century Rome had changed and, faced with censorship, arrest and severe penalties, some of the more independently-minded philosophers and clerics thought it prudent to flee across the Alps.

The drive against heresy was strengthened by the foundation of the Society of Jesus by Ignatius Loyola, a Basque who gave up his military career to become a soldier of Christ. He and his followers had gone to Rome to put themselves at the disposal of the Pope, offering total obedience, a willingness to go to all parts of the world to spread the true faith, and to fight heresy. In 1540 Paul III officially recognized Loyola's Society; within a few decades the Jesuits had established a European network of teaching institutions, a fundamental part of the Society's propaganda. By 1600 there were 372 Jesuit colleges in Europe; by 1640 around 520. A curriculum, rigidly taught according to a specified 'ratio studiorum', or method of studies, controlled the imparted education along monitored and approved lines; it was explicitly stated that any teacher opposed to Aquinas' scholastic theology would be dismissed. In Rome, where there already existed a papal university within the Vatican palace, there now arose the Jesuits' Collegio Romano, a prestigious university dedicated to the protection of Catholicism from all unorthodoxy, especially the doctrines of the Reformation. Under the influence of men like Roberto Bellarmino, professor of philosophy at the Collegio, there would be not the slightest deviation from Aristotle's philosophy, so finely woven into Catholic theology by the revered Aquinas. Apart from grammar, ethics, logic, mathematics, and theology, the new institution taught natural science, chiefly Aristotelian physics and cosmology. There was also an interest in experimental magnetism and optics, and the staff included skilful observational astronomers. But at the Collegio Romano no scientific experiment or observation was allowed to spoil the received Aristotelian orthodoxy. The Jesuit scientists at the Collegio were not conducting independent scientific research, investigating nature wherever that might lead, but always within the confines of permitted beliefs; for them science was to be applied to doctrinal and apologetic needs, 'even at the sacrifice of personal scientific ambitions' (Redondi, 1988, p.127).

Figure 4.2 The Collegio Romano, founded in 1550, showing the renovated structure of 1584. (From F. Buonanni: Numismata pontificum romanorum, quae à tempore Martini V ... *(1699). Photo: British Library Board.)*

This uncompromising dedication to orthodox Catholicism was entirely in keeping with the decisions of the Ecumenical Council of Trent (1545–63). The discontinuous sessions were held in the alpine bishopric of Trent, linguistically Italian but politically an autonomous state within the Holy Roman Empire of the German Nation. The Council's attempts to reconcile Protestant and Catholic failed but there was success in the Council's aims to reform the Catholic church and define Catholic dogma. Controversial theological questions concerning the sacraments and holy relics were settled in sharp opposition to Protestant doctrines. A year after the closing session Pius IV issued the 'Profession of the Tridentine Faith', a summary of the newly defined and approved doctrine. The Vulgate, Jerome's Latin version of the Bible, was declared a basic authoritative text; the doctrines of the early Church Fathers were endorsed as an unquestionable body of religious truth; and insistence on the Church's monopoly of interpretation of the Scriptures firmly asserted. Tradition, obedience, and a coherent body of dogma, clearly defined and officially sanctioned, governed Catholicism after Trent. And no one more than Roberto Bellarmino, the Jesuit philosopher and theologian, and theological adviser to the Pope, played a greater role in interpreting and explaining the new orthodoxy. Inevitably this role brought him to the front of the stage in heresy trials and challenges to papal supremacy.

Venice

No state in Italy was more opposed to papal authority than the Republic of Venice. Apart from a short period in the early Middle Ages when Venice had been part of the Byzantine empire, the city had never recognized any superior authority beyond its lagoon. Commerce with the Levant had been Venice's predominant interest, leading it from the thirteenth century to acquire a string of possessions which stretched from the Adriatic to the Aegean and eastern Mediterranean. Corfu, Crete, and later Cyprus were Venetian colonies which with numerous smaller islands and ports safeguarded the valuable import of spices and silks from the east. While Popes and emperors continued to call for crusades against the Turks, Venice was reluctant to participate because of the damaging effects on her privileged trade with Muslim states. With its strong fleet and colonial possessions, Venice had become a European power, and this was the fruit of a nurtured trade which would not readily be abandoned.

Venice had also expanded on the Italian mainland. At the same time that Florence annexed Pisa, Venice took another university city, Padua (1405). To promote Padua's economy the Venetian Senate in 1434 prohibited Venetians from studying at any other university; degrees obtained elsewhere were simply not recognized. The main function of the university was to train Venetian ruling nobility for the service of the state, and law was the most important subject; but foreigners also flocked to Padua to study medicine there, the most progressive discipline in the faculty of arts.

Like Florence, Venice was a city-republic of merchants, ruled by an oligarchy of the wealthiest families – in the case of Venice, trading aristocrats. Unlike Florence, Venice had no history of communal violence and revolution; but a constitution and policies which so effectively brought centuries of internal peace and prosperity that in the seventeenth and eighteenth centuries the Venetian constitution was held up for imitation in England and the Dutch Republic.

In Venice there was no princely court to patronize the arts and sciences. When Galileo came to the city he was invited to informal literary and scientific gatherings at the home of the Morosini, a noble family of Levant traders. And when he was appointed to teach at the University of Padua Galileo served as an employee of the Venetian republic.

Along with Venice's determination to preserve trade and independence went secularization and toleration. Elsewhere in Europe prelates rose to the highest offices of state: Cardinal Wolsey in England, Cardinal Granvelle in Spain and Cardinal Richelieu in France. In Venice that could not have occurred; its constitution expressly prohibited any ecclesiastic from holding any state office, even as a clerk. The republic wished to exclude all priestly influence in government thereby avoiding interfering effects of loyalties to Rome. In Venice the Church was largely organized as a department of state, the Pope reluctantly reduced to approving bishops nominated by the republic. There was an office of the Roman Inquisition in Venice but, with the utmost reluctance, the Popes had given in to the republic's unwavering insistence that the Inquisition would not function on Venetian territory unless lay citizens of the republic were admitted to its membership. Jurisdiction in matters of heresy or censorship would not be surrendered to the Papacy. When the Popes ordered that matriculation at universities should be conditional on taking an oath of acceptance of the faith defined by the Council of Trent, it was firmly resisted, because it entirely conflicted with the republic's tolerance of German Lutherans and Jews, who graduated without obstacles at Padua. The republic's policy here was eloquently stated by her official spokesman on theology, Paolo Sarpi. A friar of the Order of Servites, a devotee of the sciences, and the friend of Galileo, Sarpi clearly believed in the unhindered development of scientific knowledge:

> To confer a doctorate in philosophy and medicine is to testify that the scholar is a good philosopher and physician, and that he can be admitted to the practice of that art. To say that a heretic is a good physician does not prejudice the Catholic faith; it would prejudice it rather to say that he was a good theologian. The whole world says that Hippocrates and Galen, infidels, are most excellent physicians, and that there is not their equal among Catholics; yet our faith receives no injury from this. *(Bouwsma, 1968, p.534)*

Heretics and Jews continued to be a tolerated part of the student body at Padua; what was not tolerated was any hint of papal influence at this centre of education of the Venetian ruling elite. The Jesuits also had a college in Padua, and the academics of the university, no less staunchly Aristotelian, resented the presence of a competing institution which took away students. When in 1591 Cesare Cremonini, Padua's leading Aristotelian philosopher, appeared before the Venetian Senate, he fully exploited the republic's sensitivities towards the papacy, indicating – it was no distortion – that the Jesuits were granting degrees on Venetian territory through papal authorization. The Venetian Senate responded by suppressing Jesuit teaching at Padua.

Relations between Venice and Rome finally reached breaking point in 1605 when the republic passed a law to hinder the acquisition of land by the Church, and initiated criminal proceedings against two clergymen whom the Pope wanted to be tried in an ecclesiastical court. Paul V issued an ultimatum declaring that unless these decrees were revoked he would excommunicate the Venetian Senate and issue an interdict, prohibiting religious services throughout Venetian territory. Venice resisted and the papal interdict was declared. The Jesuits were sent packing and the republic, through its spokesman Sarpi, condemned the Pope's action as motivated by greed for power. On the papal side Bellarmino asserted that the Pope was supreme in temporal as well as spiritual affairs. But after a year's bitter quarrel, the Pope, with some loss of reputation, retracted and Venetians were relieved to have successfully fought off a threat to their independence.

Venice had been a place of tolerance, a haven for persecuted Jews, German Lutherans and a large colony of Greeks, who to the Pope's continuing irritation were allowed to retain their rituals and allegiance to the schismatic patriarch of Constantinople. Nevertheless Catholicism was a matter of deep interest to the rulers of Venice; the protection of the faith was regarded as essential to the well-being of the republic, provided the papacy was kept from interfering. And so, sporadically, heretics were prosecuted in sixteenth-century Venice, the Jews expelled for a period and their holy books burned, and other books prohibited. Freedom in Venice was not stipulated as a matter of principle in the constitution, and its continuation could not be guaranteed. There were freer places in sixteenth-century Europe: Poland, Transylvania and the Ottoman Empire; but the degree of freedom in Venice was remarkably higher than much of the rest of Europe, and any discussion of the importance of freedom for scientific development should take the case of Venice into consideration.

Naples

While Venice enjoyed tranquillity and prosperity, the Kingdom of Naples suffered from uprisings, poverty and banditry. A part of the Spanish monarchy since the beginning of the sixteenth century, when Spanish forces had defeated the French in a contest for the territory, it was ruled by a Spanish viceroy. Important offices and ecclesiastical benefices were filled by the king in Madrid. But much of the territory was governed through native feudal nobility. The traditional institutions of the kingdom were generally respected, but rioting occurred in 1510 and 1547 when the viceroy attempted to introduce the Spanish Inquisition, regarded as still more repressive than the Roman Inquisition which was eventually restored as the agency responsible for protecting the kingdom from heresy. The uprisings which occurred were usually due to the rapacity of the nobility rather than to resentment of Spanish rule. And it is because Neapolitan philosophers and clergymen became involved in this social tension, not only taking sides with the oppressed populace, but also providing attractive ideas, that for this region of Italy it is conceivable that some relationship existed between philosophical and scientific speculation, and the social context. Naples produced a crop of natural philosophers deeply interested in magic and the occult; that harmonized with the strong manifestation of popular magic and astrology in Calabria where the desperate population had turned for salvation to the occult. The urban disturbances of 1585, caused by a rise in bread prices, were blamed on Giovanni Pisano, a pharmacist and associate of the occult scientist Giambattista della Porta.

The Neapolitan clergy allied with the poor against the nobility and even supported banditry. The nobles pressed for the closure of the Dominican convent of San Domenico Maggiore in the city of Naples. This was an important centre of learning housed in the same building as the university. It possessed a large library including prohibited books and some of its friars developed bold speculative theories quite different from the received instruction in Aristotelian philosophy. This was where one of the arch-heretics of the age, Giordano Bruno (1548–1600) received his formative training, entering the convent at the age of seventeen and remaining there for over ten years. And the same convent was visited by another famous Dominican heretic, the Calabrian Tommaso Campanella (1568–1639) whose astrology, prophecy and utopian dreams offered an unorthodox road to salvation for the downtrodden; he was implicated in a plot to overthrow the state in 1634. Threatened with closure by nuncios and Popes, seen by the ruling nobility as a seed-bed of disorder, San Domenico Maggiore was preserved through popular support.

4.2 Freedom to theorize lost; the sanctions of 1576–1633

4.2.1 The suppression of Italian Platonism

The conditions of Italian intellectual life had profoundly changed with the Counter-Reformation and the Council of Trent. In the previous century Ficino's Platonic revival in Florence had presented a philosophy different from Aristotle's, mingled with oriental religious ideas of non-Christian origin, and certainly unorthodox (see Chapter 1). Ficino had raised suspicions in Rome, not as a threat to Aristotelianism but as a sorcerer. Nothing came of this and he continued to enjoy princely patronage and fame. What a contrast with the experience a century later of another Platonic philosopher, Francesco Patrizi (1529–97). He had been teaching Platonic philosophy in the Duchy of Ferrara; the Este duke had appointed him to a chair at the university there. Rejecting Aristotle's opinion on space and other matters, he argued for the existence of a vacuum. And drawing, as Ficino had done, on the neo-Platonic texts attributed to Hermes Trismegistus, whom he believed to be a prescient sage contemporary with Moses, Patrizi believed he had constructed the basis of a truer religion. These ideas were developed in his *Nova de universis philosophia* (New philosophy of the universe) (Ferrara, 1591) with a dedication to Pope Gregory XIV stating that Hermes with his glimpses of the Trinity offered a far sounder route to Catholicism than the unreliable Aristotle who denied God's omnipotence and providence. And he urged the Pope to substitute Hermetic and other Platonic doctrines for Aristotle in Jesuit and other colleges. This would strengthen weakening Catholics and perhaps even bring German Protestants back to the true faith; because 'it is much easier to bring them back in this way than to compel them by ecclesiastical censures or by secular arms ... And I beg you to accept me as your helper in this undertaking' (Yates, 1964, pp.181–4).

At first it seemed that Patrizi's offer had been heartily welcomed. Cardinal Aldobrandini praised his philosophy as harmonizing 'perfectly with Christian piety' and soon after his election as Pope, Clement VIII appointed Patrizi to a chair of philosophy at the papal university of Rome. But then in an about-turn strikingly similar to what later happened to Galileo, there were murmurs of heresy, Patrizi's book was put on the Index and in 1594 he was forced to retract the philosophy he had so loved with a statement, which must have been made with bitterness, that he had never really believed in any of it. The final touch came after his death in 1597, when Clement began to look for someone to fill Patrizi's chair. Bellarmino advised the Pope that Platonism was dangerous, tending to distract the mind from Christian truth. He recommended the suppression of Patrizi's chair; that was not done but the appointment was given to Jacopo Mazzoni, who was not hostile to Aristotle. (For more on this, see Firpo, 1970.)

The Neapolitan philosophers of nature were also hounded though with far greater persecution. The anti-Aristotelian Bernardino Telesio (1509–88) did not suffer in person for his unorthodox beliefs in a living universe and the likening of man to animals, but in 1605 his writings were totally condemned. Giambattista della Porta had a strong interest in the occult and a form of 'natural magic' which shaded into scientific experiments, notably in optics where his work with lenses led him to devise various optical instruments. Around 1580 his magical practices brought him before the Inquisition; his *Academia Secretorum Naturae* which met in his house in Naples to discuss and investigate Nature's secrets was probably closed by the Inquisition, and in 1592 further publication of his works was prohibited.

Campanella was dealt with much more harshly. Influenced by both Telesio and Porta, but most of all by the ancient Hermetic texts, his philosophy, anti-Aristotelian, prophetic and utopian, seemed to threaten religion and government. From the mythical Hermes Trismegistus he took the belief that the world was a living animal, and the importance of the Sun as the font of light, heat and life. He became a convinced Copernican, believing that the Sun's place could only be at the centre of the universe. And his astrologically guided plans for a new society to usher in a purer age were described in a book significantly entitled *The City of the Sun*, whose realization he tried to inaugurate in Calabria. In this ideal world Campanella's model citizens were opposed to Aristotle, whom they disdained as a pedant. Denounced to the Inquisition for heresy, this Dominican was tortured and spent thirty years of his life in prison in Naples and Rome. But he escaped and went to France where he benefited from the patronage of Richelieu.

Giordano Bruno did not survive his persecution. Like Campanella a Dominican, he too rejected Aristotelian philosophy for Hermes Trismegistus, and apparently believed that Catholicism could be combined with the magical wisdom of this supposed Egyptian sage. His animistic cosmology presented the Earth and other planets as living animals moving around a central Sun. He praised Copernicus for reintroducing a Sun-centred universe but thought his approach, too mathematical and insufficiently physical, had prevented him from perceiving the full truth. This was that the universe was infinite, with many other living worlds each with their own Sun.

For reasons that are not clear Bruno was already being charged with suspect beliefs during his years at the convent of San Domenico Maggiore. He fled Naples and then Rome, and began his journeys through Europe provoking quarrels wherever he went. In Geneva he encountered intolerant Calvinists; in Oxford he mocked the university's diehard Aristotelians; and in Helmstedt the Lutherans excommunicated him. Then in 1591 he accepted an invitation from a Venetian nobleman, Zuan Mocenigo, to come to Venice and teach him his art of memory. Probably confident of Venice's reputation for toleration, he must have been unpleasantly shocked when Mocenigo denounced him to the Venetian Inquisition as an enemy of religion. Worse was to come. Bruno was a Neapolitan not a Venetian, and a Dominican subject to papal authority. On these grounds the papal nuncio in Venice pressed for his extradition to Rome. At first, predictably, the Venetian Senate hesitated; but when one of the senators, Federico Contarini, delivered a harangue branding Bruno as a heretic and debauchee, the Senate saw the opportunity for some improvement in relations with the Pope. If it had been 1606 with Venice and the papacy at loggerheads Bruno would not have been handed over. But given that poor relations between the two powers were normal it still comes as something of a surprise that Bruno should have been delivered and it must remain as one of several negative qualifications of Venice's record for tolerance. Bruno was transferred to Rome, where he was tried, sentenced and burned at the stake. Some obscurity remains on the reasons for his sentence because the official document has been lost. But records of the interrogation show that he was accused of several errors of faith, theological but also touching on scientific belief: denial of the divinity of Jesus; rejection of the virginity of Mary; false views on transubstantiation; involvement in magic; and belief in the existence of multiple and eternal worlds. In the opinion of one expert 'it was probably mainly as a magician that Bruno was burned, and as the propagator throughout Europe of some mysterious magico-religious movement' (Yates, 1970, p.542). Copernicanism by itself was not the cause of Bruno's execution; but since the planetary Earth was an ingredient of his complicated heterodox philosophy, it could well be that this dramatic trial was recalled when the ecclesiastical authorities decided to move against Galileo.

4.2.2 The rise and fall of Galileo

Galileo in Pisa: university and court

Galileo Galilei (1564–1642) was born in Pisa, the son of a musician. His father sent him to study medicine at the University of Pisa but Galileo disliked the lectures and the unquestioning acceptance of Aristotle and Galen. Already he began to challenge Aristotle's assertion that bodies fell to Earth with velocities proportional to their sizes. He had noticed that hailstones, large and small, hit the ground simultaneously. Instead of attending the university he was attracted to the mathematics teaching of Ostilio Ricci at the Medici court which moved from Florence to Pisa every winter. Euclidean geometry and Archimedes' discussion of floating bodies and centres of gravity, rather than Galen's physiology, became Galileo's main interests.

He left university without a degree and taught mathematics privately in Florence and Siena. Around 1587 he discovered a new method of determining the centre of gravity of solids, and that assisted his appointment as professor of mathematics at Pisa (1589). There is no sign yet that Galileo was a Copernican; his manuscripts of these years include a commentary on Ptolemy's *Almagest*. Nor is it established that he really dropped those spheres from the leaning tower to disprove a conclusion of Aristotelian physics. But it is likely that his criticism of Aristotle was irritating Pisa's professors.

From Tuscany to Venice

In 1592 Galileo secured the chair of mathematics at Padua, a more prestigious university and paying a higher salary. He lectured on Euclid and astronomy – chiefly to medical students who were then expected to consider the stars as causes of disease and their supposedly powerful influences in the times of administering medicines. It was now that Galileo showed the first signs of interest in Copernican astronomy, producing in 1595 an explanation of the tides which depended on the assumption of a moving Earth. When Kepler sent him his Copernican treatise *Mysterium cosmographicum* Galileo replied in August 1597 that he had 'adopted the teaching of Copernicus many years ago' but had refrained from publishing because of the 'ridicule and derision, Copernicus our teacher' had received.

Galileo continued his silence until the publication of his *Starry Messenger* (1610) which presented the strongest evidence for the Copernican universe that had yet appeared. It came from observations with the newly invented

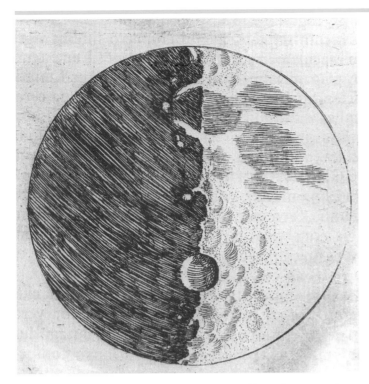

Figure 4.3 (a) The Moon drawn by Galileo. (From Galilei: Sidereus nuncius ... *(1610). Photo: British Library Board.)*

(b) The Moon near the third quarter, as seen through a modern high-power telescope. (John Sanford, Science Photo Library)

telescope. In 1609 Galileo heard that a new instrument had been invented in the Netherlands which made distant objects appear close. Realizing its value to a naval power like the Venetian Republic, his employer, he at once began trials with various combinations of lenses positioned at the opposite ends of a lead tube. Soon he had found an arrangement which made objects seem 1,000 times larger than when viewed with the naked eye. His reward was academic tenure for life in the service of the Republic. Perhaps the first ever to do so, he turned a telescope to the heavens; the results were astonishing. The surface of the Moon appeared pitted with craters. There were mountain peaks illuminated by the Sun's light. For Galileo it was just like a sunrise on a mountainous region of the Earth. Far from the perfect polish and smoothness attributed to celestial bodies by Aristotle, the Moon's surface seemed even more irregular than the Earth's. His observations of Jupiter gave added plausibility to a Sun-centred universe. Close to Jupiter Galileo saw four very bright bodies. At first he took them to be fixed stars but continued observation showed that their positions relative to Jupiter changed. Their apparent sizes also changed, an effect which could not be explained by an optical illusion caused by viewing through the Earth's atmosphere, because the apparent size of Jupiter remained constant. Also the number of those visible bodies varied on successive nights as they were hidden behind Jupiter. The only explanation was the existence of four satellites revolving about Jupiter. The assumption of Ptolemaic astronomy and Aristotelian cosmology that all celestial bodies rotated about the Earth was disproved.

Back to Tuscany

Galileo might have christened Jupiter's satellites the 'Venetian stars'. Instead he called them 'Medicean stars' and dedicated his *Starry Messenger* to Cosimo II de' Medici, once his pupil during summer vacations, and now the Grand

Duke of Tuscany. He was bidding for appointment at the Medici court. Why, when he had freedom and high pay in Venice? His letters reveal a combination of a lack of enthusiasm for university lecturing and homesickness for his beloved Florence. After his departure, his Venetian friend, government official and devotee of science, Giovan Francesco Sagredo, warned him that in abandoning the Republic for the Duchy he would face a less secure future:

> Where will you find freedom and self-determination as you did in Venice? Especially having the support that you enjoyed, which grew greater every day with the increase of your age and your authority ... At present you serve your natural prince, a great man, virtuous, young and of singular promise; but here you had command over those who govern and command others; you had to serve no one but yourself; you were as monarch of the universe. The power and magnanimity of your prince gives good hope that your devotion and merit will be welcomed and appreciated; but in the tempestuous seas of courts who can promise himself that he will not, in the furious winds of envy, be – I shall not say sunk, but at least tossed about and disquieted? I say nothing of the prince's age, for it seems necessarily that with the years he will mature in temperament and inclination and in his other tastes ... But who knows what may be caused by the infinite and incomprehensible accidents of the world? Impostures of evil and envious men, sowing and raising in the mind of the prince some false and malicious idea, may make justice and virtue themselves serve to ruin a gallant man. Princes take pleasure for a while in this or that curiosity; but then, called by interests in greater matters, they turn their minds elsewhere. I can well believe that your Grand Duke may be pleased to go about with one of your telescopes looking at the city of Florence and some nearby place; but if through some important requirement of his he must look at what goes on in all Italy, in France, in Spain, in Germany, and in the Near East, he will put aside your telescope ... And I am much disturbed by your being in a place where the authority of the friends of the Jesuits counts heavily. *(Drake, 1957, pp.67–8)*

To his cost Galileo ignored this advice. In September 1610 he became chief mathematician and philosopher to the Grand Duke. He had returned to a state where Rome's influence was markedly greater.

Soon after his return Galileo communicated his momentous discovery of the phases of Venus, nothing less than a falsification of the established Ptolemaic astronomy. Copernicus had predicted that a crucial test, involving the appearance of Venus, would decide between his heliocentric theory and Ptolemy's geocentric astronomy. For if Venus moved around the Sun it should exhibit a sequence of phases to an observer on Earth, varying like the Moon from a circle to a crescent. On the other hand if Venus revolved about the Earth as centre, as Ptolemaic astronomy maintained, the observer would never see more than a thin crescent of Venus (because Venus never appears far from the Sun). Before the telescope was available, Venus was too small for the naked eye to detect phases. But Galileo was able to see for the first time that Venus appeared to change its shape, like the Moon. He also observed the marked variation in the apparent size of Venus, showing that its distance from the Earth was altering. This was easily explained on the Copernican system, since Venus at various points in its orbit might be on the same or the different side of the Sun in relation to the Earth. But this did not establish the truth of Sun-centred astronomy because the appearances could also be explained by Tycho Brahe's theory, rejected by Galileo as absurd, which made all the planets, except the central Earth, revolve around the Sun (see Figure 3.20). And that theory had the attraction of preserving the immobile Earth in conformity with common sense, scriptural authority and academic tradition.

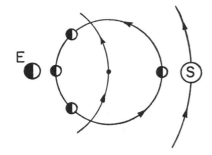

(a)

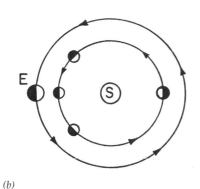

(b)

(c)

Figure 4.4 The phases of Venus in (a) the Ptolemaic system, (b) the Copernican system, and (c) as observed with a low-power telescope. In (a) an observer on the Earth should never see more than a thin crescent of the lighted face. In (b) he should see almost the whole face of Venus illuminated just before or after Venus crosses behind the sun. This almost circular silhouette of Venus when it first becomes visible as an evening star is drawn from observations with a low-power telescope on the left of diagram (c). The successive observations drawn on the right show how Venus wanes and simultaneously increases in size as its orbital motion brings it closer to the earth. (T.S. Kuhn, The Copernican Revolution, Harvard University Press/Oxford University Press, 1957, fig.44, p.223)

Figure 4.5 Frontispiece of Giovanni Battista Riccioli, Almagestum Novum *(Bologna, 1651). This Jesuit treatise shows Tycho's theory weighing more heavily than Copernicus' system. (Photo: British Library Board)*

This empirical falsification of the established cosmology was not well received by the Aristotelian professors at Pisa and Padua. Some of them would not face the evidence, refusing to look through the telescope. Others dismissed the observations as optical illusions produced by lenses. Galileo offered a large reward to anyone who could make an instrument which would display bright objects moving around Jupiter but around no other celestial body. Still other philosophers conspired to silence Galileo by inciting clergymen to condemn his work as anti-Christian. That offer was taken up by

Caccini, a Dominican, who at the end of 1614 delivered a sermon in Florence, illustrated with texts from Joshua on the moving Sun, in which he blamed Galileo for irreligious Copernicanism. He then went to Rome to denounce Galileo, but nothing came of it; the officials there could not yet see anything reprehensible.

Galileo had recently visited Rome, where he had been warmly received by Cesi's Accademia dei Lincei (see Chapter 9), the Jesuits and Pope Paul V. Once the Jesuit astronomers of the Collegio Romano acquired a sufficiently powerful telescope they reported to Bellarmino that Galileo's observations were confirmed. But this accord could not last. Cesi's Academy was an anti-Aristotelian scientific society committed to independent research; regular clergy like the Jesuits of the Collegio Romano were barred from membership. There was a strong tension between the two institutions, and it was the Academy which elected Galileo a member and would sponsor his publications. In one of these, the *Letters on Sunspots* (1613), Galileo sharply criticized the Jesuit astronomer, Christopher Scheiner, rejecting his claim to the discovery of sunspots and his interpretation, designed to preserve a perfect Aristotelian solar surface, that the apparent spots were merely small planets which periodically obstructed the view of the Sun. The quarrel soured Galileo's relations with the Jesuits.

Galileo again came to the attention of ecclesiastical authority in Rome with the report of an after-breakfast discussion at the Medici palace in December 1613. The guests included Cosimo Boscaglia, philosophy professor at Pisa; and Benedetto Castelli, Benedictine monk and former pupil of Galileo, who had just been appointed to the mathematics chair at Pisa with explicit instructions not to teach the motion of the Earth or even to discuss it. The conversation had turned on the universe and the telescope. The Grand Duchess Christina, the overbearing mother of Cosimo II, asked whether Jupiter's satellites really existed. After the meal Boscaglia advised her that while Galileo's observations were confirmed, his belief in a moving Earth was contrary to Holy Scripture. The Grand Duchess then approached Castelli quoting biblical passages against Galileo's opinions. His reply that scientific questions should be decided independently led to an argument.

Castelli informed Galileo of that morning's events, and Galileo sent his friend a letter supporting him and expounding his own similar views on science and religion. After Castelli circulated copies of the letter, it was seen by Niccolò Lorini, a Dominican, who was so shocked by its contents, by the meddling of a layman in theology, and his 'trampling on all Aristotle's philosophy which has been of such service to Scholastic theology', that he notified the Inquisition in Rome. Another of the Grand Duchy's clergy, the Bishop of Fiesole, called for the jailing of Copernicus and had to be told that this man had been dead for some years. Galileo, anxious lest his views be misrepresented in Rome, sent a copy of his letter to Rome for Bellarmino to see, and promised a less hasty, enlarged treatment of the subject.

This soon appeared as the *Letter to the Grand Duchess Christina* which circulated in manuscript. He had written this to clear himself from charges of heresy drummed up by Aristotelian academics and to guide the Church towards a wise pronouncement on Copernicanism. A sincere Catholic, he believed that the Bible and science were independent sources of truth: one was the Word of God; the other, the 'open book of Nature', was a second avenue to truth through observation and reason, and since these faculties had been given to us by God, this path too proceeded from Him. The Bible and the book of Nature both derived from God; it was therefore impossible for science and Christianity to conflict. The apparent contradictions were unreal. Following St Augustine closely here, he argued that it was necessary to go beyond the surface meaning of scriptural passages which had been written for the simple people of the time. If, contrary to common sense, the Bible had stated that the Earth moved and the Sun was stationary, the population would not have believed it, which would have raised doubts on the rest of the Scriptures. Even in contemporary society, Galileo said, not one in a thousand would deny the apparent immobility of the Earth. The Bible was not intended to be an astronomical treatise but a guide for the salvation of souls. He endorsed the remark attributed to Cardinal Baronius, the sixteenth-century historian of the church, that the purpose of the Scriptures was to teach us 'how to go to heaven, not how the heavens go'. As to Trent's ruling that all agreed opinions of the Church Fathers were binding, Galileo said this applied only to matters of faith and morals; the mobility of the Earth and Sun were not in this category.

Galileo insisted on the freedom to investigate nature unhindered by theological or other opinion. Those theologians who required natural philosophers to search for a fallacy in any conclusion which contradicted Scripture were commanding them to close their eyes. Not even the Pope could declare phenomena to be other than what really occurred in nature. Any prohibition of Copernicus' system would mean the prohibition of true astronomy because men would be commanded not to recognize the phases of Venus. Centuries before, Augustine had warned Christians to avoid firm statements on such imperfectly understood matters as the nature of animals; the properties of the Earth; and the distances of stars. If such assertions turned out to be false, non-believers would associate Christian doctrine with error and reject fundamentals like the resurrection of the dead or belief in the Kingdom of Heaven. Galileo warned that any condemnation of the Earth's motion by Rome would risk these damaging consequences. And he closed with a description of the Sun, stationary and rotating on its axis at the centre of the universe, which shows that he too sympathized with the ancient Hermetic sentiments of solar excellence befitting a Sun-centred universe:

> ... if we consider the nobility of the sun, and the fact that it is the font of light which (as I shall conclusively prove) illuminates not only the Moon and the Earth but all the other planets, which are inherently dark, then I believe that it will not be entirely unphilosophical to say that the sun, as the chief minister of Nature and in a certain sense the heart and soul of the universe, infuses by its own rotation not only light but also motion into other bodies which surround it. And just as if the motion of the heart should cease in an animal, all other motions of its members would also cease, so if the rotation of the sun were to stop, the rotations of all the planets would stop too. *(Drake, 1957, pp.212–13)*

Bellarmino's reaction was that there was no question of banning Copernicus' book, only perhaps the adding of some preface to indicate that it did not purport to give a true account of the universe. He thought that Galileo, 'that fine mathematician', would do well to follow suit.

Bellarmino was genuinely interested in science, but as the guardian of Tridentine orthodoxy he had also to be sensitive to the slightest threat to the Church's standing. He had been one of the inquisitors at Bruno's trial and had compiled heretical propositions for the philosopher to abjure. And when a Carmelite friar, Paolo Foscarini, published a treatise alleging that the Copernican system was physically true, Bellarmino's tone became stronger and he began to warn Foscarini and Galileo of dangers to the faith: it was safer to discuss astronomy hypothetically, otherwise physical assertions of a planetary Earth would threaten Catholicism by contradicting the Scriptures.

Figure 4.6 Portrait of Cardinal Bellarmino. (Mansell Collection, London)

Galileo decided he must go again to Rome to put his case in person. Amidst the debates Paul V intervened and called for an official statement on the motion of the Earth. A report was issued by the theological experts of the Congregation of the Index. On 24 February 1616 the theologians announced their decisions on the controversial matters they had been summoned to consider. The proposition that the Sun was stationary at the centre of the universe was declared 'foolish and absurd, philosophically and formally heretical, in as much as it expressly contradicts the doctrine of the Holy Scripture in many passages, both in their literal meaning and according to the general interpretation of the Fathers and Doctors'. The other proposition that the Earth was neither at the centre of the universe nor immovable, but moved as a whole and also with a diurnal motion, was adjudged deserving of 'the same censure in philosophy and, as regards theological truth, to be at least erroneous in faith'.

A week later Foscarini's Copernican treatise was put on the Index of prohibited books. The printer was imprisoned and Foscarini himself soon died in mysterious circumstances. Copernicus' *De revolutionibus* was not condemned outright. It was to be 'corrected' by a few changes which Galileo himself regarded as slight: removing the word 'star' from the descriptions of the Earth and taking a few lines from the preface which alleged that the theory was in harmony with the Scriptures. In this modified version Copernicus' work could be read again in 1620. The Church had shown that while it would not tolerate Copernican works which called for a reinterpretation of the Scriptures, the Copernican system could be discussed hypothetically. As for Galileo, the Pope instructed Bellarmino to summon him and inform him that he was neither to hold nor defend the censored astronomical propositions. An account of this meeting, later issued by Bellarmino at Galileo's request, made it clear that Galileo had received no punishment and had not been called upon to abjure anything. This document would later be the basis of Galileo's defence.

Galileo left Rome dejected. For the rest of his life he believed the Church had made a dreadful mistake; his private notes reveal his bitterness:

> On the matter of introducing novelties. Does anyone doubt that from wanting minds, created free by God, to be the slaves of others' will, most serious scandals will be born? And wanting people to deny their own senses and subject them to the rule of another; and allowing persons entirely ignorant of a science as judges over those knowing it, so that by the authority conceded to them they are empowered to have things their way. These are novelties capable of ruining republics and subverting states. *(Drake, 1980, p.65)*

It would have brought no comfort to Galileo to hear that enthusiastic support of his stand and criticism of Rome's edicts had come from Campanella, in the eyes of the Church an arch-heretic. His manuscript *Apology for Galileo*, circulated in Rome in 1616 and later published in Lutheran Frankfurt, linked Galileo with Bruno and Foscarini as the great innovators of reformed philosophy.

Back in Florence Galileo began to develop his theory of the tides which he believed to be the strongest argument for the planetary Earth. But he now had to write cautiously and when he sent the elaborated theory to Leopold, archduke of Austria, it carried a preface with the insincere description that it was just a 'poetic conceit or dream'.

In 1623 Galileo's spirits rose with the election of a new Pope. Maffeo Barberini, a fellow Florentine and a member of the Accademia dei Lincei who had written a poem praising Galileo's telescopic discoveries, was now Urban

Figure 4.7 Hopes for an alliance between the Accademia dei Lincei and the papacy. The title page of Galileo's Assayer *combines the spotted lynx, symbol of Cesi's academy, with bees, the emblem of the Pope's Barberini family. (Photo: British Library Board)*

VIII. Once in office he appointed men who favoured the policy of independent scientific research. His nephew Francesco Barberini, appointed cardinal and papal secretary of state, had recently been elected to Cesi's Accademia. And the Pope himself is alleged to have said he would never have approved the anti-Copernican edicts of 1616. Full of optimism the Accademia published Galileo's *Assayer*. Dedicated to Virginio Cesarini, member of the Academy and master of the Pope's chamber, it carried a preface, signed by all of the Academy's members, praising the Pope and welcoming the advent of a new period of enlightened patronage.

In the *Assayer* Galileo had set out 'with a most just balance' to weigh the astronomical and philosophical views of Horatio Grassi, Jesuit astronomer at the Collegio Romano. It ridiculed the Jesuit's Aristotelianism and developed an alternative corpuscular physics. But above all it was a manifesto for the

Figure 4.8 Galileo, a portrait drawn during his fourth visit to Rome in 1624. (Photo: The Trustees of the British Museum)

investigation of nature unfettered by any attachment to authority. The Pope was delighted with the presentation copy and is said to have laughed with delight at passages read to him.

The time seemed perfect for another visit to Rome. Galileo was warmly received by Urban; but already it was clear that this was not going to be a reign of unqualified freedom for scholars. When the two men discussed the theory of the tides Galileo was given permission to publish it, but only if he did not assert that the Earth's motion was demonstrated. Returning to Florence with a papal letter describing him as 'our beloved son', Galileo began work on the treatise. Five years later he was back in Rome seeking approval for the printing of the completed manuscript entitled *On the Flux and Reflux of the Sea*. The Pope thought the title gave too much prominence to a phenomenon which Galileo believed demonstrated the Earth's motion; so the title was changed to *Dialogue Concerning the Two Principal Systems of the World* (that is, the Ptolemaic and Copernican). And to bring home the hypothetical treatment Urban insisted that Galileo's discussion of the tides be weakened by the assertion that the alleged cause was only one of many possible ways that God could have produced the tides. The *Dialogue* was published in Florence in 1632.

Written in the vernacular to reach a wide readership outside the universities, Galileo cast his book in the form of a dialogue between three characters. In a Venetian setting, cosmology and physics, old and new, are debated. Salviati, in real life a Florentine friend who had observed sunspots with Galileo, represents the author's views. Against him stands Simplicio, named after the ancient commentator Simplicius; he defends Aristotelian philosophy. Listening to both and participating with comment is an intelligent layman called Sagredo, another of Galileo's friends, the Venetian patrician and dilettante scientist.

Just as Galileo had done in his visits to Rome, Salviati sets out to demolish the Aristotelian objections to the Copernican system. Since antiquity the diurnal and annual rotations of the Earth had been denied on the basis of common-sense observations. It continued to be argued that a spinning Earth would hurl buildings and animals into space; birds and clouds would be left behind as the rapidly moving Earth sped past them, and humans clinging to their speeding planetary vehicle would constantly feel the rush of wind on their faces. Projectiles thrown vertically up would never return to the thrower's feet; similarly because of the Earth's alleged rotation, stones dropped from a tower would land some distance to the east of it. Since none of these things occurred it was concluded that the Earth was at rest.

All such arguments were based on Aristotle's common-sense physics. Its premises were that bodies could remain in motion only if there was an associated mover pushing or pulling them; that bodies occupied natural places in the universe – in the case of the Earth, at the centre; and that if they were displaced, bodies returned with a characteristic natural motion: a dropped stone moved down to the centre of the cosmos, and moved rectilinearly, showing that circular motion was not natural to the Earth's material. But the Copernican system assumed a rotating Earth displaced from the centre of the universe. A different type of physics was needed; that is why Galileo's *Dialogue* was more concerned with physics than astronomy. Much of the book dealt with the motion of bodies on Earth, in a very different way from Aristotle. Salviati argues that it is impossible to tell from the fall of a stone whether the Earth was moving or at rest; the result would be the same in each case – the dropped stone would land at the foot of the tower. The reason was that the stone possessed the Earth's motion; when it fell it had not one motion but two: a rectilinear motion combined with the circular motion of the Earth. And since human observers also shared in the Earth's rotation, all they perceived was the stone's rectilinear fall. The argument depended on presenting motion as a *state* in contrast to Aristotle's idea of motion as a *process*.

Here Galileo took the first steps towards modern inertial physics, with the assertion that terrestrial bodies could persist in motion without the continued application of a force. He discussed the movement of perfectly spherical balls on a perfectly smooth plane. It was a thought experiment impossible to realize because of the occurrence of friction and material imperfections in the real world; nevertheless it persuaded Simplicio that such spheres would continue to move forever. According to Galileo the closest approach to the realization of these conditions was the movement of a ship over a calm sea. Like the ball moving on an untilted surface, the ship was at a constant distance from the centre of the Earth. Neither would have a tendency to accelerate like a falling body (or decelerate in the case of an object thrown up) since gravity was acting constantly. Galileo's conception of the way bodies persist in motion was related to the distance of the body from the Earth's centre. Consequently motion could persist only if it was in a circle. The correct law of inertia that undisturbed bodies continue in a state of rest or in a straight-line motion was never formulated by Galileo. His inertial physics still attributed some special role to the Earth, even though he had removed it from the centre of the universe.

The *Dialogue* came to its climax with the discussion of the tides, the phenomenon which Galileo had long believed was the best evidence for the Earth's motion. Salviati presented his companions with the image of the oceans and seas as vast expanses of water in a giant container, the Earth. How would the waters manifest any movement of their container? Not at all if the Earth was moving uniformly, because the seas would simply receive and follow that motion. But if the Earth accelerated or decelerated, the waters would rise and fall; tides would be observed. Salviati illustrated this effect by referring to the barges which carried water across the Venetian lagoon. Suppose a barge ran aground or hit an obstacle and was slowed down or stopped; its cargo of water, retaining impetus, would rise at the prow and fall at the stern. The reverse would occur if the barge accelerated. Provided the acceleration or retardation of the barge was not so great as to move the whole body of the contained water violently, the rise and fall of the water would be perceptible only towards the extremities of the barge. Salviati said this was exactly what occurred with the Earth's seas.

But he had yet to convince his friends how the Earth could be said to move irregularly, since the postulated annual and diurnal motions were uniform. Salviati drew a diagram in which ABCD represents the Earth, centre E (Figure 4.9). The uniform daily rotation of the Earth occurs in the direction A–B–C–D. The uniform annual revolution of the Earth is represented by the larger circle, centre K. Salviati said the combination of the two motions gave a resultant irregular motion to the Earth. At the part of the Earth A, the two motions were in the same direction and additive, so the result was greater than either; while the part of the Earth at C was moving slower, since the annual and daily motions were in opposite directions. At C, as in a barge that was slowing down, the waters would rise; at A, the waters would fall.

As Galileo appreciated, this theory predicted two tides at intervals of twelve hours, instead of the observed four at six-hourly intervals. He therefore

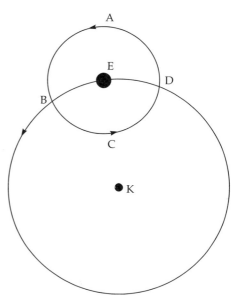

Figure 4.9

introduced the *ad hoc* hypothesis that further oscillations in the water occurred as a result of variations in the depth of water and other secondary causes.

Logically there is a fallacy in this argument; it lies in the failure to refer motion to the same point. The orbital motion of the Earth is considered relative to the fixed stars; but the motion of the water is considered relative to the Earth. Also Galileo rejected the theory, held by Kepler and later proved by Newton, that the tides were caused by the direct combined action of the Sun and Moon. Galileo regarded this as an unscientific theory of action at a distance and instead restricted the explanation of the tides to a local, terrestrial cause. Galileo's theory had elements of truth; the rotation of the Earth and the variation in ocean depths have secondary effects on the tides. But the convincing empirical evidence of the Earth's orbital motion would only come with the detection of apparent shifts in the position of the fixed stars. Like Copernicus before him, Galileo could only admit that these stellar displacements were imperceptible; Salviati predicts that one day they would be established by 'extremely accurate observations'. That did not occur until the nineteenth century; until then there was no visual confirmation of the Earth's motion.

Galileo's argument from the tides failed to convince; the theory was shaky. But Simplicio's scepticism, expressed at the close of the *Dialogue*, was based on other considerations:

> As to the discourses we have held, and especially this last one concerning the reasons for the ebbing and flowing of the ocean, I am really not convinced; ... I know that if asked whether God in His infinite power and wisdom could have conferred upon the watery element its observed reciprocating motion using some other means than moving its containing vessels, both of you would reply that He could have, and that He would have known how to do this in many ways which are unthinkable to our minds. From this I forthwith conclude that, this being so, it would be excessive boldness for anyone to limit and restrict the Divine power and wisdom to some particular fancy of his own. *(Galileo, p.464)*

This was the weakening of presentation demanded by the Pope; the Earth's motion was not demonstrated by this theory because God could have used numerous other ways of producing the tides. By using Simplicio, readily construed as signifying a simpleton, to voice the Pope's view, Galileo may well have angered Urban. In fact the Pope was furious for other reasons as well.

On trial in Rome

The Pope felt he had been tricked. Beneath its thinly veiled disclaimer Galileo's *Dialogue* appeared as a vigorous attempt to establish the physical reality of the Copernican system. The Pope was told of a document in the Inquisition's file recording that Galileo had given an undertaking to an officer of the Inquisition to 'relinquish altogether' the Sun-centred astronomy with its moving Earth, and never 'to teach or defend it in any way whatsoever' on pain of proceedings by the Inquisition. And Rome's Jesuits, realizing that their Aristotelian philosophy was in danger and perhaps seizing the opportunity of revenge for the ridicule they had received from Galileo, pronounced the *Dialogue* as more pernicious for the Church than all the works of Luther and Calvin.

The Pope rebuked Father Riccardi, who had issued the licence to print the work. Orders were given for the suspension of sales and confiscation of stock; it was too late, all copies were in circulation. The Florentine Inquisition

MELCHIORIS INCHOFER

E SOCIETATE IESV

AVSTRIACI,

TRACTATVS SYLLEPTICVS,

In quo,

QVID DE TERRAE, SOLISQ. MOTV, VEL STATIONE,
fecundùm S. Scripturam, & Sanctos Patres fentiendum,
quauè certitudine alterutra fententia tenenda fit,
breuiter oftenditur.

FIXA

HIS QVIESCIT

ROMÆ,
Excudebat Ludouicus Grignanus MDCXXXIII.
SVPERIORVM PERMISSV.

Figure 4.10 In the year of Galileo's trial (1633) this defence of traditional Aristotelian cosmology was published. The author was Melchior Inchofer, an Austrian Jesuit and member of the preliminary theological commission investigating Galileo. The frontispiece shows the Earth kept at rest by three bees, the emblem of Pope Urban's family. (Photo: British Library Board)

sent officials to Galileo's house with a summons to go to Rome within thirty days or be taken there in chains. Ferdinando II, the young Grand Duke, was pressed by the Pope to deliver his subject. Ill and approaching the age of seventy, Galileo went to Rome, where he was arrested and put on trial before the Inquisition.

The hearings began on 12 April 1633. During the interrogation conducted by the Dominican Commissary of the Inquisition, Galileo was asked to repeat what he had been told by the now deceased Cardinal Bellarmino in 1616. He replied that he had not been commanded to abjure any opinion but was only informed that the Copernican postulates had been officially adjudged contrary to Holy Scripture and that therefore it was no longer permitted to hold or defend them. And Galileo produced Bellarmino's signed statement to that effect. The Inquisitor refused to accept this as a correct record and read from another document which stated that at that meeting, seventeen years before, Bellarmino warned Galileo to abandon his erroneous opinions, and the accompanying Commissary General of the Holy Office then commanded him to relinquish these opinions altogether, and to cease holding, teaching or defending them in any way; otherwise he would face the Inquisition. This was a sterner and comprehensive ban and Galileo had not the slightest recollection of it. The document was unsigned and therefore had no legal force; yet it was deployed to bring Galileo to his knees. Historians accept that it is an authentic document but its origin remains puzzling. One explanation suggests that an official, present at the meeting with Bellarmino, was disgusted with the easy way in which Galileo had been let off and secretly told an assistant to insert a stronger account of the proceedings in the Vatican files. Another suggestion, equally unverifiable, is that the sterner orders were actually voiced by Dominicans present at the interview, but Bellarmino whispered to Galileo not to take any notice of them.

A few days after his first interrogation, three theologians, appointed to examine the text of the *Dialogue,* issued their report. They had no difficulty in showing that Galileo had not simply discussed the Copernican theory as a mathematical hypothesis, but had held, defended and taught it as a physically true account. In an attempt to get a settlement out of court, the Commissary General of the Inquisition advised Galileo that he would be treated leniently if he admitted his errors. This Galileo soon did in a statement that pride had led him to conjure up convincing novel arguments for the Copernican doctrine. And to show his disbelief in this theory he was prepared to add sections to the *Dialogue* to disprove it more effectively. This must have been a bitter retraction for Galileo; he had no choice but to throw himself on the mercy of the court. The affair could have ended there, but worse was yet to come; Galileo was to be humiliated further. This move, which seems to have originated from Urban's growing enmity, required Galileo to be interrogated again, under threat of torture. Galileo was to declare which astronomical system he adhered to. Again he retracted, stating untruthfully that he accepted Ptolemaic astronomy, and abandoned the Copernican system. Two days later Galileo knelt in a white shirt of penance before his judges in the Dominican Convent of Santa Maria Sopra Minerva. Sentence was passed. The *Dialogue* was prohibited and Galileo condemned to life imprisonment with the requirement that he repeat penitential psalms once a week for three years. Three of the examining cardinals, including Francesco Barberini, refused to sign the sentence.

The conditions of imprisonment were successively commuted until Galileo was allowed to reside in his villa at Arcetri, provided he abstained from teaching and with the restriction of visitors. Here at his villa in the hills around Florence he wrote one of his most important scientific works, the *Discourses Concerning Two New Sciences*, published in Holland in 1638. Dealing with falling bodies and the path of projectiles, it was the foundation of modern kinematics. A Latin translation of the suppressed *Dialogue* was published in 1635 in that great centre of Reformation printing, Strasbourg.

Galileo's trial and condemnation has been discussed ever since. Historians have variously blamed the Pope, the Jesuits, malicious priests, inflexible Aristotelian university professors, and Galileo himself. The latest contribution to the debate presents the astonishing interpretation that Copernican astronomy was not the real reason for Galileo's arrest but only introduced as a smokescreen in a staged trial to conceal much graver theological errors. This is how Redondi's argument goes. As guardians of Catholic orthodoxy, the Jesuits in Rome had excluded all anti-Aristotelian philosophy as incompatible with the faith, one pillar of which was transubstantiation, the supposed real conversion, during the sacrament of the Eucharist, of the bread and wine into the flesh and blood of Jesus Christ. Galileo's *Assayer* not only presented a corpuscular physics akin to heretical atomism, but threatened the cherished doctrine of the Eucharist. That book portrayed the perceived qualities of material objects, colour, taste and smell, as subjective sensations caused by the impact of invisible, moving, shaped particles on the organs of sense. For the Jesuits this undermined the Eucharist because on Galileo's theory there would be no miraculous conversion of substance; the particles of bread and wine would persist. Since the Eucharist was of such importance the *Assayer* was seen as the gravest of attacks on the faith. And it was to avoid scandal, and to protect the Pope from any association with this heresy derived from his earlier friendship with Galileo, that a charge was fabricated on the lesser error of contravening Scripture with a moving Earth theory. A mock political trial on the less serious charge would stifle scandal and silence Galileo.

Redondi's argument is based on the selective use of evidence, is fanciful and in the end several historians have found it implausible (see the persuasive criticism by Ferrone and Firpo). The cause of Galileo's trial really was Copernicanism, rightly seen as a threat to a world picture which had existed for centuries. The crisis became so acute because of the intimate association of that world picture with official Catholicism. For the Jesuits the collapse of Aristotelian philosophy meant the destruction of the Tridentine faith. What was at stake was the right of a natural philosopher to pursue his scientific research independently of authority. Galileo certainly wanted that but without any damage to the Catholic faith. In the Rome of the seventeenth century that was not acceptable; in Venice it might have been.

4.2.3 Consequences

Galileo's condemnation was the climax of a trend, which began half a century before, to restrict intellectual freedom in Italy. The effects on Italian science would be long lasting. When Giovanni Ciampoli, priest and Galilean, died in 1643, he bequeathed his scientific papers and books to Ladislao IV, the King of tolerant Poland; but just as a mule train was leaving to preserve the library from the intolerant papacy, officials of the Inquisition intervened with a sequestration order (Redondi, 1988, pp.267–8). Other followers of Galileo felt obliged to present their astronomical and cosmological treatises in the guise of hypotheses. Jesuits in the Papal States, above all in Bologna, continued to be interested in observational astronomy but carefully avoided the physical causes as hazardous ground. And in Tuscany Cosimo III de' Medici (*r.* 1670–1723) informed the University of Pisa that 'His Highness will allow no professor to read or teach in public or in private, by writing or by voice, the philosophy of Democritus, or of atoms, or any save that of Aristotle' (Hale, 1977, p.186). These were some of the ways in which Aristotelian science was preserved in seventeenth-century Italy, where so much had been done to destroy it. Not until 1757 was a Pope prepared to revoke the anti-Copernican decree of 1616; and Galileo's *Dialogue* remained on the Index until 1831.

Sources referred to in the text

Bouwsma, W. (1968) *Venice and the Defense of Republican Liberty: Renaissance Values in the Age of the Counter Reformation*, University of California Press.

Drake, S. (1957) *Discoveries and Opinions of Galileo*, New York, Doubleday.

Drake, S. (1978) *Galileo at Work. His Scientific Biography*, University of Chicago Press.

Drake, S. (1980) *Galileo*, Oxford University Press.

Ferrone, V. and Firpo, M. (1986) 'From Inquisitors to Microhistorians: A Critique of Pietro Redondi's *Galileo eretico*', in *J. of Modern History*, *58*, pp.485–524.

Firpo, L. (1970) 'The Flowering and Withering of Speculative Philosophy – Italian Philosophy and the Counter Reformation: The Condemnation of Francesco Patrizi', Cochrane, E. (ed.) in *The Late Italian Renaissance 1525–1630*, pp.266–84, London, Macmillan.

Galileo, *Dialogue Concerning the Two Chief World Systems*, transl. Drake S., (1962), University of California Press.

Hale, J.R. (1977) *Florence and the Medici. The Pattern of Control*, London, Thames and Hudson.

Redondi, P. (1988) *Galileo Heretic*, translated by R. Rosenthal, Harmondsworth, Penguin.

Yates, F. (1964) *Giordano Bruno and the Hermetic Tradition*, London, Routledge and Kegan Paul.

Yates, F. (1970) 'Giordano Bruno' in Gillispie, C. (ed) *Dictionary of Scientific Biography*, vol.2, pp.539–44, New York, Scribner.

Further reading

Drake, S. (1957) *Discoveries and Opinions of Galileo*, New York, Doubleday.

Drake, S. (1980) *Galileo*, Oxford University Press.

Redondi, P. (1988) *Galileo Heretic*, translated by R. Rosenthal, Harmondsworth, Penguin.

Iberian Science: Navigation, Empire and Counter-Reformation

Chapter 5

by David Goodman

5.1 Portugal

5.1.1 Portuguese navigation and science

In one respect Portugal occupies a position of outstanding importance in the development of early modern Europe: it was here that the first phase began of that European overseas expansion which culminated in the acquisition of colonial empires in Africa, Asia and the Americas. After an attack on neighbouring Morocco which resulted in the capture of Ceuta (1415), Portuguese ships set out on voyages of increasing duration, entering waters that were little-known or wholly uncharted. Uninhabited Madeira was colonized (*c.* 1420) and the Azores, one third of the way across the Atlantic, discovered (1427) or perhaps rediscovered. And at the same time generations of Portuguese seamen were gradually edging their way south along the coast of West Africa, reaching the equator by 1473. Then came the voyage of Bartolomeu Dias (1487) with its successful rounding of the Cape of Good Hope, so called because of the prospect of an open sea route to India. That momentous crossing was accomplished by Vasco da Gama (1498).

Soon Portuguese ships and guns conquered the most important strategic and commercial centres in the east: Hormuz (1515) at the entrance to the Persian Gulf and the centre of lucrative trade between Persia and India; Goa (1510) a prosperous emporium and now the capital of the Portuguese empire in the east; Colombo (1517); Malacca (1511), the collection centre for the lucrative commerce in spices; Macao (*c.* 1557), under its Portuguese rulers, a flourishing base controlling trade between China and Japan which Portuguese ships had reached by 1543. And in the west the Portuguese discovered Brazil (1500) and occupied an extensive area of its coasts.

By the early sixteenth century Portugal had acquired a world-wide empire, not in the form of continuous land masses with subjugated millions, but a chain of strategic outposts scattered over the oceans, protected by fortresses and tapping into the most profitable commerce in the world.

Until quite recently the entire achievement of the Portuguese voyages of discovery was credited to the inspiration of one man: Prince Henry the Navigator (1394–1460). He used to be portrayed as a great scholar and humanist of the Renaissance, so deeply interested and accomplished in astronomy and mathematics and their applications to navigation, that he retired in his early twenties to Sagres near Cape St Vincent, the south-western tip of Portugal (and of Europe), where he founded an academy of navigational science and cartography, collaborating in the study and teaching with the experts whom he attracted there. And here, so it was commonly alleged, he devised a master plan for Portuguese expeditions to Africa and

India; the subsequent successes were the fruit of the seeds he had sown. Some have attributed even more than this to Prince Henry's efforts. It is argued that when Henry sent his seamen south into the unfamiliar waters of the West African coast he 'unwittingly started the scientific revolution' because his mariners were forced to use observation and experiment, those foundations of modern science, to discover the pattern of prevailing winds in the Atlantic (Waters, 1967, p.198). But this is unconvincing; observation and experiment are centuries older than the Portuguese voyages.

Today practically nothing remains of this image of Prince Henry; even his title, the Navigator, is now seen as a romantic invention of the nineteenth century – in fact he hardly ever sailed. Nor is there any evidence of his supposed academy of navigation. But he did send pirates to raid the Moroccan coast, persistently sent ships to Africa, and sponsored the colonization of Madeira and the Azores. And he was sufficiently interested in astronomy to employ James of Majorca, an expert in navigation and geography, to teach these subjects publicly in Portugal. The revised portrayal of Henry presents him as obsessed with the conquest of Morocco and Granada and no longer the devoted student of astronomy or promoter of exploration (Newitt, 1986, p.33; Russell, 1984).

The truth behind the myth of Prince Henry is that the royal house of the fifteenth and sixteenth centuries appreciated the importance of astronomy for oceanic exploration and bestowed patronage on experts to foster sciences associated with navigation. In some cases, notably João II (*r.* 1481–95) and Manuel I (*r.* 1495–1521), the monarchs themselves directed maritime expeditions. But there is now also growing evidence that initiatives by private individuals – Portuguese noblemen and merchants – may have been even more important than monarchs as the motor of the Portuguese voyages of discovery.

In the development of astronomy for Portuguese navigational purposes, the contribution of Jews, native and foreign, was conspicuous. The Jews had strong religious reasons for studying the sciences. Since antiquity rabbis had recommended the study of astronomy to establish the Jewish calendar, an essential requirement for fixing the dates of religious festivals. Rabbis also looked to natural science to clarify difficult texts in the Torah, the basis of the Jewish religion. James of Majorca, brought over by Prince Henry, is thought to have been a converted Jew, Judah Cresques, one of the family who had raised Majorca to a leading centre of cartography in the fourteenth century. And later in 1492 when the Jews were expelled from Spain for refusing to convert to the Catholic faith, those who took refuge in Portugal included Abraham Zacut (*c.* 1452–*c.* 1515), one of the most accomplished astronomers of the time. He brought with him his *Ḥibbur hagadol* ('Great Treatise'), an astronomical manuscript in Hebrew, and part of it was soon translated into Latin and published as the *Almanach perpetuum* (Leiria, 1496), one of the earliest books printed in Portugal; it contained tables of data of great use for navigators. The translator of Zacut's manuscript was one of his students, José Vizinho, a learned astronomer and a member of Portugal's small indigenous Jewish community. Working together for the crown, Zacut and Vizinho, were of central importance in the planning of Vasco da Gama's voyage to India. But Zacut did not stay long in Portugal. In 1497 Manuel I presented Jews with the same choice that Ferdinand and Isabella of Spain had devised five years before: conversion to Christianity or expulsion. Zacut refused to convert and like many other Iberian Jews found a haven in the more tolerant Ottoman Empire, settling in Damascus.

Other Portuguese Jews decided to remain, accepted baptism and formed a new class of converts, the *conversos* or 'New Christians'. Their commitment to their new religion varied from sincere belief to expediency, assuming a public facade of Christianity to mask an unchanged attachment to Judaism. These Portuguese *converso* families continued the Jewish tradition of learning in medicine and astronomy, and the most distinguished Portuguese astronomer and mathematician of the sixteenth century, Pedro Nuñes (1502–78) was of *converso* origin. He had graduated in medicine at the university of Salamanca; Portuguese students often studied in Spain or northern Europe, because Portugal's university, oscillating between Lisbon and Coimbra, was of inferior standard. Nuñes was summoned back to Portugal by João III to serve him as court physician and tutor in mathematics and astronomy to the royal princes Luis and Enrique, a clear sign of the monarchy's continuing interest in fostering navigation. For ten years Nuñes taught at the royal palace of Coimbra and, when the court moved there, at Lisbon. Along with the princes he taught members of the nobility, notably João de Castro, a future naval commander and governor of Portuguese India – Nuñes taught him how to use some new astronomical instruments he had designed for navigational use. Nuñes' various published works on astronomy and navigation may have developed out of his palace teaching.

The Portuguese monarchs of the sixteenth century also created chairs of astronomy. In 1518 Manuel I appointed his physician Master Philip to the new chair at Lisbon; and later Nuñes became professor of mathematics and astronomy at Coimbra (1544–62). Still more immediately connected with navigation, the crown regulated the teaching and licensing of pilots and cartographers as well as the manufacture of navigational instruments and charts. In Lisbon since the end of the fifteenth century, nautical charts, regarded as of strategic importance, were deposited in the crown's department of African affairs, the Casa da Mina (House of the Mine). When Nuñes was appointed a cosmographer royal (1529) and then the cosmographer-major (1547) he was in charge of Portugal's entire navigational enterprise from teaching pilots to responsibility for maintaining a standard up-to-date master-chart of the oceans and continents.

How much then of the Portuguese voyages of discovery was due to science? The sun, moon and stars had guided seamen ever since men had begun to sail. But in the mediaeval centuries of Mediterranean voyages navigators had not relied on the observation of celestial bodies but on the technique known as 'dead-reckoning', charting a ship's course from crude estimates of its velocity, the use of a time measurer – the repeatedly inverted sand-glass – and the mariner's compass to indicate direction. In expert hands dead-reckoning is very effective. Columbus, whose seamanship was developed in Portuguese ships voyaging to Africa in the 1480s, never used astronomical observations for his voyage to America in 1492. At first the Portuguese navigators similarly dispensed with celestial observations, keeping their eyes on the compass or on distinctive landmarks of the coasts they hugged. But the coast of West Africa is featureless, and around 1460 there is the first evidence of the Portuguese resort to astronomical observation to establish a vessel's position in latitude. It was not the first time this had been done – Norsemen and Italian navigators had occasionally observed the height of the pole star above the horizon, recognizing that this varied with latitude. But the Portuguese began to take measurements systematically using a very simple graduated, sighting instrument, the quadrant, to fix latitudes as they progressed further and further south along the African coast.

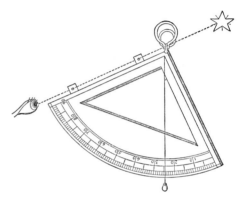

Figure 5.1 Marine quadrant of 1492. The polestar was sighted through pin-holes; a plumb-line attached to the instrument crossed a graduated scale at a certain point, and this reading gave the altitude of the pole-star.

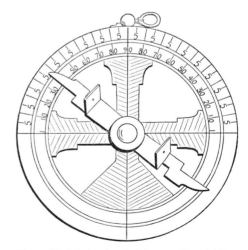

Figure 5.2 Mariner's astrolabe – another sighting instrument with pin-holes and a scale for observing the altitude of the sun at noon, so allowing the determination of the observer's position in latitude. (Figures 5.1 and 5.2 are from S.E. Morison, 1942, Admiral of the Ocean Sea: A Life of Christopher Columbus, Boston, Little, Brown and Co., pp.184 and 185.)

Reference tables were compiled which connected the altitude of the pole star to particular latitudes. The measurements were very accurate; the earliest Portuguese manual of navigation, the *Regimento do astrolabio e do quadrante*, thought to date from the 1480s, lists latitudes which are generally correct to half a degree, and in several cases to within a sixth of a degree (Taylor, 1971, p.162f).

Once the Portuguese reached the equator the pole-star could no longer be used – it disappeared below the horizon. Instead the noon altitude of the sun was taken; but the relation between this datum and latitude is much more difficult because of the complicated apparent motion of the sun, the result of the tilt in the earth's axis. Elaborate tables and rules were needed to derive latitude from solar altitude, and bringing this to a successful conclusion by 1485 was one of the principal achievements of Portuguese astronomy, due to José Vizinho and others working with him (Waters, 1967, p.204f). When Vasco da Gama discovered the sea route to India, reaching Calicut (1498), the final part of his voyage across the Indian Ocean was assisted by Muslim navigational knowledge provided by Ahmad ibn-Majid, an Arab expert in celestial navigation who joined the explorers at Malindi on the East African coast. But for the earlier phase of the voyage Iberian science had been invaluable: Zacut's astronomical tables, Vizinho's calculations and the simple copper astrolabe Zacut had specially designed for the expedition were all important in guiding da Gama's course. On the other hand Cabral's discovery of Brazil (1500) owed nothing to science; it was entirely accidental – a ship bound for India was blown far off course to the west.

The links between Portuguese navigation and the development of science are clearer when one looks at the consequences of these extended voyages. When Nuñes' pupil, the naval commander João de Castro, sailed to India in 1538, his commission instructed him to make scientific observations during the voyage on terrestrial magnetism and hydrography – the science dealing with the waters on the earth's surface: the position of rivers and seas, their depths and currents. The earth's magnetism was of central importance to navigation because the use of the mariner's compass depended on this natural phenomenon: the magnetized needle of the instrument pointed to the earth's north magnetic pole. But already in Castro's time it was well known that the compass was an inaccurate guide, because the earth's magnetic pole did not coincide with true geographical north. The difference, known as magnetic 'declination', varied with locality and in his investigation of this Castro was able to disprove the current opinion that declination was directly proportional to changes in longitude. That must have been disappointing because the much-needed determination of longitude would remain unsatisfied until the eighteenth century; the readings of the mariner's compass could not after all supply this essential datum for establishing position at sea. While cruising off Bombay, Castro observed sharp fluctuations of the compass needle, too great over the short distance he had sailed to be attributed to varying distance from the earth's magnetic pole. He therefore concluded, correctly, that the compass was being affected by an extensive ridge of rocks in the vicinity of the ship. Those rocks contain magnetic iron ore and Castro's deduction amounted to perhaps the earliest recognition of the local magnetism of rocks (Hooykaas, 1980, pp.111–12). And Castro's studies of terrestrial magnetism using Nuñes' instruments were more accurate than any others at this time.

As the Portuguese navigators sailed further into uncharted waters astronomical knowledge grew through the observation of constellations

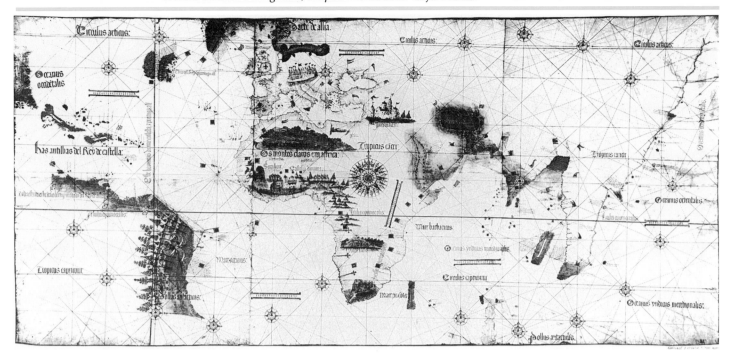

Figure 5.3 Cantino's planisphere. (From A. Cortesão and A. Texeira da Mota, Portugaliae monumenta cartographica, *vol.1, 1960. Photo: Reading University Library.)*

unknown to the ancient Greeks, notably the Southern Cross (first reported by ships in Gambia, West Africa, in 1455). But it was the improvement in maps which most clearly revealed the advances in knowledge brought about by navigation. Regions long familiar like the Mediterranean were more accurately represented than before; and many other localities were mapped for the first time. The progress in cartography is evident in the world chart now known as 'Cantino's planisphere'. This was made secretly in 1502 by an unknown Portuguese cartographer for an Italian spy, Alberto Cantino, sent by his master, the duke of Ferrara, to secure the latest information on Portuguese discoveries. This information was of strategic importance and the Portuguese monarchs issued edicts to enforce its confidentiality. But on this occasion Cantino's offer of twelve gold ducats persuaded some cartographer to break the law. A superb coloured world map soon found its way to the ducal palace in Ferrara. It was lost for years, but rediscovered in the nineteenth century in a butcher's shop where it was being used as a screen.

For the first time on any map the tropics of Capricorn and Cancer are indicated; and for the first time a good approximation of the shape of Africa – especially the West African coast north of the Congo. There is a fair representation of India; but the Malay peninsula is distorted. The longitudinal extent of the Indian Ocean is more accurate than before. Up to date with the most recent discoveries, the map indicates the position of Ascension Island (discovered 1501) and a corrected outline of the coast of Brazil based on information brought back to Lisbon by the expedition of 1501–2.

By the early sixteenth century the Portuguese had become the most accomplished cartographers in the West. Around fifty of them have been identified. Their knowledge and skills were eagerly sought by rival maritime powers, especially Spain. The rise of Seville as a leading centre of cartography and navigational instrument-making was assisted by the migration of Portuguese experts, a brain-drain which the Portuguese monarchs' legislation could not prevent. Castile benefited from the superior nautical charts prepared in Seville by Juan Dias de Solís and Diogo Ribeiro.

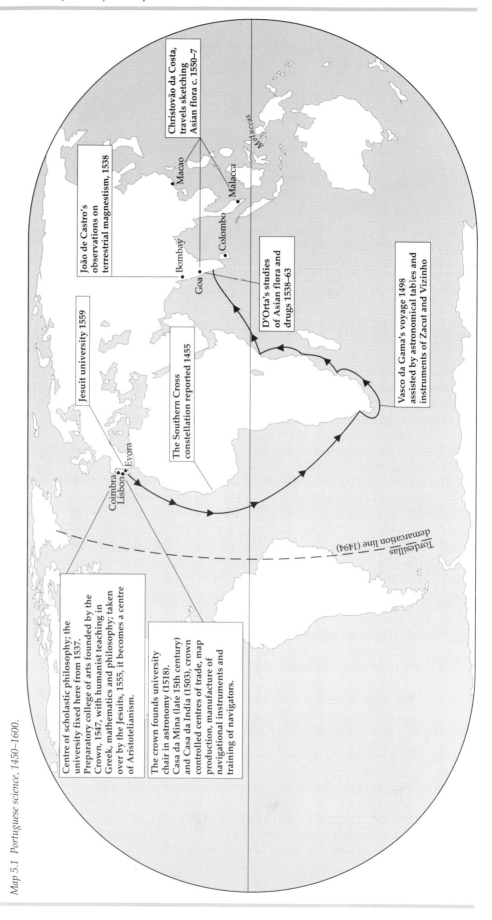

Map 5.1 Portuguese science, 1450–1600.

Jesuit university 1559

João de Castro's observations on terrestrial magnestism, 1538

Christovão da Costa, travels sketching Asian flora c. 1550–7

Moluccas

Macao

Malacca

Colombo

Bombay

Goa

D'Orta's studies of Asian flora and drugs 1538–63

The Southern Cross constellation reported 1455

Vasco da Gama's voyage 1498 assisted by astronomical tables and instruments of Zacut and Vizinho

Coimbra
Lisbon
Evora

Tordesillas demarcation line (1494)

Centre of scholastic philosophy; the university fixed here from 1537. Preparatory college of arts founded by the Crown, 1547, with humanist teaching in Greek, mathematics and philosophy; taken over by the Jesuits, 1555, it becomes a centre of Aristotelianism.

The crown founds university chair in astronomy (1518). Casa da Mina (late 15th century) and Casa da India (1503), crown controlled centres of trade, map production, manufacture of navigational instruments and training of navigators.

And one of the first manuals of navigation, the *Tratado del esphera y del arte de marear*, published in Seville in 1535, was written by Francisco Faleiro, one of two Portuguese brothers who had entered the service of the Spanish crown. He had worked on the preparation of the first circumnavigation of the earth (1519–22), an expedition led by another Portuguese in Spanish service, Magellan. The value attached to Portuguese expertise can be judged from the efforts of Philip II of Spain to secure the services of Luis Jorge de Barbuda. Philip's ambassador in Lisbon had been collecting intelligence on secret Portuguese maps and was about to return to Madrid. He was instructed to bring Barbuda with him, but King Sebastian of Portugal learned of the plot and the escaping cosmographer was arrested on the border and brought back to Lisbon in chains. After imprisonment he was released on giving certain assurances; but in 1579 he again tried to escape, this time successfully, smuggled out by Philip's agent. Soon he was working for Philip with a handsome salary of 150 ducats a year.

The voyages of discovery brought the Portuguese not only a world-wide empire but a sense of surpassing the Ancients. Portuguese writers praised the advent of a new era and an empire still greater than Rome's. In 1532 João de Barros, chronicler of the voyages, commented that if Ptolemy, the ancient authority on geography, were to return he would be ashamed by the Portuguese discoveries and embarrassed by the errors, now exposed, on his world map. Ptolemy had alleged that the torrid tropics were uninhabitable; but Diogo Gomes, a fifteenth-century pilot on a ship approaching the equator recorded 'we found the contrary'. That was the first empirical falsification of ancient authority in modern times. Its significance for the development of a scientific mentality is difficult to assess; was this really the beginning of that independent investigation of nature and challenge to revered authority that was at the heart of the scientific revolution? Some confidently assert that by demonstrating the fallibility of the Ancients, the Portuguese voyages of discovery introduced 'a new approach to nature' where 'experience, not reason nor authority should be the touchstone of truth' (Hooykaas, 1979, p.107). It is also arguable that the voyages encouraged a higher sense of accuracy: accuracy in observations with navigational instruments; exact data on latitudes; more accurate maps. This is especially noticeable in Nuñes. He devised, before Mercator, charts to eliminate sailing errors caused by the convergence of meridians on the earth's spherical surface. And for greater accuracy of observation he invented a device, called after him the 'nonius', which attached to an astronomical instrument allowed measurements to small fractions of a degree; in the refined form of the 'vernier' it remains in use today.

5.1.2 Empire, a stimulus to science

With the main centres of Asian trade in Portuguese hands, the pace of economic life in Lisbon quickened. Now the nerve centre of a world-wide empire, Lisbon's waterfront received tropical produce from three continents, and the crown and its agents became involved in the distribution of commodities like pepper and cloves throughout Europe. The profits were especially great from Indian pepper; but Portugal did not become wealthy because there were high costs to be paid in fitting out ships to collect the spices from great distances, and the expense of building and maintaining fortresses to defend isolated imperial outposts.

There is clear evidence that the rise of Lisbon to a busy emporium, the hub of a world-wide trading empire, stimulated commercial arithmetic there. In

1519 Gaspar Nicolas' *Tratado da practica darysmetica*, the earliest printed Portuguese mathematics treatise, was published with the aim of spreading knowledge of arithmetic because of its importance to Portugal's imperial commerce. The content of the worked examples in the manual reflect the transformation of Lisbon's economic activity: the reader is shown how to calculate duties on oriental merchandise imported to Lisbon; how to solve problems relating to ships carrying cargoes of pepper; how to convert different units of length and weight, and different currencies. Almost nothing is known about the author; but his book went into six editions in the sixteenth century and was still being reprinted in the eighteenth (Albuquerque, 1973, pp.99–120).

A more important stimulus for science came from the exotic fauna and flora encountered in imperial possessions in Brazil, Africa and above all India. These tropical species had been known imperfectly or not at all by revered ancient authorities, and the opportunity empire gave for direct observation of these specimens encouraged that independent and critical mentality which was to be a conspicuous feature of modern science. This development is most clearly seen in the work of Garcia d'Orta (*c.* 1501–68), the first European to give accurate descriptions of the flora of India and of the medicines prepared from them – so reliable that his book continued to be regarded as an authoritative source in the twentieth century.

His parents were Spanish Jews who had settled in Portugal after the expulsion from Spain, and had chosen baptism instead of a second forced exile five years later; their son Garcia remained a loyal Jew throughout his life beneath the mask of Catholicism. Born in Castelo de Vide, a Portuguese town close to the Spanish border, he studied philosophy and medicine at the Spanish universities of Salamanca and Alcalá de Henares and then returned to Portugal where he practised as a physician. In 1530 he was appointed professor of natural philosophy at the university of Lisbon. Perhaps because of a growing persecution of individuals of Jewish origin, d'Orta left for India, sailing in the fleet which was under the command of his friend Martin Afonso de Sousa. It has been suggested that only through this influential friend was d'Orta able to evade the recently enacted law which prohibited New Christians from leaving Portugal (Boxer, 1963). The relations he left behind were soon being harassed by the Inquisition.

D'Orta settled in Goa where he became physician to viceroys. He also traded in Asian medicinal plants, acquired specimens from far and wide, and planted their seeds in his garden. He quickly realized how inadequate were the descriptions of this flora in the ancient treatise of Dioscorides, still the West's principal source of information on medicinal plants. This tension with authority is the main theme of d'Orta's work, *Coloquios dos simples e drogas he cousas medicinais da India* (Dialogues on the simples, drugs and materia medica of India), published in Goa in 1563. And this tension is effectively brought out by the dialogue form – the book consists of exchanges between d'Orta, presenting himself as an independent investigator of nature, and Ruano, a fictitious Spaniard, a graduate of Salamanca and Alcalá, nurtured like many others on the acceptance of the doctrines of classical authorities. As d'Orta describes the parts and the habit of Asian plants, Ruano becomes disconcerted by the contradiction of what he had read in Dioscorides and other classical sources, and eventually leads to his outburst, protesting that 'it seems to me that you abolish all the writers ancient and modern ... I do not know for what reason you discredit such ancient doctors and of such high authority'.[1] D'Orta's replies to this and similar interjections reveal the extent

[1] This and subsequent quotes from the translation by Markham, 1913.

Figure 5.4 The nutmeg tree, with fruits yielding nutmeg and mace. To d'Orta's correction of Dioscorides here, Christovão da Costa added the earliest European illustration of the species.

Figure 5.5 Da Costa's illustration of the mango plant, first known in Europe through d'Orta's book.

Figure 5.6 Another plant unknown in Europe until d'Orta's description – the Indian Jack tree. (Figures 5.4–5.8 are from Garcia d'Orta, Colloquios dos simples e drogas, 2nd edn, 1872. Photos: published by permission of the British Library Board.)

to which he had liberated himself from the dependence on the classical tradition: 'Do not try to frighten me with Dioscorides or Galen, because I merely speak the truth and say what I know'; 'Does it not appear to you that Galen and Dioscorides may not have exhausted the subject, that they left many things unwritten about because they had not come under their notice'; 'Dioscorides did not see the cardamom with bark ... nor did he ever see it, nor did he know whence it came'; 'For me the testimony of an eye-witness is worth more than that of all the physicians, and all the fathers of medicine who wrote on false information'; 'I say that you can get more knowledge from the Portuguese in one day than was known to the Romans after a hundred years'. And when, after d'Orta corrects Dioscorides' erroneous description of mace as the bark of a root instead of the rind of a fruit, Ruano refers to a modern writer who alleged that the Greeks knew about this medicinal spice of the Far East, d'Orta replied: 'That was because he was afraid to say anything against the Greeks. Do not be surprised at that because even I, when in Spain, did not dare to say anything against Galen or the ancient Greeks. Yet when seen in the proper light, it is not strange that medicines should be known in one age and not in another; new things are constantly found'.

D'Orta's book contained the first description by a European of the jackfruit, mangoes, mangosteens, coco of the Maldives, camphor of Borneo and the fruit of the durian tree. And one chapter provides the first accurate account by a European of the symptoms and course of Asiatic cholera. By 1620 d'Orta's *Coloquios* had become widely distributed in Europe in five Latin editions, three editions in Italian and two in French. And much of the content reappeared in Spanish in the *Tractado de las drogas y medicinas de las Indias Orientales* (Burgos, 1578) by Christovão da Costa, another Portuguese of Jewish origin who travelled to India and China, sketched several of the plants d'Orta had described and provided much additional first-hand information.

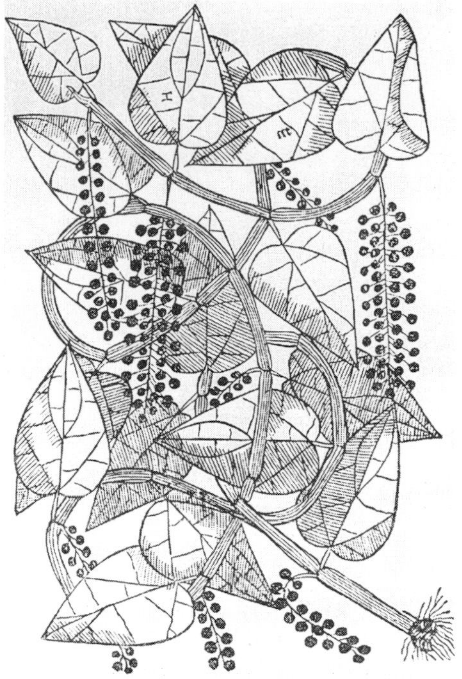

Figures 5.7 and 5.8 Da Costa's pioneering sketches of two Asian spices, cloves and pepper, whose lucrative trade the Portuguese crown sought to monopolize. On pepper d'Orta excused Dioscorides' errors as due to 'false information' gathered 'at a great distance'; but he blamed contemporary authors for not bothering to find out the facts about the appearance of the pepper tree, its fruit, and how it ripens and is gathered. (From Garcia d'Orta, Colloquios dos simples e drogas, *2nd edn, 1872. Photos: published by permission of the British Library Board.)*

Perhaps to escape the persecuting Inquisition, d'Orta's sister had joined him in Goa. But an office of the Portuguese Inquisition was established in Goa in 1560 and after d'Orta's death his sister was arrested and burned at the stake as an impenitent Jewess. At her trial she revealed that her brother had secretly observed the sabbath and other essentials of Judaism. As a posthumous punishment his remains were exhumed and burned in a public ceremony in Goa in 1580.

5.1.3 Seventeenth-century stagnation

Stimulated by world-wide navigation and empire Portuguese science blossomed but soon withered. Throughout the seventeenth century, when the sciences were being transformed by developments elsewhere in Europe, Portuguese science was stagnant. And in the eighteenth century Portuguese ambassadors and emigrés criticized the backward state of their country and the ignorance of the population, compared to what they had experienced in France, England or Italy. They campaigned to modernize their country especially through the cultivation of experimental science, and the importation of the works of Francis Bacon, Isaac Newton and other men of science which remained unknown in Portugal (Goodman, 1991).

Why this collapse in Portuguese science after such promising beginnings? Portuguese historians commonly refer to a combination of several causes: the expulsion in 1497 of the Jews, the minority who had been the most active in Portuguese medicine and astronomy; the persecution of the New Christians, those Jews who had remained after agreeing to baptism and who continued to cultivate the sciences, but whose racial origin was despised; the stifling of intellectual freedom by the establishment of the Portuguese Inquisition in 1536; the fossilizing of education from the later sixteenth century when control of higher education in Portugal and its empire was delivered into the hands of the Jesuits; and finally the decline of Portuguese navigation.

There is some truth in all of this, but much more research is needed to assess the relative strengths of these inhibiting forces. Navigation had certainly declined – to such an extent that in the seventeenth century the Portuguese commonly used foreign nautical charts and employed foreign pilots; their imperial possessions in the east had been captured by the Dutch.

The discrimination against individuals of Jewish origin also created strong tensions in Portuguese society; when in 1531 a strong earthquake occurred, the destruction was widely seen as God's punishment of the Portuguese for tolerating the presence of the New Christians who were secretly practising Judaism. And that sentiment encouraged the crown to seek papal permission for an Inquisition in Portugal, inaugurated in 1536 and given greater powers from 1547. The suspect faith of the converted Jews made them the prime target, but it is difficult to make a precise correlation here with declining scientific output; there is no record that Portugal's most eminent scientist, Pedro Nuñes, suffered any persecution from the Inquisition because of his status as a New Christian. But there is evidence that King Sebastian tried in the 1560s to weaken the prominence of men of Jewish origin in the medical profession. That seems to have been motivated by fears arising from a forged correspondence purporting to expose a Jewish conspiracy to overthrow the Christian foundations of the kingdom. The bishop of Porto Alegre had been handed letters, supposedly written by converted Portuguese Jews to their brethren in Constantinople, outlining plans to liberate themselves from persecution: they would teach their sons medicine and pharmacy in order to poison their Christian persecutors. The bishop showed Sebastian the correspondence and the king instructed the University of Coimbra to train thirty students of pure Old Christian lineage (Goodman, 1988, pp.219–20).

The Inquisition intensified the censorship of books. The first Portuguese Index of prohibited books appeared in 1547 and four other extended Indices were issued by 1625. Books on astrology were listed; but the effects on the dissemination of scientific knowledge are difficult to assess because of the uncertainties of enforcement and the degree to which clandestine literature circulated. It is, however, safe to say that Copernicanism was not acceptable in Portugal until the eighteenth century. When in 1625 Christovão Bruno, a Jesuit, was lecturing on astronomy in the Colegio de Santo Antão in Lisbon, he warned that Copernican theory should not be considered as a physical explanation and that the Roman Inquisition had rightly condemned it as a rash theory; astronomy was no more than a mathematical tool for calculations and a way to 'save the appearances' of celestial phenomena (Albuquerque, 1973, p.127f). The Renaissance humanism which to some degree was influential in the Portuguese colleges of the late fifteenth and early sixteenth centuries had been replaced by the ideals of the Counter-Reformation; the Council of Trent's rulings were implemented and, through the Jesuits, Aristotelian scholastic philosophy became entrenched and innovating scientific thought ignored.

5.2 Spain

5.2.1 The scientific traditions of the oriental minorities

Spain in the fifteenth century stood out from the rest of Europe because of its distinctive social composition. Nowhere else in Christendom were there to be found large oriental minorities, of different race and religion, the Jews and Moorish Muslims living in close contact with the Christian majority. The Jews had first arrived some time after the Roman destruction of Jerusalem, part of that diaspora which resulted in Jewish settlements throughout Europe. And Spain's Jewish communities increased during the early middle ages through further migrations from Africa. The Moors had come as conquerors in the eighth century, the westernmost thrust of that remarkable expansion of the forces of Islam. Crossing the Straits from North Africa, the invading Moors, Muslim and Arabic-speaking, soon succeeded in conquering practically the entire Iberian peninsula; only a fringe in the extreme north remained in Christian hands. From that small area began the *reconquista*, that gradual Christian reconquest which would take seven centuries to complete. Christians, Moors and Jews therefore lived at various times under Christian or Moorish rule.

For centuries the three communities had lived side by side in relative harmony; periods of peaceful coexistence and mutal admiration. There had been much intermarriage; Jews had married into the highest levels of Christian society – Ferdinand, king of Castile and Aragon, was descended from the Jewish Enríquez family. In the fourteenth century the Christian ruler of Seville, Pedro I, had built a Moorish-style palace using Moorish craftsmen loaned by the Muslim king of Granada. And at the same time in Muslim Granada a building of the Alhambra was painted with scenes depicting Christians and Moors playing chess together, hunting together and engaged in chivalrous battle. Henry IV of Castile (*r.* 1454–74), the brother of Isabella, ate like a Moor, dressed like a Moor and retained a Moorish guard. Under Christian rule Muslim craftsmen had produced the beautifully ornate interiors of mediaeval synagogues which still survive in Toledo; and later generations of Muslim craftsmen of the fifteenth century assisted in the construction of churches in Aragon, and were even permitted to complete their work with the inscription: 'There is no God but Allah'.

This intimate interaction of different peoples enriched Spanish culture, bringing oriental influences in architecture; ceramics; patterns on embroidered garments; cooking; and the Castilian language, which permanently adopted numerous Jewish words and an estimated 4,000 words from Arabic. And the character of Spanish science was also deeply affected by the presence within the peninsula of well-developed Jewish and Arabic traditions. In mediaeval Toledo scholars – Jewish, Muslim and Christian – had collaborated in the translation into Latin of Arabic and Hebrew scientific texts; that had been a process of importance not just for the development of Iberian science, but for the whole of Western science (see Chapter 1).

The continuing stimulus of the Arabic scientific tradition in Spain is clearly seen in the so-called 'Alfonsine Tables', astronomical tables compiled by Arab scholars and translated into the vernacular Castilian through the patronage of Alfonso X, king of Castile and Leon (1221–84). Based on Ptolemy's *Almagest*, they allowed the positions of the Sun, Moon and planets to be calculated for particular years or days; also the duration of lunar and solar eclipses. Circulating first in manuscript and then in modified printed editions (Venice, 1483 and 1492; Nuremberg, 1536 and 1542), the *Alfonsine Tables* remained the basis of all astronomical almanacs in Spain and throughout the West until the mid-sixteenth century. Copernicus, during his studies at Cracow, had bought a copy of the second printed edition of the *Tables*. At the Castilian university of Salamanca, students in the 1560s and 1570s who attended the lectures of Hernando Aguilera, professor of astrology, heard him discuss the same *Tables*.

But by the sixteenth century Arabic science no longer had the esteem it once enjoyed. It is too strong to speak of a decline; more correctly there was an ambiguous attitude to Arabic science. There were two reasons for this. First there was the spread to Spain of Renaissance humanism from Italy and the Netherlands. That scholarly movement, inspired by admiration of the literature of ancient Greece and Rome, sought to recover classical texts purified from mediaeval accretions; and that often meant eliminating alterations and additions introduced by Arabic commentaries. In various European universities a tension arose, notably in the medical faculties, between scholars who wished to retain established teaching texts derived from Arabic intermediaries, and a younger generation who insisted on using the original Greek texts of Hippocrates and Galen. In sixteenth-century England Thomas Linacre and John Caius were the most active of these medical humanists. In Spain the same goals were sought at the new humanist university of Alcalá de Henares, founded in 1508 by Cardinal Cisneros, archbishop of Toledo and later regent of Spain. During the 1560s the victory of humanism over Arabic at Alcalá secured the abandonment of the most famous text of Arabic medicine, Avicenna's *Canon*; it was replaced by the direct study of Greek and Latin texts.

But this humanist opposition to Arabic literature did not prevail throughout Spain. There is evidence that Avicenna's *Canon* continued to be taught at the University of Salamanca in the seventeenth century. And when Philip II built up the great library – it was intended to be accessible to scholars – in his palace of the Escorial fifty kilometres north-west of Madrid, the collections included a rich holding of mediaeval Arabic books, predominantly medical and pharmaceutical. The king brought to his palace Diego de Urrea, a professor of Arabic, to teach the language to resident monks. The purpose was in part to benefit from this concentrated stock of Arabic medical wisdom; but also to train Arabic-speaking missionaries to extirpate Mohammedanism in the peninsula.

That brings us to the second reason for the ambiguous status of Arabic science in Spain: its association with the world of Islam, a religion now regarded as heretical, corrupt and a dangerous threat to the Christian West through the might of Spain's formidable enemy, the Ottoman Turks. The last vestige of Moorish rule in the peninsula, the kingdom of Granada, had finally been conquered in 1492. After their marriage which had united the crowns of Castile and Aragon, Ferdinand and Isabella had decided to complete the centuries-old internal crusade against the Moor. As a reward for the recovery of Granada for Christendom the pope had bestowed on the victorious king and queen the title of 'The Catholic Monarchs'. But the subjugated Moors soon found they had cause for complaint. In the last stage of battle Ferdinand and Isabella had called on the city of Granada to surrender in exchange for offers of religious toleration. The offer was accepted but the promise broken. Isabella authorized Archbishop Cisneros to begin the compulsory conversion of these new subjects – some 200,000 out of Spain's six million. By 1500 Cisneros had converted all Granada's mosques into churches and, although a patron of learning at Alcalá, he organized the burning of Arabic books except for those dealing with medicine and philosophy, another sign of residual respect for classical Arabic science. The Moors were forcibly baptized.

Forced conversion, persecution and expulsion would also be imposed on Spain's other oriental community, the Jews. The reputation of Jews in Spanish society as artful magicians and possessors of knowledge secret and powerful generated both fear and admiration. Spurious texts attributed to Solomon, the Jewish prince of wisdom, circulated in the peninsula. The *Clavicula Salomonis* (Solomon's Key), purporting to be Solomon's legacy to his son

Rehoboam, gave instructions for conjuring spirits; information on the marvellous properties of herbs and stones; and revealed secret geometrical diagrams inscribed with Hebrew letters supposed to have magical powers in overcoming disease. Although the Old Testament prohibited sorcery and divination, other Jewish texts encouraged occult science. The mystical Kabbalah promised control over nature by meditation of secret script representing the signs of the zodiac; the Talmud recorded some rabbinical beliefs in the effectiveness of amulets in fighting disease and of the influence of the constellations on individuals at the time of their birth.

Probably through a combination of skill and expectations of wonder-working, the Jews became so prominent as physicians that they dominated medicine in Spain, achieving the highest positions as court physicians. But their elevated social position was made insecure by prevailing Christian beliefs that Jews had a tendency to apply their knowledge to evil ends. When Henry III of Castile (1377–1406), struggling to restore order after the massacres of Jews, died very young, his Jewish physicians were accused of poisoning him; the myth still fanned resentment in the sixteenth century. When in 1509 López de Villalobos, a convert and son of a Jewish doctor, was appointed court physician to Ferdinand the Catholic, he was imprisoned by the Inquisition on a charge of achieving his exalted position by black magic; he was later exonerated and released.

The forced conversions of the 1390s and the subsequent voluntary conversions for reasons of insurance against further persecution created a new class in Spanish society, known officially as New Christians or *conversos* ('converts') and amongst the populace as *marranos* ('swine'). Some of them or their offspring became sincere Christians (Teresa of Avila; St John of the Cross; Diego Laínez, the successor to Ignatius Loyola as general of the Jesuits). Others were Catholic only in name and to varying degrees retained secret loyalties to Judaism. And it was specifically to investigate and correct this continuing attachment of *conversos* to Jewish beliefs that Ferdinand and Isabella established the notorious Spanish Inquisition in 1480. The Inquisition had no powers over unbaptized Jews; but converted Jews were arrested, subjected to torture to secure confessions of Jewish observance, or delivered to civil authorities for execution at the stake. In the first terrible decade of its functioning the Inquisition punished an estimated 17,000 victims; some 2,000 *conversos* were burned at the stake. And the Inquisition declared that the descendants of the condemned should be ineligible to hold public office, so reinforcing ideas of racial purity. But there were still many unconverted Jews in the peninsula and there is some evidence supporting the crown's concern that they were encouraging the converted to remain faithful to Juadism. The belief that Jews were hindering the full assimilation of the *conversos* into a wholly Catholic Spanish society motivated Ferdinand and Isabella to deport the Jews from their places of maximum concentration in Andalucía to other regions of Spain; and when that didn't work to issue the edict of 1492 forcing unconverted Jews to accept baptism or be thrown out of Spain; any who returned would be sentenced to death. So ended the centuries of Jewish communal life in Spain.

Today Spanish historians commonly blame the persecution and expulsion of the Jews for Spain's scientific stagnation from the late sixteenth and seventeenth centuries, the very period when other parts of Europe were experiencing the scientific revolution. It is alleged that the arrest, murder or expulsion of Jews deprived Spain of her best scientists; and further that scientific decline followed the destruction of Spain's Jewish merchant community, because 'there is no science without the financing of science, without artisans, mechanics and merchants; yet these were the groups who were systematically eliminated' (Márquez, 1986, pp.75–6). What truth is there in this? The wider point depends on the dubious assumption that the vacuum left by departing Jewish traders and artisans was left unfilled by others who could take their place; also there is the debatable assertion that science can only flourish where there are merchants and craftsmen. But there is no doubt that through persecution Spain lost the services of distinguished scientists. The expulsion order of 1492 caused the departure of the astronomer Abraham Zacut and his expertise now benefited Portuguese navigation. Lluís Alcanyis, an enterprising physician, remained in Spain as a *converso*. He had taught in the school for surgeons which he had helped to establish (1462) in the city of Valencia. His treatise on the plague (*c.* 1490), written to combat the outbreak in the city, was the first medical text to be printed in Valencia. In 1499 he was appointed the first professor of medicine at the new university of Valencia. But in 1504 he was arrested by the Inquisition for observance of Jewish rites, imprisoned for three years and then burned alive.

But other *converso* physicians continued to practise in Spain and did well despite the risks of persecution and the obstacles of prejudice. In the cities, colleges of apothecaries exploited old fears of poisoning to exclude those of Jewish descent and secure the profession of pharmacy as a preserve for untainted Old Christians. In 1564 Philip II reaffirmed royal approval of the constitution of Valencia's college of apothecaries which prohibited *conversos* from preparing medicines, owning a pharmacy or admission to the qualifying examination in pharmacy. Similarly in Barcelona the college which supervised the practice of pharmacy there required intending pharmacists to present documents establishing a pure genealogy unspoilt by Jewish blood. And the same racial discrimination was prescribed at Zaragoza and Seville explicitly to protect the populations of those cities from *converso* poisoners.

But it is one thing to enact a strict regulation and another to enforce it. It is questionable if these regulations were effective. Genealogies were frequently forged at a price, and regulations of this type not always enforced. It is established that, despite similar statutes of purity preventing *conversos* from holding office in the tribunal of the Inquisition in Toledo, its cathedral and its municipal council, all of these in fact had *converso* members. And many *conversos*, one way or another, managed to settle in Spanish America even though that was strictly forbidden. There may have been enough Old Christian pharmacists in Spain to exclude the *conversos*. But that was not true of medical practitioners, and this explains the continued reliance on *converso* physicians. There are even some indications that Spaniards of Old Christian stock were shunning the medical profession to avoid suspicion of Jewish descent, so strongly established had the Jewish connection with medicine become. In his treatise on nobility of 1595 the Benedictine Juan Guardiola sympathized with efforts to purify the medical profession from *conversos*, but conceded that it was permissible to employ them, as Philip II did, because of their talent. When in the 1570s the tribunal of the Inquisition at Logroño needed the services of a physician, they were unable to find one of pure race. Doctor Bélez was available but he was a *converso*; could he be employed? The central office of the Inquisition in Madrid gave permission to consult him but only if his title went unrecognized. However begrudging, Spanish society in the sixteenth century could not do without its *converso* physicians.

5.2.2 Spanish navigation and science

The Spanish monarchs did not need *conversos* for navigation. Spanish historians used to allege that Columbus may well have been of Spanish Jewish descent. That is now dismissed; there is no reason to suppose he was anything other than the son of a Catholic family from Genoa. Nor does it seem that the crew who sailed with him to America in 1492 had as many *conversos* as was once supposed; but Columbus did take a *converso* interpreter, Luis de Torres, because his fluency in Arabic was thought to be invaluable for communicating with the people of China, the landfall Columbus expected to achieve by sailing west across the Atlantic. His assurance that this sea route would give access to the wealth of Asia failed to persuade the Portuguese monarch João II; but then he turned to Ferdinand and Isabella and after six years' persistence they finally agreed to support a project which many of their councillors and geographers had thought crazy, partly because they believed that the western ocean was immense and not navigable by the relatively short crossing Columbus promised.

Despite the repeated assertions that scientific knowledge was essential to Columbus' discovery of America, there is little evidence to support this. What seems to have been important were Columbus' persistence, his confidence based on a considerable underestimate of the extent of the western ocean, and above all his uncanny seamanship – not any astronomical expertise nor even occasional reliance on astronomical observations, but an intuitive mastery of winds, currents, charts and the magnetic compass which continues today to bring admiration from sea-dogs.

This absence of science may not be true for the final phase of the Iberian voyages of discovery: the expedition of 1564 Philip II sent from Mexico to the isles of the Pacific in search of spices, and which resulted in the Spanish colonization of the Philippines. The king wanted an experienced navigator skilled in geography and astronomy to guide the expedition. Andrés de Urdaneta was commissioned. He had sailed to the Moluccas before becoming an Augustinian friar. Now resuming navigation he organized the equipping of the expedition with astronomical instruments: astrolabes and cross-staves. It is not known how much use was made of these scientific instruments for the voyage because there is no record of observations. Urdaneta's achievement, one of the feats of sixteenth-century navigation, was to discover the elusive favourable winds at the high latitude of 42°N for the return journey east to the coast of Mexico. That discovery which established the route of Spain's galleons to Manila for the next two centuries owed much to Urdaneta's experience of currents and winds; it is unclear if the remarkable return journey of five months at sea without touching land was in any way assisted by his astronomical expertise.

The Spanish settlements in the Pacific had led to conflicts with Portuguese monarchs who complained they were infringements of their preserve specified by the demarcation line agreed at Tordesillas. For years the Moluccas with their valuable spices had been contested; there were battles in that remote region. Peace had come in 1529 when Charles V sold the island to the Portuguese monarch for 350,000 ducats. And now doubts were raised on the legitimacy of Spanish occupation of other islands in the same part of the world – the Philippines. It all depended on a precise determination of the agreed demarcation lines, in fact lines of longitude. And in Philip II's reign this was a matter of political as well as scientific importance. Philip had commissioned a royal cosmographer, Alonso de Santa Cruz, to survey the various methods available for determining longitude.

His resulting account is especially interesting for the clear recognition of the obstacles which prevented a contemporary solution to the problem of longitude. In theory astronomical observations could provide the answer; as Santa Cruz said, the ancient Greeks had known this. For observers stationed at different locations on earth the same lunar eclipse occurred at different times. That was because the earth took twenty-four hours for its rotation through 360°. If the recorded times of observation of the eclipse differed by six hours, the two observers were in positions separated by 90° of longitude. But in practice this collaboration was far from simple. Santa Cruz commented that a lunar eclipse could last for hours and the two separated observers would have to be sure they were timing the same phase of the eclipse. A greater difficulty ruled the method out; an accurate timekeeper was essential – even a few minutes error on the clock would result in significant errors in longitude; but Santa Cruz said the available clocks usually gained or lost as much as half an hour a day making them useless for even approximate determinations of longitude.

The same instrumental inadequacy nullified the solution recently proposed in 1522 by Gemma Frisius of Louvain, a cosmographer of the Spanish Netherlands. If a clock set at a port of departure was taken on board ship, the longitude traversed during a voyage could be found by determining local time at sea from observation of the sun; from the difference with the time on the carried clock, position in longitude relative to the port of departure could be calculated. But although he regarded this as the most promising method Santa Cruz pointed to sources of error apart from the existing inaccuracies in clocks: the variation in temperature and humidity which affected clocks and the disturbance in the clock's smooth running caused by the motion of the ship. These difficulties would not be eliminated until the eighteenth century with the invention in England of Harrison's marine chronometer.

Meanwhile the longitude data demanded by Philip II produced results far from the truth, the inevitable consequence of imprecise clocks. From recorded observations of lunar eclipses taken in Panama in 1581 it was calculated that Panama City was 49°15' west of the Canary Isles; the actual difference in longitude is 61°, an error of 11°45', or about 700 miles. And in the Philippines where Urdaneta had asked a fellow Augustinian, Martín de Rada, to take astronomical observations for calculations of longitude, the results indicated that the Philippine town of Cebu was 215°15' west of Toledo; the correct figure is 232°, a large error of just under 1,000 miles.

Navigation and the competition for territory were clearly stimulating enquiries into longitude and astronomical observation in sixteenth-century Spain. And the urgency felt by the crown for a reliable method of determining longitude is revealed by the large monetary prize of thousands of ducats offered at the beginning of Philip III's reign (1598) to anyone who could solve the problem.

The close connections between Spanish navigation, science and the crown were most apparent at the Casa de la Contratación (House of Trade) established by the Catholic Monarchs in Seville in 1503. It had been created to supervise Spain's shipping and trade with the American possessions. From 1508 it became increasingly important as a centre of scientific and technical expertise covering the whole range of knowledge required for the art of navigation. A small technical staff employed by the crown supervised the manufacture of nautical instruments, the preparation of nautical charts and the maintenance of a standard chart of the Indies continually updated as new information came in. Crown officials also taught formal courses to pilots and examined them.

In addition to these 'pilot-majors' and 'cosmographer-majors', Philip II created 'a chair of the art of navigation and cosmography' (1552). It was intended to rectify the lack of training of masters and pilots. The appointed professor was Jerónimo de Chaves, the author of works on the calendar and a commentary on Sacrobosco's *Sphere*, the standard introduction to Ptolemaic astronomy. He was instructed to teach from that elementary manual; to explain the technique of setting a ship's course by the use of marine charts; the rules for determining latitude from observation of the sun and pole star; the use of clocks; the times of tides; and the theory and practice of nautical instruments 'so that errors in them can be detected'. Before any pilot or master was licensed to operate on the Indies route this course would have to be studied and an examination passed.

All of this activity greatly impressed Stephen Borough, the English navigator who visited the Casa in 1558. He soon called for a comparable institution to be established in England. In fact Philip's repeated interventions and complaints reveal that inefficiency and corruption were affecting all aspects of the Casa's scientific work (Goodman, 1988, pp.76–8). Nevertheless Seville had become the world centre of navigational expertise. And the teaching manuals based on teaching at the Casa were translated into several languages – one of the most widely read was the *Arte de navegar* (Valladolid, 1545) of Pedro de Medina, teacher of pilots at the Casa. Spanish naval historians have some basis for their claim that other European nations learned their navigation 'from us'.

5.2.3 *The stimulus of empire*

By the 1530s Spain had acquired an empire far more extensive than Portugal's and differing from it in the possession of huge land areas populated by millions of new subjects, the subjugated Amerindian empires of the Aztecs and Incas. The crown was determined to amass detailed information about these new territories in order to secure complete control of government and full exploitation of resources. And some of the Spaniards who came to the New World were so amazed by their remarkable environment that they were led to challenge long-held opinions and generate radically new ideas. In these two ways the experience of empire stimulated Spanish science and led to real advances in knowledge. This is particularly noticeable in the middle of Philip II's reign, during the 1570s.

The Royal Council of the Indies was the central authority responsible for everything concerning Spain's overseas possessions, except finance (transferred at the beginning of Philip II's reign to the Council of Finance) and the maintenance of Catholic orthodoxy (the responsibility of the Inquisition). Created in 1524, this consultative Council met regularly in Madrid and on the basis of its advice the monarch took the decisions of government which were then executed by his viceroys in the Indies. The Council consisted of lawyers, but in 1571 Philip added to it a new scientific officer, Juan López de Velasco, the 'cosmographer-chronicler', to remedy the lack of basic knowledge of the overseas territories.

His comprehensive duties, spelled out in the king's instructions, reveal the crown's desire for precise scientific information. He was required to compile accurate geographical tables of data on the longitude and latitude of places and their distances apart. That entailed the organization of astronomical observation in the Indies and the collection of results in Madrid. The royal cosmographer was instructed to arrange for the observation of lunar eclipses in the Indies. He was to write to royal officials out there communicating the times when these were to be observed, send instructions on how to make simple instruments – a type of sundial made from paper circles with wooden or metal gnomons, and explain the technique to be employed – marking the points on the dial where the shadow of the moon crossed it at the beginning and end of the eclipse. The other essential datum for a determination of longitude with reference to Madrid was provided by López de Velasco himself: the recording of the time of the same eclipse as it occurred in Madrid. The surviving record of observations made in Panama in 1581 shows that these instructions were complied with.

In addition López de Velasco had to collect as much information as possible on the flora and fauna of the Indies. This along with other information was solicited and secured by the sending out of printed questionnaires to government officials in localities throughout the Indies, with a request for the return of completed forms to the Council in Madrid 'as quickly as possible'. Many were completed, supplying the government then and historians today with valuable descriptive information on Spanish America.

More impressive work was achieved at the same time by the crown's scientific expedition to the Indies. In 1570 Philip sent a Portuguese cosmographer, Francisco Domínguez, to conduct a full geographical survey of New Spain (Mexico); with him went Francisco Hernández, one of Philip's numerous court physicians, now appointed *'protomédico general* of our Indies'. Hernández's task was to find out all he could about the medicinal plants of the Indies. The marvellous virtues of exotic American plants had for decades raised hopes in Europe that genuine panaceas were now within reach; the newly discovered guaiacum was being employed in central Europe in the 1510s and for centuries after to treat syphilis, though its effectiveness was imaginary.

The sale of American medicinal plants had already become a profitable commercial venture and perhaps the search for greater profits was one motive behind the royal expedition. Hernández was instructed to accumulate reliable information by consulting Indian herbalists on the places where medicinal plants could be found and their supposed virtues. Then he was to test the information by observations and experiments. Viceroys were instructed to supply local artists to provide illustrations of the plants. For five years Hernández travelled by mule over the vast territory of New Spain accompanied by artists, plant-hunters and an interpreter. At various hospitals he experimented on the sick with extracts from plants he had never seen before. His own health suffered and he was never able to complete his mission with a botanical survey of Peru. But the description of thousands of species of Mexican Flora which he sent to Madrid has never been superseded. The manuscripts were deposited in the Escorial; the fine illustrations were used to decorate Philip's living rooms; and sacks of seeds and tubs of plants were transplanted in the gardens of the royal palace in Seville.

Hernández's reactions were revealing. Here were plants of great medicinal power which had never been seen in Europe. He therefore wrote to the king that he had surpassed Dioscorides, the long-established authority on medicinal plants; progress had been achieved far beyond the Greeks, because they had no inkling of this American floral treasure. And he was sure that if

Figure 5.9 Sixteenth-century engraving made from Hernández's illustration of a medicinal plant from the New World. (From F. Hernández, Nova plantarum, animalum et mineralium Mexicanorum historia, 1651.)

Alexander the Great had won fame from his patronage of Aristotle's natural history, how much greater glory would come to Philip from this expedition!

But his manuscripts were not published until the seventeenth century. And Europeans became acquainted with American medicinal plants chiefly through the *Historia medicinal* (1565–74) of Nicolás Monardes, physician of Seville, who acquired shipped specimens as part of his business ventures in the Indies. His book familiarized Europeans with Peru balm, Tolu balm, sarsaparilla and tobacco – then regarded as a medicine. In the seventeenth century American flora gave Europeans ipecacuanha and quinine; these were not imaginary panaceas but important new medicines.

The peculiar fauna of America – the iguanas, armadillos, llamas, turkeys and parrots – amazed Europeans who saw them for the first time. Some were surprised to find any form of life in equatorial America because they had learned from Aristotle and other ancient authorities that nothing could live in the scorching heat of the tropics. When the Jesuit missionary José de Acosta arrived in Peru in 1571 his ingrained trust in Aristotle, nurtured by the education of his religious order, weakened under the severe blows produced by experience of a wholly new environment. In his widely read *Historia natural y moral de las Indias* (Seville, 1590) he confessed to mocking Aristotle's philosophy because 'while his doctrine predicted everything to be ablaze I and all my companions were cold'. Acosta was experiencing the chill of the Andes, a further refutation of Aristotle who supposed that the upper atmosphere became increasingly fiery. Huge rivers and lakes falsified the Ancients' assertion that the tropics must be parched; and the presence everywhere of Indians exposed the belief in the impossibility of life in the tropics as nonsense.

But the most interesting of Acosta's reactions were his acute theoretical reflections on the origin of life in the New World. How had the Indians and animals arrived in America? His explanation began from the conviction that all humans and beasts had been generated from the mating pairs in Noah's Ark, and then liberated after the Flood's retreat on Mount Ararat where the Ark had come to rest. But that was in Asia Minor in the Old World. He rejected the idea of a separate divine creation of life in America because that would imply incompleteness in God's original creation and reduce the importance of Noah's Ark. So the Amerindians and American beasts must have crossed from the Old World. But an ocean passage seemed incredible: the Amerindians knew nothing of the magnetic compass; in fact they had been astonished at the sight of Spanish ships. Acosta therefore concluded that Amerindians and beasts had crossed from the Old World to the New by land. His prediction that future exploration would show the two continents to be joined or very close would be confirmed by the eighteenth-century exploration of the Siberian Kamchatka peninsula (no more than 56 miles of the Bering Straits separates Siberia from Alaska). As for the distinctive fauna of the New World, Acosta speculated that they had walked from the Ark to various regions of the earth, but were able to survive only in the New World. But he was left with an unresolved difficulty: why had these animals never been observed in the Old World? No answer could be given to this until the theories of evolution centuries later.

Of all American natural resources none provided the Spanish with greater wealth than the precious metals. Silver had since antiquity been important in Europe's economy for trade with the Far East. In recent centuries the chief centres of production had been the mines of central Europe (the Harz mountains, Saxony, Bohemia, the Tyrol and Hungary). All of that changed with the discovery in the 1540s of Spanish America's rich silver deposits (Zacatecas, New Spain; and the great silver mountain of Potosí, Peru). The European mines declined; they could not compete with the American scale of production and the lower price of silver. Mexico and Peru became the principal suppliers of Europe's silver, assisting the Spanish monarchy's hugely expensive military campaigns and bringing wealth to financiers in Genoa. The operation of the American mines was left in the hands of the

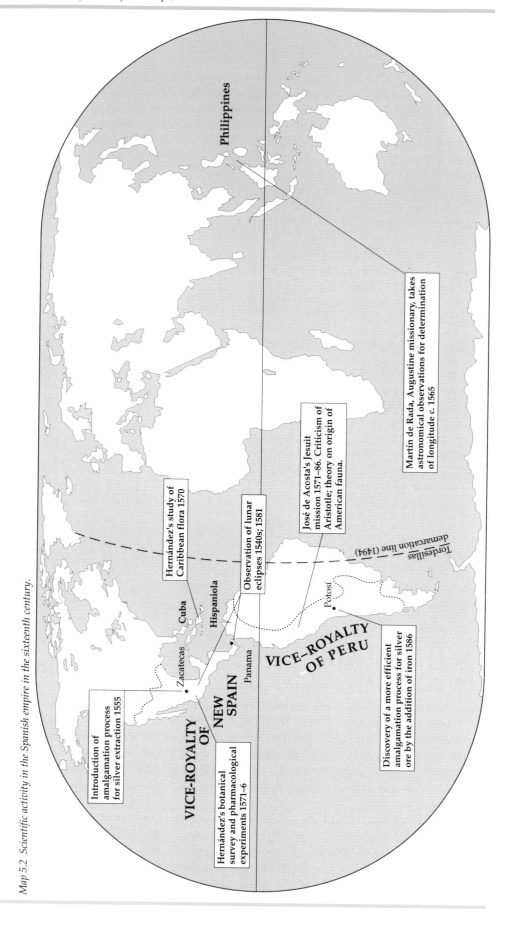

Map 5.2 Scientific activity in the Spanish empire in the sixteenth century.

Philippines

Martín de Rada, Augustine missionary, takes astronomical observations for determination of longitude c. 1565

José de Acosta's Jesuit mission 1571–86. Criticism of Aristotle; theory on origin of American fauna.

Hernández's study of Caribbean flora 1570

Observation of lunar eclipses 1540s; 1581

Tordesillas demarcation line (1494)

Cuba

Zacatecas

Hispaniola

Potosí

Panama

VICE-ROYALTY OF PERU

VICE-ROYALTY OF NEW SPAIN

Introduction of amalgamation process for silver extraction 1555

Hernández's botanical survey and pharmacological experiments 1571–6

Discovery of a more efficient amalgamation process for silver ore by the addition of iron 1586

settlers, but the crown took its share: one tenth of the yield in New Spain; one fifth in the more lucrative Peru. Viceroys departing from Spain to take up office in Mexico City or Lima were left in no doubt by the king's instructions that the fostering of American mining was to be given the highest priority.

All of this gave a strong stimulus to Spanish metallurgy. It was technological innovation which caused the silver boom in America. At first the silver had been extracted by smelting the ore; but the fuel costs were high and attempts were made to process silver ore 'without fire'. A settler from Seville, Bartolomé de Medina, successfully applied a new method of extracting the silver by mixing the crushed ore with mercury. In a slow chemical process in the cold, mercury alloyed with the precious metal to form silver amalgam. This was then heated – the only part of the process requiring fuel – and the amalgam decomposed; the mercury vaporized leaving the silver. The idea is thought to have come from Germany, but there was real achievement in Medina's successful implementation of the technique with American ores. The considerable saving in fuel was offset by additional expenses for salt, an essential ingredient, and mercury, supplied by the king's peninsular mine of Almadén and the Peruvian mine at Huancavelica. The great advantage of the amalgamation process was that it permitted the economic exploitation of the low-grade silver ore of Potosí, where production increased seven-fold in 1572–92, reaching a peak in 1592 of 900,000 marks of silver, 200 tons.

Spaniards in the peninsula and America worked on various designs of furnaces and methods of ventilation both to boost production and improve the dreadful working conditions of the Indian labour force. And a way was also discovered of accelerating the amalgamation process and reducing the consumption of mercury: in Potosí it was discovered that this could be achieved by the addition of iron filings. No one understood why this worked – twentieth-century text-books still present the amalgamation process as full of the complications that chemists call 'side-reactions'. But it did work and this addition of a metallic reagent – later a copper salt was used – remained standard practice in the twentieth century.

5.2.4 Monarch, cities, Church and science

Power, political and financial, affected the development of science in peninsular Spain. No individual had as much power as the monarch but, contrary to what is usually assumed, the power of the Spanish monarch was not absolute. In the crown of Castile, the largest and most populous part of the peninsula, the king's power was limited by the *cortes*, the parliament of the realm. That was no democratic institution but a gathering of elite representatives of the oligarchies, moneyed and aristocratic, which governed the cities. Nor did the *cortes* meet regularly; but it was not the lame duck historians have until recently supposed it to be – its ability to withhold taxes from the crown until local grievances were satisfied seriously curtailed the king's power. There were other *cortes* in the crowns of Aragon (Aragon, Catalonia, Valencia), and in Navarre. And in these regions, as well as the Basque lands, the king's power was greatly reduced by the existence of jealously guarded regional liberties. When in the sixteenth century monarchs sent *protomédicos*, royal physicians, to examine, license and control medical and pharmaceutical practice in Navarre, Aragon and Catalonia, there was in all cases resistance to what was seen as centralizing control from Castile, and the crown was forced to back down.

The extent to which an interested monarch could influence scientific development in Spain became clearer with the accession of Philip II (*r.* 1556–98), a dilettante. His strong interests in medicinal plants led him to establish a botanic garden at his palace of Aranjuez, south of Madrid, where distillers were regularly employed to extract plant essences. At the other great palace of the Escorial, an entire suite of rooms was devoted to the chemical preparation of medicines from plants and minerals. The elaborate apparatus included a giant brass distillation tower, 20 feet high, fitted with 120 glass alembics (vessels carrying distilled liquors to a receiver) which was said to produce 180 pounds of distilled essences in 24 hours. Those employed included Diego de Santiago, who designed apparatus and wrote an alchemical treatise revealing the influence of Paracelsus (see Chapter 6), and Richard Stanyhurst, a Catholic exile from Elizabethan England. The medicines prepared were stored in the associated pharmacy and dispensed to patients in the adjoining infirmary, and to the royal family. The same attraction to alchemical medicine induced Philip to bring to Madrid Leonardo Fioravanti, an Italian Paracelsian physician employed by the viceroy of Philip's kingdom of Naples. Fioravanti had written to Philip offering to prepare more effective alchemical medicines for the king's soldiers; he worked in Spain in 1576–7 disseminating Paracelsian alchemical medicine, unorthodox medical doctrine which challenged the established Galenism of university medical faculties.

Within the confines of his court Philip could establish scientific institutions, but more ambitious plans to extend them throughout Castile were less successful. That can be seen in the outcome of the planned Academy of Mathematics. Suggested by Philip's architect Juan de Herrera, and motivated by the desire to provide the kingdom with 'men proficient in mathematics, architecture and other related sciences and skills', Philip announced its establishment in Madrid in 1582. The king purchased premises, and soon nobles and military officers were attending lectures on a wide range of scientific subjects, including demonstrations with scientific instruments. The lecturers included foreign experts: João Lavanha, a Portuguese, taught geography and navigation; Giuliano Firuffino of Milan taught military engineering. But there were also Spanish lecturers: Herrera on architecture, and Cristóbal de Rojas, who gave a geometrical treatment of fortification. Apart from lectures, the staff prepared vernacular translations of classical scientific texts like Euclid's *Optics*; that was very much in keeping with the king's policy of making Castilian a dominant world language.

Philip was delighted with his new Academy and in 1587 sought to diffuse its curriculum by ordering the *cortes* of Castile to set up similar institutions in the cities. But he had neither the money nor the power to enforce this. And when several of the cities refused to provide the necessary finance, the project collapsed. In Madrid Philip's Academy continued to function until 1625, and then its scientific instruments and books were inherited by the Colegio Imperial, a Jesuit college which taught mathematical science and natural philosophy.

The cities' power was sometimes used positively to promote scientific developments. Barcelona was no longer the thriving port it had been in the thirteenth and fourteenth centuries when it was the centre of Aragon's Mediterranean empire. But it was still something of a city-state ruled by moneyed nobility and powerful guilds. Its College of Apothecaries continued to resist royal interference; in 1511 it had supervised the publication of an official pharmacopoeia, the first in Spain – only one other had yet been

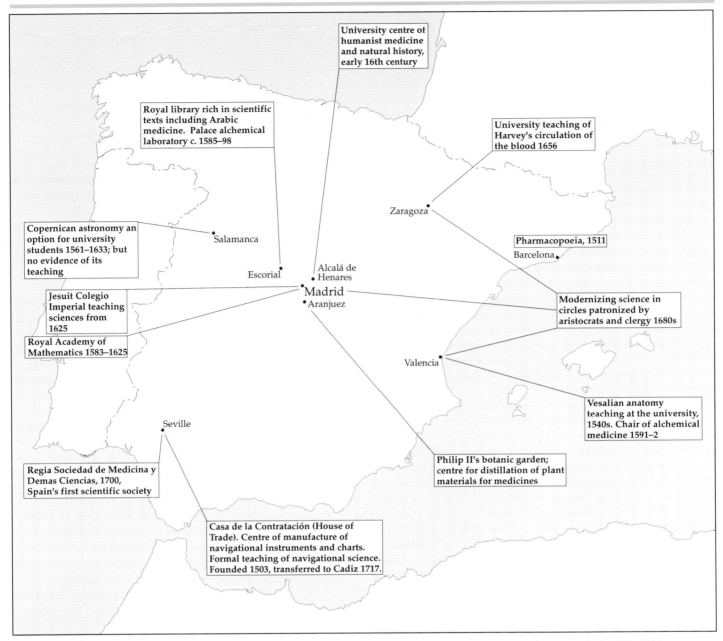

Map 5.3 *Science in Spain 1500–1700.*

published in Europe (Florence, 1498). In Valencia rich merchants controlled the city and its university. Public health was well developed and the university unusual for the dominant position of medicine. It was here that Spain's first chairs of anatomy, surgery and medical botany were created (1501). And here also that innovating teaching was introduced – Vesalian anatomy from the 1540s and, unique in Europe, a short-lived university chair in alchemical medicine (1591) given to the Valencian physician Llorenç Coçar, a follower of Paracelsus (López Piñero, 1973, pp.125–6; 1977; and 1979, pp.98–9).

What influence did Spanish religious institutions have on the development of science in the peninsula? From the sixteenth century the effective head of the Catholic Church in Spain and Spanish America was not the pope but the monarch. It was a serious offence to publish papal bulls in Spain without prior authorization by the crown. And Philip II delayed publication of the decrees of the Council of Trent until his lawyers had scrutinized them; they were then permitted to be published on the understanding that the king's power was unaffected. The Spanish Inquisition may only have been established (1480) after permission had been given by the pope; but thereafter it functioned as a royal Council directed not from Rome but by the crown. Nor is the idea tenable of a solid Catholic alliance between the popes and Spain; during the sixteenth century relations often reached breaking point for political reasons: in 1527 the troops of Charles V, emperor and king of Castile, sacked Rome and imprisoned the pope; and in the next reign Philip II's relations with Rome were embittered for years. In Spain the pope's importance was chiefly that of a paymaster – a source of badly needed revenue for the Monarchy's military campaigns. And in those campaigns of Ferdinand and Isabella against the Moors, and of Charles V and Philip II against Islam and Protestantism, the Spanish monarchs saw themselves as the divinely appointed champions of Catholicism, responsible for the protection of the Holy Faith within Spain and its overseas territories, and the overthrow of diabolical Islam and heretical Protestantism.

Since the late fifteenth century the courts of Spanish monarchs had begun to be affected by currents of Italian and Dutch humanism, widening horizons in religion and literature, influencing science, painting and architecture. In the 1520s the ideas of the Dutch humanist Erasmus spread to the court of Charles V, bringing challenging religious ideas, and in the 1530s a humanist Honorato Juan played some part in Philip II's education. But then came the reaction of the 1530s against these new currents of European thought; the intellectual atmosphere had changed and there was a return to traditional doctrines, secular and religious. The reason was the spread of the German Reformation. Spanish monarchs now became the leaders of Counter-Reformation. When he prepared Philip to succeed him, Charles V urged him to use the strongest measures to keep Spain free of the Protestant infection. Accordingly in 1559 Philip II introduced a general ban on university study to prevent contact with European Protestants; his subjects were permitted to enrol only in Spanish universities and the designated, safe, Catholic universities at Naples, Rome and Bologna.

In 1558 while Philip was in the Netherlands his sister Juana, now regent, had introduced stricter censorship laws with the death penalty for printing or importing heretical literature. The crown sent agents to inspect bookshops and universities; at Salamanca the rector and masters of the university were instructed to report to the Inquisition students in possession of suspect works or professors who taught Lutheran doctrines. And in 1559 Fernando Valdés, Inquisitor General, issued the first Index of the Spanish Inquisition. The prohibited books included the zoological treatise of Conrad Gesner, and the pioneering illustrated botanical works of Leonhart Fuchs and Otto Brunfels – all banned simply because these German authors were Protestants. The much enlarged Indices of 1583–4 issued by the Inquisitor General Gaspar de Quiroga prohibited the works of the German astronomers Peurbach and Reinhold; parts of Paracelsus' work and of his followers like Toxites and Fioravanti – Paracelsian ideas were regarded as infected with Lutheranism even though Paracelsus remained Catholic. The seventeenth-century Spanish Indices became still larger, that of 1632 listed as many as 2,500 authors, and now prohibited practically the whole of Paracelsus' work. But the astronomy of Copernicus was not prohibited until after the trial of Galileo (1633).

What was the effect of these repressive measures on Spanish science? Did they completely cut Spain off from European science and therefore cause that seventeenth-century stagnation which left Spain untouched by the scientific revolution? The evidence shows that the ban on university study abroad was imposed: at the university of Montpellier, where Aragonese students had traditionally come to study medicine, some 248 Spanish students matriculated in 1510–59; after the ban of 1559 the figure for 1560–99 is just twelve. The assessment of the general effect on Spanish science depends on one's view of the quality of university science; Spanish historians have argued that Spanish students were not missing much because the most flourishing science in Europe was produced outside the universities, which were conservative. The effects of the Inquisition on Spanish science are not easily discussed with any precision. The effectiveness of the Indices is problematic because of loopholes – the occasional official permission granted to consult forbidden works – and the impossibility of complete control; the clandestine circulation of prohibited literature is another imponderable. The Inquisition had first targeted *conversos* and Moriscos, and then concentrated on the sexual morality of Old Christians. But fear of the Inquisition may have inhibited innovatory ideas; perhaps that accounts for the marked fall in publication of scientific works in the seventeenth century (López Piñero, 1979, pp.374–5). The Spanish Inquisition was not abolished until 1834, by which time it had become moribund. But during the eighteenth century it could still spring into action after periods of dormancy, arresting scientists like the mathematician Benito Bails for consulting prohibited works or the physician Diego Zapata on a charge of Judaizing. And as late as 1790 a Spanish Index expurgated a sentence from the Castilian translation of Nordberg's *History of Charles XII* in which Copernicus was described as 'the discoverer of the true system of the world' (Domergue, 1986, p.107).

Yet Spain was not completely cut off from European developments. Vesalian anatomy was taught in peninsular universities; so for a while was Paracelsian alchemical medicine. But in the seventeenth century innovation had ceased. Not until the 1680s was a concerted effort made to modernize Spanish science with the introduction of doctrines like Harvey's circulation of the blood and corpuscular philosophy. The innovators were small groups of individuals operating outside of the universities, patronized by enlightened noblemen. The most important centres of their activity were Zaragoza, Madrid, Valencia, and Seville where in 1700 was created Spain's first modern scientific institution, the *Regia Sociedad de Medicina y demás Ciencias*, strongly inspired by Cartesian influence (Ceñal, 1945; López Piñero, 1979, pp.371–455).

Sources referred to in the text

Albuquerque, L. de. (1973) *Para a história da ciência em Portugal*, Lisbon, Livros Horizonte.

Boxer, C. (1963) *Two Pioneers of Tropical Medicine: Garcia d'Orta and Nicolas Monardes*, London, Wellcome Historical Medical Library.

Ceñal, R. (1945) 'Cartesianismo en España: Notas para su historia (1650–1750)', *Revista de la universidad de Oviedo*, pp.5–97.

da Orta, G. (1913) *Colloquies on the Simples and Drugs of India*, tr. C. Markham, London, Henry Sotheran and Co.

Domergue, L. (1986) 'Inquisición y Ciencia en el siglo XVIII', *Arbor*, 124, pp.103–30.

Goodman, D.C. (1988) *Power and Penury: Government, Technology and Science in Philip II's Spain*, Cambridge University Press.

Goodman, D.C. (1991) 'Portugal', in J. Yolton (ed.) *The Blackwell Companion to the Enlightenment*, Oxford, Blackwell.

Hooykaas, R. (1979) *Humanism and the Voyages of Discovery in Sixteenth-century Portuguese Science and Letters*, Amsterdam and Oxford, North-Holland.

Hooykaas, R. (1980) *Science in Manueline Style*, Coimbra.

López Piñero, J.M. (1973) 'Paracelsus and his work in 16th and 17th century Spain', *Clio Medica*, 8, pp.113–41.

López Piñero, J.M. (1977) *El 'Dialogus' (1589) del paracelsista Llorenç Coçar y la cátedra de medicamentos químicos de la Universidad de Valencia (1591)*, Valencia, Cátedra e Instituto de Historia de la Medicina.

López Piñero, J.M. (1979) *Ciencia y técnica en la sociedad española de los siglos XVI y XVII*, Barcelona, Labor.

Márquez, A. (1986) 'Ciencia e Inquisición en España del XV al XVII', *Arbor*, 124, pp.65–83.

Newitt, M. (1986) 'Prince Henry and the origins of Portuguese expansion', in M. Newitt (ed.) *The First Portuguese Colonial Empire*, Exeter, Exeter University Press.

Russell, P.E. (1984) *Prince Henry the Navigator: The Rise and Fall of a Culture Hero*, Oxford, Clarendon.

Taylor, E.G. (1971) *The Haven-finding Art: A History of Navigation from Odysseus to Captain Cook*, enl. edn, London, Hollis and Carter.

Waters, D.W. (1967) 'Science and the techniques of navigation in the Renaissance', in C. Singleton (ed.) *Art, Science and History in the Renaissance*, Baltimore, Johns Hopkins.

Further reading

Goodman, D.C. (1988) *Power and Penury: Government, Technology and Science in Philip II's Spain*, Cambridge University Press.

História e desenvolvimento da ciência em Portugal, 2 vols, Academia das Ciências de Lisboa, 1986.

Waters, D.W. (1967) 'Science and the techniques of navigation in the Renaissance', in C. Singleton (ed.) *Art, Science and History in the Renaissance*, Baltimore, Johns Hopkins.

Science from the Earth in Central Europe

by Gerylynn K. Roberts

6.1 Mining in Central Europe

Central Europe is a geographical region rather than a political entity. In relation to the sixteenth century, it is common to think of it as extending from the Rhineland and Switzerland in the west, through Austria and Hungary to include Silesia in the east, from the various German states in the north to parts of what is now Italy in the south. It was populated mainly by Germanic peoples in the north and west and with Slavic peoples in the south and east. Politically, it was a collection of sovereign states which formed the Holy Roman Empire, an alliance of temporal and spiritual powers dating from the time of Charlemagne which sought to express the political ideal of a spiritually unified Christendom. It was ruled, at least theoretically, by an emperor elected by seven princes who were called 'electors': the Palatine of the Rhine, the Margrave of Brandenburg, the Duke of Saxony, the King of Bohemia and the Archbishops of Mainz, Trier and Cologne. The position of the emperor was complicated by the fact that, although he was their elected head, the German lands consisted of more than three hundred feudal, ecclesiastical and urban territories, whose rulers and representatives sat in the Imperial Diet (Reichstag) with powers to constrain the emperor's actions. In practice, the emperor was only as powerful as his own territorial base and dynastic political alliances allowed.

The head of the Austrian House of Habsburg dominated the post of emperor during the sixteenth century. As was the normal pattern within the empire, Maximilian of Austria, who succeeded his father as Holy Roman Emperor in 1493, was thwarted in his efforts to establish common imperial institutions; he had instead to exercise his authority separately in each state. However, the power of his own House of Habsburg was further consolidated when his grandson became Charles V of Spain, with authority over not only the thrones of Aragon, Castile and Burgundy, but much of the Americas, the Netherlands, North Africa and key areas in northern Italy. Despite concern within the Empire about this vast territorial power base, Charles V, with timely backing from important financial interests (especially the Tyrolean mining wealth of the Fuggers), and aided by bribery of the ecclesiastical Electors, was chosen to succeed Maximilian in 1520. Charles's younger brother, Ferdinand, held great power within the Empire as head of the Austrian branch of the House of Habsburg with dynastic alliances in Bohemia and Hungary.

Quite apart from the difficulties of administering such far flung territories, dealing with the implications of his own designs on French territories and of facing continuing Ottoman challenges on both land and sea, Charles V also had to cope with the political ramifications of the Reformation. Begun in 1517 in the Saxon town of Wittenberg by Martin Luther, a priest and university professor of biblical theology, the movement spread rapidly throughout

Europe from this central geographical position, aided by the printing press. As Holy Roman Emperor and thus the chief secular defender of the Catholic faith, Charles found himself initially in the position of having an Imperial Diet which, though largely composed of Catholic rulers, would not back imperial opposition to the movement lest they were to strengthen the hand of their already powerful emperor.

Gradually, some princes began genuinely to embrace Lutheranism, while others found it expedient to do so in order to acquire secularized church lands and wealth. In 1526, the Diet of Speyer resolved that each prince should make his own decisions and, in 1529, an alliance of Lutheran princes, the Schmalkaldic League, was formed. As more princes converted, the League grew into a powerful force within the Empire with considerable military capability. Finally, almost bankrupt from campaigns on so many fronts, both internal and external, Charles V abdicated in favour of his brother in 1555. At the Diet of Augsburg in that year, the position of 1526 was formalized; two official confessions were recognized, and the choice in any particular area was to be determined by the faith of the prince, hence undermining further the ideal of spiritual unity that underlay the concept of the Holy Roman Empire.

Figure 6.1 Charles V by Titian, 1548. Painted to celebrate his famous victory at Mühlberg over the Schmalkaldic League of Lutheran princes, this equestrian portrait of Charles V expresses fully his power and splendour as the Catholic Emperor in what would be short-lived triumph over a challenge to his imperial authority. (Mansell Collection, London, and Alinari, Florence)

Map 6.1 (facing page) The Holy Roman Empire, c. 1600, showing the Habsburg territories. (Adapted from Evans, 1973, p.3.)

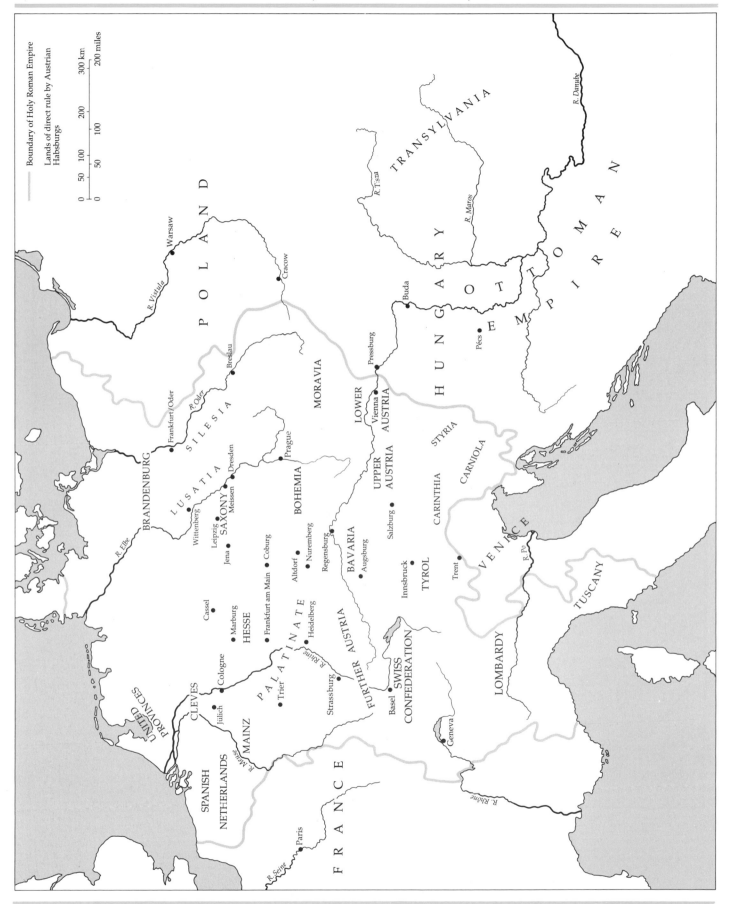

Boundary of Holy Roman Empire
Lands of direct rule by Austrian Habsburgs

300 km
200 miles

POLAND

Warsaw

Cracow

R. Vistula

BRANDENBURG

Frankfurt/Oder

R. Oder

Breslau

SILESIA

LUSATIA

R. Elbe

Wittenberg

Leipzig

Jena

SAXONY

Meissen

Dresden

Prague

BOHEMIA

MORAVIA

Cassel

Marburg

HESSE

Frankfurt am Main

Coburg

Altdorf

Nuremberg

Regensburg

BAVARIA

Augsburg

Cologne

Jülich

CLEVES

PALATINATE

Heidelberg

Trier

MAINZ

R. Rhine

Strassburg

FURTHER AUSTRIA

Basel

SWISS CONFEDERATION

Geneva

UNITED PROVINCES

SPANISH NETHERLANDS

R. Meuse

FRANCE

Paris

R. Seine

Salzburg

Innsbruck

TYROL

UPPER AUSTRIA

LOWER AUSTRIA

Vienna

Pressburg

Buda

HUNGARY

Pécs

TRANSYLVANIA

R. Tisza

R. Mures

R. Danube

OTTOMAN EMPIRE

STYRIA

CARINTHIA

CARNIOLA

Trent

VENICE

R. Po

LOMBARDY

TUSCANY

RHÔNE

R. Rhône

Map 6.2 (facing page) Mining regions of Central Europe, about 1600.

To some extent, the German princes' worries about the territorial strength of Charles V were justified. While they succeeded in tempering his direct political influence in the German lands, the economic impact of his American territories became overwhelming. In the early sixteenth century, the German states of Central Europe were internationally important centres of trade and industry that was based on linen textiles and on the production of the precious metal silver, the principal medium of exchange currency. From 1540, relatively cheaply produced silver from the Americas (see Chapter 5) dramatically affected the economies of Central Europe and began an eclipse of German economic power. Over the preceding century, the output of silver from the mines of the Tyrol, the Harz, Upper and Lower Hungary (Slovakia), Bohemia and, especially, Saxony had grown rapidly, roughly fivefold. Output from these very rich areas peaked in the 1530s (Nef, 1987, p.735).

A—FURNACE. B—STICKS OF WOOD. C—LITHARGE. D—PLATE. E—THE FOREMAN WHEN HUNGRY EATS BUTTER, THAT THE POISON WHICH THE CRUCIBLE EXHALES MAY NOT HARM HIM, FOR THIS IS A SPECIAL REMEDY AGAINST THAT POISON.

(b)

A, B—TWO FURNACES. C—FOREHEARTH. D—DIPPING-POTS. THE MASTER STANDS AT THE ONE FURNACE AND DRAWS AWAY THE SLAGS WITH AN IRON FORK. E—IRON FORK. F—WOODEN HOE WITH WHICH THE CAKES OF MELTED PYRITES ARE DRAWN OUT. G—THE FOREHEARTH CRUCIBLE: ONE-HALF INSIDE IS TO BE SEEN OPEN IN THE OTHER FURNACE. H—THE HALF OUTSIDE THE FURNACE. I—THE ASSISTANT PREPARES THE FOREHEARTH, WHICH IS SEPARATED FROM THE FURNACE THAT IT MAY BE SEEN. K—BAR. L—WOODEN RAMMER. M—LADDER. N—LADLE.

(a)

Figure 6.2 (a) A liquation and (b) a cupellation furnace at work. In the liquation process, silver was removed from the silver-bearing copper ore by heating with lead. The silver dissolved in the lead and the copper could be drawn off separately. The silver-bearing lead was then treated in a special cupellation furnace; remaining impurities were removed by being absorbed as oxides into an ash crucible, as was the lead, leaving behind pure silver on the bed of the furnace. (From Agricola, De re metallica, *Dover Publications, [1556] 1950.)*

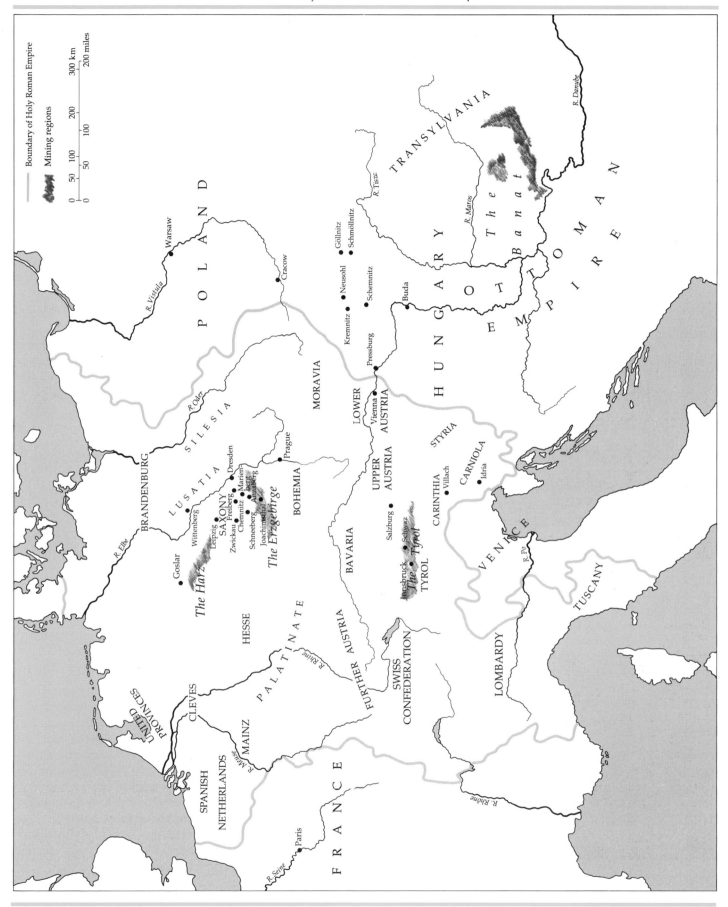

One technologically-enabling feature of this development was the introduction into Saxony in 1451 of a process to extract silver from argentiferous copper ores by means of lead. This made it worthwhile to process a whole new range of ores, but also entailed mining much deeper (600 feet was not unknown) and the consequent development of elaborate machine-driven drainage systems involving long tunnels or adits. Traditionally, mining had been undertaken by individual miners or small groups who were given rights to underground products by the territorial princes, regardless of who actually owned the land, in return for a percentage. At the same time, the princes had a responsibility to organize and administer the mines. The greater capital and more coordinated organization required for the technologically more sophisticated deeper mines of the turn of the sixteenth century led to a consolidation of mining enterprises into larger scale units. While still traditionally under the control of the territorial princes, the capital-intensive work was funded by the merchant-financiers of the great German commercial towns. This in turn brought a greater need for legal control and a more extensive system of mining officials who exercised both legal and technical supervision over individual mines on behalf of the princes. Regular inspections were part of the administrative process (Nef, 1987, pp.706, 744).

The question of the knowledge base underlying these developments arises. Only in the sixteenth century, coinciding with the rapid growth of mining and the spread of printing technology and humanist programmes of codifying knowledge, did a literature on mining and metallurgy appear. In bringing together the concerns of practical and learned men, this literature was an important strand in the development of modern science.

Development had been based on the evolving experience and craft skills of assayers, miners and metal workers. The early sixteenth-century literature shows how rich this tradition was. In particular, there was a long practical tradition of assaying which was of course fundamental to the mining of precious metals and enforced by legal standards in most of Europe by the early mediaeval period. Along with their practical techniques, assayers developed a notion of a pure substance as a substance which could be separated no further. At the same time, for obvious commercial reasons, their emphasis was quantitative (Halleux, 1986, p.281).

This practical definition of purity and quantitative approach of the assayers implied a contradiction to the prevailing learned view derived from Aristotle that the basis of the material world was a uniform prime matter manifested as four simple elements – earth, air, fire and water – which were defined by the mixture of qualities belonging to each.

All consisting ultimately of prime matter, the Aristotelian elements could change one into another by the interaction of their qualities. Furthermore, all substances consisted of a combination of the four elements; therefore, all substances could, in theory, be transformed one into another. On the one hand, as part of the classical heritage re-emphasized by the humanists of the Renaissance, this cosmology was important in learned circles. On the other hand, as developed by alchemists, this Aristotelian cosmology combined with '... an idea of the harmony and unity of the universe ... led to the belief that the universal spirit could somehow be pressed into service either through the stars or by concentrating it, so to speak, in a particular piece of matter – the philosopher's stone' (Holmyard, 1957, p.23).

The assayers' concept of a pure substance was not a theoretical definition of a simple substance, it was an operative convenience focussed on different

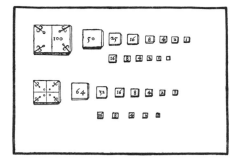

Figure 6.3 Assayers weights. (From Agricola, De re metallica, *Dover Publications, [1556] 1950.)*

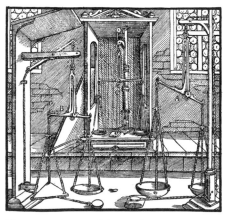

Figure 6.4 Assayers balances. (From Agricola, De re metallica, *Dover Publications, [1556] 1950.)*

The four primary qualities are the fluid (or moist), the dry, the hot, and the cold, and each element possesses two of them. Hot and cold, however, and fluid and dry, are contraries and cannot be coupled; hence the four possible combinations of them in pairs are: hot and dry, assigned to *fire*; hot and fluid, assigned to *air*; cold and fluid, assigned to *water*; cold and dry, assigned to *earth*.

In each element, one quality predominates over the other; in earth, dryness; in water, cold; in air, fluidity; and in fire, heat. None of the four elements is unchangeable; they may pass into one another through the medium of that quality which they possess in common; thus fire can become air through the medium of heat, air can become water through the medium of fluidity, and so on. Two elements taken together may become a third by removing one quality from each, subject to the limitation that this process must not leave two identical or contrary qualities; thus fire and water, by parting with the dry and cold qualities could give rise to earth. In all these changes it is only the 'form' that alters; the prime matter of which the elements are made never changes, however diverse and numerous the changes of form may be. *(Holmyard, 1957, pp.21–3)*

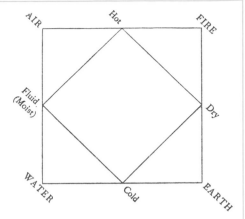

Figure 6.5 *Aristotle's four-element theory. (From E.J. Holmyard,* Alchemy, *Penguin, 1957.)*

objectives and therefore did not come into conflict with theoretical views. The assayers' aims were practical; their results were based on the experience of manipulating substances in controlled ways, rather than on theories. Their concerns were thus distinct from those of the explanatory aims of the humanists. At the same time, although they shared a practical approach, the assayers' aims were also quite distinct from the alchemists' programmes. Alchemy can be defined as:

> the art of liberating parts of the Cosmos from temporal existence and achieving perfection which, for metals is gold, and for man longevity; then immortality and finally redemption. Material perfection was sought through the action of a preparation (Philosopher's Stone for metals; Elixir of Life for humans), while spiritual ennoblement resulted from some sort of inner revelation or enlightenment (*Gnosis*, for example, in Hellenistic and western practices). *(Sheppard, 1986, pp.16–17)*

As such, the enterprise of the assayers was different from that of contemporary alchemists even though they used and developed certain tests and techniques in common (Halleux, 1986, p.291).

This is not to attribute a false 'modernity' to those on the practical side of mining, but merely to separate out distinct areas of activity. In fact, miners, for all their practical experience, commonly held animistic beliefs about the growth of metals and minerals in the womb of the earth, and associations between the fruits of the earth and the heavenly bodies were not uncommon. Often, such ideas were linked in with Christian traditions as well. Saint Anna, the patron of Saxon silver mines, was closely associated with the moon, long linked to silver in alchemical tradition, perhaps partly because of its colour. The 'silvery moon' is even now a resonant phrase for us. The divining rod also was still widely relied upon for locating profitable veins.

The mining and metallurgical literature which appeared in the sixteenth century did not mark a new phase of discovery, but it did coincide with an increase in scale, mechanization and organization of mining and metallurgical methods. These works were written in many styles and from a number of points of view; some were aimed at the growing officialdom. One style represented by the handbook known as the *Probierbüchlein* (Little book on assaying) was a set of recipes, probably long handed-down, which were collected together as an *aide-mémoire* (rather than as a manual for training). It

Figure 6.6 *Title page of* Probierbüchlein. *(From Agricola,* De re metallica, *Dover Publications, [1556] 1950.)*

appeared anonymously in several editions in the early years of the century (Agricola [1556] 1950, Hoover Appendix B, pp.612–13). Another style was that of Ulrich Rülein von Calw's description in dialogue form of mining geology and practice, *Ein nützlich Bergbüchlein* (A useful little book on mines) of *c*. 1500.

A physician to the wealthy mining town of Freiberg in Saxony, von Calw wrote his book in the vernacular German to encourage the regeneration of the town through reinvestment in the mines after a plague had destroyed four-fifths of the population (G.-R. Engewald, Freiberg Mining Academy, personal communication, April 1990).

While perhaps most striking to the eyes of the modern reader is the insight into sixteenth-century technological practices that these works provide, it is important not to lose sight of some of the authors' other aims, such as considering how prevailing theories, both classical and alchemical, related to experience derived from the mines. After all, in their concerns with the formation of minerals and metals, and the transmutation of base into noble metals, the interests of alchemists were, on one level, not unlike those of individuals wishing to understand how deposits were formed, the better to predict their location. Thus von Calw sought to systematize alchemical views with classical matter theory, arguing that the ultimate test of both would be the way they fitted with knowledge gained underground (Suhling, 1986).

Figure 6.7 Title page of Bergbüchlein. *(From Agricola,* De re metallica, *Dover Publications, [1556] 1950.)*

6.2 *Georgius Agricola*

One of the key works of the period, still recognized as a milestone in the history of technology, was the fluent Latin work *De re metallica libri XII* of Georgius Agricola (1494–1555), which appeared in 1556, shortly after his death. As a description of metallurgical practice, it was not surpassed until the middle of the eighteenth century (Smith and Forbes, 1957, p.28); Agricola's mineralogical scheme remained current for a similar period (Laudan, 1987, p.22), and the mining technology which the book described was still common for a century beyond that (Wagenbreth and Wächtler, 1983, p.52). Whilst perhaps foremost amongst the sixteenth-century technological works, *De re metallica* was also an exemplar of a classical humanist programme. Writing in Latin and devising neologisms where classical Latin lacked equivalents for the technical terminology of the German-speaking artisans whom he observed regularly, Agricola saw his work as following in the tradition of the Roman encyclopedists of the arts. Not the least important in guaranteeing its enduring influence were the numerous very accurate and detailed woodcuts with which the book was lavishly illustrated.

Figure 6.8 Georgius Agricola (1494–1555). (Published by permission of the British Library Board.)

Born George Pawr (later Bauer, which was Latinized to Agricola as was the custom) in a wealthy mining region of Saxony, his father is thought to have been a craftsman, a dyer and woollen draper. His initial education was in local church Latin schools. At the age of twenty, Agricola matriculated at the century-old university in the Saxon trading city of Leipzig to study philosophy and philology at a time when that university was becoming influenced by humanism with its emphasis on the rediscovery and examination of original Greek texts. Amongst his professors was a former student of the humanist scholar Erasmus (Engewald, 1982, p.19). Agricola was still at the university when the Reformation was launched in the nearby Saxon town of Wittenberg. After taking his first degree, he taught classical languages in schools in Zwickau near his home town for a few years where he participated in humanistic reforms of the curriculum and wrote Greek and Latin grammars for the use of his students. During this period, Zwickau was in turmoil as an active centre of the Reformation. Apparently, Agricola believed that some reform of the Catholic Church was necessary, but objected to the more revolutionary implications of the reformation movement (Wilsdorf, 1970, p.77). He remained a Catholic throughout his life.

At this time, a first degree was considered to be only a preliminary to the main goal of a university education, which was professional training leading to an advanced degree in theology, medicine or law. It was common for Bachelors of Arts to have a spell as schoolmasters before returning to university to complete their studies (see Chapter 14). Agricola returned to Leipzig in 1522 and, through friends, joined the circle of an eminent professor of medicine, another follower of Erasmus, and resolved to study medicine. During this period he met the Freiberg physician Ulrich Rülein von Calw, whose work he was later to follow up and to cite. To study in Italy, with its modern medical curriculum stressing actual practice in the healing arts in addition to a re-evaluation of the ancient medical authors, was the goal of many medical students. Agricola studied for three years in Bologna and Venice during which time he used his philological skills in helping to prepare what became the definitive editions of the Greek texts of the medical authors Galen and Hippocrates. It has been suggested that this intimate knowledge of the Greek authors introduced Agricola to the idea of using mineral substances as medicines, although common medical practice at the time concentrated on plant-based medicines. This may have motivated him to seek work in a mining area, the better to be able to examine mineral substances and the products of mining for their medicinal properties (Engewald, 1982, pp.49, 54f; Wilsdorf, 1970, p.77). After taking his MD, Agricola returned to Saxony and was elected in 1527 to the post of physician and apothecary to the town of Joachimsthal, on the southern edge of the Erzgebirge, just in Bohemia.

The Erzgebirge was at this time the most prolific metal mining district of Central Europe. Towns such as Schneeberg, Annaberg and Marienberg were approaching the peak of their silver production. Freiberg's production, on the other hand, would not peak for another hundred years (Nef, 1941, p.587). Joachimsthal was roughly fifty miles from Freiberg; most of the other mining towns mentioned by Agricola lay well within that radius. Joachimsthal itself, hardly more than a decade old when Agricola arrived, was a boom town with a population already numbering some fourteen thousand and the consequent social problems of such rapid development (Engewald, 1982, p.55). In the decade 1526 to 1535, the town was the most significant producer of silver in the area and was almost twice as productive of silver then as in any previous or subsequent decade (Nef, 1941, p.578). Thus Agricola was well placed to observe at first hand the full process of silver production, as well as the lives of the miners.

By his own account, Agricola devoted all of his time that was not required for his extensive official duties to this task and to researching possible ancient authorities on mining (Agricola, [1556] 1950, Hoover Introduction, p.vii; Engewald, 1982, pp.54, 57). He made a point of learning from all types of people involved in silver production. His aim was to bring together such descriptions as were to be found in the classics with the observations which he could make in the Erzgebirge. His first book on mining and metallurgy, *Bermannus sive de re metallica dialogus* was published in 1530. Usually referred to as the *Bermannus* to distinguish it from his later masterwork, *De re metallica*, the book was written in Latin, as were all of his major works, and in the well-understood classical form of dialogue which had already been used by von Calw and others for mining works. That it was in Latin indicates that it was intended for an international learned audience rather than one of practitioners. Publication was urged by a teacher friend who induced a highly placed mining official to send the manuscript both to Erasmus and to one of the most famous humanistic publishing houses of the day, Froben in Basel. Erasmus, who was already acquainted with Agricola by correspondence during his Italian publishing days (Engewald, 1982, p.59) and whom Agricola may have met during his Italian travels (Wilsdorf, 1970, p.77), wrote an enthusiastic letter of recommendation.

The book brought together a great deal of unsystematic, empirical knowledge from miners themselves. Agricola's aim, as developed in *Bermannus*, was to demystify and put on a more systematic basis the secret knowledge of mining and metallurgy, to uncover the laws of the crafts particularly in relation to knowledge about ore deposits (Engewald, 1982, pp.62–3). With Erasmus's introduction, *Bermannus* was very successful and established Agricola's reputation as an authority on mining.

The year 1531 found Agricola in the then quieter ambience of the smaller town of Chemnitz in Saxony (population *c.* 4,300) as town physician, looking forward to being able to continue his researches. It was here, over a period of some fifteen years, that Agricola undertook the systematic series of enquiries about the principles of geology, mineralogy, mining and metallurgy foreshadowed in *Bermannus*. Agricola furthered his reputation as a mining authority by making some very astute investments himself in new, prosperous mines (Wilsdorf, 1970, p.78) and by publishing in 1533 on the various distinct systems of weights and measures from antiquity onwards that made practical work so difficult to standardize in both pharmacy and coinage. By the 1540s he had become a very substantial citizen, one of the twelve richest men in the town (Wilsdorf, 1970, p.78), whose opinion was sought on a wide range of matters including political developments (Engewald, 1982, p.80).

The main fruits of his work in Chemnitz on geological and mineralogical topics were published together by Froben in a collection in 1546. Throughout, his approach was to examine the views of classical authors, testing them against contemporary views and against direct observations in the mining regions. Information and specimens were sent to him from around Europe (Engewald, 1982, p.83). In *De natura fossilium* (On the nature of minerals), he attempted a systematic mineralogy, establishing a scheme which was to be important for two hundred years. Minerals were very broadly defined as that which constituted the solid parts of the earth '... arranged in extended masses as rocks, veins and strata' (Laudan, 1987, p.21–2). He drew particularly on the work of Avicenna in the tenth century and Albertus Magnus's work in the thirteenth century, which sought to bring together Aristotelian and alchemical criteria for the identification of minerals by properties such as colour, smell, hardness, etc. The object of Agricola's work was to enable those minerals given different names in distinct traditions (classical, medical, alchemical, or mining) to be identified and compared. He specified some six hundred simple minerals, which he grouped into four main classes: metals, stones, earths and congealed juices (salts and sulphurs). There was a fifth class of compound minerals, which were mixtures of types from the other classes. Agricola followed Aristotle in attributing the properties of the minerals to their proportions of the four elements with associated qualities. Heat and water were the dominant agents of change.

In describing the individual minerals, Agricola included information about location and characteristics of deposits as well as their practical uses (Engewald, 1982, p.82). Agricola knew that there were traditionally thought to be six metals (gold, silver, iron, copper, tin and lead), however, he argued from practical considerations that mercury should also be considered a metal and he had previously added bismuth to the list. Contrary to the received wisdom based on Aristotle, he noted that, even in classical times, it had been suggested that many metals might be as yet unknown (Agricola, *De natura fossilium*, cited in Hoover in Agricola, [1556] 1950, p.2). Later he added antimony to his list of metals.

The other major work included in the 1546 collection was *De ortu et causis subterraneorum* (On the origin and causes of things underground). In this work, Agricola explicitly rejected the biblically based view that all deposits were formed at once as well as the views of mediaeval alchemists, who believed that metals were composed of varying proportions of quicksilver (mercury), sulphur and salt (Agricola [1556] 1950, Hoover notes, p.44). He objected to the alchemical linking of the planets with the metals, which argued that the metals were formed through the agency of a heavenly force on stones. Agricola also opposed the idea that metals contained the seeds of their own development rather like plants, another common mediaeval view (Adams, 1938, pp.84–90). Some alchemical writers from the fifteenth century elaborated a traditional miners' view that metals were transmuted from the baser ones into the more noble ones underground by an organic process of maturation within the womb of earth's crust (Adams, 1938, p.298; Halleux, 1974, p.215). The task of the alchemist in search of gold then was to reproduce and hasten this process artificially.

Agricola's own ideas on the formation of ore deposits were not based merely on a speculative scheme, but on testing various theoretical suggestions against detailed observations in the Erzgebirge (Agricola [1556] 1950, Hoover notes, p.52; Halleux, 1974, p.217; Adams, 1938, pp.93, 309). He required evidence. A century later, some 4,500 copies of Agricola's works in this collection were still in circulation.

By 1546 when this series of works was published, Agricola had begun to undertake certain important public duties. The territory and leadership of Saxony had been divided in 1485 between two branches of a family, the larger part becoming an electorate and the smaller a dukedom, subject to the former. The electorate, in which Wittenberg was located, nurtured the Reformation. On the Meissen side of the House of Saxony, the Duchy, in which Agricola worked, remained Catholic until the accession of the Protestant Maurice in 1541. Maurice joined the Schmalkaldic League of Protestant Princes. Anxious to become the elector himself, however, Maurice struck a deal with Charles V, abandoned the League of his coreligionists in 1546 and joined the Catholic side, taking the opportunity of their attack in that year to invade electoral Saxony. After the final defeat of the League at Mühlberg in 1547, Maurice was rewarded with the electorate. Maurice thus obtained his goal, but the Emperor Charles V, though victorious in battle, could not achieve his goal of imposing a single religion on the Empire as he could not gather sufficient support, even among the Catholic princes, for the establishment of a general imperial league (see Figure 6.1). Their sovereignty was seen by the princes as too great a price to pay for religious uniformity. Meanwhile, Maurice changed sides again, offering an alliance to Charles's French enemy while Charles and his Catholic Austrian brother, Ferdinand, made a final attempt on the French territories. Maurice died of battle wounds in 1553 and the religious issue was decided politically at the Diet of Augsburg in 1555.

It was Maurice who as Duke, and later Elector, of Saxony called most on Agricola's energies; Agricola took care to dedicate his works of the 1540s to Maurice. The Duke appointed Agricola Burgomaster of Chemnitz in 1546, when he was also made a member of the Diet of Freiberg and a councillor to the Court of Saxony at Dresden. As a Catholic, Agricola was sent on numerous diplomatic missions to liaise with his coreligionists, Charles V and King Ferdinand of Austria, when Maurice was allied with them. His previously published views in favour of an imperial solution to problems of the Holy Roman Empire would also have made him a suitable diplomat. Agricola carried on various public duties right up to the time of his death. He also continued his medical duties, finding the black plague years of 1552–3 particularly demanding as town physician. Some attribute Agricola's own death in 1556 to the severe physical and emotional pressures of this period, coupled with his demoralization over the Augsburg settlement, which brought the defeat of his personal hopes for a unified faith (Wilsdorf, 1970, p.78). Chemnitz itself had become aggressively reformed, so the final symbol of failure was that Catholic Agricola was refused interment in the parish church.

During this period of active public service, Agricola saw through the press a number of works. He was also preparing his masterwork, *De re metallica*, for which he had conceived the programme some twenty years before. The manuscript was completed in 1550, but supervising the preparation of 292 woodcuts meant that the book did not reach Froben until 1553 and did not appear until after Agricola's death. *De re metallica* covers mining, assaying and metallurgy. It brings together the ideas covered in his other works and follows their pattern of exploring the ideas of classical, mediaeval and contemporary authors in relation to his own observations and experiences gained in the mines. The basis of his work was: 'Those things which we see with our eyes and understand by means of our senses are more clearly to be demonstrated than if learned by means of reasoning' (Agricola [1556] 1950, quoted in Hoover Introduction, p.xiii). For Agricola, the woodcuts reinforced this observational basis.

> ... for with regard to the veins, tools, vessels, sluices, machines, and furnaces, I have not only described them, but have also hired illustrators to delineate their forms, lest descriptions which are conveyed by words should either not be understood by men of our own times, or should cause difficulty to posterity, in the same way as to us difficulty is often caused by many names which the Ancients (because such words were familiar to all of them) have handed down to us without any explanation. (*Agricola, [1556] 1950, p.xxx*)

It was while Agricola was working on such subjects that the Saxon mining civil service was initiated by the Elector Maurice in 1542 and consolidated by his successor, August. This would coordinate the next great phase of investment in mining, the building of water-powered energy systems for entire mining areas, rather than just individual mines. Agricola's achievement in *De re metallica* was not one of discovery or invention, but of bringing together effectively the state of knowledge both theoretical and practical about mining and metallurgy in the sixteenth century at a time when Central European mining output was at its peak and just beginning to be challenged by American mining. Agricola showed not only the highly sophisticated and differentiated technology which permitted that output, but also the conditions of work of those who produced it. His combining of a humanist programme of classical scholarship with actual observation and the knowledge of practical workers made him famous in his own day, and also gave his work lasting significance.

6.3 Paracelsus

It is difficult to imagine a figure more thoroughly contrasting with Agricola than his near contemporary, Theophrastus Bombast von Hohenheim, generally known as Paracelsus (1493/4–1541). Their lives have a number of common features. They both trained as physicians in Italy, were interested in the diseases of miners, remained non-Protestant in Protestant areas of post-Reformation Germany, favoured mineral remedies, promoted empiricism to some extent, travelled widely in Central Europe and indeed had personal links with Erasmus and the publishing house of Froben. Yet there is no evidence that they ever met, nor did they cite each other's work. Paracelsus's thinking derived from a quite distinct Renaissance tradition from that of Agricola: '... his ideas were moulded in an atmosphere of gnosticism, neo-Platonism, and cabbalism; he was a Hermetic medical magus' (Rocke, 1985, p.39).

Figure 6.9 Paracelsus (1493/4–1541). (Mansell Collection, London)

Paracelsus's personality was difficult:

> [He] was a provocative, extravagant, inflammatory figure (literally: he once made a public bonfire of the printed works of Avicenna). He was contemptuous of authority, a forceful champion of his own merits, a trouble-maker of the most energetic kind in words and writings ... [H]e was an acute observer of diseases, and enjoyed some notable professional success especially in the treatment of wounds and chronic sores. But it was the legacy of his writings, many of which became known only after his death, which made this strange man one of the most influential thinkers of renaissance Europe. (*Hall, 1983, p.80*)

It is widely agreed that, however influential his writings, they were and are difficult to understand.

> His neologisms have often been judged barbarous – certainly they are nonclassical – and Partington cannot be faulted for believing they reflected 'the intentions of puzzling and irritating the conventional physicians and making his writings impressive' [Partington, 1961, p.127]. His coarse language was often filled with invective, and he despised both scholastic and humanistic conventions. He probably knew no Greek, [he] lectured and wrote in [Swiss German] ... (*Rocke, 1985, p.39*)

Hall refers to him as '... by no means a lucid author or one whose message was easily grasped' (p.80). At the same time, his most authoritative biographer, Walter Pagel, suggests that even the details of Paracelsus's biography are shrouded in mystery.

> The large volume of extant books and papers on the life of Paracelsus is out of proportion to the scarcity of well-documented facts. Even the few data which seemed solid enough to be transmitted from book to book for centuries have recently been challenged for good reason. We are not even on safe ground when quoting the famous names: Phillippus Aureolus Theophrastus Paracelsus. The latter, by which he is commonly known, was a nickname given to him at a later period in his life. That he used it himself at any time cannot be demonstrated. (*Pagel, 1958, p.5*)

The name 'Paracelsus' is thought to be meant to indicate both 'superior to Celsus' (a Roman writer on medicine) and the paradoxical nature of his work. His father was a well-respected doctor who studied at Tübingen and practised in Villach in the mining province of Carinthia (south-eastern Austria) in the Holy Roman Empire from 1502 until his death in 1534. Paracelsus's early education was in the hands of his father and probably consisted of an exposure to mining, mineralogy, botany and natural philosophy. He also studied with a number of eminent churchmen, one of whom was a noted exponent of the occult and who was in touch with Agrippa von Nettesheim (1486–1535), the contemporary cabbalistic philosopher known for his association with the occult and defence of natural magic. Agrippa, although ultimately sceptical about claims of transmutation and the search for the philosopher's stone, saw the operational aspects of alchemy as fundamental to the understanding of nature (Müller-Jahncke, 1975, p.150). Another phase of Paracelsus's education consisted of practical work in the Fugger mines near Villach and Schwaz. It is likely that he studied medicine at a number of Italian universities, including Ferrara where Agricola took his medical degree, but the evidence that Paracelsus himself took a degree is vague. Pagel (1974, p.304) suggests that he may have taken what was then seen as a lower degree, qualifying himself as a surgeon rather than a physician. Certainly in his early adult years he was a military surgeon in the employ of the Venetian state.

Paracelsus's career and his character display a certain restlessness. He spent some ten years wandering about Europe in his capacity as a military surgeon and seldom settled anywhere for long. This was partly due to what is called *Wanderlust* in German, but also to the fact that he seems to have had a knack for coming into conflict with authority wherever he stopped. Indeed his early career showed the tension between an academic and an artisanal approach to medicine. Paracelsus found it difficult to accept the current distinctions between the practice of medicine and that of surgery. Physicians, being defined as scholars, were not meant to engage in activities that were purely manual; while surgeons, as skilled artisans, were thought of as curing on the basis of their skills. Paracelsus objected to this mediaeval guild distinction, arguing in a book of 1529 that theory and practice in medicine should be inextricably intertwined (Pagel, 1958, p.17). Such a challenge to contemporary professional codes courted hostility wherever he attempted to practice.

Early in 1527, through the influence of the publisher Johannes Froben, whom he cured, and Erasmus, to whom he gave medical advice, Paracelsus was made town physician at Basel by the Protestant town authorities. This entailed a responsibility to lecture at the university, which was still Catholic, although the professoriate was not consulted about the appointment. His position lasted only for a year. Paracelsus alienated the local academic medical community by challenging traditional Galenic medical teachings and threatening to introduce his own syllabus based on his observations as a naturalist and his experiences at the sickbeds of patients. He lectured in German, instead of the scholar's Latin, and in defiance of guild traditions, admitted barber-surgeons to his university lectures. The famous book-burning incident of what was a canonical text in academic medicine took place as a St John's day rag stunt during this period. Paracelsus was also at odds with the apothecaries of Basel on account of his distinctive views on the nature of medicines, over which they as a guild had formal control. By early 1528, he was on his way out of town for a further decade of wandering having aroused not only the hostility of his medical colleagues, but also fallen foul of a local magistrate who refused to pay for his medical services. It is thought that Paracelsus never really practised again except perhaps for a few years at the end of his life.

Most of Paracelsus's works were written during his second ten years of wandering in Central Europe. During this time, he seems to have had access to chemical furnaces and to have observed at first hand the miners' diseases of silicosis and tuberculosis, while continuing his studies of the occult. He cast himself as a man of the people and claimed to have picked up on his travels a great deal of knowledge from peasants, barbers, chemists, old women, quacks and magicians (Partington, 1961, p.117). An amanuensis of his once reported a debauched life-style characterized by drinking bouts interspersed with periods of lucid dictation. His reputation was retrieved somewhat when he had an audience with Ferdinand of Austria in about 1537 (Pagel, 1974, p.305). Paracelsus's grave, however, was in the grounds of an almshouse; it became a place of pilgrimage for the poor and sick.

Paracelsus was a self-consciously pugnacious reformer of both medicine and alchemy. One of his most enduring achievements was in bringing to medicine a new chemical orientation which was reflected in the pharmacopoeias published from the seventeenth century onwards (Pagel, 1958, p.347). Indeed, he referred to himself as an 'iatrochemist', one who knows both chemistry and medicine. He seems to have been the first sixteenth-century writer to have used this name (Partington, p.135). Undeniably, Paracelsus drew heavily from the work of his predecessors and his influence is due primarily to the publishing labours of subsequent 'Paracelsians', rather than to his own efforts. Yet, for all that and for all his complexities, he himself stood for a certain spirit of an age in transition, not yet modern, but in the process of reorientation towards the natural world which would later be characterized as the Scientific Revolution.

Unlike the humanist programme in which Agricola was engaged, Paracelsus himself repudiated explicitly the teaching of the ancients, preferring to turn to nature alone as his source of knowledge. At the same time, and in a typically contradictory way, the humanistic revival not of Aristotle, but of '... the Hellenistic blending of Jewish, Christian, Greek and Oriental ideas ... as expressed in Neo-Platonism, Gnosticism and Kabbala, Alchemy, Astrology and [natural] Magic' was central to his work (Pagel, 1958, p.39). He also shared the humanist dislike of mediaeval scholasticism and the hierarchical mediaeval world-view (see Chapter 2), promoting instead an anthropocentric world in which individuality was important. This is perhaps clearest in his doctrine of correspondence between microcosm and macrocosm, that is a system of analogies between the workings of nature in the human body and in the universe as a whole. For Paracelsus, nature is a visible reflection of the invisible work of God.

Paracelsus thought that man was the centre of creation and a small replica of the world around him, uniting in himself all powers of the constituents of that surrounding world – minerals, plants, animals, celestial bodies. Knowledge existed within things as their essence or virtue. The planets, the organs of the human body, minerals and plants were all linked by their shared virtues, and it was thus fruitful to seek out sympathies between them. Because of this unity, individuals could intuit via an experience, a sort of trance or a dream, rather than relying on reason operating on objects viewed as external to them. Pagel argues that knowledge gained in this way through the mutual sympathy between individuals and natural objects rather than through book-learning and logical deduction marks a move toward scepticism and empiricism which were to be so important in the development of modern science (1958, p.51). In other words, it was the microcosmic nature of man that made the study of nature possible. The person who gains this 'transcendental' knowledge of all things is nothing less than the Magus. This is how Paracelsus saw himself.

In relation to medicine, Paracelsus's cosmology led to a rejection of the classical humoural conception of disease. He replaced it with the view that specific diseases have specific causes. In Aristotelian medicine as transmitted by Galen and later by Avicenna (the object of Paracelsus's notorious bonfire), the body of a healthy individual had a balance of four humours, each of which was related to one of the four Aristotelian elements and its associated qualities. Indeed, the individual's nature, or personality, was determined by this balance. Disease occurred when the balance was upset; appropriate treatment consisted of whatever was necessary to restore the balance to normal. However, the correct balance for any person was particular to that individual. Diseases therefore were specific to individuals rather than isolatable separate entities with specific causes and specific effects. By contrast, Paracelsus sought an external cause for each disease and was particularly concerned with the specific location of diseases within the body. These external causes, which he identified as coming directly from all parts of the macrocosm (terrestrial and celestial), could be linked to specific parts of the body. Each external (or foreign) cause was an entity in its own right and the disease caused by it also had a specific existence, threatening the body perhaps to the point of death. In terms of his cosmology of macrocosm and microcosm, just as metallic seeds grew in the womb of the earth and developed into metallic veins, so the seeds of disease developed within specific organs in the body.

This new concept of disease implied as well a new form of treatment. The agents of specific diseases should be countered by specific treatments. One aspect of this was his 'doctrine of signatures'. He did not completely reject herbal remedies, but argued that plants which resembled an affected organ could be the source of efficacious remedies for it.

> And then we have the thistle: do not its leaves prick like needles? Thanks to this sign, the art of magic discovered that there is no better herb against internal prickling ... And the *Syderica* bears the image and form of a snake on each of its leaves, and thus according to magic, it gives protection against any kind of poisoning. *(quoted in Hall, 1983, p.86)*

Most importantly, in relation to plant remedies, Paracelsus reversed the Galenic practice of compounding medicines, that is of mixing several plants, but rather he sought to isolate the specific active agent of each substance, its *quinta essentia*, or quintessence. A really radical therapeutic departure, however, was his introduction of the use of mineral substances as internal remedies. Mercury had already been introduced as a treatment for syphilis, reportedly a new disease of the time, but in very heavy doses. Paracelsus counselled moderate doses on homeopathic principles rooted in German folk medicine of 'like cures like', not unrelated to the doctrine of signatures. For Paracelsus, the laboratory was the true home of the doctor.

> In the meantime, I extol and adorn, with the eulogium rightly due to them, the Spagyric physicans ... These do not kill the time with empty talk, but find their delight in their laboratory ... Like blacksmiths and coal merchants, they are sooty and dirty, and do not look proudly with sleek countenance. In presence of the sick they do not chatter and vaunt their own medicines. They perceive that the work should glorify the workman, not the workman the work, and that fine words go a very little way towards curing sick folks. Passing by all these vanities, therefore, they rejoice to be occupied at the fire and to learn the steps of alchemical knowledge. *(quoted in Leicester and Klickstein, 1965, p.18)*

Interestingly, he defined 'the steps of alchemical knowledge' in practical terms, like the processes of distillation, extraction, separation and reduction. For Paracelsus, alchemy was the technique of making medicines, rather than the illusive quest for the philosopher's stone or transmutation. He himself devised a number of practical techniques including the process of concentrating alcohol by freezing and invented a number of preparations, including some ethers and possibly tartar emetic (Pagel, 1974, p.308). His practical efforts to isolate the active ingredients in substances were underpinned by his theoretical beliefs.

The 'essences', which the alchemist/magus aimed to extract from material substances such as plants and minerals, were invisible spirits, the very life substances of objects. They were seeds which emanated from God and existed preformed in a nonmaterial prime 'matter' of water, earth, air and fire. These are not the Aristotelian four elements, but rather '... wombs that give birth to groups of objects, each specific to its source; thus, minerals and metals are the "fruit" of water, and plants and animals including man – the "fruit" of earth' (Pagel, 1974, p.309). Paracelsus did not reject the four-element theory of Aristotle, but superimposed a three-element system of a different nature upon it, the three principles (*tria prima*) of salt, sulphur and mercury. For Paracelsus, these were not elements in any material sense, but rather principles within matter that determine how it develops. Sulphur and mercury had long been considered by certain alchemists to be the basic constituents of all metals, and the addition of salt had been thought to affect the degree of nobility or baseness of any particular metal.

Paracelsus built his three principles into his wider cosmological system by identifying salt, sulphur and mercury with body, soul and spirit respectively. All substances have the three in some proportion, salt being responsible for solidity, sulphur for inflammability and mercury for fluidity. However, the nature of the three principles varies from substance to substance. They are not then elemental constants. Nature produces individual substances out of them by means of the preformed 'seeds'. Thus the alchemical doctrine of the 'seeds' of metals is extended to the whole of nature (Pagel, 1958, p.103). Besides giving alchemists a new, respectable agenda of seeking new medicines by means of the practical task of isolating the active ingredient in substances, Paracelsus's system, for all its complexity and obscurity, was

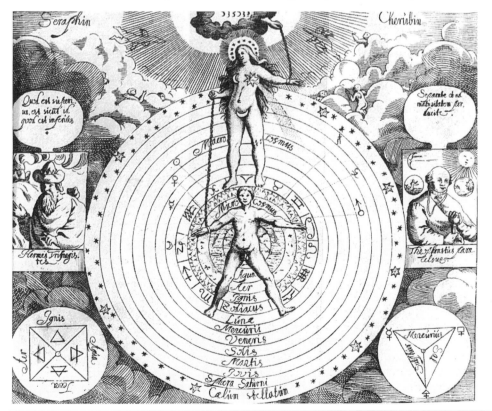

Figure 6.10 Man is seen here as a microcosm in the middle of an Earth-centred universe. He is linked to his creator via the chains of nature, the macrocosm, represented here as a young woman. Note how man occupies the region below the sphere of the Moon (Luna), which is where the four Aristotelian elements (earth, shown as the planet; water, acqua; air, aer; and fire, ignis) also exist. To the left is a portrait of Hermes Trismegistus and a diagram of the four-element system. To the right is a portrait of Paracelsus and a diagram of the system of three principles. (From Tobias Schütz, 1654. By permission of the University of Chicago Library, Department of Special Collections.)

important for the subsequent development of chemistry in stressing the very concept of the activity of substances. But then, a whole new range of active substances which had not been available to the ancients, such as soap, gunpowder, distilled alcohol and some acids, were the very stuff of sixteenth-century 'chemical' experience (Hall, 1983, p.83).

The problem then remains of what to make of Paracelsus and the contradictions in his life and thought. It is difficult to do other than recapitulate the assessment of Walter Pagel. On the one hand, Paracelsus was clearly skilled in the practical aspects of the alchemy of his day and used them to good effect in the laboratory; he established an alternative medical programme in asserting a chemical basis for the processes of life and disease; and he attempted some sort of system of what we would now call inorganic chemistry. But 'however much inspiration and actual addition to chemical knowledge may be due to him, Paracelsus was neither a scientist nor a chemist in the modern sense' (Pagel, 1958, p.344). His chemistry was only part of a much wider philosophy far removed from anything which might be termed 'scientific'. Similarly in medicine he had various specific, enduring achievements such as the identification of certain occupational diseases of miners. Most notably, his theory of disease, with its emphasis on specific, external causes, challenged in what we might wish to term a 'modern' way the humoural systems of the ancients. At the same time, however, his medical system as a whole was based on his theory of microcosm/macrocosm analogies: 'In this, observation and protoscientific elements are widely overgrown by a farrago of speculations which strike us as fantastic' (Pagel, 1958, p.345).

In the end, what Paracelsus did accomplish by attacking the contents, methods and privileges of academic medicine was to stir things up, and to initiate a fruitful cross-fertilization of ideas and practices derived from a wide spectrum of artisans and healing practitioners. It is possible to find mediaeval sources for most of Paracelsus's ideas and for the practical techniques he advocated in the laboratory. His achievement, however, was to bring together various strands which had been outside the scholastic mainstream and to apply them to medicine. He drew on what had survived from far older traditions of neo-Platonism, gnosticism and cabbalism which, together with more than a little folklore, he fused into a unique synthesis of medicine, alchemy, chemistry, religion and cosmology.

Perhaps the most significant fact about Paracelsus's accomplishment is that he spawned an avid following. During the second half of the sixteenth century (most of his works were published late in the 1580s) and throughout the seventeenth, the new chemical medicine was an active enterprise. Furthermore, his world-view had wide influence in Central Europe.

6.4 The emergence of chemistry

It was from the late sixteenth century onwards that a separate chemical science began to emerge. The metallurgical tradition, particularly assaying, and the Paracelsian iatrochemical tradition both contributed to the differentiation of specifically chemical techniques from more general technical practices (Hall, 1983, pp.272–3). During the seventeenth century, a vigorous pedagogic tradition emerged in the field which helped to define the subject of chemistry as a discipline (Hannaway, 1975, p.ix). At the same time, the chemical world-view of the Paracelsians was an alternative to that of the mathematically based mechanical philosophy in the Scientific Revolution. It was believed at this time that there was a chemical key to the understanding of nature as well as a mechanical one (Debus, 1978, chs 2 and 8).

One strand of development can be seen in the work of Lazarus Ercker (1530–94), who was born in a Saxon mining town and educated at the University of Wittenberg. Like Agricola, he was a respected public figure. His early career in the 1550s was as an assayer within the Saxon Mining Civil Service under the patronage of the Elector August. After a spell as Master of the Mint at Goslar in the Harz mining district, Ercker settled in Prague where he received the patronage of the Habsburg Bohemian Court, first of Maximillian II, and later of Rudolf II, who was also well known as a patron of alchemists. He became chief inspector of mines and was knighted for his technical services. Thus he had a broad experience of assaying and metallurgy, which he recorded in a number of practical books. The most substantial of these, published in 1574 in the vernacular German, was his *Beschreibung allerfürnemisten mineralischen Ertzt* (Treatise on ores and assaying). Drawing in part from Agricola's *De re metallica*, it is a comprehensive review of known assaying and refining techniques, but it also includes methods of obtaining other substances needed for these processes (such as acids, salts and saltpetre) and introduces a number of new methods. Ercker was hostile to alchemical ideas and only included in his *Treatise* processes that he had actually tested (Hubicki, 1971, p.394).

The practical aspects of his work were very influential, and it was often borrowed from wholesale by subsequent authors. Ercker's work was certainly relied upon to some extent by Andreas Libavius (1560–1616). Libavius too was born in Saxony. He studied at the universities of Wittenberg and Jena, and ultimately took a medical degree at a centre of humanism, the University of Basel. At Basel, he came under strong anti-Paracelsian influence and his career, principally as founding rector of a school at Coburg, was marked by this outlook. As an orthodox Lutheran institution, the school failed to receive an imperial charter to become a university from the pro-Paracelsian Rudolf II (Hubicki, 1973, p.309). In 1597, Libavius published what has long been recognized as a key book in the development of a chemical discipline, his *Alchemia*, effectively the first chemical text-book. The definitive edition appeared in 1606. In it, he brings together the techniques and preparations of all the chemical arts and crafts and attempts to present chemistry as a subject for independent study.

> This work is the first text which conceives of chemistry as an independent and integral discipline divorced from its applications and which seeks to organize the techniques and prescriptions of the subject in such a way that they can be taught. As such, it is much more than a late sixteenth-century encyclopedia of the operations and recipes of the various chemical arts. It marks the appearance of a new scholarly discipline on the horizon – the discipline of chemistry. (*Hannaway, 1975, pp.142–3*)

Libavius's *Alchemia* was written in a scholarly Latin and clearly not intended itself to be a practical treatise, although it drew on wide-ranging practical traditions. His method was to organize the subject according to a series of definitions with subdivisions, elaborating with examples proceeding from the general to the specific.

His primary divisions were techniques or operations on the one hand and the preparations made by them on the other. Known operations and preparations were thus grouped together, rather than treated separately according to their practical origins in, for example, pharmacy or metallurgy.

Libavius's main division is between *Encheria* (techniques and operations) and *Chymia* (the chemical species prepared by these operations); these two principal divisions are treated in separate books. Thus chemical species are defined in relation to the operations which produced them rather than in terms of their inherent chemical composition. Focussing on the first part of the second book, the principal dichotomy is between *simplex* and *composita*. The former is in turn broken down into the fundamental divisions of *extracta* (those species which derive from some sort of extraction process such as distillation, precipitation etc.) and *magisteries* (those species which are perfected without any separation of an inner essence from the starting material). The end point of the subdivision of the *magisteries* is itself dichotomized – the (Aristotelian) elements and the (Paracelsian) principles. (Hannaway, 1975, pp.146–50)

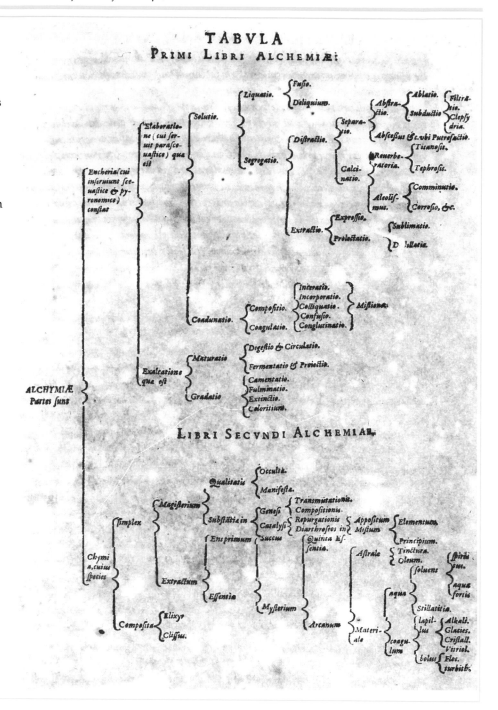

Figure 6.11 Libavius's diagrammatic representation of chemistry. (From Alchymia, *1606. Published by permission of the British Library Board.)*

This was an important innovation, as Libavius tried to clarify in words what different substances were, quite independently of the different names attaching to them from their diverse practical origins. Thus where a substance or operation might previously have been known by one name to metallurgists and another to dyers, Libavius identified it as the same substance/operation and gave it a common name. It is not that Libavius wished to discount the importance of the practical traditions; indeed, he was an exponent of the current view that chemistry should be applied to medicine. In order to be applied, it had, however, first to be understood; and in order to be understood, it had first to be organized in a manner that would make it teachable. This has been identified as an essentially Ramist programme (see Chapter 3) rooted in Lutheran humanism. It avoided first principles based on authority in order to seek clarity through the systematic differentiation of subject-matter based on the collective experience of its practitioners (Hannaway, 1975, ch.6). Following Libavius, a flourishing chemical text-book tradition emerged. Whatever theoretical principles subsequent authors chose to espouse, Libavius's organization of the subject, as a method of teaching and defining it, remained influential until well into the eighteenth century.

Ercker and Libavius may be taken to represent an influential non- or even anti-Paracelsian trend in the German states of the late-sixteenth and early-seventeenth centuries. However, Paracelsianism remained tremendously important there throughout the period, as well as in France and England. The career and ideas of Oswald Croll (1560–1609) emphasize this. Right at the end of his life, Croll also bought together his ideas in what would be an influential text-book, the *Basilica chymica*, which was published in 1609. Croll was educated at the University of Marburg and subsequently at Heidelberg, Strassburg and Geneva, probably taking an MD in 1582. He eventually practised as a physician in the Holy Roman Empire, settling in Prague from about 1597. Croll was in the employ of Christian I of Anhalt-Bernburg, a Calvinist, who was a leader in the cause of the Protestant Princes against the Catholic Holy Roman Emperor Rudolf II, whose court was in Prague. Croll's Paracelsianism gave him access to the court, which was sympathetic to such ideas, so Christian I used him as an informal emissary in political negotiations (perhaps even as a spy). In return, Croll received financial support for his iatrochemical work (Schröder, 1971, p.471).

The *Basilica chymica*, which received the imperial imprimatur and went through many editions, was probably one of the main sources through which a knowledge of Paracelsian medicines and teachings were disseminated (Partington, 1961, p.175). It is known for its comprehensive exposition of the role of chemical techniques and preparations in pharmacy and especially for its (un-Paracelsian) descriptive clarity. Not only did Croll introduce a number of new mineral remedies, but he has been credited with influencing the development in a 'chemical' direction of the German pharmacopoeias of the seventeenth century, and hence a major reform of pharmacy. For Croll, the setting of this work was his enthusiastic and mystical religious and ethical interpretation of Paracelsus's ideas. Croll believed that:

> The true Physick whereof by the Devine assistance I intend here to treat, is the mear gift of the most high God; it is not to be sought for or learned from the Heathens, but from God alone ... (*quoted in Hannaway, 1975, p.11*)

Clearly anti-Aristotelian, Croll emphasized a neo-Platonic, Christian hermetic philosophy and also found inspiration in a christianized version of the magical number philosophy of the Cabala. It has been argued that Croll's particular synthesis was to bring together Paracelsian thinking with Calvinistic ideas (Hannaway, 1975, ch. 2). Because of his position in the centre of the universe in relation to the microcosm–macrocosm analogy, man (indeed the physician) could gain understanding.

> Illumination led not only to a purely passive knowledge of God's spiritual designs for mankind in Scriptures and a pious appreciation of His wisdom through the study of nature; but through this knowledge man could acquire spiritual and natural powers ... Man illuminated by the light of nature could gain knowledge and control of the natural forces with which God had endowed His universe; man illuminated by the light of grace could attain to mystical experience and exercise supernatural powers. (*Hannaway, 1975, p.11*)

Paracelsians in a similar way argued, that there were two divine books given by God to man, namely that of Revelation (the Scriptures) and that of Creation (the Book of Nature). Thus they were part of the tradition concerned both with biblical exegesis and with a fresh observational and experiential approach to nature. The Book of Nature was one written in secret signs which had to be interpreted through sympathetic understanding. At one level, the medical man had to be a practitioner in the sense of a practical man whose effectiveness is proved by the cures he works. At another level, the Paracelsians valued practical craft knowledge as knowledge which had been 'experienced'. Chemistry as the practical science of the transformation of material substance was to provide the key for a new understanding of nature as a whole. Indeed the biblical account of the Creation was interpreted as a chemical process and the earth was seen as a vast chemical laboratory. Such well-known phenomena as volcanoes, hot mineral springs, the 'growth' of metals underground and the beneficial effects of manuring were all explained chemically during the early years of the seventeenth century. Paracelsus's three principles, although they did not directly replace the Aristotelian four elements, influenced discussions of element theory in the seventeenth century. Some iatrochemists were attracted to the *tria prima* by the trinitarian analogy of body, soul and spirit; others simply used the three elements in describing the products of their practical operations such as distillation (Debus, 1978, p.25).

The influence of Paracelsianism on official medicine – that taught and practised by university-trained physicians and that controlled by legally incorporated guilds – varied from place to place. Nearly everywhere the introduction of chemically prepared medicines provoked controversy. In France, there was a particularly divisive debate between the Galenists, who upheld official therapy, and the iatrochemists, who argued for the introduction of chemically prepared medicines of a Paracelsian kind. In many places official medicine took a pragmatic approach, extracting what it found useful from Paracelsian prescription while rejecting the underlying theory and ideology. Thus the London Royal College of Physicians incorporated some chemical medicines in its great pharmacopoeia of 1618. In several of the newly founded Protestant universities in the German-speaking territories, various strands of Paracelsianism found their way into the academic curriculum. The fact that text-books existed like those of Croll and Libavius, which had systematized the operations and preparations of the new chemical medicines, led to formal instruction in this aspect of Paracelsian practise. Thus in 1609, a professorship of *chemiatria* (or iatrochemistry) was founded at the University of Marburg to teach the preparation of chemical medicaments. Its practical instruction was based largely on the *Basilica chymica* of Oswald Croll.

6.5 Science at the court in Central Europe

The establishment of the Marburg chair highlights another aspect of the development of science in Central Europe in this period, the important role of the princely courts. Especially in the Protestant territories of Germany, the sphere of the court was not entirely separate from the sphere of the university. Some Protestant princes who were anxious to ensure the confessional uniformity of their territories according to the authority gained from the Augsburg settlement of 1555 took a personal interest in the appointments at their territorial universities and supervised the teaching given there to the future schoolmasters, pastors, lawyers and physicians of their lands (see Chapter 14). Thus the Marburg chair of *chemiatria* was established at the direct command of Prince Moritz of Hessen-Kassel (*r.* 1592–1632), who just at this time was imposing Calvinism on his formerly Lutheran subjects. Moritz was both a Calvinist and a Paracelsian, interpreting the religious doctrine of the elect as consistent with the Paracelsian idea of a privileged access to knowledge given to the alchemical adept, possibly influenced by Croll's version of Paracelsian doctrine as published in the introduction to the *Basilica chymica*. He amassed a large 'chemical' library, performed experiments to develop new medicines (which had to be tried by unsuspecting courtiers) and was at the centre of a circle of Paracelsians brought to his court. His creation of the chair of *chemiatria* should not be seen as an effort to be 'modern'. It was just part of his larger effort to impose a unified world-view and theology on his territory. He appointed his own personal physician, Joannes Hartmann, to the post (Hannaway, 1975, pp.50–7; Moran, 1985).

Foremost amongst the court centres in Central Europe was the Catholic Habsburg court in Prague of the Holy Roman Emperor Rudolf II (*r.* 1576–1611). Although they represent quite different strands in the development of 'chemical' practice, Lazarus Ercker and Oswald Croll had in common their involvement with the Rudolfine court. They were both admitted for their expertise, one in metallurgy and the other in medicine. This chemical expertise fitted particularly well with the intellectual climate of Rudolf's court with its strong interest in all things scientific, especially those involving alchemy and the occult. It was there, some fifty years after the lonely death of the outcast Paracelsus, that his followers brought his teachings to the very centre of courtly life in the empire.

> ... the Holy Roman Emperors maintained at their court some of the most notable figures in the intellectual life of the age. The Habsburgs were a cosmopolitan focus for the wide-ranging interests of the time, for collectors, historians, antiquarians, natural philosophers, students of minerals, plants, stones, and the rest; and the endeavours thus focused on the court mirrored the schemes of unity in diversity which so engaged contemporary thinkers. The culmination of the process came, like the seeds of its collapse, with Rudolf II. (*Evans, 1973, pp.116–17*)

Following family tradition, Rudolf's father, Maximilian II, had brought together a group of scholars with humanist leanings and international reputations. Natural history, especially botany and plant collecting, was a particular interest. (It was one of Maximilian's entourage who, during diplomatic travels, introduced the tulip to the West from Turkey.) Maximilian's court historiographer, whose job was to extol the glories of the Habsburg dynasty, subsequently held a similar position under Rudolf. He amassed a vast library which encompassed a wide range of scientific and occult books, including Copernicus's and Paracelsus's works.

Rudolf II was a curious personality who was eccentric at the very least and said to have had occasional periods of madness. It was he who moved the imperial court from Vienna to Prague shortly after beginning his reign, partly out of consideration to his other title as King of Bohemia and partly to remove the imperial centre further from the reach of the ever-threatening Turks. It was a very cosmopolitan court both in terms of the flow of ideas and the flow of people. Rudolf reigned at a time when political conflicts between confessional interests were coming to a head. As Holy Roman Emperor, he was formally aligned with the Catholic cause, but the

Figure 6.12 Rudolf II as the gardener's god, Vertumnus, by Arcimboldo. Arcimboldo was Rudolf's Master of Ceremonies, and Rudolf was apparently well-pleased with this painting, imbued with correspondences and signs, which is an extraordinary commentary on the world-view of his court (Evans, 1973, pp.173–4). This representation of a Holy Roman Emperor could scarcely be more contrasted to that in Figure 6.1. (Skoklosters slott Balsta, Sweden)

confessional allegiances of the various territories of his empire were very varied. During his reign too there was a marked resurgence of Catholicism as a result of the Counter-Reformation.

Rudolf himself remained something of a recluse in the midst of his intellectually lustrous court. His interest in the occult has been seen as characteristic of an age anxious to penetrate beyond the realm of everyday experience to an underlying spiritual reality. Paracelsianism was of course part of this orientation as natural philosophers '... studied the world around them, not as discretely observed patterns of cause and effect, but as motive spirits acting through a divine scheme of correspondences' (Evans, 1973, p.197). As we have seen, such knowledge was meant to provide not only understanding but the power of control and the ability to effect applications as well. It is not perhaps difficult to understand the appeal of such a world-view to an emperor whose empire was conceived as a unified spiritual entity but who confronted daily the political reality of an ever-fragmenting number of states increasingly in conflict with each other.

Alchemy, and in particular the search for the philosopher's stone, was a central preoccupation of the closing years of the sixteenth century in the Germanic lands. Prague was a particular centre of alchemical activity. The alchemists were concerned not only with the question of the regeneration of metals, but also with the regeneration of the spirit. Rudolf's most direct contact with alchemists was through his court physicians, who were seen by contemporaries as an elite. Perhaps the best known of these is Michael Maier (1568–1622) who, after his period in Prague, published a well-known series of symbolic works on alchemical topics that remain one of the best expressions of that age's yearning for spiritual renewal through the learned study of ancient myths and the chemical transformation of the base metals.

Figure 6.13 Michael Maier, Atlanta Fugiens, *1618. Emblem 8 illustrates the separative function of the sword. The alchemist wields the (fiery) sword above the Philosopher's Egg and he is instructed 'to attack it cautiously (as is the custom), with the fiery sword; let Mars (iron) lend his aid to Vulcan (fire), thence the chick arising from it will be the conqueror of iron and fire' (Sheppard, 1986, p.16). (Published by permission of the British Library Board.)*

And it was not only alchemists who were attached to the Rudolfine court. The Danish astronomer Tycho Brahe became imperial mathematician in 1599, having been in correspondence for some years with the astrologically minded principal court physician there. It is significant and revealing of the age, that Brahe, too, prepared chemical remedies, and in an underground laboratory had practised alchemy. This he called 'terrestrial astronomy' in recognition of the correspondence he believed existed between the stars in the heavens and the metals and minerals under the earth (Hannaway, 1986). The presence of Brahe in Prague drew Johannes Kepler there after he had been ejected from the Austrian town of Graz by the forces of the Counter-Reformation. And Kepler succeeded Brahe as imperial mathematician in 1600. Kepler's mystical inspiration resonated with the atmosphere of the Rudolfine court and his achievements over the next twelve years were the high point of courtly science in Prague (see Chapter 3).

Kepler's last and, for him, most significant work, *The Harmonies of the World*, was published in 1619, one year after the outbreak of the Thirty Years War (1618–48), a most unharmonious period in Central Europe, which would alter forever the vision of political and spiritual unity of the Holy Roman Empire. Rudolf's reign proved to be the culmination of an era in Central Europe which witnessed the final throes of a cosmpolitan humanist culture in the service of a universalist political ideal. It saw, however, the first shoots of a new scientific vision emerge from that fertile soil.

Sources referred to in the text

Adams, F.D. (1938) *The Birth and Development of the Geological Sciences*, Baltimore, Williams and Wilkins.

Agricola, G. ([1556] 1950) *De re metallica*, tr. with Introduction and Epilogue by Herbert Clark Hoover and Lou Henry Hoover, New York, Dover Publications.

Debus, A.G. (1978) *Man and Nature in the Renaissance* (Cambridge History of Science Series), Cambridge, Cambridge University Press.

Engewald, G.-R. (1982) *Georgius Agricola* (Biographien hervorragender Naturwissenschaftler, Techniker und Mediziner), vol.61, Leipzig, BSB B.G. Teubner Verlagsgesellschaft.

Evans, R.J.W. (1973) *Rudolf II and His World: A Study in Intellectual History, 1576–1612*, Oxford, Clarendon Press.

Hall, A.R. (1983) *The Revolution in Science, 1500–1750*, London, Longman.

Halleux, R. (1974) 'La nature et la formation des métaux selon Agricola et ses contemporains', *Revue d'histoire des sciences appliqués*, 27, pp.211–22.

Halleux, R. (1986) 'L'alchimiste et l'essayeur', in Christoph Meinel (ed.), *Die Alchemie in der europäischen Kultur- und Wissenschaftsgeschichte*, Wiesbaden, Otto Harrassowitz.

Hannaway, O. (1975) *The Chemists and the Word: The Didactic Origins of Chemistry*, Baltimore and London, Johns Hopkins University Press.

Hannaway, O. (1986) 'Laboratory design and the aim of science: Andreas Libavius versus Tycho Brahe', *Isis*, 77, pp.585–610.

Holmyard, E.J. (1957) *Alchemy*, London, Penguin (1968 reprint).

Hubicki, W. (1971) 'Ercker', in C.C. Gillespie (ed.), *Dictionary of Scientific Biography*, vol. 4, New York, Scribner.

Hubicki, W. (1973) 'Libavius', in C.C. Gillespie (ed.), *Dictionary of Scientific Biography*, vol. 8, New York, Scribner.

Laudan, R. (1987) *From Mineralogy to Geology: The Foundations of a Science, 1650–1830*, Chicago, University of Chicago Press.

Leicester, H.M. and Klickstein, H.M. (1965) *A Source Book in Chemistry, 1400–1900*, Cambridge, Mass., Harvard University Press.

Moran, B.T. (1985) 'Privilege, communication, and chemistry: The hermetic alchemical circle of Moritz of Hesse-Kassel', *Ambix*, 32, pp.110–26.

Müller-Jahncke, W.D. (1975) 'The attitude of Agrippe von Nettesheim (1486–1535) towards alchemy', *Ambix*, 22, pp.134–50.

Nef, J.U. (1941) 'Silver production in Central Europe', *J. Political Economy*, 49, pp.575–91.

Nef, J.U. (1987) 'Mining and metallurgy in mediaeval civilization', in M.M. Postan and Edward Miller (eds), *The Cambridge Economic History of Europe*, vol. II *Trade and Industry in the Middle Ages*, 2nd edn, Cambridge, Cambridge University Press.

Pagel, W. (1958) *Paracelsus: An Introduction to Philosophical Medicine in the Era of the Renaissance*, Basel and New York, Karger.

Pagel, W. (1974) 'Paracelsus', in C.C. Gillespie (ed.), *Dictionary of Scientific Biography*, vol. 10, New York, Scribner.

Partington, J.R. (1961) *A History of Chemistry*, vol. 2, London, Macmillan.

Rocke, A.J. (1985) 'Agricola, Paracelsus, and "Chymia"', *Ambix*, 32, pp.38–45.

Rosen, G. (1943) *The History of Miners' Diseases: A Medical and Social Interpretation*, New York, Schuman.

Schröder, G. (1971) 'Oswald Crollius', in C.C. Gillespie (ed.), *Dictionary of Scientific Biography*, vol. 3, New York, Scribner.

Sheppard, H.J. (1986) 'European alchemy in the context of a universal definition', in Christoph Meinel (ed.), *Die Alchemie in der europäischen Kultur- und Wissenschaftsgeschichte*, Wiesbaden, Otto Harrassowitz.

Smith, C.S. and Forbes, R.J. (1957) 'Metallurgy and assaying', in C. Singer, *et al.* (eds), *A History of Technology*, vol. III *From the Renaissance to the Industrial Revolution, c. 1500–c. 1750*, Oxford, Clarendon Press.

Suhling, L. (1986) '"Philosophisches" in der frühneuzeitlichen Berg- und Hüttenkunde: Metallogenese und Transmutation aus der Sicht montanistichen Erfahrungswissens', in Christoph Meinel (ed.), *Die Alchemie in der europäischen Kultur- und Wissenschaftsgeschichte*, Wiesbaden, Otto Harrassowitz.

Wagenbreth, O. and Wächtler, E. (1983) *Technische Denkmale in der Deutschen Demokratischen Republik*, Leipzig, VEB Deutscher Verlag für Grundstoffindustrie.

Wilsdorf, H.M. (1970) 'Agricola', in C.C. Gillespie (ed.), *Dictionary of Scientific Biography*, vol. 1, New York, Scribner.

Further reading

Debus, A.G. (1978) *Man and Nature in the Renaissance* (Cambridge History of Science Series), Cambridge, Cambridge University Press.

Evans, R.J.W. (1973) *Rudolf II and His World: A Study in Intellectual History, 1576–1612*, Oxford, Clarendon Press.

French Science in the Seventeenth Century

by Noel Coley

7.1 Science and religious authority

During the first half of the seventeenth century, Europe was ravaged by the Thirty Years War (1618–48), a series of political conflicts between Catholics and Protestants which began in Bohemia and mainly concerned German states ruled by the Habsburgs, but also involved other European countries, including France. The war ended with the Peace of Westphalia in 1648, but in France, where the cost of the war had caused massive increases in taxation, public opposition to the continuing military campaigns had grown, and civil unrest developed, especially in Paris. Immediately after a service of thanksgiving at Notre Dame Cathedral for the end of the war, Cardinal Mazarin arrested some of the civic leaders of Paris for their support of the popular revolt. Civil war, the *Fronde*, ensued, during which the secular powers accrued by Cardinals Richelieu and Mazarin were challenged by the French nobles who sought to reduce the influence of the Church in political affairs. There was a long period of unrest until 1659, during which the government was frequently ineffective, and it was not until Louis XIV (r. 1643–1715) assumed personal rule in 1661 that a degree of political stability was achieved.

One important consequence of the Thirty Years War was that the international power of institutional religion was seriously undermined. The treaty which ended the war recognized the authority of the secular ruler within each sovereign state. The influence of the pope had declined to the point where he could no longer play an effective role in the political affairs of Western Europe, and after 1648 wars were not fought solely on the grounds of religious differences. The authority of the Church was also challenged in other ways, but nowhere more powerfully than in the intellectual sphere, as the foundations of received wisdom shifted from revealed knowledge to new ideas and discoveries arising from observations of nature and the exercise of reason. This movement affected all European countries in various ways, but in France, where social and religious unrest coincided with the disintegration of the intellectual consensus, the challenge to religious authority was widespread and acute (Cragg, 1970, pp.9ff).

Within the Church there were several factions, of which the Jesuits were the strongest (Treasure, 1981, pp.113ff). In their efforts to liberalize the Church so as to make Catholicism easier to practise and more popular among ordinary people, the Jesuits condoned practices which seemed to some to debase true Christian belief, such as the use of indulgences. Fierce criticism came from the Jansenists, followers of Cornelius Jansen (1585–1638), Bishop of Ypres in Spanish Flanders, who had sought a purer, simpler Christian faith. Like Calvin (1509–64), Jansen believed in redemption through grace, which was given only to the elect, but he differed from Calvin in denying any direct relation between the soul and its maker – this could come only in and through the Roman Catholic Church. The Jesuits and their supporters, anxious to proceed with their modernization of the Church, regarded the Jansenists as an infuriating group of niggling precisionists who could not see the wood for the trees. Nevertheless, Jansenism became an important movement in seventeenth-century France. Their belief in the responsibility of the individual, their critical attitude and desire to reconsider traditional doctrines made them receptive to the new ideas in philosophy and science. Strongly opposed within the Church, Jansenism found support among *parlementaires* (elected representatives to local government), members of the upper classes and Gallicans, who were hostile both to the ecclesiastical hierarchy and to the Jesuits (Briggs, 1977, p.162).

Unquestioning obedience to the authority of the pope, still demanded by the Church and supported by the Jesuits, was anathema to Gallicans, loyal French Catholics who insisted that although ultimate control of the Church should be in the hands of the Vatican Councils the direction of the Church in France must be the concern of the king. Gallicanism became a powerful movement in France by the mid seventeenth century, especially at court, and conflicts between king and pope over questions of authority persisted throughout the century. Veering between indecision and the use of excessive force, Louis XIV's repeated interventions against the Jansenists stiffened their resistance and helped to broaden the appeal of Jansenism in opposition to the Jesuits.

While the authority of the Crown and Church were thus challenged, traditional learning, based on Aristotelian philosophy, also came under critical scrutiny. Though Aristotelianism remained enshrined in church teaching throughout the seventeenth century, its influence gradually declined. At a time when intolerance and repression marked the conduct of public affairs in France, secular authorities supported the Church as it sought to impose its rites, dogmas and teaching, including the Aristotelian philosophy, on all French people. From the mid seventeenth century, Cartesianism (see Section 7.7) gained support among French scholars, but it did not secure the official recognition of the Church, which resisted all efforts to re-interpret the physical world in the light of reason and experiment. In the 1680s the government, acting under Jesuit influence, sought to arrest the drift away from traditional learning by enforcing the teaching of Aristotelian philosophy in French universities, but the challenge of the Cartesians, the revival of alternative philosophical systems (e.g. neo-Platonism and Epicurean atomism) and the discoveries of experimental science ensured the eventual decline of Aristotelianism.

Aristotelianism

Aristotle (384–322 BC) was a skilled biologist who also applied his observational and descriptive techniques to physics and astronomy. He regarded the universe as geocentric, entirely filled with matter and limited by an outer sphere; stars and planets moved round the Earth with uniform, circular motion in perfect crystalline spheres showing the changeless, eternal order of the heavens. Influenced by Pythagorean ideas,

Aristotle mechanized the mathematical spheres of Eudoxus (see Chapter 2). In the region below the Moon he saw imperfection, change and chaos: earthly things consisted of the four elements – earth, water, air and fire – each of which sought to move towards its 'natural place', with earth at the centre surrounded by water, air and fire. Natural motion in the terrestrial region was therefore rectilinear (vertical), and until Newton the difference between

celestial and terrestrial motions remained a mystery. The four-element theory, unquestioned before the seventeenth century, survived to the end of the eighteenth. Aristotelian philosophy was syncretized with Christian doctrine in the thirteenth century by St Thomas Aquinas and had become the accepted world view of the Church. Thus any attack on Aristotelian ideas was interpreted as a challenge to religious authority.

The classical inheritance also provided antecedents for sceptics seeking to re-interpret physical phenomena. Certain classical works, re-discovered since the fifteenth century, offered exciting alternatives to Aristotelianism. Of these, Greek philosophical atomism, rejected as atheistic during the Middle Ages, was to become one of the most important for the development of scientific ideas in the seventeenth century. With the growing interest in mechanical explanations based on physical interactions between material particles, ancient atomic theories were revived. Free-thinkers and amateurs of science, such as Pierre Gassendi, who embraced the Epicurean philosophy, sought rational explanations for physical phenomena and showed their hostility to superstition and credulity. This critical attitude to established beliefs shocked the orthodox in France and led to a gradual tightening of censorship. The condemnation of Galileo in 1633 demonstrated the potential hostility of the Church to a wide range of advanced thinking, and counselled caution, although in France a great deal of tolerance was shown to unorthodox views so long as their holders were not too vociferous.

> **The mechanical philosophy**
>
> According to the mechanical philosophy of Descartes, Gassendi, Boyle and others, matter consisted of particles of various sizes and shapes in random motion and collision. This idea was then used to explain all physical phenomena. Matter and motion were the underlying, quantifiable, primary properties of bodies, while colour, roughness, taste or smell, which depend on sense-impressions, were regarded as less important, qualitative, secondary properties. The mechanical philosophy explained *observed* phenomena, but was unable to predict *new* phenomena.

7.2 *Chemistry and botany at the Jardin des Plantes*

The formation of a botanic garden in Paris was first proposed in the sixteenth century by Henri IV (*r.* 1589–1610) and his minister Sully, but it was not established until 1626, during the reign of Louis XIII (*r.* 1610–43). The rise of French chemistry and botany during the succeeding two centuries was closely bound up with the activities of the professors and demonstrators at this institution, generally known as the Jardin du Roi. The first professor of chemistry and botany was William Davisson, a Scot who had migrated to Paris, where he became councillor and physician to Louis XIII and was appointed to the Jardin du Roi in 1648. Davisson's chemistry was governed by its usefulness in medicine, and in common with many other seventeenth-century French chemists he adopted the three hypostatical principles – sulphur, mercury and salt – out of respect for Paracelsus (see Chapter 6). By contrast, Estienne de Clave, a contemporary who also worked at the Jardin du Roi, favoured a theory of five elements – water or phlegm, earth, mercury or spirit, sulphur or oil, and salt. In his *Cours de chimie* (1646), de Clave described the preparation and properties of many chemical substances, but like Davisson, he was particularly concerned with those useful in medicine.

Davisson left Paris in 1650, and soon afterwards Nicolas le Fèvre was appointed demonstrator at the Jardin du Roi. Le Fèvre was an experienced pharmacist who had come to Paris to give public lectures and demonstrations on chemistry and at the same time conduct a business as an apothecary. He gained a very high reputation and after almost ten years at the Jardin he moved to London as Royal Professor of Chemistry to Charles II and Apothecary-in-Ordinary to the royal household with a small laboratory in St James's Palace. Soon after his arrival in London, his *Traité de la chymie* was published in Paris. Although it contained a discussion of chemical theory, it was largely a compendium of medicinal preparations for apothecaries. Le Fèvre's chief authorities were Paracelsus, van Helmont and the German chemist Glauber. For the next hundred years le Fèvre's book ranked as a standard chemical treatise.

When le Fèvre resigned from the Jardin du Roi in 1660, he was succeeded by Christopher Glaser, another apothecary, who also wrote a treatise describing his work at the Jardin du Roi. Glaser's *Traité de la chymie* was published in

Figure 7.1 Title page of Nicolas le Fèvre, Traité de la chymie, *2nd edn, Paris, 1669. Note the close connection between chemistry and medicine; Le Fèvre claims to have increased the book, and therefore its value, by the addition of a good number of excellent remedies. (Published by permission of the British Library Board.)*

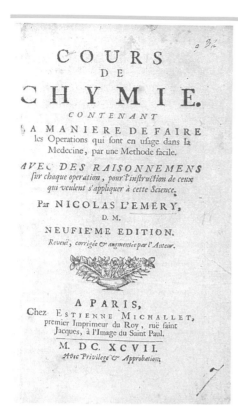

COURS

DE

CHYMIE.

CONTENANT

LA MANIERE DE FAIRE
les Operations qui font en ufage dans la
Medecine, par une Methode facile.

AVEC DES RAISONNEMENS
fur chaque operation, pour l'inftruction de ceux
qui veulent s'appliquer à cette Science.

Par NICOLAS L'EMERY,
D. M.

NEUFIE'ME EDITION.

Reveuë, corrigée & augmentée par l'Auteur.

A PARIS,
Chez ESTIENNE MICHALLET,
premier Imprimeur du Roy, ruë faint
Jacques, à l'Image du Saint Paul.

M. DC. XCVII.
Avec Privilege & Approbation.

Figure 7.2 Title page of Nicolas Lémery, Cours de chymie, *Paris, 1697. The subtitle reads 'Chemistry devoted to the practice of medicine'. (Published by permission of the British Library Board.)*

Paris in 1663 and became very popular during the next fifty years. It contained very little chemical theory, dealing mainly with chemical operations and preparations under the three categories of mineral, vegetable and animal substances. Glaser adopted Boyle's theory of the absorption of igneous particles from the air during calcination (the process of strongly heating metals) and de Clave's five-element theory of chemical composition. In 1666, a young apothecary named Nicolas Lemery came to Glaser in Paris from his home town, Rouen. He hoped to enlarge and perfect his knowledge of chemistry, but soon found Glaser uncommunicative, and after a couple of months Lemery left for Montpellier, where there were freer attitudes towards medicine. There he worked with pharmacists, gave lessons in chemistry and also practised medicine. In 1672, Lemery returned to Paris to set up a manufacturing laboratory. He also gave lectures illustrated by experiments, which by their clarity and attractive style, established his fame.

In 1675, Lemery's famous textbook, *Cours de chymie*, was published in Paris. This book brought chemistry to the popular notice, and no fewer than ten French editions were published during the seventeenth century. Although Lemery was never employed at the Jardin du Roi, his book was strongly influenced both in its general plan and in some of its details by Glaser's *Traité*, but its popularity stemmed from his easy style and his attempt to make chemistry useful and plain by removing as much of the mystery from it as possible. He was an opponent of alchemy, and exposed many of its tricks. He often introduced anecdotes to explain points of detail and became mildly discursive, even argumentative. Lemery was not bound by any one chemical theory, but was a free-thinker willing to try any promising line of argument. He adopted a rudimentary atomic theory similar to that of Boyle, in which the properties of substances depended on the shapes of their particles. Thus, besides its popular appeal, Lemery's book also exerted an influence on chemical thought.

Another theory which demonstrated the current confusion about the fundamental nature of matter was put forward by Lemery's friend and co-worker Wilhelm Homberg, a German from Saxony who spent much of his life in France. He identified four species of sulphur – vegetable, animal, bituminous and metallic – and claimed to have shown that these forms of sulphur were made up of the sulphur principle combined with a salt and an earth in roughly equal proportions and tiny amounts of a metal (see also the discussion of phlogiston in Chapter 15). Consequently, the analysis and synthesis of common sulphur seemed quite feasible.

In addition to chemistry, botany was also taught at the Jardin du Roi, though little progress was made in developing this subject in France until the late 1680s when J.P. de Tournefort occupied the chair of botany. De Tournefort established a new system of classifying plants in his *Éléments de botanique* (1694), which although somewhat artificial, was to dominate botany until it was supplanted by the Linnaean system in the eighteenth century. Among flowering plants de Tournefort distinguished between those with petals and those without, and the petalous plants he further subdivided into those with only one petal (e.g. Canterbury bell) and those with more than one (e.g. wallflower). He also introduced the term 'pistil' for the central organ of the flower, comprising ovary, style(s) and stigma(s), and combined his distinctions with others, such as the popular one between trees, shrubs and herbs, to produce a system of 22 classes of plants. Many of these genera – there were 698 of them by 1700 – were recognized and re-defined by Linnaeus.

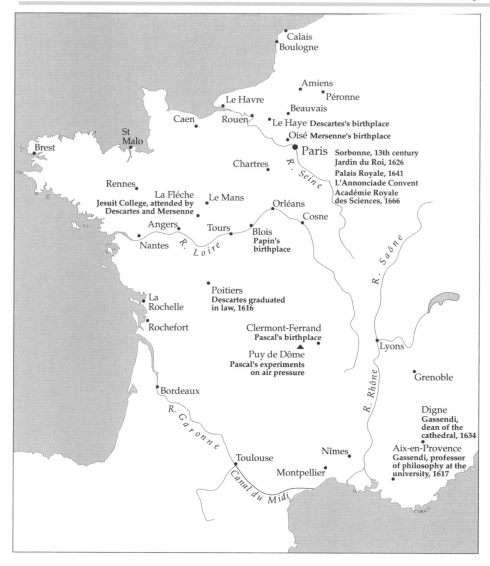

Calais
Boulogne

Amiens
•Péronne
Le Havre
Beauvais
Caen   Rouen
Le Haye **Descartes's birthplace**
Oisé **Mersenne's birthplace**
St
Malo
Brest
Paris
Sorbonne, 13th century
Jardin du Roi, 1626
Chartres
Palais Royale, 1641
L'Annonciade Convent
Rennes
Académie Royale
des Sciences, 1666
La Fléche   Le Mans
Jesuit College, attended by
Descartes and Mersenne
Orléans
Angers   Tours   Cosne
Nantes   Blois
Papin's
birthplace

Poitiers
Descartes graduated
in law, 1616
La
Rochelle
Rochefort
Clermont-Ferrand
Pascal's birthplace
Lyons
Puy de Dôme
Pascal's experiments
on air pressure
Grenoble
Bordeaux
Digne
Gassendi,
dean of the
cathedral, 1634
Aix-en-Provence
Nîmes   **Gassendi, professor**
Toulouse   **of philosophy at the**
university, 1617
Montpellier

R. Seine
R. Loire
R. Saône
R. Rhône
R. Garonne
Canal du Midi

Map 7.1 Science in seventeenth-century France.

7.3 The Académie Royale des Sciences

The Jardin du Roi served to establish a tradition of royal patronage for science in France, and in 1666, under the guidance of Jean-Baptiste Colbert (1619–83), Louis XIV's finance minister, the Académie Royale des Sciences was established in Paris with financial support from the king (see Chapter 9). The academicians were paid a regular salary, and some leading foreign mathematicians, astronomers and natural philosophers were attracted to Paris. Among the first of these was the Dutch physicist and inventor Christiaan Huygens (1629–95), who came to Paris in 1666. Huygens's early work was concerned with optics, the wave theory of light and the invention of the pendulum clock, but he later became interested in contemporary experiments on the vacuum. He began to investigate the use of atmospheric pressure to drive a pumping engine. His first suggestion involved the explosion of gunpowder in a cylinder fitted with valves. The heat generated by the explosion caused the gases in the cylinder to expand, some of which would escape through the valves. On cooling down, a partial vacuum would

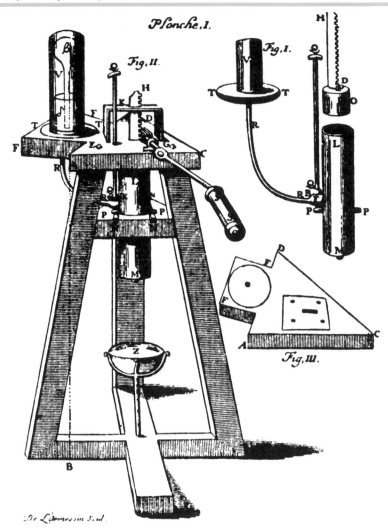

Figure 7.3 This air-pump was designed by Huygens and Papin in Paris in 1674; it clearly owes much to the design of Boyle's second air-pump (see Section 8.7.2). (From Huygens, Oeuvres complétes, 22 vols, The Hague, 1888–1950, vol.XIX, p.217. Photo: British Library Board)

be formed in the cylinder, which was closed at the top by a movable piston. The weight of the air above the piston would cause it to be forced into the cylinder in a way which could be used to operate a pump. Unfortunately this device was both dangerous and impractical.

Denis Papin, Huygens's assistant, turned his attention to steam instead of gunpowder. In 1679, he invented a 'steam digester', the forerunner of the autoclave, a closed vessel with a tightly fitting lid in which steam could be generated under pressure well above the normal boiling point of water. In 1690, applying similar principles, Papin showed how, by alternately heating and cooling water in a cylinder, a piston could be made to move up and down, exerting a force on the down stroke due to air pressure above and a partial vacuum below. Although Papin's invention would not have worked in practice, it introduced the idea of an engine driven by steam, which was developed by Savery and Newcomen in England within a few years. These early stages in the invention of the steam engine illustrate the international character of late seventeenth-century science and technology. In 1684, Papin was appointed temporary curator of experiments at the Royal Society of London, where he worked with Robert Boyle. In 1687, he became professor of mathematics at the University of Marburg in Germany and later moved to Cassel. In about 1707, he relinquished this post and returned to London where he died in obscurity ten years later.

During the reign of Louis XIV, France became so politically influential in Europe that other powers began to fear its ascendancy. Finance Minister Colbert succeeded in stabilizing the government, building up the navy, increasing national prosperity and generally strengthening the position of the king, who was suspected of having grandiose plans for political and military domination in Europe. Colbert persuaded the king to provide financial support and patronage for various cultural and intellectual enterprises, including the Académie Royale des Sciences. Royal pensions for writers, artists and savants were established, which caused a great deal of sycophantic adulation of the king, although the distribution of favours was often somewhat arbitrary. In any case the cost of Louis's support for culture and the arts was dwarfed by the enormous sums lavished on Versailles and its satellite palaces – magnificent temples of the cult of monarchy.

One consequence of the provision of pensions and sale of offices was an increase in the number of groups of prosperous and educated men with time on their hands, who were the staple of intellectual and cultural life. Many writers and artists of the time came from this background, which also provided an informed public capable of supporting cultural activity independent of the court. After 1680, the salons frequented by such groups of intellectuals in Paris became more significant than the official academies and more important for setting cultural standards than the court. The last decades of the century saw the establishment of local academies in many provincial towns and cities, adding to those already in existence since the early years of the century (Briggs, 1977, pp.196ff). Natural philosophers found themselves addressing interested amateurs for whom it was necessary to explain the new ideas, discoveries and inventions in simple, direct language. Personal libraries also grew in size and diversity, while the possession of *cabinets de curiosités* by many dilettanti showed an increasing general interest in science.

7.4 Gassendi's revival of Epicurean philosophy

Pierre Gassendi (1592–1655) was born in Provence, a region of south-eastern France with its own language and laws. Until 1639, Provence had its own regional elected assembly, which met annually and included representatives of the Church, the nobility, 22 towns of the region and fifteen other privileged places. Gassendi was educated at Digne, a cathedral city which was at the centre of the religious wars in the region during the sixteenth and seventeenth centuries, and at Aix-en-Provence, the capital of the region. He took holy orders and became professor of philosophy in the university at Aix in 1617. He was critical of Aristotelianism, suggesting that many of Aristotle's ideas were open to question and pointing out that rival philosophical schools had once competed with the Lyceum in Athens. He accepted contemporary scepticism about the limitations of human knowledge and was dissatisfied with the dialectical style of Aristotelian discourses. In 1622, a new Jesuit administration of the university, unwilling to allow him freedom to express such criticisms, removed him from the chair of philosophy. This gave him both the time and the motive to develop his critique of Aristotle and in 1624 he published his doubts in a collection of lectures. He travelled extensively in Flanders and Holland, returning to Digne as dean of the cathedral in 1634. Later he accompanied the Duc d'Angouleme on a tour of his domains in Provence, but in 1645 he was appointed to the chair of mathematics at the Collège Royale in Paris, where he died in 1655.

Figure 7.4 Pierre Gassendi (1592–1655). Philosopher, scientist and humanist, he opposed the blind acceptance of Aristotle, revived Epicurean atomism and advocated an empirical realism, although he admitted that the intellect attains notions and truths of which the sensations can give no indication (e.g. general ideas and universals such as the notion of God). (Roger Viollet, Paris)

In Provence, Gassendi's friends included Joseph Gaultier, Vicar General of Aix, an astronomer and early observer of the satellites of Jupiter, and Louis-Emmanuel de Valois, Comte d'Alais and Governor of Provence. Gassendi also wrote a biography of Nicolas-Claude Fabri de Peiresc, counsellor to the king in the *Parlement* of Provence, the leading bibliophile and collector of manuscripts of his generation. Beyond this immediate circle there were others who assisted him, including the following: Jean Chapelain, a critic of Aristotelianism and secretary of the Académie Française; Gabriel Naude, writer of history and librarian or secretary to Cardinals Richlieu and Mazarin and to Queen Christina of Sweden; and Marin Mersenne, mathematician, philosopher and founder of the most influential group of scientific and philosophical savants in Europe, whose work is discussed in Section 7.8.

In the seventeenth century, various efforts were made to revive ancient Greek atomic theories and amongst these Gassendi's attempt to revive Epicureanism was one of the most important. Inspired by Daniel Sennert in Wittemberg, Sebastian Basso and David van Goorle in Leiden – all of whom published works in the 1620s advocating particulate theories of matter – Gassendi sought to re-instate Epicurean atomism as an alternative to Aristotelian physics.

Gassendi explained his theories about atoms in footnotes to his *Observations on the Tenth Book of Diogenes Laertius* (1649). Dating from the third century AD, Diogenes' work is the earliest known source for the ideas of Epicurean atomism. Gassendi brought it to the attention of his contemporaries in a critical edition which contained the original Greek text, a Latin translation and his own comments, whose main purpose was to divest the Epicurean philosophy of the charges of atheism and materialism, which had caused it and the atomic theory in general to be rejected by early Church fathers and mediaeval philosophers alike. The revival of atomism in the seventeenth century required it to be reconciled with Christian thought, and it was in this respect that Gassendi, a Catholic priest, was most successful. In general, he approved of Epicurean *atomism*, but disassociated himself from Epicurean *materialism* by denying that the atoms were eternal. Instead, he held that they were created by God in limited numbers and endowed at their creation with forces which gave them motion and powers of mutual attraction. Since physical changes were caused by the motion of the atoms, Gassendi thus re-introduced the concept of divine purpose into the material universe. Quoting from the Roman poet Lucretius, whose poem *De rerum natura* expresses the nature of Epicurean atomism, Gassendi agreed that atoms join together to

---

Epicurean atomism

In the fourth century BC, Epicurus adopted the atomic philosophy of Democritus and Leucippus. He held that matter was made up of minute immutable particles, or atoms, moving at random in empty space and that chance rearrangements of these atoms gave rise to random changes in the physical world. The Epicurean philosophy was entirely materialistic: there was no hint of any divine plan, and objections to it on grounds of atheism had been raised since the beginning of the Christian era. The atoms were recognized by the senses, but space, the other essential ingredient of the theory, was a necessary inference of reason. Without space, the atoms would be immovable and change would be impossible. Epicurus also argued that the universe must be infinite since its boundaries would only be observable in contrast to something else and as nothing existed apart from the atoms and space, there could be no limit to the observable universe. The atoms, which were in perpetual random motion, were not all of one shape or size, and although the number of different kinds was limited, the number of atoms of each kind was infinite. Epicurus even extended his materialism to the soul, which he held to consist of fine particles dispersed among the atoms of the body. He considered the senses to be reliable, and thought that errors of perception arose when the mind formed an incorrect opinion which went beyond the evidence of sensation.

form corpuscles (molecules) and larger bodies which form the 'seeds' of things, just as letters may be joined together to form words, phrases and sentences. Gassendi later extended his ideas in the *Syntagma philosophicum* (published posthumously in 1658), where he advocated the Baconian method of observation and induction (see Chapter 8).

As a humanist, Gassendi situated scientific ideas in a broad context of history and philosophy gained during a lifetime of study. In redefining the atomic theory, he also sought to establish cultural continuity between ancient Greek and modern European ideas about the nature of the physical world. At the same time he argued that science should be independent of cultural influences (despite the fact that Epicurean atomism was a product of early Greek thought). In his efforts to reconcile these incompatible notions, Gassendi's aim of demonstrating cultural continuity became obscured (Joy, 1987, p.5). He wrote as a historian of science and philosophy at a time when history, philosophy and science were not clearly differentiated from each other, but as the seventeenth century advanced, the aim of reviving classical learning which had inspired the humanists of earlier decades went out of fashion and Gassendi's attempts to demonstrate the continuity of culture no longer seemed worthwhile.

The atomic theory was clearly capable of fulfilling most of the demands of the seventeenth-century mechanical philosophy, though in a different way from the Cartesian system. Rather than revive an ancient philosophy as Gassendi had done, Descartes preferred to set all ancient learning aside and begin anew. He abandoned the historical mode altogether and set up his system as a way of judging all other claims to knowledge, especially those of the Aristotelians. Descartes and Gassendi were therefore rivals. Both proposed ideas which were important for the rise of European science and both contributed to the ideas of men like Robert Boyle, John Locke and Isaac Newton, but an important transformation in the organization of learning began in the seventeenth century which determined the fate of their respective systems. That transformation was the severing of ties between areas of study such as history, physics and philosophy, which in the ancient world had been undifferentiated and which Gassendi had tried to re-work *as a whole*. Descartes was among those who began the process of severance; Gassendi could not accept it. Thus, while Descartes's work provided a basis for the early stages of the Scientific Revolution, Gassendi's was largely forgotten.

7.5 Descartes's idea of completed scientific knowledge

René Descartes (1596–1650) was undoubtedly the most influential French philosopher of the period. In common with many of his contemporaries, Descartes was critical of Aristotelianism. He questioned all that he had been taught on Aristotelian principles at the Jesuit College at La Flèche, yet he accepted the idea of a philosophical system which would encompass the whole of nature and provide solutions to all physical and metaphysical problems. In fact he aimed to devise an alternative to the Aristotelian system that was just as complete but more reliable – a science of nature about which he could be absolutely certain (Hooker, 1978, p.114). Rejecting Aristotelian arguments, Descartes developed his own physical explanations based on his

Figure 7.5 René Descartes (1596–1650). Apart from his contributions to cosmology and science, Descartes's most important advances were in 'co-ordinate geometry'. By applying algebra to geometrical figures it becomes possible to explain any curve in terms of a moving point whose position at any instant may be identified in terms of its relation to two fixed lines, or axes, as in a graph. Descartes's analytical method first appeared in his Geometry, *1637. (Hulton Deutsch Picture Collection)*

corpuscular theory and the mechanical hypothesis. Cartesian cosmology and physics were later challenged by Newton's theory of universal gravitation (1687), but Cartesians opposed Newton's world-view, and controversy raged between the two camps far into the eighteenth century.

Prior to the seventeenth century the most far-reaching attack on Aristotelianism had come from Pierre Ramée, professor of philosophy at the Collège de France. In 1543 Ramée published a critique of Aristotelian logic based on his premiss that the whole of Aristotle's teachings were false. Ramism enjoyed considerable popularity for a time in France, Holland, Germany, Sweden and England. Its orientation lent support to the pursuit of solutions to practical problems and to a utilitarian outlook in science such as that adopted at Gresham College in London (see Chapter 8), but Ramée's critique was premature and insufficient to undermine permanently the whole Aristotelian corpus. Other early workers like Paracelsus, Telesio or Campanella also attacked certain parts of Aristotelian philosophy, but none before Descartes had sought to replace it entirely with a comprehensive alternative philosophy.

Born into an influential family, Descartes inherited private means from his mother's estates in Poitou and was thereby enabled to spend the greater part of his life on philosophy without undue concern for his livelihood. Feeling dissatisfied with his education at La Flèche, and thinking that the fault lay not in his own abilities but in the sciences themselves, he set himself the task of improving them. After spending a year in Paris, he entered the University of Poitiers where he graduated in law in 1616. He then joined the army of Prince Maurice of Nassau as a gentleman volunteer, hoping to find time for study as well as opportunities for travel and adventure. At Breda in Holland he met Isaac Beeckman, the Dutch scholar, who introduced him to some interesting problems in mechanics and acoustics while at the same time encouraging him to develop his interest in mathematics and science. Descartes later reported his invention of analytical geometry to Beeckman.

In 1619 Descartes came to two conclusions: first, he decided that to discover true knowledge he must construct a complete system just as an artist creates a work of art; and, second, he was convinced that he must begin by questioning the truth of all the tenets of contemporary philosophy while looking for self-evident principles on which reliable science could be constructed by the application of logic. He believed that he was warned in a dream against past errors and that the 'spirit of truth' had opened up to him the path to true knowledge, a vision which remained with him for the rest of his life. In the 1620s Descartes came into contact with the group which had formed around the Minim friar Marin Mersenne in Paris where he met Pierre Gassendi, among others. In about 1628 Descartes finally became convinced that he was destined to found a new universal science and withdrew to a solitary life in Holland where he found greater tranquillity and more intellectual freedom than in his native France. He lived in Holland for the next twenty years until 1649 when he moved to Stockholm at the request of Queen Christina, who wished him to instruct her in mathematics (see Chapter 12). After only about four months in Sweden Descartes caught a severe chill and died there on 11 February 1650.

Descartes began to construct his philosophical method in relation to medicine and mechanics, the former contributing to the conservation of health and the latter aimed at reducing the labours of mankind. He attempted to solve problems of all kinds – mechanical, biological and cosmological – and Fontenelle said of him that he tried to put himself at the source of everything, to make himself master of the first principles ... so that he could then simply descend to the phenomena of nature as to necessary consequences of these principles. *(Fontenelle, 1729, pp.404–6)*

Thus, Descartes sought the physical causes of what he observed by applying the deductive logic of mathematics to what he already knew. Mathematical logic became for Descartes the basis of a scientific method which could be applied to *all* parts of science. When science had become as precise and certain as mathematics he believed that he would have attained his goal of completed knowledge.

In his *Discours de la méthode*, published in the Netherlands in 1637, Descartes gave an autobiographical account explaining how he came to his method of reasoning. He wrote in French rather than Latin, being one of the first, along with Galileo, to present philosophical and scientific ideas in the vernacular to make them more accessible to a wider public. He outlined four stages in this method of inquiry: first, he accepted as true only those things of which he felt certain; next, he resolved problems into small parts; then, he took each part separately, beginning with the simplest and moving systematically to the most complex; and last, he undertook periodical reviews to make sure that he had not overlooked anything. By this method Descartes thought he could not fail to reach the truth. Rejecting as unreliable all that he had previously learned, he concluded that the one thing he could not doubt was his own capacity for critical thought. He must therefore exist as a thinking, rational being and the famous maxim 'I think therefore I am' (*Cogito ergo sum*) followed. Having found reason to doubt so much that he had previously accepted, he concluded that his mind could not be perfect. Yet, he thought, it was possible to imagine a perfect being. Such an idea could neither have come from nothing, nor was it part of himself; it could only have been put into his mind by the perfect being, God, who must therefore also exist. Descartes felt certain that God would not deliberately deceive his creatures, and with this assurance he thought he had a guarantee that clear and distinct mental concepts were true and that reason could act as a reliable check on the senses.

Descartes accepted the Aristotelian view that matter and space were co-extensive. For Descartes, as for Aristotle, there could be no vacuum. On the other hand, while Aristotle allowed the division of matter into minima only as a possibility, Descartes argued in line with the mechanical philosophy that matter was actually made up of minute corpuscles whose shapes, sizes, motions and interactions were the cause of all natural phenomena. The whole universe was completely filled with these minute particles of matter which, unlike the Epicurean atoms, were capable of being changed in shape and size – by abrasion, for example. Descartes held that the only properties of external material things which are clear and distinct are extension and motion, the primary qualities of matter. He compared these measurable properties with such indefinable qualities as colour, smell and taste, which depend on subjective sense-impressions and vary from person to person and from time to time in each individual. Thus he accepted Aristotle's distinction between primary and secondary qualities, which was also revived and refined in the seventeenth century by some of his contemporaries, including Galileo, Bacon, Gassendi and Boyle, all of whom, like Descartes, made significant contributions to the development of the mechanical hypothesis.

Descartes also suggested that motion, created separately, was associated with matter in accordance with certain physical laws and that the total quantity of motion in the universe was constant. The Cartesian universe was therefore a great machine consisting of minute particles of matter in continual motion. For Descartes, matter was infinitely divisible, and consequently particles must also be infinitely divisible; space was therefore completely filled with matter. The motion of one part gave rise to motion in adjacent parts and this resulted in the circulation of particles as in a whirlpool or vortex.

Vortex theory

At the Creation, according to Descartes, innumerable vortices were set in motion. Friction between the contiguous particles of matter rubbed off finer particles from them, making them smooth and spherical. The finest particles tended towards the centre of the major vortices where they formed suns and stars while the spherical particles moved towards the circumference of each vortex, where they composed the transparent heavens. Grosser particles formed opaque bodies such as the Earth and other planets, each of which was at the centre of a smaller vortex carrying satellites such as the Moon. Comets pursued a tortuous path between the different vortices.

Descartes's descriptive Copernican cosmology owed more to speculation than to astronomical observations. The details of the vortex theory were fanciful: it was neither compatible with quantitative measurements, nor susceptible to mathematical analysis. Moreover, it continued to assume circular planetary motions in conflict with Kepler's elliptical orbits. Descartes completed his *Le Monde* in 1633, but he withheld the book from publication after learning of Galileo's trial and condemnation. It was published in its entirety only after Descartes's death, though some of the less controversial parts appeared in the *Discourse* (1637). By the time that his *Principles of Philosophy*, a general system of physics, appeared in 1644, caution led him to present it as hypothetical and not a picture of the real world. Nevertheless, he wrote as though he had discovered the true philosophical methods by which completed scientific knowledge could be achieved, making much of the ideal of a physics based on mathematical deduction.

Descartes also extended his physical and cosmological speculations to living things and applied the mechanical model to animals. He even treated the human body as a machine and suggested that the mind was an unextended thinking substance which inhabited the body. He distinguished sharply

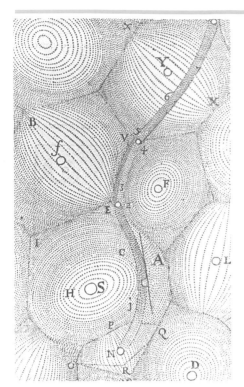

Figure 7.6 Diagram of vortices from Descartes, Principia philosophiae, *Amsterdam, 1644. (Ann Ronan Picture Collection)*

Descartes's view of sense-perception

Descartes suggested that sense-impressions were received by the pineal gland, which is situated at the base of the brain, and were transmitted from it to the mind, seated in the brain itself. This placed a material barrier between sense-impressions and the external material world from which they originated, and raised fundamental questions about the nature and extent of human knowledge, which lie at the root of Cartesian 'dualism'.

Figure 7.7 Illustrations from Descartes, Tractatus de hominie, *Leiden, 1664 (published posthumously), showing his views on the reception of sense-impressions. The pineal gland, seated at the rear of the brain, receives physical sensations, conveys them to the mind, and through the central nervous system co-ordinates muscular movements in response. (Ann Ronan Picture Collection)*

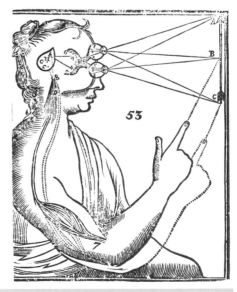

between mind and matter to the extent that any property attributed to one was denied to the other: mind being essentially active; and matter, inert. How these two mutually exclusive categories could interact, as they appeared to do in human beings, was an important problem for Cartesian physiology and it remained a fundamental difficulty for the Cartesian system as a whole. The problem was debated by Gassendi, Hobbes and Descartes in 1641 and discussed at length by later Cartesians.

7.6 Descartes and the circulation of the blood

It has sometimes been said that Descartes paid little attention to experimental observations, though he criticized other philosophers who neglected them. This is not entirely true: Descartes conducted experiments in physics and chemistry; he also studied anatomy, performing dissections and vivisections on various animals and their organs. He clearly recognized the applications of experiment and hypothesis as the basis for his completed scientific knowledge, but in his biological work his dualistic philosophy and devotion to mechanism led him to propose formalized physiological models which obscured his knowledge of anatomy. Although he spent much time on his own experiments, Descartes was always sceptical of experimental results reported by others, as shown by his reactions to Harvey's discovery of the circulation of the blood.

In *De motu cordis* (1628) Harvey described the action of the heart and blood vessels by reasoning from his own experimental observations (see Chapter 8). News of Harvey's discovery reached Mersenne in Paris in 1629, and he quickly informed others, including Descartes. The latter accepted the general notion of a closed circulation system at once, but showed his own prejudices in the obstinate way in which he clung to ancient beliefs about the function of the heart itself. Harvey observed that the heart acted as a muscular pump, expanding as it filled with blood and contracting as it pumped the blood out, whereas the traditional explanation of its action had been that it was dilated by the sudden vaporization of small quantities of blood seeping into the ventricles from the auricles (a modern analogy for this view of the heart's action would be the internal combustion engine). The vaporization was thought to be due to the great heat of the heart, and it was this idea that Descartes continued to support by pure reasoning and the authority of traditional beliefs. He never accepted Harvey's account of the action of the heart despite incontestable experimental evidence (Whitteridge, 1971, pp.151–7).

Thus Descartes fell into the old errors of scholasticism, invoking common sense and traditional opinions. Relying on his analytical method, Descartes claimed that as it was founded on simple, elementary evidence it must lead to the true explanation of the motion of the heart. He was so certain of this that he wrote to Mersenne saying that if anyone should think his ideas on this subject false, he would be ready to accept that the rest of his philosophy was also worthless. Descartes's ideas about the motion of the heart were indeed false. He accepted the traditional view that the heart was emptied as it expanded and filled in contraction, despite Harvey's experiments on the beating heart having established the exact opposite. Yet, in spite of his apparent preference for reasoning over observation, Descartes praised the experimental method and regretted the delay to his own investigations due to lack of assistance. He would have done better to pay proper attention to Harvey's careful experiments instead of merely judging the conclusions theoretically on the basis of his mathematical method. Here, then, is a case where the Cartesian method not only failed to yield the correct explanation of the heart's action, but was *shown* to be at fault by carefully devised and conducted experiments.

7.7 Science and Christian belief

7.7.1 Cartesianism

Cartesianism had become the most fashionable school of thought in France by the 1660s. Seen as a challenge to Aristotelianism, it was a comprehensive system which offered a new synthesis to replace outmoded Aristotelian ideas, although there were some similarities between the two systems. The most fundamental differences lay in Descartes's use of mathematical reasoning and the corpuscular philosophy to provide mechanical explanations for physical phenomena without reliance on occult forces or teleological arguments. In this respect his system was in line with the contemporary fashion for mechanisms. Cartesianism established the authority of reason while raising questions which stimulated argument and further work. Its most intractable problems in terms of Christian belief arose from the separation of mind and body and the relationship of each to God. Descartes was unable to explain how mind and body could form a unity. He merely contented himself with the doubtful conclusion that states of mind could not cause bodily changes, nor could any physical change affect the mind.

In an effort to overcome this difficulty the theory of 'occasionalism' was proposed in 1670 by Nicolas Malebranche (1638–1715), a priest of the Oratory in Paris. Impressed by Descartes's *Traité de l'homme*, Malebranche resolved to devote himself to the study of Cartesian philosophy. For Malebranche, to be a Christian was to be a philosopher and to be a philosopher was to be a Cartesian. In 1674–5 he published *De la recherche de la verité*, in which he adopted Cartesian principles, including the following: the separation of mind and matter; the idea that animals, having no souls, can feel neither pain nor pleasure; and the vortex theory. As a Roman Catholic priest, however, Malebranche was more concerned with the nature of God than Descartes had been and he extended Cartesian ideas in some interesting ways. His concept of occasionalism invoked the direct intervention of God in all cases of sense-perception. It offered a way of resolving the mind–body problem using and extending Cartesian principles; it made Cartesianism respectable in some clerical circles while introducing it to a wider intellectual public.

According to Malebranche, ideas, being spiritual rather than physical, are modifications of the soul. When we see an object – since the mind does not leave the body and material objects cannot affect the mind directly, nor can they send out immaterial images – we do not see the object itself, but form a mental picture of it. This cannot come either from the mind or from the object and must therefore come from God, who is present in our minds. Malebranche argued that as all physical sensations occur in the same way, we are utterly dependent on God for all our knowledge of the external world. A moving object cannot move itself and it cannot communicate its motion to another object, nor can finite minds move matter – the real cause of all motion is God. Thus, if a ball strikes another and seems to set it into motion, it does not do so in reality, but is only the 'occasional' cause of the motion, which in fact comes from God. Again, external objects are not the cause of pleasure or pain; they are neither agreeable nor disagreeable. Pleasure and pain come from God; the human mind can know, feel or will only as God intervenes. We cannot control our bodies by mental activity, because there is no interaction between body and mind; our will is only the 'occasional' cause of our movements – the real cause is God. As an extension of Descartes's thought, occasionalism clearly affects the nature of experimental observations and therefore of physical science. Malebranche argued that his version of Cartesianism was in complete conformity with the Christian religion. It was widely read and played an important part in the spread of Cartesianism in France.

7.7.2 Mersenne, the Baconian friar

A different approach to the relationship between science and Christian belief was taken by the Minim friar, Marin Mersenne (1588–1648). In contrast to many of his contemporaries in France, Mersenne was critical of Cartesian thought and instead modelled his scientific work on Baconian lines (see Chapter 8). Unlike either Descartes or Galileo, he aimed for precision in observation rather than certainty of knowledge, and it was in Bacon that Mersenne found a method for acquiring what he considered to be true scientific knowledge, although he criticized Bacon for not keeping up with progress in science and for suggesting that it would be possible to discover the essential nature of things. In Mersenne's view the latter was out of the question since only God knew the essences of things. He concluded that the only true knowledge of the physical world available to humans came from the quantification of observed effects, which science could explore by means of experiments and hypotheses.

Figure 7.8 Descartes (centre right) and Mersenne (extreme right) at the court of Queen Christina of Sweden. The armillary sphere in the foreground is based on the heliocentric (Copernican) system. (Cliché des Musées Nationaux de Paris)

Like Descartes, Mersenne was educated in the Jesuit College of La Flèche, after which he spent two years at the Sorbonne in Paris. In 1611 he joined the monastic order of Minims and in 1619 entered the Convent of l'Annonciade near the Palais Royale in Paris, where he remained until his death in 1648. The leaders of his order, recognizing that he could best serve the community by the use of his intellect, allowed him to spend his life in study, teaching and writing. In 1623 Mersenne set up a salon at the convent where many savants met, including Gassendi, Descartes, Hobbes, Beeckman, van Helmont, Fermat and Pascal. He also maintained a correspondence with many other natural philosophers all over Europe and as far afield as Tunisia, Syria and Constantinople – a widespread scientific network institutionalized in 1635 when Mersenne organized the Academia Pariensis, which later formed a nucleus for the Paris Académie Royale des Sciences.

Mersenne's approach to experimental science was founded on two important criticisms of Cartesian thought. In the first place he questioned the validity of Descartes's search for fundamental physical axioms derived from scientific demonstrations and, in the second, he doubted whether nature could be fully explained by any *strictly* mechanistic theory. Mersenne illustrated his grounds for the first of these criticisms by referring to contemporary debates about the Copernican system, especially in regard to the question of a moving Earth. He had first rejected this idea, but later accepted it, while pointing out that the Copernican theory could be neither refuted nor positively demonstrated. For this reason, according to Mersenne, those who accepted the Copernican theory could not justifiably claim to base their physics on truly scientific demonstrations, and this included Descartes. He was also strongly opposed to those extrapolations of Copernican cosmology which extended it into theologically sensitive areas such as the plurality of worlds, in which life on other planets was imagined, or the infinity of the universe, which threatened to remove God far beyond the reach of humankind. The notion of a rational and knowable nature – a mechanism governed by quantitative laws – appealed to Mersenne, who wrote a defence of Christian theology against atheists, magicians, deists and others. He criticized atomism as an atheistic theory while condemning the materialism of Bernardino Telesio and Tomaso Campanella, who held that everything which existed was produced by the effects of heat and cold on formless matter, but above all Mersenne condemned the ideas of his contemporary Robert Fludd along with the whole range of hermetic, cabalist and naturalist doctrines about occult powers and the harmonies of creation. These criticisms, coupled with the defence of rationality in nature, attracted the attention of Gassendi who, despite Mersenne's fierce opposition to atomism, became his closest friend.

In 1629 Mersenne wrote to Galileo offering his services in getting his work on Copernican cosmology published. Galileo made no reply, but Mersenne persisted since he saw in Galileo's work a supreme example of the rationality of nature governed by mechanical laws. In 1633 Mersenne published his first critique of Galileo's *Dialogues Concerning the Two Chief World Systems* (1632) and in the following year, after Galileo's trial and condemnation, he published a French version of Galileo's early work on mechanics with summary accounts of the first two 'days' of the *Dialogues* and of the trial. While he was not prepared to risk a schism for the new astronomy, he planned a defence of Galileo in 1634, though he later gave up the idea. He did not accept Galileo's 'necessary demonstrations' and was unconvinced by his arguments for the motion of the Earth. But, though he was sceptical about the Earth's motion ever being demonstrated, he encouraged the search for more experimental evidence to support the theory.

His own experiments were concerned with the nature of light and sound, both of which he considered to be corporeal, in line with the mechanical philosophy. He studied ancient and modern theories of music and optics, always trying to eliminate magic and the irrational. His chief intellectual preoccupation after 1623 was the composition of a major work on music, which appeared in 1636–7 as *Harmonie universelle, contenant la théorie et la practique de la musique*. In experiments on the physics of sound Mersenne investigated the relationship between the length of a vibrating string and its frequency of vibration, based on propositions stated by G.B. Benedetti, Vincenzio Galilei (Galileo's father) and Beeckman. He also discovered the relation between the time of swing and the length of a pendulum in 1634, one year before Galileo. He investigated the upper and lower limits of audible frequencies, harmonics and the speed of sound in air and also discovered the inverse square law, relating the intensity of a sound to the distance from its source. In experiments using the vacuous space above the mercury column in a barometer tube (see Figure 7.13), Mersenne showed that the air is the true medium of the propagation of sound.

Mersenne made an important contribution to experimental methods in science by insisting that all the details of each experiment should be carefully specified and recorded, that experiments should be repeated to obtain consistent results and that any approximations should be identified. In acoustics he showed the need to identify the limits of accuracy of any experiment, to quantify observations and link experimental results with mathematical theory. His ideas about the nature and purpose of experimental work differed from those of Descartes and Galileo in some important respects. While Descartes aimed to achieve certain knowledge and Galileo used experimental results to confirm mathematical deductions, Mersenne, who disbelieved in the possibility of achieving physical certainty and distrusted mathematical explanations of nature, aimed instead for accurate experimental observations. Recognizing the difficulty of making accurate measurements, he repeated Galileo's experiments with the inclined plane, using a seconds pendulum to measure the times of fall. His results showed such discrepancies from the exact relationship Galileo claimed to have obtained that Mersenne wondered whether Galileo had really performed the experiments at all.

Yet, significant as these developments were, Mersenne contributed even more to the rise of scientific Europe in quite another way: through his extensive international correspondence and the provision of a forum for discussions in his salon in Paris. He was among a small number who demonstrated the importance of scientific communication, and as a consequence scientific ideas and discoveries were widely disseminated throughout Europe before the era of the scientific journal. Mersenne informed those with whom he was in correspondence about the latest developments in science and encouraged them to make further experiments. He promoted the Baconian ideal of collaboration in experimental science and in this respect his approach was in direct opposition to Descartes's belief in a system of completed scientific knowledge created by a single worker. Mersenne's salon, like the earlier academies in Italy, was a forerunner of European scientific academies such as the Royal Society in London and the Académie Royale des Sciences in Paris. Both the publication of experimental results in scientific journals such as the *Philosophical Transactions* of the Royal Society or occasional publications of the Académie Royale des Sciences and the establishment of scientific societies were essential elements in the rise of European science later in the seventeenth century, and Mersenne played an important part in pioneering these developments.

Figure 7.9 Title page of Marin Mersenne, Harmonie universelle, Paris, 1636. Mersenne studied the structure and performance of music from early times to his own day, covering so wide a range of musical instruments that this book became an essential source for musicologists. He also sought to explain the effects of music on the soul, aiming to discard the magical and occult powers supposedly attached to words and sounds. This led to the analysis of language and a popular seventeenth-century pursuit – the invention of a universal language. (British Library Board)

7.7.3 Pascal: science and Jansenism

The desire to reconcile science with Christian beliefs, a common characteristic of many seventeenth-century natural philosophers, is shown most clearly in the life and work of Blaise Pascal (1623–62). Pascal's mother died when he was 3 years old and he was brought up by his father, Etienne, a prominent official in Clermont-Ferrand and a respected mathematician. In 1631 the family moved to Paris, where Blaise began his scientific studies under his father's direction. A mathematical prodigy, he read Euclid's *Elements* in 1635 when he was only 13 years old and, according to a biography written by his sister, he also began to attend Mersenne's Academia Pariensis at about this time. Pascal began with a study of the conics, based on a difficult mathematical treatise published by Desargues in 1639. As he did not use Cartesian algebraic symbols and vocabulary, Desargues's work baffled most of his contemporaries, including Descartes, but Pascal was able to appreciate the richness of the work which laid the foundations of projective geometry and of a unified theory of conic sections. In 1640 he published a pamphlet setting out a plan for further work in this field, including an outline for a work on conics. In the same year his connections with Mersenne's salon were broken when he moved with his two sisters to join his father, who was then working as a royal tax official in Rouen.

In 1646 his father fell ill and as a consequence of seeking medical advice Pascal came into contact with two Jansenists from Port Royal. Originally a Cistercian abbey, Port Royal had come under Jansenist influence through the reforming zeal of Angelique Arnaud. Pascal found himself in sympathy with the Jansenists' desire to purify Roman Catholic practices and he won over his family to this point of view. Throughout his life he retained this reforming spirit and although his religious feelings ebbed and flowed, his desire to live a devout life and to cleanse the Catholic Church of some of its less attractive ways remained strong. Pascal's talents were very varied, indeed he might justifiably be considered a universal genius. His literary and theological

Figure 7.10 Blaise Pascal (1623–62). Pascal developed a mathematical theory of probability and created new geometrical conceptions. His experimental work, done c. 1650, was published posthumously in 1663 and was criticized by Boyle in Hydrostatical Paradoxes, *1666. Note Port Royale in the background, a reference to Pascal's important connections with the Jansenists there. (Photographie Giraudon, Paris)*

Figure 7.11 Port Royal was a celebrated Cistercian abbey founded in 1204 on a low-lying site in the valley of the Yvette near Marly, a few miles south-west of Paris. The church (K) was built in 1229. In 1598 Angelique Arnaud sought reforms, and the abbey became linked with the Jansenist revival. In 1626 ague (malaria) drove the nuns to Paris where they settled at Port Royal de Paris, at the end of the Faubourg St Jacques. In 1648 some of the nuns returned to the abbey where they set up a little school (D) for the sons of Jansenist parents. Racine received his education there. In 1653 Pope Innocent X condemned Jansenism and in 1656 the school was disbanded. There followed half a century of changing fortunes until in 1709 the remaining nuns, all of whom were then over 60, were forcibly removed and the buildings were pulled down. (Bibliothèque Nationale, Paris)

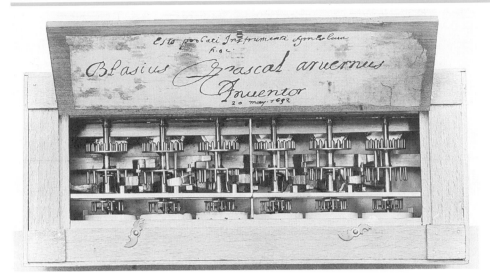

Figure 7.12 Two views of Pascal's calculating engine (this example dates from 1652). The dials allow sums up to 999,999 to be calculated by straight-forward addition and subtraction. Multiplication would be achieved by multiple additions and division by multiple subtractions. (Science Museum, London)

works are as highly esteemed in their fields as are his contributions to mathematics, philosophy and experimental science. It is only the latter which concerns us here.

In about 1642 Pascal began to direct his attention to the problems of constructing a calculating machine to assist his father with his accounting. Beginning with the fundamental arithmetical operations of addition and subtraction, he showed how, in theory at least, these operations could be carried out using gears. When it came to constructing an actual machine, however, the wheel arrangements turned out to be so complex that it was very difficult to make a satisfactory machine with the rudimentary and inaccurate gear-cutting techniques then available. Nevertheless, during a period of about two years he devised and constructed more than fifty different machines until in 1645 he finally produced his definitive model and organized its manufacture and sale himself. He was given a monopoly by royal decree in 1649 and three years later he demonstrated the machine to Queen Christina of Sweden. By that time, however, his interests were turning to the physical problem of the vacuum.

In his *Discourses on Two New Sciences* (1638), Galileo remarked that it was impossible to raise water with a pump more than about 30 feet, a fact confirmed by experiment in 1642. Two years later Galileo's pupils Viviani and Torricelli, experimenting with mercury, a much denser liquid than water,

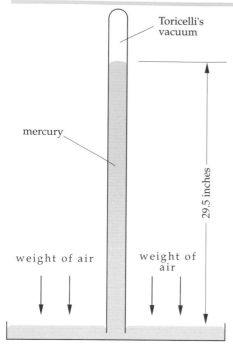

Figure 7.13 *The mercury barometer. The weight of the air (i.e. air pressure) supports the column of mercury.*

found that the height of the column was about 29.5 inches. Torricelli explained the effect in terms of the weight of the air and realized that the space above the mercury column must be a vacuum (see Figure 7.13). In 1646 Petit, a French government official in charge of fortifications, who had learned of Torricelli's work from Mersenne, visited Rouen and being an able experimenter, he repeated Torricelli's experiments assisted by Etienne and Blaise Pascal. The latter repeated them again in the presence of important citizens of Rouen. In his most spectacular experiment he used a glass tube 46 feet long filled with red wine. He had this set up vertically with the open end stopped and immersed in a vessel of water to the depth of a foot. When the stopper was withdrawn the wine in the tube dropped to a height of about 30 feet above the surface of the water in the vessel. When the tube was inclined water entered at the bottom, until when the vertical height of the top was about 30 feet above the water level in the vessel the tube filled again, but now there was wine from the top to 13 feet above the water level and water in the bottom 13 feet (wine being lighter than water). By this experiment and others in which he used tubes of various widths and shapes (see Figure 7.16), as well as different liquids, including mercury and oil, Pascal showed conclusively that the critical factors in every case were the nature of the liquid and the *vertical* height from the level of the liquid in the vessel to the top of the column, not the length, width or shape of the tube (Barry, 1973, pp.3ff).

Mersenne, in his *Reflexions physico-mathematiques* (1647), suggested the idea of comparing the height of the mercury column in a barometer tube at different altitudes as a means of confirming the weight of the air. In the same year

---

The principle of the hydraulic press

Pascal investigated the forces exerted by columns of liquids, basing his experiments on the principle that the pressure at a point in a liquid is the same in all directions. He considered the relationship between the total force exerted on pistons and their cross-sections. He showed that if two cylinders of different cross-sections containing a liquid and closed by pistons were put into communication, equilibrium was established when the pistons were loaded with weights proportional to the areas of their cross-sections. He realized that this principle could be used in a machine to multiply force. Thus if the area of B is one hundred times the area of A, W_2 will be one hundred times the weight of W_1.

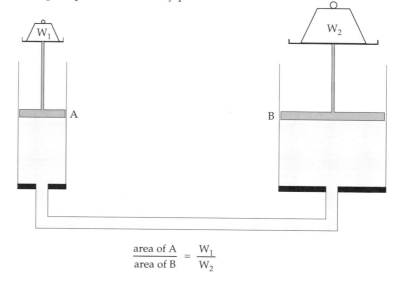

$$\frac{\text{area of A}}{\text{area of B}} = \frac{W_1}{W_2}$$

Figure 7.14 *Pascal's experiment showing the multiplying effect of the hydraulic press.*

Pascal wrote a pamphlct on his experiments, *Expériences nouvelles touchant le vide*, which he distributed in France, Sweden, Holland, Poland, Germany, Italy and throughout the whole European scientific community. At about the same time, Pascal heard of Torricelli's idea of the weight of the air and that this might be the true cause of the effects which had hitherto been ascribed to the Aristotelian belief that 'nature abhors a vacuum'. As there was no experimental evidence to support this conjecture Pascal made several attempts to test it. In the most sophisticated of these he set up a mercury barometer in the vacuous space above the mercury in a much larger barometer. No mercury column was formed in thc inner barometer tube, bul when the air was let into the larger tube the mercury in it fell while the mercury column rose to its normal height in the inner tube.

Still not entirely satisfied, Pascal conceived the famous experiments which were carried out in 1648 by his brother-in-law Perier, who lived at Clermont-Ferrand near the Puy de Dôme, a mountain of about 5,300 feet. Pascal asked Perier to observe the height of the mercury column at the foot of the mountain and again at various altitudes, noting any variations. On 19 September 1648 Perier and a group of observers measured the height of the mercury column in the gardens of the Minim fathers, which were at just about the lowest point in Clermont-Ferrand. After repeated trials using freshly distilled mercury in two barometer tubes each 4 feet long the height of the mercury column in each tube was found to be 26 inches and 3.5 lines. Then, leaving a reliable observer to record the height of these barometers at intervals throughout the day and to note any changes in the weather, Perier and his group ascended the mountain taking the other barometer tube to measure the height of the mercury column at the summit. They found it to be 23 inches and 2 lines, a difference of 3 inches and 1.5 lines. On returning to the garden of the Minims in Clermont-Ferrand they found that no change in height of the mercury column there had occurred during the day, despite unsettled weather (Barry, 1973, pp.103ff). After repeated trials Perier was able to show that for every foot he climbed on the lower slopes of the mountain the height of the mercury column fell faster than it did a thousand feet up the mountain. Pascal immediately published a short account of these experiments with Perier's measurements, confirming Torricelli's notion about

Figure 7.15 View of Puy de Dome from Clermont-Ferrand. (Académie de Clermont-Ferrand)

the weight of the air and showing that it declined steadily with increasing altitude. The success of the experiments led Perier to repeat them at the base and top of various towers, and also in the attic and cellar of a house. Pascal later repeated similar experiments on various tall buildings in Paris.

Figure 7.16 (a) Pascal's apparatus for measuring the weight of fluids. The receptacles I–IV are of different shapes and sizes, but if placed in contact with the straight cylinder (VI) by being fixed into the box, water poured into the apparatus will reach the same height in all cases.
(From The physical treatises of Pascal, *tr. I.H.B. and A.G.H. Spiers, New York, Octagon Books, 1973.)*

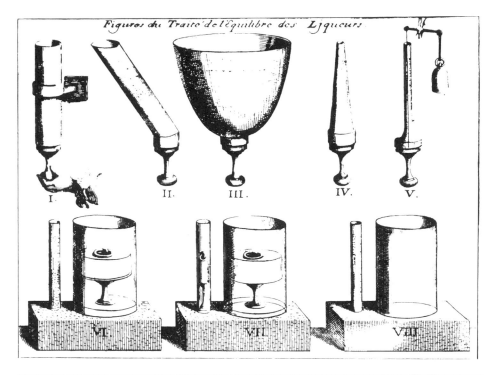

Figure 7.16 (b) Pascal's apparatus for measuring the weight of air. To demonstrate the weight of the air, Pascal described a number of phenomena (figs I–VII show some of these):
I It is impossible to operate bellows which are sealed up because the weight of the air prevents it.
II It is difficult to separate two polished surfaces in close contact for the same reason.
III & IV Water will not enter the flask in III or the sealed tube in IV due to the air inside the apparatus.
V If the bent tube is filled with water and then sealed at the top of the straight leg, water will not run out due to the weight of air pressing on the open end.
VI Represents a lift pump.
VII Represents the siphon.
(From The physical treatises of Pascal, *tr. I.H.B. and A.G.H. Spiers, New York, Octagon Books, 1973.)*

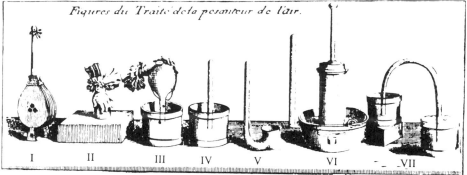

His results showed that the experimental method could be used not only to demonstrate specific points but also to lead on to more general conclusions. Like Mersenne, he applied the experimental method in a modern sense. Beginning with a theory, he tested it quantitatively and then used his results to reach wider generalizations by inductive reasoning. Thus, Pascal's physical experiments are important not only for their specific results, but also for their contribution to the development of the scientific method. In spite of the care he took to ensure that his experimental results were reliable, there was much controversy over his ideas, and he planned to publish a large treatise in which he would explain his work and answer all the objections that had been raised. This never appeared; instead, with his passion for brevity he condensed his accounts into two small works written between 1651 and 1653, though only published after his death, in 1663. These were *Traité de l'equilibre des liquers* and *Traité de la pesanteur de la masse de l'air*. Both are important works in the history of science.

In 1654 Pascal returned to his work on conics, and it was at this time that he encountered the problems of prediction connected with the throwing of dice which led him to studies of probability and the theory of numbers. He wrote a little work on the cycloid curve as applied to the roulette wheel under the pen-name of Ettonville and in his *Traité du triangle arithmetique* (1654) he laid the foundations of the modern theory of probabilities. From the end of 1653, however, he again began to feel religious scruples concerned with fears about personal ambition. The 'night of the fire', a powerful religious experience which Pascal had in November 1654, began a new life of Christian devotion. He entered Port Royal and although he never became one of the *solitaires* (brothers) he wrote thereafter only at their request. His best known religious work, *Les Provinciales*, was written in 1656–7 in defence of Antoine Arnauld, the Jansenist Cartesian, on trial before the faculty of theology in Paris following his criticisms of the Jesuits' practices, especially their methods in the confessional. He was unable to save Arnauld from losing his post at the Sorbonne, but *Les Provinciales* was an immediate success and was to achieve lasting popularity. Thus, while Pascal is important for his scientific discoveries, he is also a very important figure in French literature. His theological writings mark the beginning of modern French prose and are still intensively studied. From 1656 he spent time in meditation on miracles and other proofs of Christianity in preparation for a work of Christian apologetics. This was never published, but during the next two years he put together most of the notes and fragments he had collected for this work. This collection was later edited and published posthumously as *Pensées* (1657–8); it is the best known of all Pascal's literary works.

7.8 Baconianism in the Académie Royale des Sciences

The Académie Royale des Sciences, established in December 1666, was organized in two sections: mathematics (including mechanics and astronomy) and physics (including chemistry, botany, anatomy and physiology) (see Chapter 9). Concerned with experimental work, the academicians modelled themselves on the Royal Society, and their programme followed Baconian ideals. Experiments previously reported by the Accademia del Cimento in Florence and the Royal Society were repeated. These included tests on the freezing of water and experiments with the air-pump. Metals like lead and antimony were found to increase in weight on calcination and it was thought that something had been absorbed during heating, either from the air or from the containing vessel. Some mineral waters were also analysed. A large number of dissections were carried out, including the bodies of a panther and an elephant from the Versailles menagerie. These dissections were not done according to a plan, and differences were sought rather than similarities. As in the Royal Society, some attempts at blood transfusion were made, but without success. Plants were examined by chemical and mechanical methods; juices were squeezed out and tested with chemical solutions or allowed to evaporate so that the salts which crystallized out could be examined. Plants were subjected to destructive distillation and the distillates were tested for acids or 'sulphureous' particles. No recognizable botanical studies seem to have been attempted.

In mathematics the academicians worked on problems identified by Descartes, while in hydrostatics they investigated the relation between the speed of flow of a fluid from an orifice and the head of pressure in the containing vessel. Like the Royal Society the Academy appointed several of its members to investigate the construction and working of tools and machines in common use with the object of improving or simplifying them. Ingenious mechanical devices – including frictionless pulley systems, pumps and automatic saws – invented by members of the Academy were published in an illustrated catalogue. All these activities show the influence of Bacon, both in the random collection of data and in the efforts of the academicians to make their science useful.

The most successful work was done in astronomy. At first, astronomical observations were made from a garden at the rear of the meeting place of the Academy in the Royal Library, but as this was hemmed in by houses, an appeal was made to the king to found a proper observatory, which was built to a design of Claude Perrault in the Faubourg St Jacques. It was in use by 1672. From 1669 astronomical work was carried out under the direction of G.D. Cassini, an Italian astronomer invited to Paris by Colbert. Huygens, who had already made improvements to the telescope whilst in Holland, contributed to the work of the observatory not only by his astronomical observations, but also with his pendulum clock for measuring time more accurately and with his wave-theory of light. Huygens, Picard and Auzout, together with Cassini, improved the precision of astronomical observations and measurements by introducing the practice of attaching graduated circles to their telescopes for measuring angles. The object lens of the telescope was also fitted with cross-wires to make the measurements more accurate and micrometers were used to measure small angular separations between two objects seen simultaneously in the telescope. The first successful study of optical refraction was also made in Paris by Huygens, who communicated his ideas to the Académie Royale des Sciences in 1678.

Several foreign expeditions were organized by the Academy for astronomical purposes. In 1671 Picard went to Denmark to determine accurately the position of Tycho's observatory at Uraniborg, which was already in ruins. He brought back with him Olaus Romer, who became a member of the Academy and later made important observations on the velocity of light. Another group led by Richer, went to Cayenne in French Guiana to observe Mars in opposition to the Sun. From comparisons of the observations made simultaneously in Cayenne and by Cassini in Paris, more precise calculations of the size of the solar system resulted. Richer also noted that a pendulum to beat seconds must be made shorter at Cayenne than in Paris, indicating a difference in gravity between the two places – a discovery which marked the beginning of speculations about the shape of the earth. All of these activities were stimulated by Colbert's enthusiastic support, and after his death in 1683 the Academy declined to the end of the century, until its reorganization and enlargement in 1699.

7.9 Conclusion

The authority of the Roman Catholic Church in France remained strong throughout the seventeenth century in spite of internal quarrels. Jesuits, Gallicans, Jansenists and others all maintained a stout loyalty to the Church while disagreeing amongst themselves on theological issues. Under Richelieu and Mazarin, who achieved enormous political power, intellectual freedom in France was limited, censorship was strict and the threat of censure before ecclesiastical courts was ever present. The Church also exerted its authority in intellectual matters in its efforts to maintain the Aristotelian traditions. It was to escape from this repressive environment that Descartes felt obliged to leave France for the relative tranquillity and greater intellectual freedom of Protestant Holland. Similar reasons also account for his decision to present his *Principles* as hypothetical and not a true physical account of nature, contrary to the aims which he stated in the *Discourse.*

Descartes took the attack on Aristotelianism further than any of his contemporaries, apart from Bacon, and devised a philosophical system which was capable of deposing Aristotelianism from its dominant position in natural philosophy. Using the new mechanical philosophy based on physical interactions between material particles, Descartes was able to account for most observable phenomena, but falling into the same errors as the Aristotelians themselves, he was tempted to imagine more than the observations alone would allow and his description of the physical universe turned out to be highly speculative. Cartesian science was full of such philosophical notions, yet it proved remarkably resilient and was stoutly defended on the continent, especially in France, until well into the eighteenth century.

With regard to the experimental method, neither Descartes nor Gassendi can be said to have made significant advances. Though Descartes made many experiments, few were quantitative and he seems to have used his observations mainly to confirm his theoretical deductions. For Descartes, as for Galileo, mathematical argument and philosophical deduction came before experimental observations. Moreover, he was reluctant to accept observations reported by others, no matter how convincing. By comparison, both Mersenne and Pascal show a more empirical approach to experimental techniques. Mersenne, who favoured Bacon's methods, derived his theory of musical harmony from his experimental observations, while Pascal used quantitative measurements to investigate ideas suggested earlier by Galileo, Torricelli and others. The utility of science does not figure prominently in the work of either Mersenne or Pascal, however. Possible practical applications of experimental discoveries, although sometimes suggested, were rarely followed through. Thus, Pascal's work on pressure in liquids did not progress beyond the experimental stage, though it is clear that he recognized the potential of his discovery for the construction of a hydraulic press. On the other hand, Pascal persisted with great fortitude in perfecting his invention of the calculating engine. Later in the century, the members of the Académie Royale des Sciences showed a growing interest in experimentation, especially in astronomy, and a desire to demonstrate the utility of science. It is true of course that the academicians may have felt an obligation to show some practical return for the pensions they received, but the example set by the Royal Society clearly influenced the work of the Academy, where a similar Baconian programme was followed.

Sources referred to in the text

Barry, F. (ed.) (1973) *The Physical Treatises of Pascal*, tr. I.H.B. and A.G.H. Spiers, New York, Octagon Books.

Briggs, R. (1977) *Early Modern France 1560–1715*, Oxford, Oxford University Press.

Cragg, G.R. (1970) *The Church and the Age of Reason 1648–1789*, Harmondsworth, Penguin.

Fontenelle, (1729) *Ouvres diverses*, new edn, vol.3, Paris.

Hooker, M. (ed.) (1978) *Descartes: Critical and Interpretative Essays*, Baltimore and London, Johns Hopkins University Press.

Joy, L.S. (1987) *Gassendi the Atomist: Advocate of History in an Age of Science*, Cambridge, Cambridge University Press.

Treasure, G.R.R. (1981) *Seventeenth Century France*, 2nd edn, London, Murray.

Whitteridge, G. (1971) *William Harvey and the Circulation of the Blood*, London, Macdonald.

Further reading

Aiton, E.J. (1972) *The Vortex Theory of Planetary Motion*, New York, Elsevier.

Briggs, R. (1977) *Early Modern France 1560–1715*, Oxford, Oxford University Press.

Brundell, B. (1987) *Pierre Gassendi: From Aristotelianism to a New Natural Philosophy*, Dordrecht, D. Reidel.

Cottingham, J. et al. (eds) (1985) *The Philosophical Writings of Descartes*, 2 vols, Cambridge, Cambridge University Press.

Cragg, G.R. (1970) *The Church and the Age of Reason, 1648–1789*, Harmondsworth, Penguin.

Dijksterhuis, E.J. (1961) *The Mechanisation of the World Picture*, Oxford, Oxford University Press.

Greene, M. (1985) *Descartes*, Brighton, Harvester Press.

Gillispie, C.C. (ed.) (1970–6) *Dictionary of Scientific Biography*, 16 vols, New York, Charles Scribner's Sons. On Descartes, see vol.4, pp.51–65. On Gassendi, see vol.5, pp.284–90. On Mersenne, see vol.9, pp.316–33. On Pascal, see vol.10, pp.330–42.

Joy, L.S. (1987) *Gassendi the Atomist: Advocate of History in an Age of Science*, Cambridge, Cambridge University Press.

Leoffel, H. (1987) *Blaise Pascal 1623–1662*, Basle, Birkhauser.

Smith, N.K. (1963) *New Studies in the Philosophy of Descartes*, London, Macmillan.

Treasure, G.R.R. (1981) *Seventeenth Century France*, 2nd edn, London, Murray.

Williams, B. (1978*) Descartes: The Project of Pure Enquiry*, London, Penguin.

Science in Seventeenth-Century England

by Noel Coley

8.1 England at the beginning of the seventeenth century

As the reign of Queen Elizabeth I (*r.* 1558–1603) drew to a close, the prosperous in England were reaping the benefits of half a century of geographical exploration, commercial development and expanding knowledge. The economic policies of the Lord Treasurer, William Cecil, had established the monopoly of particular trades in the hands of great companies and had given the Merchant Adventurers fuller powers over commerce with the Netherlands and Germany. Trade with Muscovy and Persia had been extended, and John Hawkins's ventures to West Africa and the West Indies had been supported by the Crown and Privy Council. These moves had caused a remarkable increase in maritime and commercial activity as merchants were encouraged to venture further afield. By 1588, at least a hundred English ships called at Baltic ports each year, while trade with Spain, Mediterranean countries and the Levant flourished – seaborne trade which called for better maps, charts and navigational techniques. At home, industrial development flourished: new industries were often backed by patents, such as those for the manufacture of copper and brass, salt and glass; there was a remarkable expansion in coal-mining and the coal trade from the late sixteenth century; and many skilled workers fleeing from religious persecution in the Netherlands and France settled in England.

The religious attitudes which were the chief cause of the English Revolution in the seventeenth century had their origins in Elizabeth's reign. Presbyterian sympathizers in the Anglican Church tried to limit Roman Catholic influence, and up to the early 1580s the Presbyterian movement flourished. However, after 1583 under John Whitgift, archbishop of Canterbury, the power of the episcopal courts and the court of high commission increased. Their severity, reminiscent of the Inquisition, broke the Presbyterian movement, but may have strengthened the independent spirit of Puritanism in a wider context. The term 'Puritan' was coined in the 1560s and applied contemptuously to those in the Anglican Church who sought a more drastic reformation than the Elizabethan religious settlement allowed. This demand for reform arose from an underlying spirit of religious and moral earnestness, and the Puritans were identified with strict and closely regulated habits of life, every aspect of which was viewed in the light of God's demands.

From 1574 there was a revival of Roman Catholic activity in England, and after the arrival of the Jesuits in the 1580s, anti-Catholic penal laws were passed by a parliament fearful of a return of Catholic domination. By 1600 more than two hundred English Catholics had been executed as traitors while heavy fines were imposed for saying or hearing mass and for not attending Anglican services. This persecution led some to plot against Elizabeth's life with the aim of bringing Mary Queen of Scots to the throne with Spanish help – plots which culminated in Mary's death on the scaffold in 1587 and the attempted invasion of England by Spain in the following year. After the defeat of the Armada there were no further conspiracies, but patriots tended to associate Catholicism with treason and the threat of Spanish domination. The war with Spain dragged on to the end of Elizabeth's reign, creating a serious drain on resources, both of labour and money: armies were maintained in France and the Netherlands while Spain threatened further naval attacks; taxes rose and outlets for trade decreased with the closure of markets in France, Spain and the Netherlands, causing an increase in unemployment. Discontent with the queen's handling of affairs grew. Parliament had learned how to seize the initiative in legislation, common lawyers were beginning to harass the lesser conciliar courts, and while the Star Chamber (the High Court) had not lost its hold, Lord Chief Justice Sir Edward Coke and others criticised some aspects of the law. Thus, as the seventeenth century opened it was becoming clear that the constitutional powers of the monarch in England would no longer remain unchallenged.

Throughout Elizabeth's reign the desire of Puritans to reform the Anglican Church was thwarted by the queen's determination to retain her prerogative to power in both Church and State, but with the accession of James I (*r.* 1603–25) the Puritans hoped for a more favourable hearing. Although James also firmly rejected their demands, he allowed a degree of freedom which ensured some stability during his reign, and it was not until Charles I (*r.* 1625–49) came to the throne that anti-Puritan attitudes began to harden. Archbishop Laud adopted rigorous measures to impose his Anglo-Catholic

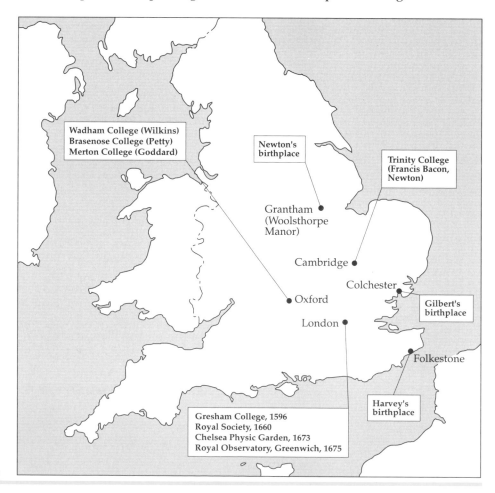

Map 8.1 Science in seventeenth-century England.

reforms on the Anglican Church; the Puritans in Church and Parliament objected and Charles, dissolving the latter, embarked on a period of personal rule which lasted through the 1630s. When an attempt was made to impose Laud's liturgy on the Calvinist Church in Scotland, the Scots rose in revolt and invaded England. Charles, lacking the finance to quell the unrest, was forced to recall Parliament in 1640, which denied him the right to impose additional taxes until various abuses of his personal reign were corrected. At the same time Parliament moved to abolish the episcopacy, creating opposition which enabled the king to rally the support of many noble families of Anglo-Catholic leanings, and with the outbreak of civil war between the royalists and Puritan supporters of Parliament in 1642, the religious and political unrest which marked the middle years of the seventeenth century in England began.

8.2 *Gilbert's experiments on magnetism*

In the realm of scholarship during the second half of the sixteenth century, the Copernican theory attracted attention as astronomers and mathematicians strove to improve the accuracy of planetary tables. In the last quarter of the sixteenth century, ideas about the size of the universe and distribution of the fixed stars also began to change. In 1576 Thomas Digges suggested that there was no reason why God should have limited the sphere of the stars: the universe might tend towards infinity with stars scattered throughout unbounded space. In the following years the Danish astronomer Tycho Brahe (1546–1601) observed several new comets and in 1588 boldly suggested that the crystalline spheres did not exist (see Chapter 3). Yet, if the heavenly bodies were not held in place by their attachment to the crystalline spheres, there must be another cause to keep them moving in their circular orbits and Brahe had no alternative explanation to offer. The problem remained unresolved until Newton's theory of universal gravitation was published in 1687, though one possible cause, the force of magnetism, which might explain the motions of the heavenly bodies was suggested by Queen Elizabeth's physician William Gilbert (1544–1603) in 1600.

Born at Colchester in Essex and educated at Cambridge, Gilbert obtained his MD in 1569. Although he was a physician, he showed a keen interest in the practical applications of science and frequently associated with mathematicians and navigational instrument makers in his home town and in London, where he lived from the 1570s. In *De magnete* (1600), prefaced by the leading English applied mathematician, Edward Wright (1558–1615), Gilbert showed how an experimental investigation of the behaviour of magnets could be made. His aim was to improve methods of navigation at sea. Navigators had long been aware of the variation of the magnetic compass needle from true north, and it had been suggested that this might be used to determine longitude. Gilbert was doubtful, but he did think that the angle of magnetic dip, discovered in about 1580 by Robert Norman, a London instrument maker, might be used to determine latitude at night and in cloudy weather and suggested a grid and special quadrant for measuring the angle of dip.

Yet, useful as Gilbert's suggestion was as an aid in navigation, the part of *De magnete* devoted to this purpose has turned out to be the least important. Gilbert's experiments on the nature of the loadstone, the natural magnet, and

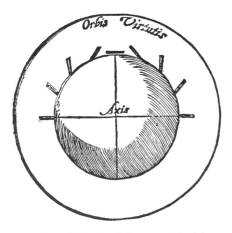

Figure 8.1 Gilbert found that a small loadstone on the surface of a terella would be parallel with the surface at the equator and vertical to it at the poles. The angle of dip would increase regularly with the position of the loadstone between these two extremes. He suggested that accurate measurement of the angle of dip could be used to determine latitude at sea when the sun was obscured by clouds, or at night. (From William Gilbert, De magnete, 2nd edn, London, 1628. Photo: Science Museum, London)

Figure 8.2 *Title page of Gilbert's* De magnete, *2nd edn, London, 1628. Note at the top left the diagram showing Gilbert's recognition of the angle of dip and at the bottom left his dip circle. The sailing ship at the bottom centre reminds us of Gilbert's aim to improve marine navigation. (Science Museum, London)*

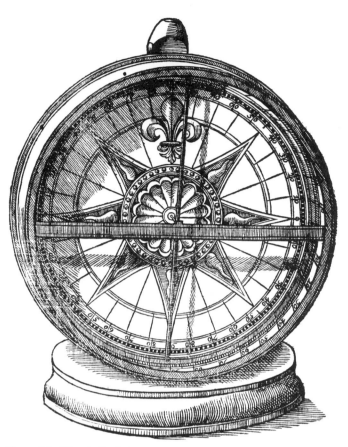

Figure 8.3 *Gilbert's dip circle instrument. (From William Gilbert,* De magnete, *2nd edn, London, 1628. Photo: Science Museum, London)*

its five movements (coition, direction, variation, declination and revolution) have proved his most valuable contribution to experimental science. He distinguished between electric and magnetic attraction and, using a sphere of loadstone (*terrella*, or 'little earth') with small bar magnets, he demonstrated the magnetic movements. The angles of dip and variation of a suspended magnet depended on its position on the surface of the *terrella*, and Gilbert argued that the same forces would apply to a compass needle on the Earth's surface if, as he believed, the Earth itself were a large magnet. This was confirmed by the fact that blacksmiths could produce magnetism in a hot iron bar by hammering it as it lay in a north–south line. It followed that the compass needle was affected by the Earth's magnetic field and not by the celestial poles as had previously been thought. The observation that a freely suspended *terrella* revolved led Gilbert to suggest a theory based on the Earth's magnetism to explain its daily rotation.

In spite of his use of experiments and observations to construct a general theory of magnetism and explain the diurnal rotation of the Earth, it is certainly not justifiable to ascribe to Gilbert a modern approach to physical science. He mentioned, often with derision, earlier writers who had thought the magnetic force occult, but his own world-view was strongly influenced by belief in a world soul.

> ... we deem the whole world animate, and all globes, all stars and this glorious earth too, we hold to be from the beginning by their own destinate souls governed.

Figure 8.4 *Blacksmith hammering a piece of hot iron held in a north–south line. The iron becomes magnetized showing the effect of the Earth's magnetic field. (From William Gilbert,* De magnete, *2nd edn, London, 1628. Photo: Science Museum, London)*

Moreover, the stars have

> ... reason, knowledge, science, judgement, whence proceed acts positive and definite from the very foundations and beginnings of the world. *(Gilbert, [1600] 1958, book 5, ch.12)*

Thus, in these respects Gilbert's cosmological ideas were in line with those of men like John Dee, Giordano Bruno and Sir Walter Raleigh, whose conception of nature was fundamentally occult and magical.

8.3 *William Harvey and the circulation of the blood*

The experimental method led to another important discovery when the physician William Harvey (1578–1657) applied physical measurements and arithmetical calculations to his observations of the movement of blood through the heart. Born at Folkestone in Kent, Harvey studied medicine at Caius College, Cambridge, and at Padua. On returning to England he became a prominent physician in London and was appointed Lumleian professor at the Royal College of Physicians in 1616. At Cambridge Harvey had been immersed in Galen's theories, but Fabricius at Padua introduced him to a more mechanical and experimental approach to physiology. Fabricius replaced Galen's primacy of the liver with Aristotle's preference for the heart as the chief organ of the body. Following Andreas Vesalius's methods of dissection and first-hand observation, Harvey investigated the structure and functions of the heart. He asked himself why the 'arterial vein' nourished only the lungs while the 'venous artery' served the whole body; why the lungs should need so much more nourishment than the rest of the body; why when the lungs moved, the right ventricle of the heart also moved and why it should exist solely to serve the lungs.

Figure 8.5 *Portrait of William Harvey (1578–1657). Harvey studied at Padua (1600–2) where he came under Galileo's influence. Harvey's demonstration of the circulation of the blood, published in 1628, depended on observation coupled with Galileo's principles of measurement and calculation. (From W. Pagel,* William Harvey's Biological Ideas, *Basel, S. Kager, 1967.)*

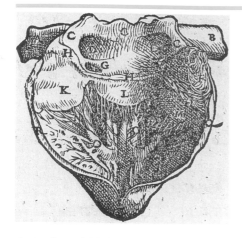

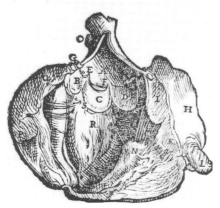

Harvey first set out his theory of the motion of the blood in 1618. For the next ten years he continued to experiment on the motions of the heart and blood in animals until in 1628 he published *De motu cordis*, in which he treated the motions of the heart and blood from an anatomical point of view. To observe the heart in motion, Harvey dissected live cold-blooded animals such as fish, toads, frogs, snakes and lizards as well as a mammalian foetus, establishing a tradition which was to be continued in the Oxford Experimental Philosophy Club and later in the Royal Society. Harvey found that the heart is a muscle, active in systole (i.e. when it contracts to expel blood) rather than in diastole (expansion), as had always previously been thought. He also found that he could correlate the dilation of the arteries with the pulse and heartbeat. He recognised that the heart performed a mechanical function in transmitting blood by means of the arteries to the extremities of the body, but he also thought it had a vital function in maintaining life. He observed the pulmonary circulation in fish and showed that the blood passed from the right side of the heart through the gills to the left side. This led him to conclude that the blood follows a circular path from the left side of the heart through the arteries to the rest of the body and back to the right side of the heart via the veins. He calculated that the weight of the total quantity of blood passing through the human heart every twenty-four hours was greater than the weight of the whole body. As this could not all be new blood produced by the liver as Galen had taught, it followed that the *same* blood must be moving round the body via the heart. Harvey concluded that the circulation of the blood was a logical necessity (Whitteridge, 1971, p.142)

Harvey's discovery was framed within the Aristotelian tradition.

I began to think whether there might not be a motion as it were in a circle ... in the same way as Aristotle says that the air and the rain emulate the circular motion of the superior bodies [the planets].

Figure 8.6 Dissections of the heart from Andreas Vesalius, De humani corporis fabrica, *Basel, 1543. Harvey was also influenced by Vesalius's anatomical work. (Reproduced by permission of the British Library Board.)*

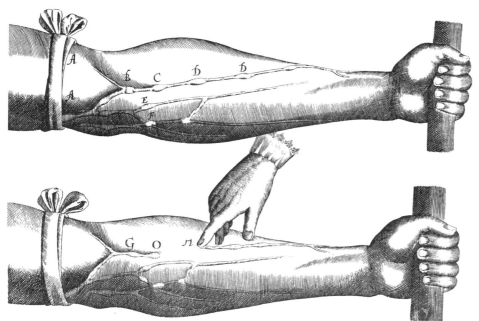

Figure 8.7 Membranes in the veins and heart which allowed the blood to pass in one direction only had long been known. Harvey showed that they acted like valves which kept the blood moving in a circle from the arteries to the veins throughout the body. (From William Harvey, De motu cordis, *London, 1628. Reproduced by permission of the British Library Board.)*

Figure 8.8 Title page of the first edition of William Harvey, De motu cordis, *London, 1628. (Reproduced by permission of the British Library Board.)*

He also used neo-Platonic arguments to support his new theory (see Section 2.4.3).

> The heart ... is the sun of the microcosm, even as the sun in its turn might well be described as the heart of the world. *(Harvey, [1628] 1907, pp.56–7)*

The sun and the heart both provided living creatures with warmth and life; the heart ruled the body just as the sun governed the heavens. Thus, behind Harvey's empiricism, as with Gilbert's work on magnetism, there lay a mysticism typical of the period, shared by Dee, Fludd, Bruno, Digges, Gilbert and Copernicus (Kearney, 1971, p.86). Nevertheless, Harvey's discovery was an important step in the transition from scholasticism to experimental science. His investigation of the heart and circulatory system created a precedent which could readily be applied to other areas (Webster, 1975, pp.315f). Harvey's theory appealed to younger physicians, but caused heated arguments with older members of the Royal College of Physicians, who remained staunch adherents of Galenism. In the treatment of disease, traditional ideas prevailed and Harvey's discovery remained no more than an intellectual diversion (Whitteridge, 1971, *passim*).

8.4 Francis Bacon's scientific method

Harvey's contemporary Francis Bacon (1561–1626) is usually regarded as a pioneer of seventeenth-century scientific method. Bacon recognised the need to reform knowledge, but rather than repeating the mistakes of the past, he argued that new information should be sought, not from books but by the direct observation of nature. Since it would never be possible to observe every instance of each phenomenon, conclusions would have to be drawn from the available observations, and Bacon showed how improved explanations of physical phenomena could be achieved by collecting and classifying experimental results. But Bacon criticised the mere empiricist as much as the dogmatic theorist. In *Novum organum* he wrote

> The men of experiment are like the ant; they only collect and use; the reasoners resemble spiders, who make cobwebs out of their own substance. But the bee takes a middle course; it gathers its material from the flowers of the garden and of the field, but transforms and digests it by a power of its own. *(quoted in Kearney, 1971, p.91)*

Figure 8.9 Francis Bacon (1561–1626). (National Portrait Gallery, London)

Statesman, lawyer, essayist and philosopher, Bacon had no time to compose more than a small part of the comprehensive work he envisaged. Like Descartes, he aimed to reform all knowledge and create a new learning which would stem from the application of scientific method. His ideals were never fully realised, but his vision inspired others and his advocacy of the experimental philosophy, together with his great personal prestige provided moral support for those who pursued similar goals later in the century. He censured scholasticism for being concerned with arguments largely unconnected with observations of the physical world and for perpetuating subjective errors arising from the unreliability of the senses, the search for final causes and the insidious influence of prejudice. He was aware of the attraction of false ideas and superstitions as well as the tendency to construct whole systems on the basis of a few observations in one particular area, as Gilbert had done in magnetism. These pitfalls Bacon called 'idols', since he thought they had often become the unchallenged tenets of belief.

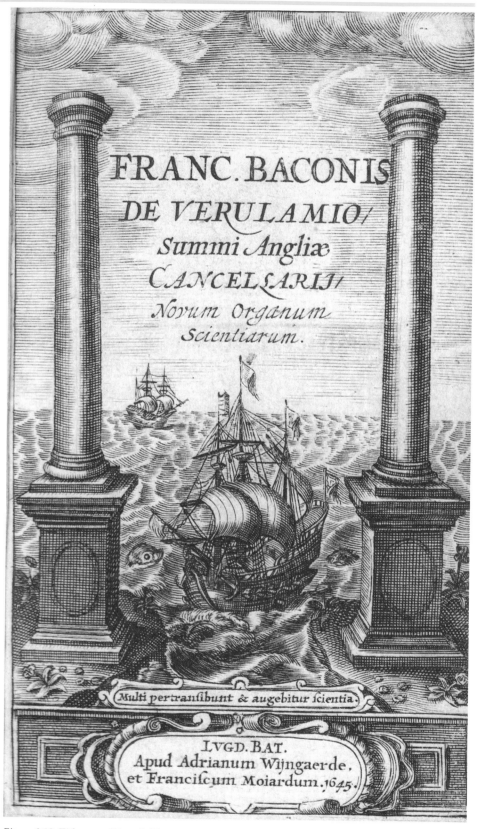

Figure 8.10 Title page of Bacon's Novum organum, *London, 1620. The ship sailing through the 'pillars of Hercules' from the Mediterranean Sea to the Atlantic Ocean is an allegory of Bacon's intellectual 'voyage' from the relative safety of traditional ideas to the uncharted waters of new knowledge. (Reproduced by permission of the British Library Board.)*

Bacon's contribution to the rise of science was threefold. First, his writings made him a successful advocate for science. He had a gift of exposition which enabled him to present ideas in lively terms, ensuring a demand for his works. Second, he taught the inductive method by which conclusions are drawn from large numbers of observations. He also stressed the importance of well-designed experiments, though he was not himself an experimenter. Third, he countered the danger that the new science would be condemned by the Church by seeking to separate science from religion. He argued that increased knowledge about nature need not lead to disbelief in God, but that being purged of fancies, the understanding would become more submissive to the divine oracles. The new knowledge would reveal the glory of God and improve the human condition. Yet, he was unable to free himself from tradition entirely: he accepted occult beliefs in astrology and alchemy; he gave credence to the interpretation of dreams, to divination and some forms of natural magic. Together with most of his contemporaries, he accepted the alchemists' dream of the philosopher's stone which would change base metals into gold and prolong human life. He also sought a new theory of the structure of matter based in part on Paracelsian ideas (Rees, 1977, pp.110–25). Convinced that the experimental method was the only way to improve knowledge of nature, Bacon recognised that the development of science by this means would require corporate effort and he drew up a code of procedure in his *Novum organum* (1620). He believed that almost any industrious person of ordinary common sense could make useful discoveries. In this he underestimated the need for originality, the importance of mathematics and the difficulty of reducing experimental methods to simple rules of procedure.

With the growth of industry and overseas trade, it was desirable to improve manufacturing techniques and Bacon thought that science could provide the means of doing this. He believed that artisans would be as able to contribute to the advancement of science as scholars – the lack of a classical education need not preclude contributions to experimental knowledge. For seventeenth-century English Puritan dissenters, deprived of access to a university education, this offered hope and encouragement. During the Civil War and Cromwellian Interregnum (*c.* 1642–60), Baconianism was embraced by many leading English Puritans who played a prominent part in the rise of the new experimental philosophy and helped to demonstrate the usefulness of science.

In *The New Atlantis* (1626), Bacon discussed the organisation of the fictional Solomon's House, where experiments would be conducted by 36 academicians working in small groups of three or four. He envisaged caves for tests on refrigeration, working in metals and curing diseases, high towers for observing meteors and the weather, lakes, fountains, walls, orchards and parks, each with its own programme of experiments. There would be furnaces, brew-houses, kitchens, dispensaries, perspective houses for optical work, engine houses for work on motion and a mathematical house with geometrical and astronomical instruments. He drew up a long list of topics to be investigated, many of which reveal the importance he gave to utility. Yet, despite this corporate activity and his disclaimers, Bacon's ideas still owed much to Aristotle. His proposed coverage was encyclopaedic in the Aristotelian tradition and he still used terms such as 'humours' and the contrast between 'natural' and 'violent' motion (Webster, 1975, p.333). Later in the century, Bacon's followers in the Royal Society adopted his idea of corporate action in the hope of preparing the ground for a technological revolution in which science would play an important role.

8.5 Precursors of the Royal Society

8.5.1 Gresham College

In London at the beginning of the seventeenth century, there were already opportunities to learn about science by attending public lectures given at several institutions. Lectures on anatomy and surgery were given at the Royal College of Physicians and the Company of Barber-Surgeons, and at Gresham College, public lectures on practical topics such as surveying, land measurement and navigation were available. Founded under the terms of Sir Thomas Gresham's will (*d.* 1579), Gresham College opened in 1598 with seven professors (of rhetoric, divinity, music, physics, geometry, astronomy and law), who were appointed to live and lecture in his former house on Bishopsgate Street in the City. Most of these men were Puritans who advocated the practical applications and usefulness of science. They also held other posts or were frequently absent from the College for various reasons, but in science and mathematics the Gresham professors excelled. They concentrated on mechanics, statics, anatomy, chemistry and navigation, all of which were studied experimentally with an eye to utility.

The Gresham professors of geometry and astronomy were especially successful in establishing their College as a centre for applied mathematics and navigation. They became prominent for their work on magnetism, the construction of dials for compasses and other navigational instruments and the use of logarithms. Henry Briggs, first Gresham professor of geometry, was a successful teacher whose work was continued by Edmund Gunter, while Henry Gellibrand, the first professor of astronomy, was followed in 1636 by Samuel Foster – all four had strong Puritan ties. With their experimental approach and emphasis on utility the Gresham professors formed close links with the naval dockyard across the river at Deptford. Here, in collaboration with naval architects they became involved with the design and construction of ships and the determination of displacement

Figure 8.11 Founded in 1598, Gresham College was well established by the 1620s as a centre of useful mathematical and experimental learning. Henry Briggs, professor of mathematics at Gresham College, popularised logarithms calculated to the base 10 for practical use. He also established the College as a meeting place for discussing problems in trigonometry, nautical measurement and navigation. (Mary Evans Picture Library)

tonnage, all related to the stability of fighting ships at sea and in battle. Gellibrand also tried to use the variation of the magnetic declination as a means of improving navigation at sea. In these ways Gresham College established a tradition for collaboration between scholars and artisans in the application of scientific ideas to technical problems. By the 1630s the College began to fall into decline, but its fortunes revived in the 1640s when a group of natural philosophers began to meet there, and in the late 1650s under the leadership of Christopher Wren, Gresham professor of astronomy from 1657, it provided the first home for the group who founded the Royal Society in 1660.

8.5.2 Samuel Hartlib's projects

Another perspective on the organisation of science and learning in England can be found in Samuel Hartlib's work. Son of a Polish merchant, Hartlib came to England in 1625. His efforts to improve agriculture and his interest in useful learning in general were recognised by parliament, who granted him a pension of £100 in 1646, trebled in the following year. From then on he devoted all his energies and most of his financial resources to promoting the growth of useful knowledge for the public good. His circle of friends and supporters included John Milton, Robert Boyle, John Beale, Seth Ward and William Petty. Hartlib also knew Marin Mersenne in Paris and, like him, carried on an extensive correspondence, acting as an 'intelligencer' for new ideas, discoveries and inventions from all over Europe. Hartlib was enthusiastic about the experimental philosophy, and his dearest wish was to establish an institution to promote the new learning in England, backed by the State. His works are littered with references to usefulness, the public good, relief of the poor, the advancement of learning and always with reference to the applications of the experimental method. He constantly employed Baconian terms and agreed with Bacon's views about the role of the artisan and the potential of corporate enterprise. Hartlib espoused the utilitarian aspects of Baconian science not only in the search for useful knowledge through experiment but also in the array of inventions and social projects he sponsored. Though these were not themselves scientific, they used the scientific method and were founded on the observation of nature rather than meditation, speculation or the study of books. Improvement in the human condition was sought not in abstract philosophy, theology or ancient lore, but in the efficient deployment of practical expedients.

Among Hartlib's many proposals was an idea for an Office of Public Address, which was to be a state-regulated institution founded on Baconian principles (Webster, 1975, pp.67–77). As an international correspondency for the interchange of ideas, the Office would formalise and extend what Hartlib was already doing: it would act as a bureau to direct the efforts of inventors and make available information on new inventions for the common good. Hartlib also hoped that by having state patronage the Office would be able to persuade inventors to remain in England. Plans for this Office were discussed in 1646 and made public in the following year. It was to be organised in two sections: an Office of Accommodations, a kind of labour exchange and advertising agency; and an Office of Address for Communications, a national agency for research. Hartlib, who would be the agent for communications, aimed to encourage English scholars, entertain foreign visitors and maintain an international correspondence. He proposed to contact the librarians of all the major English libraries, requiring them to supply useful information from their collections of books and manuscripts. The enterprise was enormous; for more than a decade plans for this Agency were discussed, modified and refined. By 1655 its essence had changed and it had become chiefly concerned with the advancement of Baconian science. Efforts were made to secure parliamentary support, but the scheme never received official recognition.

Hartlib was patron to the mathematician John Pell, to whom – along with Theodore Haak, an exile from the Palatinate – he delegated international correspondence on scientific and mathematical topics. Haak initiated a correspondence with Mersenne in Paris (see Chapter 7) which gave some impressions of scientific work in France and helped to awaken the desire to encourage similar activities in England. Discussions and experiments on scientific or technical matters were often carried on in private houses, the consulting rooms of physicians and surgeons, apothecaries' shops or printers' workshops, and such activities usually went unrecorded except in letters. Sir Charles Cavendish, who lived much abroad, often reported the latest European developments in science to Pell, Hartlib, Boyle or Petty, whilst Mersenne mentioned the latest discoveries and theories in letters to various individuals. Such correspondence was very important in the absence of more formal means of disseminating information and ideas.

8.5.3 Wadham College

During the 1640s in England several groups of scholars and 'virtuosi' became interested in the new experimental science. Their activities were to be combined when the Royal Society was founded in 1660. One group began to meet regularly at Gresham College from about 1645. Another was formed at Oxford around John Wilkins. After the surrender of Oxford by Charles I in 1646, changes were made in senior posts in the university as Puritans received preferment under Cromwell: Wilkins was made Warden of Wadham College in 1648; John Wallis became Savilian professor of geometry in the following year; and in 1651 Jonathan Goddard was appointed Warden

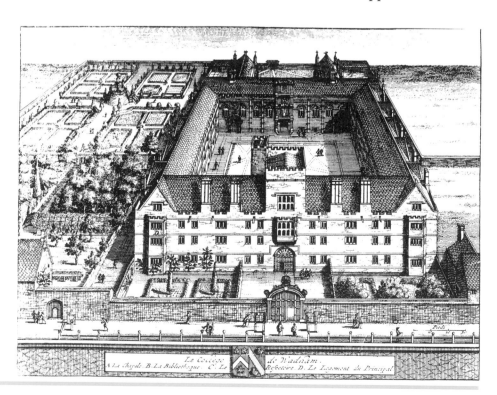

Figure 8.12 Wadham College, Oxford, from the west. The scholars who gathered around John Wilkins at Wadham College in the late 1640s formed the nucleus of the original Royal Society. (Mary Evans Picture Library)

of Merton College. Once established in their new posts Wilkins and Wallis began to meet together with others as they had become accustomed to do at Gresham College in London. A sizeable group was soon engaged in collecting information, discussing new ideas and conducting experiments on problems in natural philosophy, mathematics and astronomy. Like the London group, those meeting in Oxford were marked for their political and religious toleration, though Wilkins, Wallis and Seth Ward, a royalist member of this group and Savilian professor of astronomy and a royalist, were all theologians. Their scientific ideas were coloured by natural theology: the wish to demonstrate by purely rational arguments the existence and attributes of God and the main tenets of Christian belief.

From 1650 the Wadham College group was supplemented by another, led by William Petty, who arrived in Oxford in 1649 as assistant to the professor of physick (medicine). Petty studied medicine in Holland and Paris, where he had met Thomas Hobbes and the members of Mersenne's circle. In Oxford he began to hold meetings on chemical and medical subjects, complementing the interests of the Wadham group, in his lodgings above an apothecary's shop. Petty became professor of anatomy in Brasenose College, but in 1652 he left academic life to become physician-general to Cromwell's army in Ireland. Here he organised a comprehensive land-survey in 1655–56 for which he received extensive Irish estates in payment. He was later a member of the first council of the Royal Society.

While chemistry was among the most popular of the sciences at this time, optics also attracted a great deal of attention. Experiments were carried out with reflectors, mirrors and lenses, while improvements were made to the telescope and microscope. The techniques of lens grinding and polishing attracted much attention, and Wren is said to have designed micrometer scales to be fitted to telescopes and microscopes so that they could be used to make accurate measurements. Both were considered noble instruments which provided ways of studying the 'Wisdome of the Great Architect of Nature'. Ward and Wren wanted to establish an astronomical observatory at Wadham as had been envisaged by Henry Savile. This project was begun in 1650 when £25 was set aside for a portable observatory on the tower of Wadham College. The collection and construction of telescopes seems to have been a major activity at Wadham, and in 1655 Wilkins and Wren are said to have been building an 80-foot telescope, which would have been the largest of the time. In the end this was scaled down to a more modest 24 feet. Wren used this instrument to examine the surface of the Moon. He also collated the observations of Saturn made by Oxford astronomers since 1649. From this he produced a new theory of the phases of Saturn and of an elliptical corona which he announced in his lectures on astronomy at Gresham College.

Wilkins's garden at Wadham was devoted to experiments in horticulture, husbandry and the cultivation of plants, including 'Indian wheat', fruit trees and flowering plants. Wren invented a beehive with a glass wall so that the activities of the bees could be observed. Other inventions were concerned with farm machinery, ploughs, seed drills and so on, revealing a serious interest in the improvement of husbandry by the Oxford group. Wilkins's passion for machines was well known; he had a collection of devices invented by himself, Wren and Petty, and he tried to show that the operations of complex machines depend on the basic laws of mechanics. In other experiments inspired by Harvey's work on the circulation of the blood, Wren injected ale and medicines into the veins of living dogs, beginning the series of experiments which culminated in the blood transfusion tests carried out later by Richard Lower. In these and other ways the Oxford group

Figure 8.13 John Wilkins (1614–72). On the Parliamentary side in the Civil War, he became Warden of Wadham College, Oxford, in 1648. He established the Experimental Philosophy Club at Wadham College, and was connected with the group meeting at Gresham College from 1654. He married Cromwell's sister, and later became bishop of Chester. He was one of the first secretaries of the Royal Society along with Henry Oldenburg, in 1660. (Courtesy of the Warden and Fellows of Wadham College, Oxford.)

demonstrated the potential utility of science. Both Sprat, in his *History of the Royal Society* (1667), and Wallis give the impression that the Oxford group broke up in about 1658 when its leading members moved to London and revived the meetings at Gresham College which culminated in the establishment of the Royal Society, but in fact the group was still in existence long after the Restoration. Led by Robert Boyle, it included Wren, Lower, Willis and Mayow, all of whom were engaged in important experimental work.

8.6 Puritanism and science

After the execution of Charles I, England was governed for the next eleven years along strict Puritan lines under Cromwell. Yet despite the uncompromising attitude of the Puritan rulers, principles of religious toleration were slowly established as a matter of expediency. Moreover, by allowing debates in the army and in the self-governing churches, and by responding to demands for liberty and the denunciation of arbitrary power, the Puritans also contributed to the rise of democracy and an intellectual climate which favoured independence of mind and self-reliance. Seventeenth-century English science with its practical emphases on experiment, invention and utility contrasts sharply with the philosophical and mathematical system-building which marked contemporary French science (see Chapter 7), and it has been suggested that one reason for the difference can be found in the influence of Puritanism. Gresham College, where theory and practice went hand-in-hand, was a hot-bed of Puritanism while many, though by no means all, of the leaders of the Oxford group had Puritan sympathies or depended on Cromwell for their university posts.

The rise of the new learning in England certainly coincided with the political ascendancy of Puritanism, but the coincidence is not easy to explain. The nature of the relationship between science and religious belief has exercised scholars ever since the sociologist Max Weber (1814–1920) first postulated a connection between capitalism and Calvinism. Robert Merton, proposing a similar connection between Puritanism and science in 1938, suggested that Puritan values – such as self-restraint, orderliness, simplicity, attention to detail and single-mindedness – promoted an aptitude for scientific and technological research. But both Weber and Merton recognised that similar attitudes could also be found among other religious sects such as Quakers, Independents, Mennonites and Pietists (Hooykaas, 1973, pp.135f). Moreover, as the membership of the Royal Society in the 1660s shows, single-minded devotion to science does not depend upon particular religious beliefs. Among the Fellows there were men of every political and religious persuasion, united by a common intellectual interest in science (Hunter, 1989, pp.58–60). Most shared the renewed biblical ethic and the new understanding of creation.

Nevertheless, Baconianism did appeal to Puritans for several reasons. It offered a practical way of studying nature and extending knowledge of God's creation without book learning and was therefore more egalitarian than classical scholarship had ever been. It was anti-authoritarian, an attitude which reflected Puritan independence of mind. Moreover, Christian theology permeated Bacon's philosophical works and this fitted perfectly the Puritan ideal of Christianizing every part of life. Reason being in accord with divine truth, learning must lead to 'sublime knowledge'. The discovery of the New World during the preceding century seemed likely to lead to the revelation of a new intellectual world. Wilkins even postulated a new world in the heavens with his idea of inhabitants on the moon. He regarded the arts that conquer and surpass nature as having the highest aim – to take away the curse of labour and restore human dominion over nature. These ideas and qualities remained important in the activities of the Royal Society, but they were not the sole prerogative of Puritans. Others who had no such pretensions shared the same ideals, notably the Society's two most prominent Fellows, Boyle and Newton. Both were nominally Anglican, but both had deep religious convictions and lived strict, ascetic lives.

8.7 Early scientific work of the Royal Society

The Royal Society was founded on 28 November 1660 and its first meeting place was Gresham College, where Christopher Wren was professor of astronomy (see Chapter 9). In its scientific activities the Society followed fashionable lines of enquiry – observations with the microscope, experiments with the air-pump, and anatomical and physiological experiments, including vivisection. Medical researchers in the Society carried out tests on the blood and made attempts at blood-transfusion as a method of treatment in sickness and old age. Experimental techniques were developed under the influence of Robert Boyle and Robert Hooke, the Society's first curator of experiments. Hooke remarked that new instruments like the microscope and telescope extended the range of the senses, while axioms and theories aided the memory and provided a theoretical basis for reason. Using the cyclical method of observation, experiment, hypothesis and verification, he argued that there was nothing in nature which could not be analysed. He demonstrated many new observations, instruments and experiments at the weekly meetings of the Society. The experimental demonstrations were chosen partly for their potential value in scientific research, partly by trial and error in the hope of hitting upon some useful discovery and partly it must be admitted, for their entertainment value.

8.7.1 Microscopy

The simple microscope, or magnifying glass, a single convex lens giving a magnified image, was well known long before the seventeenth century from the use of eye-glasses and spectacles, but it was not until some time after Galileo's telescopic discoveries that the microscope began to be turned on the 'inner' world of the minute. In the seventeenth century some important discoveries were made using simple microscopes, especially by Marcello Malpighi at Bologna, and Anthoni van Leeuwenhoek at Delft. Robert Hooke and Nehemiah Grew, in London, used the more complex compound microscope. Their observations were diverse; they did not initiate new lines of investigation, but they did extend the knowledge of the structures and functions of living things. In 1661 Malpighi confirmed Harvey's theory of the circulation of the blood when he observed the capillary vessels linking veins and arteries in the lung tissue of a frog. This was further confirmed when Leeuwenhoek observed blood circulating in the tail of an eel. These observations finally established Harvey's theory while at the same time supporting the idea of the living body as a mechanism.

Hooke's own observations showed the great complexity of the world beyond the range of unaided human vision. His discoveries were presented in

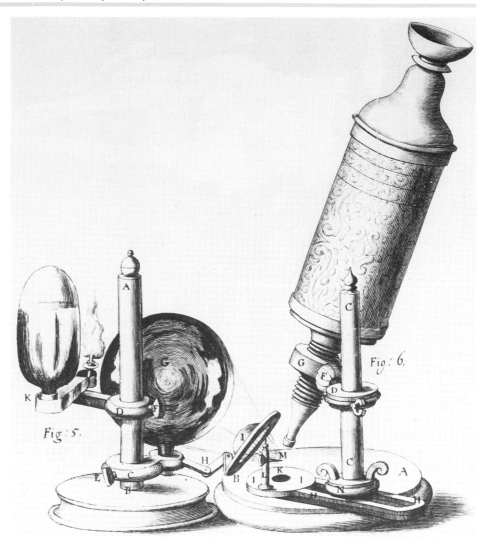

Figure 8.14 Hooke's microscope. This was a compound microscope with objective and eyepiece lenses. Light, either from the sun through a small aperture, or from a small oil lamp, was concentrated onto the specimen on the stage by means of a condenser (a sphere filled with water or brine) and a convex lens. (From Hooke's Micrographia, *London, 1665. Reproduced by permission of the Trustees of the British Museum.)*

Figure 8.15 Title page of Hooke's Micrographia, *London, 1665. (Science Museum, London)*

lectures delivered to the Royal Society and were published, together with illustrations mostly drawn by Wren, in *Micrographia* (1665). They range from the structure of a fly's eye, a feather, flea, louse or ant, to the point of a needle, various crystals and the cellular structure of cork. But Hooke was not content merely to describe these discoveries, he also used them to explain other physical phenomena such as the force of gravity, the freezing of water to form ice, how crystals are formed, or iridescence – the interplay of bright colours in the light reflected from substances such as mother of pearl, oil films and soap bubbles. Hooke tried to use his microscopic discoveries to explain the essence and behaviour of the objects he examined, and his microscopic observations were almost as disturbing as the revelations of Galileo's telescope in the previous generation. The telescope and the microscope served to extend human knowledge of creation from the immensity of the universe to the world of minute things invisible to the eye. Those, like Boyle, who sought to justify the pursuit of science on the grounds of its utility could cite telescopic and microscopic evidence to support religious beliefs and provide physical proof of God's omnipotence.

From the outset, the Royal Society had been in contact with foreign scholars through the extensive correspondence of its first secretary, Henry Oldenburg, and the Society was held in high regard by virtuosi throughout Europe. The fellowship was also offered to carefully chosen foreign scholars: Christian

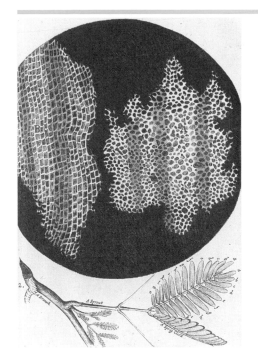

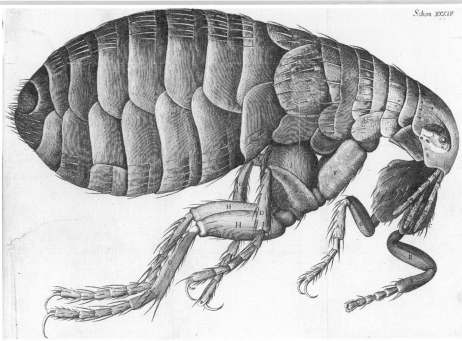

Huygens was elected in 1663; and Leeuwenhoek, a Dutch draper, in 1680. Using simple microscopes of his own invention and construction, Leeuwenhoek discovered spermatozoa, infusoria in pond water and the red and white corpuscles in blood, among many other minute bodies. He communicated his discoveries in letters to Oldenburg and Hooke at the Royal Society and many of them were published in the *Philosophical Transactions*. Yet, in 1700 when he had been demonstrating the scientific possibilities of the microscope for 25 years, Leeuwenhoek was still the only serious microscopist in the world. He had no rivals and few imitators; his observations aroused keen interest, but no one seriously tried to repeat, challenge or extend them.

This increased interest in microscopy revived ancient beliefs in spontaneous generation, as the microscope revealed minute bodies which seemed to appear from nowhere. Leeuwenhoek observed organisms in infusions of hay which within a few days or even hours became turbid as active microscopic forms multiplied, apparently spontaneously. William Harvey had stated the true position in the frontispiece of his book *On the Generation of Animals* (1651) when he added the words *Ex ovo omnia* (All living things come from an egg). In the late 1660s, Francesco Redi, a Florentine physician, showed that blowflies developed from eggs deposited on meat. Others showed that the same was true for other insects, fleas and lice, but the question of spontaneous generation still remained open for the minute organisms found in infusions. It was a question which was not to be resolved until the late nineteenth-century work of men like John Tyndall, Louis Pasteur and Robert Koch.

Figure 8.16 Illustrations from Hooke's Micrographia: *(left) the cellular structure of cork and (right) the flea. Some of the illustrations in* Micrographia *were drawn by Christopher Wren, others by Hooke; they all show first-hand observations through the microscope. (Reproduced by permission of (left) the Science Museum, London, and (right) the British Library Board.)*

8.7.2 Robert Boyle

Experiments with the air-pump

In the 1650s, Robert Boyle (1627–91) learned of experiments exploring features of the vacuum above the mercury column of a barometer tube that were carried out in Florence at the Accademia del Cimento. The vacuous

Figure 8.17 *Robert Boyle (1627–91), natural philosopher and chemist, son of the Earl of Cork. One of the founders of the Royal Society, he championed the corpuscular and experimental philosophies and used his observations to promote Christian belief. Boyle was a careful experimenter and a prolific writer. (Mansell Collection, London)*

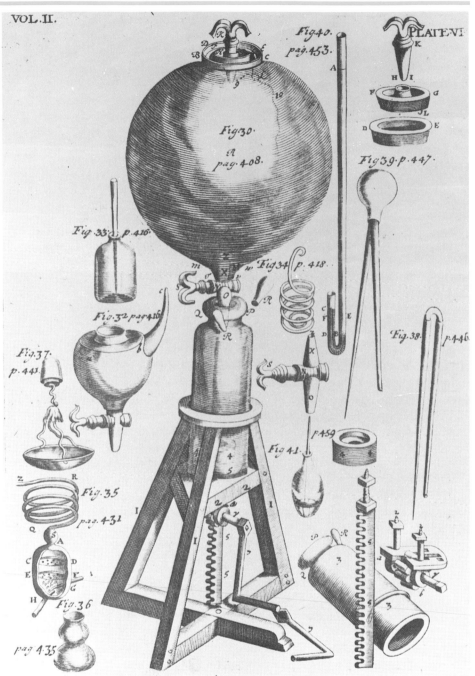

Figure 8.18 *Illustration of Boyle's first air-pump showing its component parts. (Ann Ronan Picture Library)*

space was, however, limited by the width of the tube and it was difficult to pass objects up through the mercury column. What was needed was a device for removing the air from a glass globe in which pieces of apparatus or small animals and birds could be placed. In Germany the military engineer Otto von Guericke (1602–86), mayor of Magdeburg, had devised an air-pump capable of creating a vacuum inside a closed vessel. This too came to Boyle's attention and later, in the Royal Society, he worked with Hooke to build an air-pump which would be convenient for laboratory experiments.

Boyle's first air-pump is described in his *New Experiments Physico-Mechanical, Touching the Spring of the Air, and Its Effects* (1660). It consisted of a glass globe that had a 3-inch hole at the top sealed by a stopper and which was connected through a stop-cock at the bottom to a cylinder in which a piston could be moved to withdraw air from the globe. Objects and pieces of apparatus could be introduced into the globe through the hole at the top before withdrawing the air. Boyle's second air-pump, also constructed with Hooke's assistance, was even more convenient as the glass vessel to be evacuated was sealed with wax over a hole in a metal plate through which the air could be withdrawn. With these devices Boyle observed many facts related to the functions of the air. He showed that if a barometer was arranged with its reservoir in the glass globe, the height of the mercury column fell as the air was removed from the globe. Then he demonstrated the truth of Mersenne's suggestion that air is the medium through which sound passes by demonstrating that a bell struck inside the evacuated globe cannot be heard. Lastly he found that light objects such as feathers fell at the same rate as heavy ones in a vacuum.

Mechanical theory

Boyle remarked that there were at least two ways of accounting for the 'spring' of the air. One was to imagine each particle of the air as compressible like a spring and the whole mass of air as resembling a fleece of wool. The other was to think of the particles as being in random motion and elastic collision with each other. In common with many of his contemporaries, he favoured the mechanical theory, according to which physical phenomena were considered to be caused by minute particles of matter in random motion. From Gassendi and Descartes (see Chapter 7), the revivers of corpuscular theories in France, Boyle took the notion of particles with particular size, shape and motion, but while he accepted the Cartesian corpuscles, which were capable of being changed in size and shape by attrition, he did not accept the Cartesian plenum. Instead Boyle's corpuscles moved at random in vacuous space.

Corpuscular theories allowed the possibility of measurement, quantification and the application of mathematics to experimental observations. Boyle welcomed this, but he also saw the need for the corpuscular theory to account for chemical phenomena as well. He therefore suggested that the corpuscles must carry both chemical and physical properties. Thus, the imaginary shapes and sizes of Boyle's corpuscles enabled the chemical properties of matter to be described. The corpuscles of acids, for example, were said to be spiky to account for their sharp taste and their ability to break up and dissolve the corpuscles of alkalis and metals. Oils, on the other hand, consisted of slippery corpuscles, while salts and metals had corpuscles which could pack together to form appropriate crystal shapes. But Boyle's corpuscles, like those of Descartes, were capable of being abraded and so changed in shape, and since they were all made of the same basic matter and the differences between substances were dependent on the different sizes and shapes of corpuscles, abrasion resulted in the conversion of one substance into another. The chemist's task was to discover the means by which the appropriate kinds of abrasive action could be brought about, and thereby achieve the transmutations the alchemists sought.

Although Boyle's air-pump extracted the air from a vessel, there was still the possibility that the space remaining was filled with a subtle matter, the fine particles of which could pass through pores too small to allow the passage of the air. Such a material (the aether) was postulated by Descartes to explain how the whole of space could be full of matter. On the other hand, atomists like Gassendi thought that the space between the atoms of matter was utterly void. Boyle made some ingenious experiments to test the Cartesian theory by observing the motion of this subtle matter. He failed to observe any such

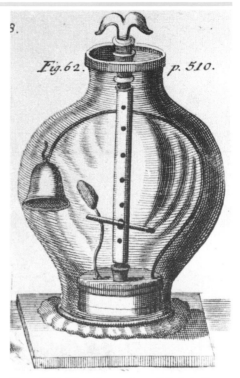

Figure 8.19 When the vessel was full of air the bell could be heard to ring as the clapper struck it. As the air was withdrawn the sound became fainter until it was no longer audible although the clapper was still seen to strike the bell. This experiment shows that sound is transmitted by the air. (Ann Ronan Picture Library)

motion and was sceptical about the existence of the aether. Instead he agreed with those who postulated empty space between the particles of matter. While Boyle's corpuscular theory was derived from Descartes and Gassendi, he developed it in his own way so that his ideas did not resemble either very closely, but by his consistent use of the corpuscular theory he contributed significantly to the seventeenth-century concept of mechanism.

Boyle's law

One objection to the existence of a vacuum above the mercury in the barometer tube was put forward by a little-known Aristotelian, Franciscus Linus (1595–1675), who suggested that the mercury column was held up by an invisible membrane, or thread (funiculus), just strong enough to support a mercury column of 29.5 inches. To refute this fantastic notion Boyle devised experiments with a J-shaped tube of which the short limb was sealed and the long one open (see Figure 8.20). The results were only approximate due to the crudity of the scales, the unevenness of the tube and the fact that Boyle ignored temperature changes. Yet they were good enough to lead him to the relation between pressure and volume in a gas which bears his name.

This result was obtained by carefully planned experiments and not by philosophical argument based on theory alone, although both traditions were important for the development of science in the seventeenth century. Comparing the work of Pascal (see Chapter 7) with that of Boyle in the field of pneumatics illustrates these contrasting methods. Pascal was primarily a mathematician, and Boyle, an experimentalist. It is not always clear whether Pascal actually carried out the experiments he described in his work on hydrostatics and pneumatics or whether they were merely thought-experiments by which the principles could be explained. His arguments proceeded by deduction from a few postulates with here and there an experimental check, as in the case of the Puy de Dôme observations. Yet even here it is possible that the apparent precision and consistency of the results was due to prior belief in the principle being tested. Boyle, on the other hand, used experiment as a means of discovery rather than mere confirmation of previously argued positions. He described in great detail the practical difficulties of his experiments and explained the steps he took to overcome

> **Boyle's law**
>
> For a given quantity of a gas at a constant temperature,
> Pressure x Volume = Constant

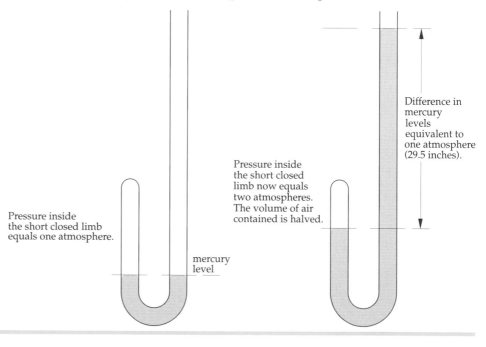

Figure 8.20 Diagram of Boyle's apparatus. Each limb was marked with a scale, and the mercury in the two limbs was first equalised so that the air in the closed limb was at atmospheric pressure. Mercury was then poured into the long limb until the difference in height of the mercury in the two limbs reached 29.5 inches, when the pressure in the closed limb was two atmospheres. The volume of the air in the closed limb was found to have been halved.

Pressure inside the short closed limb equals one atmosphere.

mercury level

Pressure inside the short closed limb now equals two atmospheres. The volume of air contained is halved.

Difference in mercury levels equivalent to one atmosphere (29.5 inches).

them. He gently poked fun at Pascal for writing about experiments which he could never have carried out because they were not possible and for failing to give sufficient details in other cases so that the experiments could not be successfully repeated. By contrast, given proper resources and sufficient practical skill, anyone could repeat Boyle's experiments guided by his lengthy instructions.

8.7.3 Experiments on the chemical properties of the air

In his experiments with the air-pump Boyle showed that air is necessary for respiration and combustion: small animals and birds placed in the receiver of the air-pump expired when the air was removed; and the flames of candles or spirit-lamps were extinguished. It seemed to Boyle that the air contained a 'vital quintessence' needed to support life and flames. He performed experiments in which he dropped combustible materials onto a red-hot iron plate in the exhausted receiver and found that whereas pure sulphur would only melt and produce vapours, gunpowder (a mixture of sulphur with charcoal and saltpetre) would still burn in a vacuum. At first he thought that some air had become mixed with the saltpetre as it crystallized, but when he found that saltpetre that had been crystallized in a vacuum was just as effective, he concluded that nitre gave off some 'agitated vapours which emulate air', though he was not clear about the composition of the air. He thought it contained 'celestial effluvia', exhalations from the stars and planets along with fiery particles from the Sun and many other things in a confused jumble. Boyle did not grasp the true role of the air in combustion and he certainly missed the reason for the gain in weight during the calcination of metals. He thought that igneous particles were absorbed by the metal causing the increase in weight, and both he and Hooke realised that only a part of the air was used up in respiration and combustion. This was clearly stated by Hooke in *Micrographia*, who called this part 'a substance inherent and mixt with the Air, that is like, if not the very same, with that which is fixt in Salt-peter' ([1665] 1961, p.103).

Hooke also speculated about combustion in his *Micrographia*. He gives no details of experiments, although he made many unpublished observations in his work for the Royal Society. Instead, he put forward a number of propositions, among which he says the air is the universal dissolvent of all sulphureous (i.e. inflammable) bodies and in the process of dissolution great heat is evolved, which we call fire. He thought there must be a substance in the air similar, if not identical, to that fixed in saltpetre which gives rise to the vigorous reaction of burning, when not only heat but also light is released. During combustion, part of the burning body is united with this part of the air and dissolved, turning into air as a result. Hooke's theory is therefore very close to that later put forward by Mayow, but he never succeeded in isolating the common constituent of the air and nitre.

The nature of air was further investigated by John Mayow (1641–79), a London physician who had been educated at Wadham College in the 1660s. Mayow published the results of his experiments in Oxford between 1668 and 1674 (see Figure 8.22). He was aware of Boyle's work and he mentions Malpighi's discoveries in some detail, but his own experiments were overlooked and their significance was not recognised until the end of the eighteenth century.

The difference in colour between arterial and venous blood seemed problematical after Harvey's discovery that they were both part of the same

Figure 8.21 John Mayow (1641–79). (Science Museum, London)

217

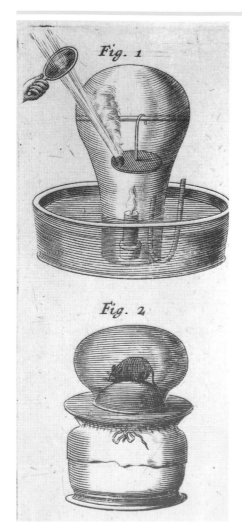

circulation. Arguments about the nature of the blood and its relation to the air had been going on in Oxford since the 1630s, first among Harvey's circle and later in the Oxford Experimental Philosophy Club. Thomas Willis, an eminent Oxford physician and anatomist, became interested in the role of the blood during respiration, while in 1667–8 Hooke and Richard Lower at the Royal Society showed that dark venous blood became florid when exposed to air. Lower also found that venous blood became florid as it passed through the lungs where, he thought the 'nitrous spirit' of the air saturated it.

> On this account it is extremely probable that the blood takes in air in its course through the lungs, and owes its bright colour entirely to the admixture of air. Moreover ... it is equally consistent with reason that the venous blood, which has lost its air, should forthwith appear darker and blacker. *(Lower, [1669] 1932, quoted in Boas, 1970, p.168)*

Lower also attempted blood transfusions between sheep and dogs and dreamed of what might usefully be done to relieve diseases and old age in humans by this means.

Figure 8.22 Mayow's experiments on combustion and respiration, 1674. Mayow burned phosphorus, candles and gunpowder in closed vessels of air inverted over water. He also confined mice and other small animals in the closed container. In all cases he found that only a fraction of the air was used up, and he gave this fraction the name 'nitro-aerial particles', which he thought were also found in the nitre used in gunpowder. (From John Mayow, Tractatus quinque medico-physici, *Oxford, 1674. Photo: Science Museum, London)*

Figure 8.23 Seventeenth-century blood transfusion. Richard Lower became one of the most noted physicians in London. His attempts to carry out blood transfusions, though extremely dangerous, do not appear to have resulted in serious injury to the subjects, possibly because little if any blood actually passed from one to the other. (Mary Evans Picture Library)

8.7.4 Boyle, the founder of modern chemistry

Boyle has often been considered the founder of modern chemistry, and he has some claim to that title as he showed that chemistry was worthy of study for its own sake. Besides his experiments on the chemistry of the air, Boyle was also interested in a wide variety of other chemical topics. He succeeded in preparing a solution containing phosphorus from urine, having learned about the discovery of this element from the German chemists Johann Kunckel and J.D. Krafft. In 1677 Krafft had shown Boyle a sample of phosphorus and gave him some hints about its preparation, but Boyle was unable to obtain the solid element until 1680. He observed some of its properties, including the phosphorescent glow which made the element such an object of curiosity to alchemists and chemical operators. Boyle showed that the glow occurs only in the presence of air and found that only very small quantities of phosphorus were needed to produce it.

Boyle was always concerned to show how useful science could be and in this connection he devised a scheme for the analysis of mineral waters commonly used in medicine. Published in 1684, Boyle's scheme was a great improvement on the haphazard methods previously used. In this respect too Boyle introduced more rigorous experimental methods into chemistry than had been in use before his time. In his scheme for mineral-water analysis, perhaps his most important innovation was the introduction of test papers soaked in strong solutions of vegetable colours and dried in the air. With these the acidity or alkalinity of a mineral water could readily be tested and different samples could be compared. He also used papers soaked in tannic acid from oak galls to test for the presence of iron, and in the course of his experiments he recognised the presence of elementary sulphur in certain waters, distinguishing it from bituminous or oily matter which had often previously been confused with sulphur.

Boyle's most famous book, the *Sceptical Chymist*, was published in 1661. In it he complained about the obscure language of the alchemists. He cast doubt on the four Aristotelian elements (earth, water, air and fire) and on the three Paracelsian principles (sulphur, mercury and salt) because none of these things could be isolated from bodies with any certainty. Instead Boyle proposed a new definition of the chemical element which would be capable of experimental investigation.

> I now mean by Elements ... certain Primitive and simple ... bodies; which not being made of any other bodies, or of one another, are the Ingredients of which all those call'd perfectly mixt Bodies are immediately compounded and into which they are ultimately resolved.

(Boyle, [1661] 1911, p.187)

This is in effect the modern definition of a chemical element. However, since Boyle's corpuscular theory postulated that chemical properties depended on the shapes and sizes of the corpuscles, it follows that the collection of chemical properties which define a chemical element requires a similar number of differently shaped corpuscles. The idea that an element could be made up of *identical* corpuscles is untenable on Boyle's corpuscular theory and he was himself led ultimately to doubt whether any such chemical elements existed. Nevertheless, his critical approach was valuable and his attempt to define the chemical element stimulated a useful experimental approach to chemical analysis.

8.8 *Limitations of the mechanical philosophy*

Throughout the seventeenth century, as we have seen, the mechanistic view of the universe was closely linked with the experimental philosophy. At Oxford and Cambridge, the new ideas developed outside the mainstream of the undergraduate curriculum. At Gresham College in London, the practical applications of science and mathematics which had been promoted from the beginning of the century were mainly concerned with naval science, navigation, surveying and the improvement of mathematical and measuring instruments. In medicine the new developments in anatomy and *materia medica* (substances used in medical treatments) gave rise to heated arguments between traditional physicians and followers of Paracelsus (see Chapter 6). In 1673 the Society of Apothecaries established a Physick Garden at Chelsea for the cultivation and study of plants, especially those used in medicine. Two years later the Royal Observatory was founded by Charles II at Greenwich to revise the astronomical observations, solve the problem of determining longitude and consequently improve navigation at sea. Fellows of the Royal Society were involved with all of these bodies.

Thomas Hobbes (1588–1670) was introduced to the mechanical philosophy through contact with members of Mersenne's circle in the 1640s during his eleven years of exile in Paris. He was favourably impressed and, influenced by continental system-builders like Descartes and Gassendi, he based his system of ethics and psychology, human behaviour and politics on the mechanical philosophy, extending its influence to areas which had always been the exclusive preserve of the Aristotelians. In *Leviathan* (1651) Hobbes described the State as a machine, governed by the will of the Sovereign. He thought movement and change were fundamental features of human life and he dismissed the fancies of an animistic world along with ghosts, fairies and witches as figments of the imagination. Miracles were relegated to the margins of creation and the Deity was dismissed from his scheme. For Hobbes the laws of nature did not reveal any underlying rationality in the universe, they were merely theorems which people agreed for their own peace and quiet.

Hobbes was thought to have taken his mechanistic view of society too far and his ideas were greeted with hostility. Critics like Seth Ward and Edward Hyde, who embraced the mechanical philosophy in science, retained an Aristotelian view of ethics and politics in the educational and religious traditions of English gentlemen. Even in France, where Hobbes enjoyed more success than in England, the aristocratic code of behaviour and hierarchical view of society proved a strong barrier against his mechanistic system. Within the Royal Society too the new learning was far from universally accepted. Many Fellows were at least as interested in spectacular events, monstrous births and natural catastrophes as in genuine scientific discoveries. Ancient beliefs had often suggested that creation was a mystery which could only be explored by occult methods, and while Fellows like Boyle and Hooke were engaged in rational experiments leading to useful discoveries, the intellectual milieu in which they worked was riddled with ideas of magic, astrology, alchemy, witchcraft and other esoteric traditions. These ideas lent credence to wonders such as the weapon salve and Sir Kenelm Digby's powder of sympathy (1658). Digby, by describing his powder as strictly chemical and explaining its action in mechanical terms, ensured its acceptance by Boyle, who was open-minded about any claim provided it could be accounted for by the corpuscular theory.

Figure 8.24 Sir Kenelm Digby (1603–65), writer, naval commander and diplomatist, an Aristotelian interested in alchemy and the transmutation of metals, famous for his 'powder of sympathy', a quack remedy said to cure wounds by sympathetic action when applied at a distance to a bandage taken from the wound. (Ann Ronan Picture Library)

Alchemy also figured prominently in the activities of some of the Fellows. Boyle himself was actively involved in alchemical experiments as was Newton. In 1652 Boyle claimed to have so purified quicksilver (i.e. mercury) that the product approached in purity the mercurial spirit of the Paracelsian *tria prima*. Boyle was not a traditional alchemist, but as we have seen he was more than half convinced of the possibilities of transmutation, which his corpuscular theory explained. In *The Origin of Forms and Qualities* (1666), he described the artificial transmutation of bodies as 'one of the noblest and usefullest effects of humane skill and power'. A decade later in 1676, he published a vague description of his method of purifying quicksilver in the *Philosophical Transactions*, although he kept secret most of the details.

8.9 Sir Isaac Newton

Born at Woolsthorpe Manor near Grantham in Lincolnshire on Christmas Day 1642, Isaac Newton (1642–1727) entered Trinity College, Cambridge, on 6 June 1661. Despite his love of book-learning, his undergraduate career was undistinguished. He graduated BA in 1665, but the outbreak of plague in that year forced him to seek refuge in the relative safety of his home and for the next two years he lived in seclusion at Woolsthorpe. It was during this period that Newton's genius became evident. In mathematics he discovered several new theorems and methods, including the differential calculus, the single most important mathematical innovation since the time of the ancient Greeks. In May 1666 he discovered the principle of the integral calculus by which the areas under curves could be found. These discoveries alone would have ensured for Newton a pre-eminent position in the history of science, but they were accompanied by two others of equal importance in physics. He showed by experiment the compound nature of white light and worked out a theory of colours. During the same period, he also discovered the force of gravity which holds the Moon in its orbit and attracts bodies to the Earth, leading to his theory of universal gravitation based on astronomical observations and Kepler's laws of planetary motion. He did not publish the latter discoveries for twenty years, although the problems involved were widely discussed among astronomers and mathematicians and Newton's prowess in the field was well known.

Soon after returning to Cambridge in 1667, Newton became a Fellow of Trinity College and in 1669 his tutor Isaac Barrow, who had long recognised Newton's talents, resigned as Lucasian professor of mathematics in favour of his protege. Newton's lectures at this time were on the nature of light, a subject which had interested him since 1663 when he began to experiment with lens-grinding techniques in an effort to correct chromatic aberration, the coloured fringes seen around the edges of images produced by spherical lenses. This work led to the discovery that white light could be analysed into the colours of the rainbow when passed through a glass prism. It appeared that the faults of refracting telescopes lay not in the lenses but in the nature of light itself, and Newton erroneously thought that it would never be possible to devise a lens system which would obviate this effect. He therefore constructed a new kind of telescope, a reflecting instrument, using a curved mirror instead of the objective lens. Following the presentation of the telescope to the Royal Society together with an account of his researches on the nature of light, he was elected a Fellow of the Society in 1672. There Robert Hooke and others were already working on optics and Hooke

Figure 8.25 Isaac Newton (1642–1727). Newton's scientific work included observational astronomy, mathematics, dynamics, optics, astrophysics and chemistry. Newton became Lucasian professor of mathematics at Trinity College, Cambridge (1669), Fellow (1672) and later president (1703–27) of the Royal Society, warden and later master of the Mint (1695–1727) and two brief spells as a Member of Parliament (1688–9; 1701–2). His discoveries in astronomy, mathematics and physics consolidated the work of Copernicus, Galileo and Kepler and set new objectives for the physical and mathematical sciences. (University of London, Autographs Collection)

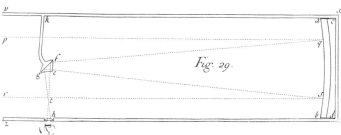

Figure 8.26 *Newton's reflecting telescope. Light enters the open end of the tube and is reflected from the concave mirror at its base to the prism, where it is turned through an angle of 90° by total internal reflection to pass through the eyepiece. (Science Museum, London)*

challenged Newton's priority in the discovery of the nature of white light. While admitting the accuracy of Newton's experiments, Hooke criticised his theory of colours. This caused strained relations between Hooke and Newton, who was always extremely sensitive to criticism.

8.9.1 *Universal gravitation and the* Principia

In the 1680s there were many unresolved arguments about the nature of planetary motions and the orbits that would result from certain combinations of forces. In January 1684, Hooke boasted to Halley and Wren that he had demonstrated all the laws of the celestial motions – though he did not produce any evidence. In August 1684, Halley, who admitted his own inability to solve the problem of planetary motion, visited Newton in Cambridge and on asking him what would be the path of a body moving under the action of a central force which varied as the inverse square of the distance from the centre, was told at once that it would be an ellipse. Newton's calculations in the 1660s had shown him that this must be the case, but after his initial burst of inspiration he had set the work aside. This was in part because he was dissatisfied with the reliability of his results. By 1684 Newton had lost his original calculations, but he was able to reconstruct the solution and Halley, himself a talented mathematician, persuaded Newton to return to his work on planetary motion and prepare it for publication. With Halley's encouragement and support Newton completed his *Philosophiae naturalis principia mathematica* (Mathematical principles of natural philosophy), which was published by the Royal Society three years later. By accounting for all the discoveries in astronomy and mechanics which had been made from the time of Copernicus, the *Principia* is the most powerful expression of the mechanical philosophy in the seventeenth century. It sets Newton apart as a giant among his contemporaries and ensures his place as probably the most powerful mathematical and scientific intellect of any age.

The central idea of the *Principia* – that the force of gravity which attracts a stone to the Earth is the same as the force which holds the Moon in its orbit round the Earth and all the planets in their orbits about the Sun – led Newton to the three laws of motion that are the basis of Newtonian mechanics, and which proved so successful in astronomy and physics that he has since appeared to many as the doyen of mechanists. Nevertheless, in the seventeenth century the idea of gravitational attraction was not universally welcomed in spite of the fact that it explained the cohesion of matter as well as Kepler's elliptical planetary orbits. That it was rejected by Aristotelians is easily understood, but Cartesians, dominant in France and Holland since the 1580s, were equally hostile. They dismissed Newton's theory on the grounds that it re-introduced the ancient concept of action-at-a-distance, or occult forces, and that in that respect it was a retrograde step. Huygens, the Dutch Cartesian, dismissed Newton's theory of gravitational attraction as absurd and in no way explicable by any principle of mechanics, while Leibniz, who also quarrelled with Newton over priority in the discovery of the differential calculus, classed him with Aristotle as a believer in 'sympathies' and 'antipathies'. The reasons for this opposition become clearer in the light of the neo-Platonic influences apparent in Newton's world-view and his inability to offer any physical explanation for the cause of gravity.

Although his cosmic system operated like a gigantic mechanism obeying natural laws, Newton refused to accept mechanical forces as the sole cause of motion. The image of Newton as the arch-mechanist, whose God was a 'Divine Mechanic' intervening as required to correct faults in the machine, is an eighteenth-century aberration. Newton's world-view owed much to the cosmology of Johannes Kepler, a neo-Platonist and something of a mystic. Also, in the 1660s at Cambridge, Newton came into contact with Henry More and Ralph Cudworth, both of whom held neo-Platonic notions of the world as a living organism with a soul. These Cambridge Platonists were fiercely critical of purely mechanical theories and, influenced by their ideas, Newton agreed that the universe must be maintained under the control of an Intelligent Being. Even the disposition of the stars, set at immense distances from each other in space lest gravity should cause them to coalesce, showed the benevolence of the Creator, in Newton's view. Recent scholarship has suggested that Newton's scientific insights were based on his religious beliefs and world-view. Seen in this way the *Principia* is not just the product of mathematical and astronomical research, but is part of a larger intellectual and religious synthesis. Newton insisted that God was involved *continuously* in preserving his Creation; space, the sensorium of God, and time were part of the Divine Presence.

Another aspect of Newton's attempt to establish a universal world-view was his persistent interest in alchemy. This may have owed much to Boyle's work on the structure of matter and the search for a new comprehensive matter-theory. But Newton's interest in alchemy also included the neo-Platonic writings of the hermetic school which, like the ideas of the Cambridge Platonists, looked back to antiquity for their inspiration. Newton too was sympathetic to the idea of *prisca sapienta* (ancient wisdom) which had been lost. A characteristic concept of Renaissance neo-Platonism, it was thought that *prisca sapienta* had been expressed in obscure, symbolical language to protect it from the attentions of the vulgar and that it was given only to the

Newton's laws of motion

1 A body will continue in its state of rest or of uniform motion in a straight line unless acted on by an external force.

2 Any change in the state of rest or uniform linear motion of a body is proportional to the impressed force and occurs in the direction in which the force acts.

3 To every action there is an equal and opposite reaction.

These laws define the effects of forces on the condition of rest or motion of a body and the first law indicates that the universe must necessarily be infinite. Newton was in no position to investigate this law experimentally; it was just the logical consequence of his calculations. Gravity, too, was a necessary constant in the calculations, but as Newton himself admitted, he could give no physical explanation for it.

initiated to understand its mysteries. This may well explain Newton's persistent interest in alchemy, his huge collection of alchemical books and his own extensive writings on the subject. He rejected the atheistic implications of the mechanical philosophy and sought the occult sources of 'active principles' as a corrective. The same aims may also have inspired the mathematical and philosophical work by which he attempted to penetrate to the heart of lost knowledge using his powerful new mathematical tools together with the experimental method (Hoppen, 1976).

8.9.2 Experimental method and the Opticks

In his approach to experimentation, Newton was an innovator. He began each investigation, not with a traditional theory, plausible hypothesis or shrewd guess, but with experimental observations which he then used as the basis for mathematical calculations and philosophical deductions. The theories which he then propounded were therefore based on positive experimental evidence. Thus, besides his important scientific discoveries, Newton also made a significant contribution to the development of the scientific method. His approach was distinctly more 'modern' than most of his contemporaries. In his *Opticks* (1704), Newton again described the experiments he had carried out in the 1660s which led to his discovery of the compound nature of white light. He shows how he placed a glass prism in the path of a ray of sunlight coming through a small hole in the shutter of his window. The resulting band of coloured light was thrown onto a white board. It had often been suggested that glass had a 'staining' effect on the light, but Newton showed this was not so by a critical experiment in which he showed that no further analysis of the seven spectral colours was possible (see Figure 8.27). Newton's experiments demonstrate another side of his genius – his grasp of experimental techniques and his use of experiments both to make primary observations and to confirm deductions arising from them.

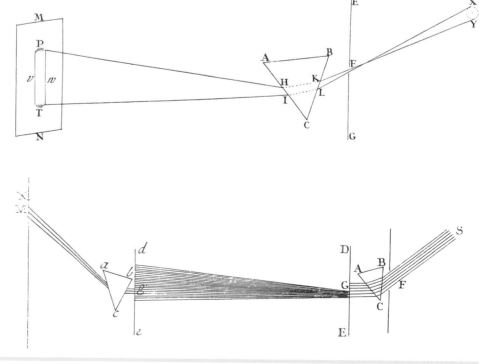

Figure 8.27 Newton's experiments on the composition of white light. These diagrams illustrate Newton's use of the experimentum crucis. *In the top diagram, sunlight passes through a small hole (F) in the window shutter (EG) and then through a glass prism (ABC) which separates the colours by differential refraction to form a spectrum (PT) on the screen (MN). The bottom diagram shows that by manipulating the prism (ABC) each part of the spectrum can be made to pass through a hole in the screen in turn. By a further selection at a second screen (de) a monocoloured ray is obtained which passes through a second prism (abc). It is found that although the ray is refracted to the screen (MN), no further colour dispersion takes place. Each coloured ray is differently refracted but no new colours appear, confirming that white light consists of the seven colours of the spectrum and no others. (Science Museum, London)*

Since *Opticks* was written in English, the experiments described in it were readily understood in England. The book presented a new experimental approach to the study of optical phenomena which was as important for its method as for its content. The method was later applied to the investigation of other phenomena in the eighteenth century, as Newtonianism spread across Europe. Newton also added 31 'Queries' at the end of the *Opticks*: ideas about the nature of the physical world which raised important questions and invited further investigation. These included his corpuscular theory of matter and of light, chemical change and the forces of nature. Thus the book offered the outlines of a research programme in science and many of the problems identified by Newton were to form the starting point of scientific investigations throughout Europe from the eighteenth century onwards (see Chapter 10).

Sources referred to in the text

Boas, M. (1970) *Nature and Nature's Laws*, London, Macmillan.

Boyle, R. ([1661] 1911) *The Sceptical Chymist*, London, Dent.

Gilbert, W. ([1600] 1958) *De magnete*, tr. P. Fleury Mottelay, New York, Dover.

Harvey, W. ([1628] 1907) *De motu cordis*, London, Dent.

Hooke, R. ([1665] 1961) *Micrographia*, New York, Dover.

Hooykaas, R. (1973) *Religion and the Rise of Modern Science*, Edinburgh, Scottish Academic Press.

Hoppen, K.T. (1976) 'The nature of the early Royal Society', *British Journal for the History of Science*, IX, 1–24; 243–73.

Hunter, M. (1989) *Establishing the New Science: The Experience of the Early Royal Society*, Woodbridge, Boydell and Brewer.

Kearney, H. (1971) *Science and Change 1500–1700*, London, Weidenfeld and Nicholson.

Lower, R. ([1669] 1932) *Tractatus de corde*, tr. K.J. Franklin, in R.T. Gunter (ed.), *Early Science in Oxford*, vol. IX, Oxford University Press.

Newton, I. ([1704] 1952) *Opticks*, New York, Dover.

Rees, G. (1977) 'Matter theory: A unifying factor in Bacon's natural philosophy', *Ambix*, 24, pp.110–25.

Sprat, T. ([1667] 1959) *History of the Royal Society*, Washington University Press.

Webster, C. (1975) *The Great Instauration*, London, Duckworth.

Whitteridge, G. (1971) *William Harvey and the Circulation of the Blood*, London, Macdonald.

Further reading

Dijksterhuis, E.J. (1961) *The Mechanisation of the World Picture*, Oxford University Press.

Frank, Jr, R.G. (1980), *Harvey and the Oxford Physiologists*, Berkeley, California University Press.

Gillispie, C.C. (ed.) (1970–6) *Dictionary of Scientific Biography*, 16 vols, New York, Charles Scribner's Sons. On Boyle, see vol.2, pp.377–82. On Gilbert, see vol.5, pp.396–401. On Hartlib, see vol.6, pp.140–2. On Harvey, see vol.6, pp.150–62. On Hooke, see vol.6, pp.81–8. On Mayow, see vol.9, pp.242–6. On Newton, see vol.10, pp.42–101. On Oldenberg, see vol.10, pp.200–2.

Hall, A.R. (1983) *The Revolution in Science 1500–1700*, London, Longman.

Hunter, M. and Schaffer, S. (eds) (1989) *Robert Hooke: New Studies*, Woodbridge, Boydell and Brewer.

Manuel, F.E. (1968) *A Portrait of Isaac Newton*, London, Frederick Muller.

Vickers, B. (1987) *English Science, Bacon to Newton*, Cambridge University Press.

Westfall, R.S. (1980), *Never at Rest*: *A Biography of Isaac Newton*, Cambridge, Cambridge University Press.

Scientific Academies Across Europe   *Chapter 9*

by Gerrylynn K. Roberts

9.1 The organization of science in the seventeenth century

The academy emerged as the typical form of organization for the pursuit and communication of science during the seventeenth century.[1] Indeed, the new form of scientific organization rose with modern experimental science; by the end of the seventeenth century, most individuals who had an active interest in scientific knowledge were affiliated to such a body, however loosely (Hall, 1983, p.210). From the mid seventeenth century to roughly 1800, some seventy official scientific academies were formed in Europe and America, plus a number of smaller private bodies (McClellan, 1985, p.xx). The institutions founded in Italy, England, France and Prussia during the second half of the seventeenth century shared a number of features and indeed helped to foster in Europe a pursuit of scientific knowledge for its own sake and, rhetorically at least, for the sake of its possible applications for the benefit of humankind. Their establishment marked the public recognition of experimental knowledge and, importantly, of the enterprise of those who pursued it. The reasons for the rise of this new form of organization as a locus for, and expression of, the new scientific knowledge are complex, rooted in the social and intellectual circumstances of the distinctive national cultures of the period as well as having to do with the nature of the new science itself.

Some historians have argued, not without reason, that these new institutions developed in opposition to the traditionalism of the universities, whose curricula were rooted in a conservative scholasticism based on textual and ecclesiastical authority. While it is undoubtedly the case that the pursuit of the new science found little support through the formal structures of the traditional universities, it should be remembered that such activities lay largely outside their programme (see Chapter 1). Their prime aim was teaching law, theology and medicine for the traditional professions and generally educating future participants in society, not extending knowledge. Although, of course, circumstances varied from country to country where science was taught, it was for the most part taught in relation to the requirements of the seven liberal arts or of medicine or of theology, and then often in the spirit of handing on set knowledge. Where scientific investigation was undertaken in the universities, it was generally done by individuals as what we might call 'extra-curricular' work. The importance of this intellectual activity outside the formal structures should not be underestimated, although it is difficult to determine the extent to which it

---

[1] The term 'academy' will be used to refer to a whole range of bodies established in the seventeenth and eighteenth centuries. Strictly, the term refers to organizations with limited, appointed (sometimes salaried) hierarchical memberships and formal state or princely sponsorship. Such bodies are sometimes called 'closed' academies. 'Open' academies, or learned societies, had no fixed membership categories, but were open to applicants who were then admitted by election and required to pay a fee. The activities of learned societies were financed mainly through members' contributions.

influenced teaching and undergraduate experience. It is also important not to overlook the fact that many active members of the scientific academies throughout Europe were themselves educated in precisely these traditional institutions (Curtis, 1959, ch.9; Heilbron, 1982, ch.2; Webster, 1975, ch.3, part ii). In that sense, the foundations of their future work were laid there.

Quite apart from what was or was not going on in the universities, from the sixteenth century, the scope for intellectual activity and leadership broadened beyond the narrow confines of academic life. At one level, princely patronage helped intellectual life outside the universities to flourish (see Chapter 6). Though perhaps themselves university-educated, physicians, lawyers and others, such as some merchants, began to participate in intellectual life in settings other than the academic. At another level, the rapidly growing trend to the vernacular, which began in the sixteenth century and was fostered by the growth of printing, made the work of the new learning accessible to people outside the universities who would not previously have explored such ideas. In addition, there was a growing tendency for those with practical skills (engineers, pharmacists, metal workers, surveyors, navigators, etc.) to write about their work and for their writings to be consulted by those interested in wider problems. The new natural history consequent on the voyages of discovery stressed the importance of the direct experience of observation and its precise recording (Hooykaas, 1987, p.472). Similar interests brought people together in informal, and largely ephemeral, association (Hall, 1983, pp.210–11; Webster, 1975, ch.5).

In discussing the new organizations for science, the nature of the new science has to be considered. For, though pursued mainly by individuals, the furthering of the new science depended on its communication to others who were in a position to judge it. Therefore, interchange between the practitioners was crucial; hence the importance of the seventeenth-century correspondence circles of Mersenne and Oldenburg. In so far as the new science was experimental, the audience depended not only on the niceties of the written arguments to make their judgements, but more importantly, on the verification of the experiments themselves, hence on repetition before the community. Again, though experimental science was pursued mainly by individuals, it was quite clear already in the seventeenth century that access to relatively expensive instruments and apparatus would come to be important to experimental science – access that could be provided through communal ownership. In so far as the new science was vast and largely undefined in scope, it was thought that following a Baconian model of communal decision making would help combine individual expertise and breadth of outlook. From the 1660s, this approach was increasingly promulgated through the academies' own official publications.

The academies, of course, were not the only institutional expression of the new knowledge: for example, in the sixteenth century, the anatomy theatre, which allowed students direct observation of dissection, became a part of continental medical teaching. The observatory, too, was a feature of the period in a number of places, as was the botanical garden. The academies, however, gave expression to a new general scientific culture in which the stress was not only on observation, but on experiment as well; they began to serve as a locus for the consolidation of that collection of developments that we have come to call the Scientific Revolution. They served science as a whole, as well as individual scientists and programmes. So by the end of the seventeenth century, most participants in science were affiliated to an academy; but at the same time, the academies gave science and scientists a place in society.

9.2 *The Accademia dei Lincei, Rome, and the Accademia del Cimento, Florence*

In Italy, where what is generally agreed to have been the first formal scientific academy was established in 1657, there was a long tradition of setting up special societies for intellectual and social activities. In fifteenth-century Italy, the 'Renaissance academy' had developed as a new form of organization for the discourse of humanists outside the universities. During the sixteenth century, such academies were common loci of intellectual activity: some seven hundred arose then (McClellan, 1985, p.42). They should not be thought of as being in the same league as the famous and enduring learned societies of the later seventeenth and eighteenth centuries: their importance is as an early stage of extra-university activity. They were small, informal and, for the most part, ephemeral bodies dependent upon the fortunes of an individual patron and generally attached to a Renaissance court. Collectively, Renaissance academies focussed on a very wide range of cultural affairs, but there are records of only a very few, such as della Porta's Accademia dei Segreti in Naples (see Chapter 4), that were devoted to what we would call 'science'.

One conspicuous example was the Accademia dei Lincei (for the lynx-eyed, or clear-sighted ones), founded in Rome by Prince Federico Cesi in 1603, which Galileo joined in 1611 on coming to Rome. Cesi was interested in natural history, and the programme of his academy was a typical humanist one (Redondi, 1988, pp.80–8). At its heart was the establishment of a great library. Libraries in this period were important centres of traditional culture as well as the elaboration of new ideas; Cesi's library, however, was to be of works devoted to science, including classical authors – works covering the explosion of developments in botany, zoology, anatomy and alchemy of the previous century and brand new books, some of which the Academy would publish. Indeed, the Academy was Galileo's publisher in Rome. Its grand project was the publication of a 'humanist' encyclopaedia of scientific knowledge which would replace classical sources – an elaboration of Galileo's work. At the same time, adopting a tactic of the Jesuits, it urged the creation of scientific libraries in a number of Italian cities to attract scholars.

By implication, of course, this was an anti-Aristotelian programme. But it was anti-Aristotelian, not in the spirit of opposing the Church, but in the spirit of opposing the dominance of the Jesuit-controlled Collegio Romano (see Chapter 4). It was a programme seeking cultural reform, including the reform of Catholicism. Its illustrious clerical members were in a position to work toward that end; as such, although it never had more than thirty members, the Academy's support for Galileo was important in promulgating the new knowledge, if ultimately unsuccessful in its objectives. The Accademia dei Lincei ceased with Cesi's death in 1630. As an organization, it was typical of many of the preceding century, but with scientific knowledge as its focus.

In much the same tradition, but following a rather different model, was the Accademia del Cimento, which was set up in Florence in 1657. Its founders and patrons were two sons of Cosimo de' Medici, who had been so influential in the career of Galileo, Grand Duke Ferdinand II and Prince Leopold. They had come under Galilean influence and performed experiments in their own laboratory long before setting up the Academy. The Accademia del Cimento had nine hand-picked members, some with direct or indirect contact with Galileo. They were the cream of Italian science of the time.

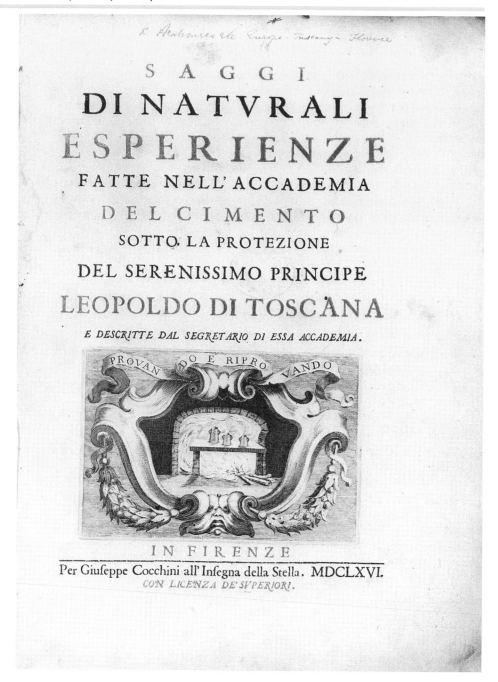

Figure 9.1 Frontispiece to Saggi, *1666. (Photo reproduced by permission of the British Library Board.)*

The word *cimento* means experiment. What distinguished the Accademia del Cimento from the traditional academies, and what makes it generally regarded as the first modern scientific academy, was its devotion to experiment. In a well-founded laboratory provided by the Medici brothers, it conducted over a period of roughly ten years a series of experiments dedicated to confirming the scientific work of Galileo and aiming to undermine the still officially influential Aristotelian physics and astronomy. Much attention was devoted to measurement and the devising of apparatus. The work was collaborative in that it was performed at meetings of the group and it was recorded anonymously. In effect, it was an early physics research laboratory.

Perhaps the most important innovation of the Academy was its final act, the publication of its activities in the *Saggi di naturali esperienze* in 1667. That was, in effect, its official scientific journal. The *Saggi* recorded in great detail the labours of the Academy. The motto on its frontispiece indicated the Academy's aims – *provando e riprovando*, [proceeding] by trial and error. The preface describes its methods.

> As one may take a heap of loose and unset jewels and seek to put them back one after another into their setting, so experiment fitting effects to causes and causes to effects – though it may not succeed at the first throw, like Geometry – performs enough so that by trial and error it sometimes succeeds in hitting its target. (*quoted in Middleton, 1971, p.90*)

In addition to trying some new experiments, they felt it important to repeat the work of others in order to test conclusions handed down with authority and to consolidate that work with more proof. The final paragraph of the preface shows their astuteness in avoiding confrontation, but also their limitations.

> Finally, before everything else we protest that we would never wish to pick a quarrel with anyone, entering into subtle disputes or vain contradictions; and if sometimes in passing from one experiment to another, or for any reason whatever, some slight hint of speculation is given, this is always to be taken as the opinion or private sentiment of the academicians, never that of the Academy, whose only task is to make experiments and to tell about them. (*quoted in Middleton, 1971, p.92*)

The academicians expressed the hope that others would pick up and repeat their work and that there would be '... free communication from the various societies, scattered as they are today throughout the most illustrious and notable region of Europe' (quoted in Middleton, 1971, p.91). This hope proved to be ill-founded as far as the Accademia del Cimento was concerned. The fact that the *Saggi* was in Italian limited its immediate accessibility. Although a copy of the *Saggi* was sent to the fledgling Royal Society of London in 1668, it was not translated into English until 1684. Not until 1731 was it translated into Latin and thus made more universally accessible. Although the Academy was dissolved in 1667, scientific organizations of the Renaissance academy type continued to be established in Italy well into the eighteenth century.

9.3 The Royal Society of London

In marked contrast to the Accademia del Cimento, the Royal Society of London, which was founded in 1660, chartered by Charles II in 1662 and re-chartered in 1663, was not a private ephemeral body dependent upon a single powerful, wealthy patron. It was established as a corporate, national institution for the pursuit of scientific research. Located in the English metropolis, it was funded (though inadequately for the achievement of its ambitious objectives) by its membership. The Society endured and is now one of the most prestigious scientific institutions in the world. Founded in the year of the Restoration, its early decades coincided with an astonishingly fruitful period for English science, recognized even among school children today through the eponymous laws of such luminaries among its early membership as Boyle, Hooke and Newton (see Chapter 8). However, neither the Society's present-day status nor the flowering of experimental science in Restoration England should determine our view of the role and significance of the early Royal Society.

9.3.1 Origins of the Royal Society

At a general level, the question of why a unique body such as the Royal Society should have emerged at this time is of interest: a body devoted to experimental natural philosophy and indeed participating in its definition; a body devoted to research rather than teaching or the certification of professionals; a public body, legally incorporated along the lines of a number of formal English institutions with particular rights. With publicly guaranteed rights went certain responsibilities, as the Society's first president stressed when seeking (unsuccessfully) public funds in 1662.

> '... the interest of private persons, though very deserving', should properly be subordinated to 'the public concern of a society, whose designs, if protected and assisted by authority, may so much conduce to the greatness and honour of their prince, the real good of his dominions, and the universal benefit of mankind'. *(quoted in Hunter, 1989, p.4)*

Many features of the context of the establishment of the Royal Society have been considered already (see Chapter 8), including the methodological and organizational ideas of Francis Bacon. Though founded in a year of great change in London resulting from the Restoration, the new Society had a number of roots in a variety of small circles of keen investigators and others interested in the new knowledge and experimental methods which had been developing from the 1640s and, most particularly, during the Interregnum after the death of Charles I in 1649. Many were the proposals for harnessing the new science to the reform of education and society during this period of ferment in English life. Groups such as those which met at Gresham College in London, the circle of Samuel Hartlib, the College of Physicians in London, and certain activities at Oxford are all part of the context out of which the Royal Society evolved. Indeed it was after a lecture to one of these informal circles by Christopher Wren at Gresham College on 28 November 1660 that what became the Royal Society was first formally proposed.

> ... it was proposed that some course might be thought of to improve this meeting to a more regular way of debating things, and according to the manner of other countries, where there were voluntary associations of men in academies, for the advancement of various parts of learning, so that they might do something answerable here for the promoting of experimental philosophy. *(Note in Royal Society Journal Book, vol.1)*

News of foreign precedents in Italy and in France was quite likely to have reached Britain via travellers and correspondents. More locally, not only was there interest in the new science, but the ambience of London was changing too. Already a centre of trade and commerce as well as politics, London of the 1650s saw the rise of a new moneyed, leisured class characterized by the club-like atmosphere of the new coffee houses. Some were a focus for people with common scientific interests. In certain respects, the early Royal Society undoubtedly had a clubbable ambience. However, it is also important to recognize that the Society was established during a period when the founding of organizations to deal with various issues was a significant trend. Part of this trend was toward more public and formal, rather than private and ad hoc, activity. The possible relation of this to Puritan ideas has been mentioned in Chapter 8. However, as noted there, among the early active members of the Royal Society were individuals with a wide range of religious and political sympathies, and many of the ideas underlying the Society were widely shared by royalists. Contemporary polemics expressing a desire for moderation, to achieve a sense of permanence and balance in the midst of the atmosphere of instability which characterized the aftermath of the Civil War, have also been cited as part of this trend, which it has been argued in turn, had an effect on the kind of science which was done, enhancing the appeal of the laborious, co-operative inductive methods of a wide-ranging Baconianism (Hunter, 1981, p.29).

Whether or not all of its ambitious goals were in fact achievable, it is important not to lose sight of the fact that the Royal Society's focus was the establishment of the new science, both in the sense of furthering its development and achieving its wider recognition as an enterprise in society.

9.3.2 The working of the Royal Society

That the Society did ultimately succeed in helping to further the new science has been amply demonstrated in Chapter 8, although by no means all of the scientific activity in the period was its product. Indeed, the impressive scientific output rather belies the fact that the Society's fortunes in its opening decades were often parlous and it was not overly successful in achieving the full range of its ambitious goals.

Figure 9.2 Frontispiece to Sprat's History of the Royal Society, *1667. This frontispiece encapsulates the spirit of Restoration science at the Royal Society. Beneath the Society's centrally placed coat of arms is a bust of Charles II being crowned with a laurel wreath by the figure of Fame. The pedestal on which the bust stands is flanked by the seated figure of Francis Bacon in his Lord Chancellor's robes on the right and by the President of the Society, Viscount Brouncker, on the left. The table to the left of Brouncker holds the Society's mace, and the bookshelf contains works by Bacon, Copernicus and Harvey as well as various Fellows of the Royal Society. Many of the scientific instruments surrounding the figures were quite up to date ones used by Fellows for experiments at the Society. The telescope in the background may be the 35-foot instrument used at Gresham College, while the building on the horizon of the right-hand side may perhaps represent Solomon's House (Hunter, 1981, pp.194–7). (Science Museum, London)*

In the first place, many initial members were interested for social, as much as for intellectual, reasons. From the granting of the first charter in 1662, all new members were elected by the existing membership, which tended to perpetuate a characteristic membership profile. There was a predominance, throughout the period to 1700, of members from the professional and landed classes, as well as from the government and the court. To a certain extent, the founders deliberately cultivated an aristocratic strand in the membership in order to give the new institution and the new learning which it wished to promote a high standing, commensurate with the status of the royal patronage that it hoped to secure. The required payment of fees also restricted the membership. Indeed, some of the very sections of the scientific community which were most active at this period, such as various groups with a professional interest in applied areas, were effectively excluded by this combination of election and fees. Furthermore, despite the Society's deliberate posture as a national institution, there was a *de facto* restriction of active membership to those who were able to attend meetings at Gresham College in London, although this did not discourage a considerable number of foreigners from seeking membership (10 per cent of the total).

Such relative social homogeneity, however, did not entail a uniformity of intellectual beliefs. While it made the development of policy and activities in the early years somewhat difficult, this diversity of views has retrospectively been seen as a great strength, increasing the number of those who would be satisfied by the Society's activities – especially important in the absence of the hoped-for public funding. Overcoming the difficulties of developing policy in view of this diversity resulted in the development of a totally new type of institution, through a combination of considerable internal negotiation and trial and error.

The core activity of the Society was to be its weekly meetings in London, the initial plan being that the Society's scientific work should consist of experiments performed and observations made in front of its membership and therefore witnessed by them. Volunteers were relied upon and an extraordinary range of natural phenomena was covered, with no coherent plan. Key problems of natural philosophy were interspersed initially with the examination of curiosities and the exploration of various practical, technological problems; some topics would seem to us to fall more in the area of magic. It was as if it was thought that the sheer weight and diversity of empirical investigation in a formal setting would be significant.

This tendency was not unremarked and, along with the great volume of work in organizing such activities for so many meetings, was probably a factor in the Society's appointment of Robert Hooke as curator of experiments late in 1662. This was a salaried scientific appointment, an innovation in itself. Hooke's job was to organize, under the direction of Council, a programme of experiments and observations for the meetings, as well as to examine topics sent in by Fellows. In addition, it was laid down that, in the reporting of experiments, their significance should be clearly stated and fact carefully distinguished from speculation. Although his range was still wide – covering the sciences, arts and inventions, plus the histories of natural and artificial things – Hooke brought a new degree of coherence to the Society's deliberations, devising series of experiments to be pursued over several months. His *Micrographia* of 1665 was the result of such a series, but it would be unwise to overstress the amount of systematic work that Hooke was able to introduce.

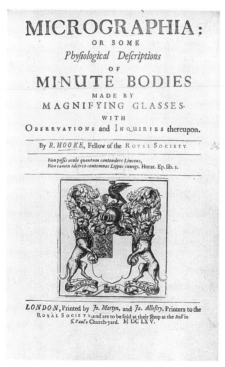

Figure 9.3 Title page of Hooke's Micrographia *showing the Royal Society imprimatur. (Photo reproduced by permission of the British Library Board.)*

Indeed, it was the Society's publications that were perhaps the most successful aspect of its activities in the early years. With access to participation in meetings limited by geography, it was the publications, under its own imprimatur granted by charter, that gave the Society an external identity. The publications were as varied as the Society's activities: John Evelyn's *Sylva, or a Discourse of Forest-Trees*, published in 1664, focussed on practical matters; by contrast, Hooke's elaborately illustrated *Micrographia* gave a literally larger-than-life picture of aspects of the natural world hitherto unknowable (see Figure 8.16).

Most important was the inauguration of the scientific journal *Philosophical Transactions* in 1665 by the Society's Secretary, Henry Oldenburg. It was his own enterprise, not to be officially taken over by the Society for a hundred years, but it had the Society's approval and reported its activities. Oldenburg was already serving as an important London-based focal point for international scientific correspondence, but the multi-copy publication of the new journal extended his effectiveness and became an important international medium for disseminating the new science. It was a short, quick, public form, totally different in style from books or correspondence and ideal for reporting focussed experimental findings. The *Philosophical Transactions* soon gained an international following and many foreigners contributed to its pages. The contents page to volume 1, number 1, gives an idea of just how wide the Royal Society's coverage was in its early years.

In addition to the formal institutional arrangements for promoting experimental work and publication, attempts were made to institutionalize other aspects of the Society's work. Perhaps the most public of these was its Repository, which came to be a popular attraction in seventeenth-century London (Hunter, 1989, ch.4). The Repository arose from the collecting and classifying aspects of Baconian science. The rather indiscriminate collecting of curiosities had long been a tradition of the well-to-do, and cabinets of curiosities were not uncommon possessions among the virtuosi in the Society's membership. In one sense, the Society's Repository was a public, institutional form of this private activity. From its early days, the Society received donations of such items, and in 1663, Hooke was ordered to look after the collection formally at Gresham College; it was much expanded by purchase in 1666. There was a tension between the virtuoso element in the Society, which felt that such a public collection should be of curiosities in the traditional manner, and those who felt that it should serve a more systematic scientific purpose in the Baconian manner and be a comprehensive natural history collection requiring the inclusion of quite mundane items if they were important for classification or understanding. This latter trend was exemplified by the publication in 1681 of a lavish and famous catalogue of the Repository on behalf of the Society by an erstwhile curator Nehemiah Grew. The catalogue was called *Musaeum Regalis Societatis. Or a Catalogue & Description of the Natural and Artificial Rarities Belonging to the Royal Society and Preserved at Gresham College*.

Some hoped that such a collection would help to promote internationally the collaborative study of natural history and contribute materially to the Society's programme for the reform of knowledge. The Repository was maintained by the Society, though seldom with adequate administrative input and thus with varying degrees of enthusiasm and success, for over a century until 1779 when it was offered to the British Museum.

Equally important, but also ultimately unsuccessful, was the Society's ambitious Baconian programme for applying science to the improvement of

Figure 9.4 Contents page from Philosophical Transactions, *vol.1, no.1. (Reproduced by permission of the Syndics of Cambridge University Library.)*

techniques in arts and manufactures. In the spirit of corporate enterprise and to complement the investigations which took place at meetings, committees were set up among groups of Fellows with common interests and appropriate expertise to run projects that were beyond the scope of the meetings. Certain of these committees were devoted to technical matters. However, the ability to influence technical change, particularly in the absence of practitioner input, proved elusive. In the longer term, the system failed partly because Fellows volunteering effort in their private time were unable to achieve such ambitious agendas. In fact, only a small core group of already active Fellows gave any support. Not surprisingly, it proved difficult for them to maintain the high pitch of activity of the early years. Furthermore, there is the question of whether such normally individual inventive activities were actually amenable to corporate activity. The Society's emphasis gradually shifted more towards an achievable pure science.

9.3.3 The Society's fortunes

Throughout the final decades of the seventeenth century, the Royal Society's financial circumstances were precarious. In addition to membership fees, which were unreliable, it depended on occasional donations to keep going. The hoped-for benefactions from those traditional patrons of intellectual activity, the aristocracy, were not forthcoming. Furthermore, the Society suffered from a remarkable lack of unanimity on just which projects to pursue; neither a research laboratory nor permanent premises were secured despite considerable discussion and sentiment in their favour. In part the Society's early problems are a reflection of the fact that its administration relied almost totally on voluntary effort. Despite its early difficulties and misfortunes, it was clearly seen by contemporaries abroad as an important institution; the *Philosophical Transactions* was highly esteemed. Ironically, the Royal Society became very quickly the very symbol of the new science.

In 1671, its membership numbered 199, but that fell by a quarter by the end of the century. Its fortunes revived during Newton's presidency from 1703, and towards the middle of the eighteenth century, the number of members was nearly 350; it was 545 in 1800. The size of the foreign membership was significant throughout. For the whole of the eighteenth century, there were still few permanent staff and a great reliance on voluntary administration. Many of the problems of the early years persisted and the Society enjoyed mixed fortunes, losing after Newton's death in 1727 even some of the lustre which it had had in foreign eyes. Public funding was still not forthcoming, although the Royal Society did help to administer government funds for Captain Cook's voyages. Towards the end of the century, the institution revived somewhat under the presidency of Joseph Banks from 1778; some further measure of royal recognition in the form of accommodation at Somerset House was granted.

At the same time as the national institution was thus less than flourishing, there was a remarkable flowering of provincial institutions throughout Britain as metropolitan, Enlightenment values of rational enquiry into the problems of society were espoused. The emergence of the Royal Society of Edinburgh, established in 1783, is discussed in Chapter 11. The Royal Irish Academy was formed in 1785 in Dublin. Most particularly, the Literary and Philosophical Society movement of the final decades of the eighteenth century gave remarkable scope for scientific discourse throughout the country. At the same time as the British national society was in eclipse, the

Figure 9.5 Emblem of the Royal Society of Edinburgh. (From Transactions, *1788. Reproduced by permission.)*

Académie Royale des Sciences in Paris, the French national society which had been founded contemporaneously but on a very different basis, positively basked in Enlightenment France.

9.4 *The Académie Royale des Sciences, Paris*

Founded in December 1666 in Paris and, in some respects, stimulated by the existence of the Accademia del Cimento and the Royal Society, the Académie Royale des Sciences could hardly have been more different from those institutions. For while it too was a formal organizational expression of the developments of the Scientific Revolution, the French institution was also seen from the start as an instrument of State: the Crown was its patron. It was designed to be a setting for the intellectual munificence of Louis XIV (*r.* 1643–1715) and, in turn, was meant to be a contributor to the welfare of the State. The Academy's membership was closed. It had a small (only fifteen members initially), State-appointed, hierarchically structured membership consisting of representatives of various branches of the sciences; all members received income from the Crown and were expected to devote part of their time to work on problems of interest to it.

9.4.1 *The early years of the Academy*

In France, too, there had been a number of private, informal 'precursor' institutions, but more like the Italian situation and unlike the independent English gatherings, the French ones tended to be more reliant on individual patrons. Such institutions were inevitably linked to the fortunes of their patrons. The importance of Mersenne as an international correspondent and of his Academia Pariensis was mentioned in Chapter 7. Not long after Mersenne's death in 1648, the Montmor Academy began to meet at the home of Habert de Montmor, patron of Gassendi. Though a private body, its establishment has to be seen against the background of the literary Académie Française, which had been founded by the Chief Minister of France, Richelieu, as a cultural manifestation of the absolute monarch. Montmor and others felt that similar encouragement ought to be given to cultivators of the sciences. From 1660, it kept in close touch with the Royal Society of London, whose promotion of experimental methods it admired.

In 1663, when the Montmor Academy was nearing collapse, its secretary, Samuel Sorbière, pointed out some of the problems of running a voluntary organization dedicated to experimentation.

> ... to build an arsenal of machines to carry out all sorts of experiments is impossible ... Think of the space needed for the observation of the stars, and of the size of the apparatus necessary for a forty-foot telescope ... Was not Tycho Brahe forced to build his Uraniborg, a castle not so much for lodgings as for the making of celestial observations?

> Truly, gentlemen, only kings and wealthy sovereigns or a few wise and rich republics can undertake to erect a physical academy where there would be constant experimentation. A special structure must be built to order; a number of artisans must be hired; and considerable funds are necessary for other expenses. *(quoted in Hahn, 1971, p.8)*

One proposal for dealing with this problem was for the establishment of a company of arts and sciences – a proposal elaborated by Thévenot, the wealthy polymath inventor of the spirit-level, and two astronomers, Auzout and Petit. The proposed company was of the nature of what we might call a research institute, with such features as laboratories, observatories, data-collecting expeditions, translators and assigned foreign correspondents, and committees of experts. Quite apart from implications for funding, the proposal implied that science should be pursued by those properly dedicated to it rather than by the dilettante amateur.

Sorbière and Auzout brought such ideas to Colbert, the then chief minister, who had long displayed a personal interest in scientific research, both for its possible utilitarian benefits and for its rational methodology. At the same time, it was he who saw the possibility of bringing together all French cultural and intellectual activities around the absolute monarch for mutual enhancement. He had already started to seek out individual scholars, writers and scientists who might like to receive a government subsidy. Despite this very favourable conjunction of views, it took almost three years to finalize plans before the Academy was established at the end of 1666.

The long gestation period was due in part to the complexities of setting up a royally funded institution, a problem clearly not encountered by the Royal Society of London. If they were to accept state pensions for their scientific work, what was to be expected in return from the academicians? How would they balance their own inclinations with those of the State? Furthermore, a certain tension between a possible Baconian, utilitarian programme of researches, calling for specialists, and a more general cultural programme, calling for natural philosophers, had to be resolved before the membership could be selected. Indeed, there was even a proposal that science be one of a number of subject areas in a great general academy.

The general academy plan failed, but members eventually chosen by Colbert for the Académie Royale des Sciences could easily have belonged to it. They were a well-educated and therefore socially exclusive group chosen for their scientific knowledge and orientation toward experiment; artisans did not fall within those categories, nor was technical prowess a criterion of selection (Hahn, 1971, pp.14–15). At the same time, the academicians had a responsibility to advise the State on practical problems too. The new Academy was to be composed of two sections: the mathematical section, including the 'exact' sciences such as mechanics and astronomy; and the physical section, including the 'experimental' sciences such as physics, chemistry, anatomy and physiology.

Twice-weekly meetings, closed to all but academicians and a few of their students, were held in the King's Library, pending the construction of proper facilities. Minutes were kept, but the early proceedings of the Academy were characterized by secrecy, though reports of their activities were published without attribution to individuals in the recently founded *Journal des savants*, a private periodical devoted to a range of subjects which included science. The academicians attempted to be systematic in their approach to the study of nature, although the record of their activities indicates that they did not necessarily achieve this aim (Hall, 1983, p.224). In a Baconian manner, they prepared annual programmes of important problems which would profit from a corporate experimental approach. Significantly, Colbert directed their attention to a range of problems of long-term interest to the State, such as the determination of longitude at sea, mapping the country, studying gunpowder and artillery questions. But they also responded to a number of specific technical requests, such as commenting on new substances or inventions. The Academy was involved in various scientific expeditions, and had an

Figure 9.6 Visit by Louis XIV and Colbert to the Paris Academy of Sciences. This rather fanciful scene of a visit which probably never took place shows the Paris Observatory under construction in the background. (Frontispiece of C. Perrault, Mémoires pour servir à l'histoire naturelle des animaux, *1671. Photo reproduced by permission of the British Library Board.)*

agreement with travelling Jesuits to receive any information they might collect on natural history, topography, etc. Colbert saw to it that their researches at the Jardin du Roi were subsidized.

Perhaps the most significant event symbolically in the early years of the Academy was the building over the period 1667 to 1672 of what became, in effect, its permanent premises, the Paris Observatory. Though ultimately less ambitious than originally planned – it should have housed chemical laboratories, natural history specimens and collections of models in addition to astronomical facilities – the Observatory was funded by the Crown and provided tangible evidence of both the status of the Academy and the level of commitment of the Crown to science.

Colbert's death in 1683 is sometimes supposed to have led to a period of decline at the Academy until an important series of reforms in 1699 launched it into the next century with renewed vigour. However, the Academy's fortunes during this period cannot be dissociated altogether from important issues in the political life of France, such as the revocation of the Edict of Nantes in 1685, which resulted in the loss of a number of important Protestant foreigners, including Huygens. Furthermore, the State was preoccupied with military crises, and consequent expenditure plus some bad harvests disrupted the economy in the 1690s. It is also necessary to recognize that the last two decades of the seventeenth century was a period of consolidation and maturation for the Academy as the role of the academicians evolved in relation to the State and their science (Hahn, 1971, pp.19–21; Stroup, 1987, p.9). It was during this period that the academicians moved away from their initial Baconian orientation, involving original collective work, towards a position of acting collectively as adjudicators of work produced by individuals, both within and outside the Academy. At the same time, certain meetings became open to the public. This change of orientation, which was codified in 1699, came about gradually as the academicians sought to deal with the various problems brought to their attention. In the sphere of technology, *approuvé par l'Académie* became a much sought-after and valued designation. The models of inventions deposited by those seeking approval subsequently became the nucleus of the Museum of the Paris Conservatoire des Arts et Métiers. Similarly, the designation by the Academy of scientific work as worthy of publication came to signify work of a high standard. Association with the Academy was highly valued. Furthermore, the Academy's prestigious publication, its *Histoires* and associated *Mémoires*, was instituted in 1699 and soon became a standard bearer for science. Like the short-lived *Saggi*, it was a direct institutional publication.

9.4.2 The Academy in the eighteenth century

The reforms of 1699 were developed in the constitution of 1716. The structure of the Academy became more elaborate and, importantly, allowed for a much broader membership. Under the new regulations, there were to be 44 regular scientific members, all of whom had to live in Paris and none of whom could be a member of a religious order. These members were divided into the same six scientific classes as at the outset. Within each class there was also a set of hierarchical subdivisions which formed a sort of career ladder: two adjuncts, two associates and three *pensionnaires* (so called because they were entitled to a salary). In addition the secretary and the treasurer were *pensionnaires*. Candidates for regular membership had to display scientific excellence. As members, they were required to attend meetings regularly and to report on their researches to the Academy from time to time, as well as to deal with formal requests to comment on matters of technology or science submitted to the Academy. There were also new categories of 'irregular' members: free associates, who were generally individuals of scientific eminence in the provinces; honorary members, chosen more for their position in society or government than for their science; and a considerable number of correspondents from the rest of France and around Europe, who had the right to communicate their researches to a full member. Although only the *pensionnaires* were entitled to a salary, occasionally members of other categories did receive some income from the Academy. However, even the senior *pensionnaires'* salary was not adequate to provide a living in Paris; either a private income or a further post or consulting work for government firms was necessary. The status of being an academician of course assisted in obtaining employment (Heilbron, 1982, pp.107–9).

Even with the expansion, the Academy was still quite small and select, with an average membership in all categories of 153 at any particular time in the eighteenth century and a core active membership of only 44. Yet in that period it was perhaps the premier scientific institution in Europe; almost every important discovery of the time was announced or discusssed there. Its scientific achievements have been discussed in Chapter 7. According to McClellan (1985, p.20), the Academy achieved such eminence because it was so carefully adapted 'to the Society in which it was created; like the rest of French society in the ancien régime, the Paris Academy was based on privilege, centralized power, and hierarchical stratification'. Part of that congruence was due to the changing cultural environment of the period of the Enlightenment which gave a new emphasis to the value of knowledge and a new status to the detached intellectual with a responsibility toward society (see Chapter 11). Birth alone could no longer define status: talent and merit were also to be criteria. In the realm of science, the Academy served to differentiate those intellectuals equipped to make sound judgements on science and technology; it simultaneously defined an elite and that elite's relationship to the community and to the bureaucratic absolutist State (Hahn, 1971, p.43). Hahn sums up its significance thus:

> The scientific academician appeared as the prototype of the citizen of the eighteenth-century Republic of Letters ... By submitting his creation to the collective judgment of his colleagues, he protected everyone – including himself – from the vagaries of his imagination and the subjectivity of individual creation. Composed as it was of specialists, the Academy was the impartial guarantor of the validity of individual discovery. Through its communal procedure of consensus, the new institution was an instrument for separating sound knowledge from conjecture and for certifying the 'way of truth'. By its publications, it could communicate the new-found truth to others and aid in turning the Enlightenment's dream of progress into a reality.

> ... the Royal Academy of Sciences [was] the model institution of its time. Beyond this, the enterprise upon which it was embarked was held in the highest esteem during the era. To many, it was the most promising undertaking of the century. Science seemed destined to revolutionize navigation, warfare, agriculture, the arts and crafts, health, and all the varied material amenities of life on earth. It was expected to be of immense service to the state, and above all to civilize mankind by offering it a new secular religion. *(Hahn, 1971, pp.56–7)*

The congruity between the Academy and the ancien régime was so strong that as the century progressed the former became closely identified with the régime itself, rather than with the advancement of science. Thus, having become somewhat inflexible, the Academy did not adapt readily to the societal changes culminating in the events of the final decade of the century. Ever-increasing demands for practical results, while reinforcing the importance of the Academy as a central arbiter of technology, threatened to upset its equilibrium between research and consultation. Furthermore, a growing trend toward egalitarianism was in direct conflict with the Academy's elitism and hierarchical organization. Also a rising Romantic individualism brought into question the Academy's system of collective authority.

In fact, the Academy survived the early years of the Revolution, establishing links with the new National Assembly while simultaneously maintaining them with the Crown and its ministers. With freedom of the press and freedom of association, the Academy lost its monopoly of scientific publication in France, and many other voluntary societies dealing with science were set up. Yet at the same time considerable demands for consultative work continued to be forwarded to it, especially after the outbreak of war in 1792. The National Assembly initially sought to use the Academy much as previously and assigned it a major national task, the reform of weights and measures. One of the Academy's members became the first mayor of Paris. However, the fundamental nature of the Academy idea, rooted in early eighteenth-century aristocratic principles, was ultimately at odds with the new democratic politics. Its level of activity declined as members became otherwise occupied with events, although the legislature's demands for scientific activity remained high and arguably more scientists were paid for working directly for government agencies as public servants than had been paid by the Academy. The voluntary associations were cultivated as centres of scientific strength, and the rise of a new educational system created important niches for scientists working in their specialities.

The Académie Royale des Sciences was closed in 1793, to be reconstituted two years later as the scientific wing of the new Institut de France, the National Institute of Arts and Sciences. Ironically, in relation to the Academy, this was in some measure an attempt to restore an institution whose functions had been taken over by other bodies. The primary role left to it became that of recording scientific achievements. Despite the loss of its functions, it remained important as a symbol.

9.5 The Prussian Royal Academy of Sciences and Letters (the Berlin Academy)

Between them, the Royal Society of London and the Académie Royale des Sciences in Paris provided the models for the academies established throughout Europe over the eighteenth century. An academy in Berlin, which was finally established in 1700, certainly had the English and French precedents in view and ultimately chose to follow the French model. Although established at the very start of the century, its founder, Gottfried Wilhelm von Leibniz (1646–1716), had been working towards its establishment for many years. In emphasis it really belonged to that late seventeenth-century generation of organizations which were seeking to consolidate the work of the Scientific Revolution.

The Berlin institution was not without predecessors devoted to science in Germany, sometimes based on Italian models (Ornstein, [1913] 1975, pp.167–77). However, even though one of those founded in the seventeenth century is still in existence, it then had the status more of a club for like-minded individuals with medico-scientific interests than a learned society actively promoting the development of science. In contrast to Italy, Britain and France, the new science did not flourish in a general way in Germany in the seventeenth century, although of course there were exceptional individuals. Some have attributed this variously to Germany's territorial divisions into numerous units, the ravages of the Thirty Years War, and backward social and economic conditions which served to undermine its schools and universities (Hall, 1983, p.228). Leibniz considered that the fact that Latin and French were the scholarly and polite languages, rather than the vernacular German, forestalled the kinds of fertile developments which drew on the knowledge of the wider, non-scholarly community.

Indeed, Leibniz was the prime mover in the establishment of an academy for Berlin. A polymath – historian, philosopher, mathematician, author – with legal and diplomatic skills, he had in view nothing less than the reconciliation of Catholicism and Protestantism, the reform of education and the regeneration of the national economy and German imperial status through the establishment of an academy which would study, communicate and apply science. He saw science as a means of unifying the German states and, when wedded to economic policies, as a prestigious instrument of the State (see Chapter 14). Above all, science should be useful. His efforts over some twenty years in this direction were unproductive until, towards the end of the century, he came into association with the then Elector of the Palatinate, later King Frederick I of Brandenburg-Prussia (*r.* 1701–13).

In the 1690s, for reasons of State, Frederick I embarked on a programme of cultural expansion, including the establishment of the University of Halle and various colleges. Among the reforms sponsored was the reform of the Prussian calendar in 1700. Seeing his chance, Leibniz suggested a new academy to deal with the technical aspects of that reform. The rather clever and innovative deal was that, in return for producing accurate almanacs and calendars, the new institution would receive all income from them, and thus be self-financing (McClellan, 1985, p.70). Actually called the Societas Regia Scientiarum, or Royal Society of Sciences, this academy was at first loosely modelled on the Paris Academy, although it was governed by a council system similar to that of the Royal Society of London. It had official status within Brandenburg-Prussia and received privileges in return for its services; membership was by royal appointment. Leibniz's influence is clear in that the membership fell into four classes: mathematical sciences, physical sciences, German language and history, and literature. It was therefore a much more broadly based cultural institution than its title would imply, but that is how Leibniz saw science.

The Society did not flourish in its early years, partly owing to the State's siphoning off of the almanac income for other purposes. This reflected more fundamentally a certain bureaucratic lack of will to follow through with the initial plans, as well as a new, unsympathetic king, King Frederick William I (*r.* 1713–40). Following foreign precedents, the Society did begin to publish a series of memoirs in 1710, the *Miscellanea Berolinensia*, but neither it nor the Society made much international impact. One historian commenting on the seven volumes produced to 1744 suggested that 'The new spirit of science was not too forcefully felt in these volumes' (Harnack, quoted in McClellan, 1985, p.72).

Frederick II's coming to the throne of Brandenburg-Prussia in 1740 transformed the Society. Exiled abroad in his youth, Frederick II identified strongly with French culture and aimed to open up narrow Prussian society to wider international influences. He made deliberate moves to establish an academy on the French model with the aim of rivalling Paris and London in matters scientific. In 1746, the Society was reconstituted as the Berlin Academy, formally the Royal Academy of Sciences and Letters of Prussia (Académie Royale des Sciences et des Belle-Lettres de Prusse), with classes for experimental philosophy, mathematics, speculative philosophy and literature. There were set numbers of members in various grades for each class, plus an administrative structure for each class, all royally appointed and funded by the restored almanac income.

Most importantly, the orientation of the new court was able to attract eminent scientists from elsewhere. The mathematician Euler was brought in for the mathematical class from St Petersburg, and Maupertuis, a Newtonian, came from France as president, thus putting the leadership into the hands of

Figure 9.7 Emblem of the Berlin Society of Sciences. (From Miscellanea Berolinensia, *1710, frontispiece.)*

a scientist. The salaries offered were somewhat better than those in Paris, and living in Berlin was cheaper. Like their Parisian counterparts, the academicians in Berlin had access to additional employment related to their expertise (Heilbron, 1982, pp.111–12). Very quickly, Maupertuis introduced changes to bring the Academy into line with French precedents and into prominence: he instituted the publication of the *Mémoires* in French; established categories of foreign members; and set up an international prize competition. The upshot of all this activity was that Berlin did come to equal Paris in the stature of its premier scientific institution. It was a success deliberately sought as a matter of state policy.

9.6 The spread of the academies in the eighteenth century[2]

During the eighteenth century, there was a rapid increase in the number of scientific societies and academies as can be seen in Figure 9.8. A scientific academy became, especially after 1750, a not unusual feature of a city striving to make its mark as an intellectual and cultural centre. Though all were established in distinctly local circumstances, it cannot be denied that the movement itself generated a certain momentum. Indeed, this institutional expansion from mid-century was an aspect of the Enlightenment (see Chapter 11). The renewal of the Berlin Academy at about the same time should be viewed in a similar light. The academies became the embodiment of the harnessing of rational knowledge, as against ignorance and superstition, both for its own sake and for political reasons to do with the application of such knowledge to the improvement of economic life.

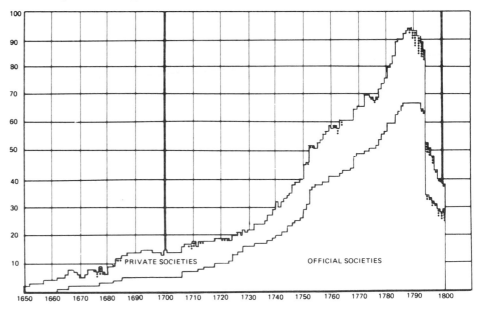

Figure 9.8 Growth of academies and societies of science, 1650–1800. (From J.E. McClellan, Science Reorganized: Scientific Societies in the Eighteenth Century, *New York, Columbia University Press, 1985, pp.67.)*

During the period to 1750, important academies were established in a number of European cities: Berlin, St Petersburg, Stockholm, Bologna, and in some of the larger provincial cities of France. From mid-century, there was a change of gear. A number of these continental societies which had had slow beginnings, such as the Academy in Berlin, achieved a new stability and,

[2] This section is largely based on McClellan (1985), chs 3–4.

along with London and Paris, were to be the premier bodies of the second half of the century. Recognizing a new era, the secretary of the Berlin Academy could proclaim in 1745 that societies needed no further justification:

> No one can ignore how the idea of learned companies was born in Europe during the past century and how, vying with one another, they have multiplied in the principal kingdoms in this part of the world. *(quoted in McClellan, 1985, pp.110–11)*

Map 9.1 Scientific societies in Europe in 1789. (Adapted from J.E. McClellan, Science Reorganized: Scientific Societies in the Eighteenth Century, New York, Columbia University Press, 1985, pp.6–7.)

Figure 9.9 Emblem of the Imperial Academy of Sciences and Arts, St Petersburg. (From Novi Commentarii, *1750, frontispiece. Published by permission of the British Library Board.)*

Throughout the rest of the century, this initial framework of major institutions was enlarged by the establishment of secondary and tertiary institutions. The motives for the establishment of these were varied. Often they drew on the work of the major bodies to set their own agendas, and many worked on specifically local problems. In the second half of the century, different types of activity were undertaken. The extent and self-awareness of the movement became evident in a new level of activity between societies. They began not only to imitate each other, but to interact. Systems for exchanging publications were set up and there were some co-operative observational projects among a number of societies.

It is appropriate to speak of these institutions as a group because they had many features in common. Indeed, they saw themselves as related. However, they also varied greatly as each was unique, created in particular national circumstances.

9.6.1 Russia

In Russia, the Imperial Academy of Sciences and Arts, St Petersburg, established in 1724 on the Paris model, was created by Peter the Great as part of his westernization programme. His motive was clear: to bring an economically backward, vast empire into the mainstream of eighteenth-century life through the promotion of science (see Chapter 13). Although based on the Paris model, the institution proposed for Russia was in fact far more complex. The plan called for nothing less than the creation *de novo* of an entire scientific infrastructure to be funded by revenues from the new Baltic ports. The original scheme included, as well as research and the provision of expert advice to the government, two teaching institutions and a publishing arm which was to help popularize science. The academicians were thus to have a treble role as researchers, teachers and popularizers. There was also to be an observatory, a botanic garden and a department for co-ordinating the mapping of the Empire.

That there was no scientific tradition in Russia to draw upon, and effectively no educated middle class, had a number of implications for the Academy. In the first place, it had to be a thoroughly government institution and hence dependent on a changing imperial climate; there was no other source of support. At the same time, everything for the new academy, from apparatus to academicians, had to be imported. Latin was its *lingua franca* and later French was used as well. In effect, then, as was the case of the activities of the Russian educated elite, it was a foreign island, an implant that took some considerable time to become Russianized.

9.6.2 Sweden

Sweden, by contrast with Russia, had a well-developed scientific structure and, though on the northern edge of Europe, had numerous contacts with activities in London and in Paris (see Chapter 12). Two scientific societies were founded there in the early years of the eighteenth century. The Royal Society of Sciences (Societatis Regiae Scientiarium), Uppsala, made royal and official in 1728, grew out of a private society founded in 1710 by a group of mathematicians and physicians from the University of Uppsala. Cartesian in its emphasis, it took London as its organizational model. Though it had a wide range of interests, its emphasis tended to be strictly academic and it used Latin as its official language. It included many of the luminaries of Swedish science.

The second Swedish institution to be devoted to science was founded in 1739 in Stockholm and given its official identity as the Royal Swedish Academy of Sciences (Kungliga Vetenkapsakademien) in 1741. It also was modelled on the Royal Society of London, but was quite distinct from the Uppsala Society in orientation. Its adoption of the vernacular Swedish as its official language is one indicator of difference. The results of its researches were to be made accessible to the general Swedish public. A number of its scientific initiators were in fact members of the Uppsala Society who wanted Swedish scientific activity to move in a different direction with the explicit utilitarian goal of harnessing science to economic development for the benefit of the State. The fact that a second society was founded owes much to the political prominence in the 1730s of a party dedicated to economic and industrial development, whose top members had close links with scientific circles. A third of its members were aristocrats or high civil servants; a fifth were professors. The stimulus then was almost the inverse of that in Russia; the Swedish Academy was promoted by interested citizens working through the parliamentary aspects of their political system.

Figure 9.10 Emblem of the Royal Swedish Academy of Sciences. (Reproduced by permission of Kungliga Vetenskapsakademien.)

Though explicitly modelled on the Royal Society in London, the Swedish Society enjoyed better funding, partly through generous bequests, but also by adopting the Berlin Academy strategy of running a highly profitable almanac business under royal license from 1747. This freed it from dependence on members' subscriptions and altered the level of its activities. Some twenty years later, its income was half that of the Paris Academy and it could afford to increase its profits further by lending money to members (Heilbron, 1982, p.118). It also varied from London practice in having a limited number of members, although these were chosen by open election. Its journal, the *Handlingar*, was published in Swedish, but quickly translated into German on a regular basis, as at least half the material was devoted to economic development or agriculture. This combination of service, utility and high science communicated through a regular publication and eager, voluminous correspondence put the Royal Swedish Academy at the forefront of eighteenth-century scientific institutions.

9.6.3 Italy

The most notable Italian development of this period was the establishment in 1714 of the Bolognese Academy of Science (Accademia della Scienze dell'Instituto). It evolved in fact from a 'Renaissance' type academy modelled on the Accademia del Cimento, which flourished in Bologna from the 1690s and was indeed its most vigorous successor. In 1704, this private body, the Accademia degli Inquieti, adopted a constitution based on that of the Paris Academy of Sciences and unofficially constituted itself as a pressure group for the reform of scientific education in the University of Bologna. Winning crucial support from a Vatican diplomat, the reform plans were approved by the Bologna town Senate in 1714. A new Institute of Bologna was formed as the scientific wing of the university. It was incumbent upon the professors to use demonstration, experiment and other practical aides in their teaching, as well as to do research in the laboratory or observatory provided. The Bolognese Academy of Science was created as an appendage to the Institute, organized around a core of its academic members. It had a structure of

Figure 9.11 Emblem of the Institute of Bologna. (From Commentarii, *1731, frontispiece. Published by permission of the British Library Board.)*

classes and subject specialities like the Paris Academy, and was funded jointly by the Bolognese Senate and by the Vatican, under which Bologna, as a Papal State, came at this time.

So here is yet another example of a scientific society, having goals in common with other international bodies, responding to specific local circumstances. These circumstances also affected the publication of its journal, the *Commentarii*: its volumes were much delayed by the necessity of seeking the approval of the Inquisition before publication. The Academy underwent important reforms in 1745 at the behest of Pope Benedict XIV, who had Frederick II's cultural reform programme for Berlin very much in view. He endowed a major library at the Academy and created a new class of paid members, who had a responsibility to attend meetings regularly and to publish research papers. Whether or not this early 'publish or perish' policy was in the best interests of the long-term quality of work at the Academy, in the short term it made it a very active body. Yet despite this support, the Bolognese Academy did not achieve international recognition as a major institution.

In Turin, when a group of active scientists came together at the university from the 1740s, they immediately looked to organization on an international level as a means of furthering their work. Initially, as a private society, they sent their publications to the major academies elsewhere. The very positive response to this effort from abroad induced the king-to-be Victor-Amadeus (*r.* 1773–96), again with the Enlightenment programme of Frederick II in view, to offer patronage to the Turin group, which constituted itself as the Royal Society of Sciences of Turin in 1759. Almost from the start, with its eyes on the international community, the society sought to be upgraded by royal sanction to an official academy.

> [We] observed the good effect that learned societies and academies produce in the interior of those countries where sovereigns have judged it appropriate to establish them, and by their constitutions they tend thus to form, despite the distances between them, a worthy tie between the gifted men of all nations ... [We seek] the title and prerogatives of a Royal Academy, based roughly on the model of some establishments of this nature which exist in various states of Europe in order that, working on the same basis that has given rise to their institution, the new Academy will tend toward the same goal and contribute in its turn to the attempts and discoveries that are made daily in foreign countries; experience has shown in these last years how much the concourse of nations multiplies the advantages that have begun to be forthcoming from nearly a century of cooperation among learned men. (*quoted in McClellan, 1985, pp.128–9*)

With several members of international stature, the Turin Society was a major scientific centre despite its lack of academy status. So, except in financial terms, the eventual upgrading to an academy in 1783 made little difference. It closed in 1792.

9.6.4 *France*

Even a cursory glance at Map 9.1 shows the extraordinary multiplication of societies and academies in provincial France, where more than twenty-five highly varied societies were established during the eighteenth century. Though many of these adopted certain forms from the Paris Academy, they more often had the characteristics of an open society like the Royal Society in London, with no fixed memberships and no prescribed topics. Among the most prominent of these were the ones at Lyon, Montpellier, Bordeaux, Dijon and Toulouse. Effectively, by mid-century, the Academy in Paris was backed up by a provincial network of academies throughout the country in cities of the second tier. Science became a central concern for many of them for the first time during this period. For the most part, these academies were 'grass-roots' institutions, arising from the interests of local individuals who formed themselves into private associations for which they then sought official (that is, central government) recognition. With no central funding, the precariousness of their finances rather limited their activities to those of local interest, especially of relevance to the local levels of government. Local mapping and natural history were typical foci. In the second half of the century, a shift has been noted in the activities of the provincial academies, away from pure science and toward more utilitarian projects specifically applying science (McClellan, 1985, p.134). At the same time, they took on more of a direct service function, and especially began to undertake more formal roles in the teaching of science.

Although not perhaps scientifically illustrious, the provincial societies in France did play an important role in local communities. They made science an ordinary feature of the institutional life of the country and introduced it to a very wide audience. The fact that they were informally linked among themselves gave the work that was done in the provinces greater prominence, and activity in provincial academies proved an effective stepping stone to Paris for many an aspiring scientist.

9.6.5 *Elsewhere in Europe*

In Northern Europe, too, during the second half of the eighteenth century, there was a flowering of academies, again highly varied in type and local motives. That in Göttingen, founded in 1752, was actually planned in relation to the university which opened there in 1734. In fact, in terms of subject coverage, it divided its efforts with the university medical faculty. An almanac monopoly secured its finances and it was able to publish a journal. However, it never achieved independent eminence as a society itself; its reputation was tied to that of one eminent professor of medicine.

The Academy in Erfurt was established in 1754 deliberately outside the local university by Protestants who were isolated from certain activities at that Catholic institution. It was well equipped with practical facilities and its orientation was explicitly utilitarian. Indeed, it has been linked with the growing movement of societies throughout Germany wishing to use science to improve economic life and national fortunes. In Munich, too, the Bavarian Academy of Sciences, which opened in 1759, was part of a broader movement of cultural regeneration when a variety of academies were established, though not without Jesuit opposition. In Mannheim, the Academy, founded in 1763, was part of an orchestrated local renaissance seeking to make the town a scientific and cultural centre. It had international pretensions and was well supported by the Elector of Mannheim.

Examples of such institutions can be multiplied in Poland, in Bohemia, and in the Austrian territories. Conspicuously, Vienna itself lacked a society, despite Leibniz's early efforts to promote one there and despite a number of developments elsewhere under Austrian patronage. In Vienna, Jesuit opposition, arising from a tension between a fixed curriculum and the freedom of enquiry likely to characterize a society, was effective in forestalling mid-century plans (Heilbron, 1982, p.122). The Dutch, too, were less vigorous in the promotion of academies than their counterparts elsewhere. In part this was owing to the long-established strength of science in their universities and a consequent concern that that would be undermined by independent institutions. The establishment of societies there had more to do with developing provincial culture and status than with science per se. Spurred by foreign example, an academy was formed in Haarlem in 1756, with the eminent electrical experimenter van Marum as its permanent secretary.

9.7 The role of the academies

If nothing else, the catalogue in the previous section, coupled with Map 9.1, shows how vital and extensive the academies movement was in the eighteenth century. However various and whatever the particular local circumstances, the academies as a body did serve an important purpose in promoting science. In the first place, whether eminent or humble, the societies served their members though mutual encouragement of their activities. Also not to be overlooked is their importance as publishing bodies. During the eighteenth century, the scholarly article in a journal of a learned society came to be the principal mode of introducing new material – books and commercial journals becoming less significant for this purpose. Furthermore, the international prize competitions introduced by a number of academies served to promote science in a very particular way, as did their collaborative projects. Neither should their efforts in popularization and, in many cases, their efforts to be directly useful be overlooked. Insofar as many enjoyed at least some measure of public recognition, if not finance, they were important in mediating between science and the State.

In addition to their significance for science, it is essential to consider their significance for scientific practitioners. During the latter part of the seventeenth century, these practitioners no longer needed fora where the principles of the new philosophy could be debated. The scientific movement had, as it were, prevailed. The function of the societies then became to consolidate the Scientific Revolution, to provide fora where developments in the new science could be validated and communicated. People who worked on the sciences, then, gained an identity related to this type of organization. The eighteenth-century institutional developments have tended to be presented as a long extension of this function until the next major institutional changes resulting from the professionalization of research in the context of nineteenth-century universities.

Recently, it has been argued persuasively that the eighteenth-century institutional developments marked a distinctive and important transitional stage in the professional development of the practitioner of science into the 'scientist' from the seventeenth to the nineteenth centuries (McClellan, 1985). In the first place, in however embryonic a form, the eighteenth-century societies did begin to provide some career opportunities based on scientific knowledge. Even where the participants were amateurs, the academies created certain expectations of standards and role, which were judged by the scrutiny of peers. The person advancing scientific knowledge came to be expected to behave in certain ways through the institutions which became the 'natural' outlets for such knowledge.

Quite apart from other considerations, as noted in Section 9.1, from the end of the seventeenth century, the leading practitioners of science were all attached in some way to societies, even if their bases were elsewhere, such as in a university. Most eminent practitioners took very active administrative roles in the societies and indeed were important initiators of institutions. The institutions served as a sort of fixed centre for their activities. The academies in particular provided a specified social role for practitioners, that of academician. It was not just the money – few academicians in any country could support themselves with salaries alone – but the status of the identity it afforded. In some of the bigger, more prestigious academies, 'academician' was a career position. Even in the French provinces, it was possible to create a career of sorts using an academy appointment as a nucleus, around which to build teaching and consulting activities.

At the same time, these bodies were able to impose standards on their members. Election or appointment was on the basis of certain criteria, though admittedly not necessarily always strictly that of scientific excellence. More importantly, they subjected the work presented to stringent cross-examination. Many societies reserved the power to impose sanctions, fines or dismissal on miscreants. The societies were also the principal vehicle for international communication, which became formalized in this period. At the same time foreign memberships came under stricter control by the various national bodies. In establishing and enforcing criteria for native and foreign membership, the societies were de facto commenting on the nature of the scientist's role.

While it would be anachronistic and indeed irrelevant to speak in terms of professionalization in the eighteenth century, it is none the less the case that that period did see the emergence of more defined roles for those involved in scientific activity. The fact that the number of practitioners was increasing so rapidly should also not be overlooked. It is through the institutions which proliferated in this period that they furthered their work and gained an identity.

Sources referred to in the text

Curtis, M.H. (1959) *Oxford and Cambridge in Transition, 1558–1642: An Essay on Changing Relations Between the English Universities and English Society*, Oxford, Clarendon Press.

Hahn, R. (1971) *The Anatomy of a Scientific Institution: The Paris Academy of Sciences, 1666–1803*, London, University of California Press.

Hall, A.R. (1983) *The Revolution in Science, 1500–1750*, London, Longman.

Heilbron, J.L. (1982) *Elements of Early Modern Physics*, London, University of California Press.

Hooykaas, R. (1987) 'The Rise of Modern Science: When and Why?', *British Journal for the History of Science*, 20, pp.453–73.

Hunter, M. (1981) *Science and Society in Restoration England*, Cambridge, Cambridge University Press.

Hunter, M. (1989) *Establishing the New Science: The Experience of the Early Royal Society*, London, Boydell Press.

McClellan III, J.E. (1985) *Science Reorganized: Scientific Societies in the Eighteenth Century*, New York, Columbia University Press.

Middleton, W.E.K. (1971) *The Experimenters: A Study of the Accademia del Cimento*, Baltimore, Md, Johns Hopkins University Press.

Ornstein, M. ([1913] 1975) *The Role of Scientific Societies in the Seventeenth Century*, New York, Arno Press Reprint (based on her 1913 Columbia University Dissertation).

Redondi, P. (1988) *Galileo Heretic*, tr. R. Rosenthal, London, Allen Lane.

Stroup, A. (1987) 'Royal Funding of the Parisian Académie des Sciences during the 1690s', *Transactions of the American Philosophical Society*, 77, part 4.

Webster, C. (1975) *The Great Instauration: Science, Medicine and Reform, 1626–1660*, London, Duckworth.

Further reading

Gascoigne, J. (1990) 'A reappraisal of the role of the universities in the Scientific Revolution', in D.C. Lindberg and R.S. Westman (eds) *Reappraisals of the Scientific Revolution*, Cambridge, Cambridge University Press.

Hahn, R. (1971) *The Anatomy of a Scientific Institution: The Paris Academy of Sciences, 1666–1803*, London, University of California Press.

Hunter, M. (1981) *Science and Society in Restoration England*, Cambridge, Cambridge University Press.

McClellan III, J.E. (1985) *Science Reorganized: Scientific Societies in the Eighteenth Century*, New York, Columbia University Press.

The Reception of Newtonianism in Europe — Chapter 10

by Colin A. Russell

10.1 The complex legacy of Isaac Newton

If any one theme is central to the rise of scientific Europe it must surely be the movement forming the subject of this chapter. And if any one person can be said to have dominated the international scientific scene in the eighteenth century that was the English progenitor of this movement, Sir Isaac Newton. So, at least, we are often led to believe, and not only by those who might be over-anxious to press the claims of their countryman for nationalistic or even jingoistic reasons, be they Newton's contemporaries or British historians of science of the twentieth century.

In fact, as we have seen (Chapter 9), Sir Isaac Newton was a titanic figure in science by any standards. Of that fact many of his contemporaries were complacently aware. The Scottish poet James Thomson described him as

> Newton, pure Intelligence, whom God
> To Mortals lent, to trace his boundless Works
> From laws sublimely simple.
>
> ('The Seasons: Summer' (1727), lines 1160–2)

Hardly less eulogistic is the oft-quoted couplet in which Alexander Pope declaimed:

> Nature, and Nature's Laws, lay hid in Night;
> God said, *Let Newton be!* and All was *Light*.
>
> (Epitaph intended for Sir Isaac Newton)

It is certainly the case that a great deal of the light of science which streamed all over Europe for the next hundred years emanated from this one man and bears his own name, *Newtonianism*. However, before we slip into the easy view that here is a, or perhaps *the*, true explanation for the later progress of scientific Europe, it is salutary to add several cautionary notes.

In the first place, Newtonianism was not the only movement contending for European loyalty and support; it had several important rivals. These included Hutchinsonianism in Britain and Cartesianism in France.

Secondly, Newtonianism was not therefore accepted immediately or at the same rate in different areas. In the present chapter we shall be looking particularly at England, the Low Countries and France; later chapters touch on the subject for Scotland (11), Sweden (12), Orthodox Europe (13) and Germany (14).

Thirdly, and perhaps most importantly, Newtonianism is in fact a complex movement, with several distinct components which are, as one might nevertheless expect, related to each other. It is a cluster of ideas whose rich diversity reflects the thinking of a man in whom science, philosophy and theology were inseparable. Newton was not afraid to think about nature and

God at one and the same time, or to speak about them in the same breath. He lacked the compartmentalism of modern man and the two-decker mind-set that is a legacy of the so-called 'Enlightenment'. So we must not be surprised to find Newtonianism to be quite a mixed bag of ideas (see box) (Schofield, 1978).

Newtonianism: the main items

1 Gravitational attraction

This is the belief that between material bodies there exists an attraction, which acts across a distance without any identifiable material intervention; thus gravitation could act across a vacuum, as is nearly the case in outer space between (say) sun and planet.

2 Laws of motion

These three famous laws constitute the basis of the whole science of dynamics, terrestrial and celestial. They stress (amongst much else) the tendency of bodies to conserve their state of rest or momentum (quantity of motion) unless compelled by other forces (such as gravitation) to change that state.

3 Inverse square law

This law states that the force between two bodies is inversely proportional to the square of the distance between

them. It can be expressed in mathematical terms as:

$$F = km_1m_2r^{-2}$$

(where F is force, k a constant, m_1 and m_2 the masses of the two bodies, r their distance from each other).

Together with the first two propositions this was thought to give a complete explanation of all celestial motion (as indeed it does if speeds do not approach the velocity of light).

4 Experimental method

Belief in the validity of the experimental method was not new to Newton, but was an integral part of his creed, and was especially visible in his work on optics.

5 Uniformity of nature

Although others had come near to articulating this idea, and Descartes had

proposed it, it was Newton who explicitly applied his laws to the whole of the material universe. There was not, for instance, one set of laws operating below the sphere of the Moon and another above it, but *one* universal set to be discovered on or from Earth, perhaps, but to be applied as far as eye or telescope could see. Of course this was an act of faith, but it was essential to the scientific enterprise as Newton conceived it.

6 The universe upheld by God

This was no optional theological extra but as much part of Newton's scientific thinking as the uniformity of nature. He regarded God as not merely the Creator of the visible universe but also its Upholder, from time to time intervening in physical events and thereby temporarily overruling the laws that were otherwise inviolable.

To understand the subsequent fortunes of this package of ideas it is helpful to recall what a mixture it was. First, there was a (deceptively) simple set of generalized *scientific laws* (2), together with a specific law (3) set in mathematical terms. Then there was a *world-picture* (1, 5) into which these laws were deemed to fit. There was a philosophy of science, concerned with *methodology* (4), and finally a *world-view*, particularly articulated in 6 but arguably including 5 as well.

10.2 Newtonianism in the land of its birth

10.2.1 'Mathematics have engrossed all'

This remark by a Cornish naturalist, Walter Moyle, is a rueful reflection of one class of people in England, many of them unable to follow the detailed arguments of the *Principia* yet overawed by the sheer majesty and comprehensiveness of Newton's cosmology. In 1719 Moyle wrote of the still youthful Royal Society, long identified with Gresham College in London, its former meeting place:

> I find that there is no room in Gresham College for Natural History: Mathematics have engrossed all; and one would think the Gentlemen of that Society had forgot that the chief end of their Institution was the advancement of natural knowledge. *(quoted in Raven, 1953, p.214)*

In fact Moyle was expressing a common English view of that time. He had to wait another ten years before the *Principia* was available in English, and following Newton's arguments in Latin must have required more than common dedication. Moreover many people are frightened by mathematical presentations and early eighteenth-century England was no exception. How many Englishmen actually understood Newton's mathematics must be a matter for conjecture, for it depended in part on his own discovery (in 1665) of the method of fluxions. An alternative tool, the differential calculus, and more suitable for the development of the Newtonian science of mechanics, was developed in Germany by Leibniz about ten years later. But the celebrated feud between the two mathematicians, with the English patriotically siding with Newton, rather inhibited them from taking advantage of the method of his Continental rival.

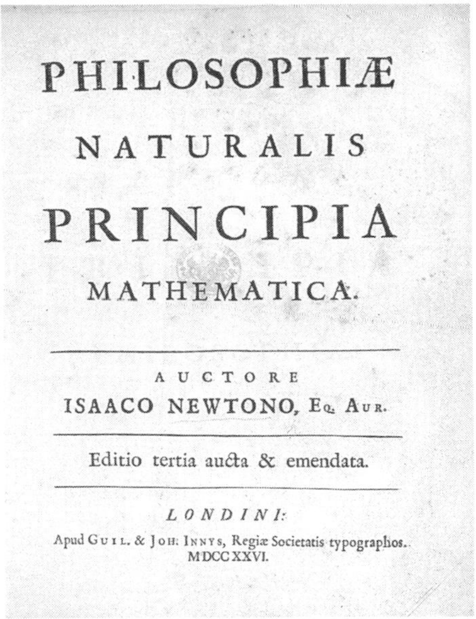

In England, as elsewhere, the mathematical education needed to comprehend Newtonianism in its fullness was generally lacking. For most of the populace little training in formal mathematics was available, except in the Dissenting Academies towards the end of the century. Throughout Europe most of the work in pure mathematics, including dynamics and astronomy, took place within the Academies (see Chapter 7). England was no exception, though as time passed the Royal Society was less inclined to such pursuits than its Continental counterparts, nor did it conform to the perceptions of Moyle earlier in the century. Nor were those observations untypical. Echoing the complaint of Moyle, members of the Coffee House Physical Society objected to a paper on astronomy in the following terms:

> The constitution of the Society does not extend to the consideration of subjects that may lead to mathematical disquisition, but is confined to experimental philosophy. *(quoted in Heilbron, 1980, p.364)*

This strain of mathematical philistinism occurs elsewhere, but in England is particularly obvious, one reason being the strength of an opposed tradition, also mediated by Newton: the tradition of experimental science.

10.2.2 *Experimental philosophy*

Newton did not only write *Principia* (and numerous works on pure mathematics). He also wrote a treatise on light, the *Opticks*.[1] This book reflects his consummate skill as an experimenter, a practical worker with lenses, prisms and so on.

> As in Mathematicks, so in Natural Philosophy, the Investigation of difficult Things by the Method of Analysis, ought ever to precede the Method of Composition. This Analysis consists in making Experiments and Observations, and in drawing general Conclusions from them by Induction, and admitting of no Objections against the Conclusions, but such as are taken from Experiments, or other certain Truths. For Hypotheses are not to be regarded in experimental Philosophy.

In other words Newtonianism embraced empirical as well as mathematical techniques: induction from experimental data as well as deduction from mathematical theorems. And this kind of approach was much more comprehensible to the ordinary citizen, who had only a dim awareness of what the *Principia* was really all about.

In England the domination by mathematics was to be short-lived as electrical, chemical, botanical and other sciences flourished in their pleasant, homely,

PHILOSOPHIÆ

NATURALIS

PRINCIPIA

MATHEMATICA.

AUCTORE
ISAACO NEWTONO, Eq. Aur.

Editio tertia aucta & emendata.

LONDINI:
Apud Guil. & Joh. Innys, Regiæ Societatis typographos.
MDCCXXVI.

Figure 10.1 Principia, *3rd edition, 1726. (Reproduced by permission of the British Library Board.)*

---

[1] This work is also famous for the 31 'Queries' in the later editions (especially of 1721 and 1730), where Newton engages in much speculation on scientific phenomena.

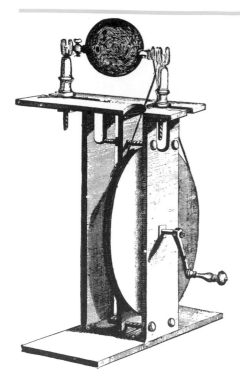

Figure 10.2 Hauksbee's electric machine. The glass globe is rapidly rotated against 'rubbers' (not shown), and acquires a high charge of static electricity. (From P. Benjamin, The Intellectual Rise in Electricity, *1895.)*

non-mathematical way. To some extent this reflects the pursuit of science by wealthy amateurs, so characteristic of England at this time. A new interest in the study of electricity was generated by the curator to the Royal Society, who needed spectacular effects for scientific demonstrations. Starting in 1705, Francis Hauksbee published 43 papers on the subject within eight years, and invented a primitive machine for generating large charges of static electricity. He was introduced by, and worked for, Newton himself. Other research in this area included the discovery of electric conduction by Stephen Gray in *c.* 1730, and numerous investigations by Fellows of the Royal Society into atmospheric electricity. Even these, however, were impeded by the English climate, with a scarcity of thunderstorms in the 1750s, and an excess of rain which tended to ruin the apparatus. But although the spirit of Newtonian empiricism inspired some English work, and even Newton himself had participated in electrical experiments and speculation, electricity was not a very obvious manifestation of English Newtonianism, at least for most of the eighteenth century.

The same cannot be said for chemistry. Here again Newton had led the way. He had long been preoccupied with alchemical studies, and as Master of the Mint had been directly concerned with the extractive chemical processes of metallurgy. In the early years of the century the Reverend Stephen Hales, Perpetual Curate of Teddington, began to collect 'airs' (gases) from numerous experiments and so laid the foundations of a distinctively English branch of chemistry, that concerned with gases. It exuded Newtonian empiricism, and, in Hales's case, actually built on specific proposals of Newton. This 'pneumatic chemistry', as they called it, was later taken much further by Cavendish and Priestley. Moreover, Newton had spoken in the *Opticks* of atoms as 'solid, massy, hard, impenetrable, movable particles' and frequently alluded to the short-range forces that held them together. The idea of quantifying chemical forces, as Newton had quantified those of gravitation, gave to some English chemists almost a new agenda, or what has alternatively been styled a 'Newtonian dream'. In so far as this is true it represents a blending of both elements in Newtonian methodology, the empirical and the mathematical.

Several English workers accepted this agenda at the beginning of the century, even before the 1717 edition of the *Opticks* had formally announced that 'it is the business of experimental philosophy' to discover 'those agents in nature able to make the particles of bodies stick together by very strong attractions'. They included John Freind and the brothers John and James Keill (disciples of that ardent Oxford Newtonian David Gregory, as was Archibald Pitcairne in Scotland), Hauksbee, Hales and Brook Taylor. Although the lead in this field passed to the Continent by mid-century, English works continued to expound these principles until, early in the nineteenth century, they found partial and unexpected realization in the chemical atomic theory of John Dalton in Manchester.

It would be stupid to attribute all the developments in English chemistry to Newtonianism alone. The science had a mixed ancestry, and owed at least as much to the empirical tradition of soap-boilers, dyers, brewers and other manufacturers who had probably never heard of Newton, and also to the pharmaceutical practice of extracting drugs from plant materials. That practice was indeed one reason for a continued interest in botany, but it cannot explain the veritable explosion of specimen-collecting, nature poetry and appreciation of the natural world that swept Europe in the latter half of the eighteenth century. That characteristic feature of Enlightenment values may also owe much to a third aspect of Newtonianism, especially in England.

10.2.3 Natural theology

From Newton to Darwin there reigned a 'holy alliance between science and religion', aptly described as 'that typically English phenomenon' (Willey, 1972). Much of it is directly attributable to Newton himself, and therefore a legitimate component of Newtonianism.

In a vast outpouring of works, much in manuscripts that are still awaiting detailed examination, Newton confided his thoughts about God and nature, doubts about conventional views of the Trinity, and his hopes for the future in the light of biblical prophecy. On Newton's theology, and its intimate relationship to his science, an immense amount has been written. It is sufficient to say that he saw science as a servant of religion and an antidote to atheism. He hoped that 'considering men' might find in his *Principia* grounds 'for the belief of a Deity, and nothing can rejoice me more than to find it useful for that purpose'. And in perceiving a pathway 'from nature up to nature's God' (as was commonly said) he was not only helping to set a trend, but was also expressing one that, in England at least, already had a considerable popularity. The use of arguments based on natural phenomena to demonstrate the wisdom and goodness of God became known as 'natural theology', or sometimes 'physicotheology'. It was specially prominent in the works of Francis Bacon and Robert Boyle. A set of lectures endowed by the latter had the twin effects of popularizing not only natural theology but also Newtonianism. The first lecture, by Richard Bentley, consisted of eight sermons, the last three devoted to Newtonianism, delivered after Newton himself had offered the preacher the reassuring words just quoted (see Figure 10.3). Other lectures developed in greater detail the 'argument from design' that is the keynote of natural theology, and for the rest of the century there flowed a stream of English books on the same theme, culminating in the *Natural Theology* of William Paley in 1802 (see Figure 10.4).

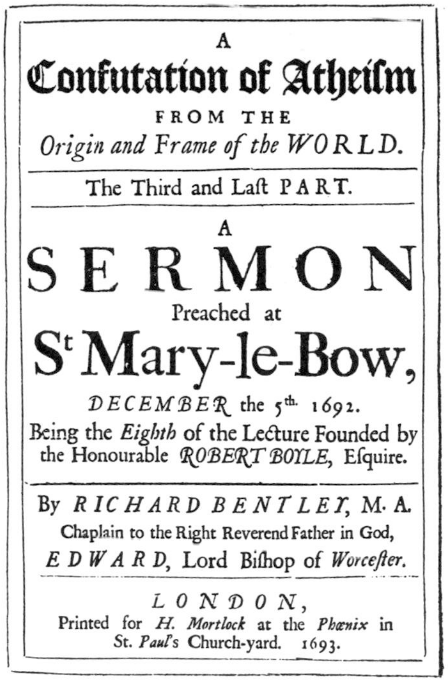
A
Confutation of Atheiſm
FROM THE
Origin and Frame of the WORLD.
The Third and Laſt PART.
A
SERMON
Preached at
St Mary-le-Bow,
DECEMBER the 5th. 1692.
Being the *Eighth* of the Lecture Founded by the Honourable *ROBERT BOYLE*, Eſquire.
By *RICHARD BENTLEY*, M. A.
Chaplain to the Right Reverend Father in God, *EDWARD*, Lord Biſhop of *Worceſter*.
LONDON,
Printed for *H. Mortlock* at the *Phœnix* in St. *Paul's* Church-yard. 1693.

Figure 10.3 Title page of the eighth sermon of the first Boyle Lecture, by Richard Bentley, 1692.

10.2.4 'The world natural and the world politic'

This phrase of Isaac Newton points to an analogy, and possibly a connection, between science and society. It has been argued that the reception of Newtonianism in England owes something to this analogy. The key lies in natural theology, which was a very English phenomenon, though not exclusively so. It represented a strong alliance between Newtonianism and the Church of England, which was then fighting a whole range of atheistic sects and glad to have an extra weapon with which to attack them. Unlike many of its Continental contemporaries, England had given birth to numerous and powerful rivals to the national Church. Nonconformists like John Bunyan flourished during the Commonwealth and made their presence felt even in the years between the Restoration of the monarchy (1660) and its reform in the Glorious Revolution of 1688. After that momentous event who was to say what would happen next? If the monarchy itself were to be in transition, what might not befall the national Church? It was in the interests of public order (and of the Church of England) that people should be persuaded of the unchanging realities of life and of order in the *natural* world which might well be imitated in the *national* one. And so, the argument goes, the real purpose of the Boyle Lectures and similar propaganda was to use Newtonianism as a means of social control.

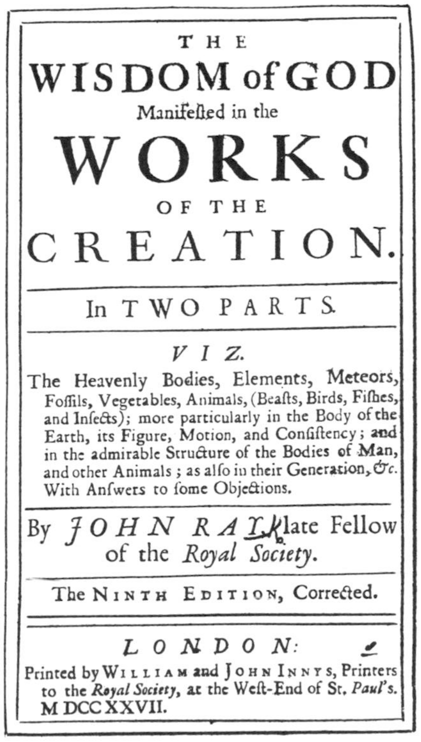
THE
WISDOM of GOD
Manifeſted in the
WORKS
OF THE
CREATION.
In TWO PARTS.
VIZ.
The Heavenly Bodies, Elements, Meteors, Foſſils, Vegetables, Animals, (Beaſts, Birds, Fiſhes, and Inſects); more particularly in the Body of the Earth, its Figure, Motion, and Conſiſtency; and in the admirable Structure of the Bodies of Man, and other Animals; as alſo in their Generation, &c. With Anſwers to ſome Objections.
By *JOHN RAY*, late Fellow of the *Royal Society*.
The NINTH EDITION, Corrected.
LONDON:
Printed by WILLIAM and JOHN INNYS, Printers to the *Royal Society*, at the Weſt-End of St. *Paul's*. M DCC XXVII.

Figure 10.4 Title page of Ray's Wisdom of God, *a typical and important work of eighteenth-century natural theology. (9th edition, 1727).*

This contention has been argued (Jacob, 1976) and challenged (Russell, 1983, and many others). What is not in dispute is the fact that the most visible proponents of natural theology were the party of Anglican clergy who

Figure 10.5 William of Orange (1650–1702), who ascended the English throne, with Mary his wife, in 1689. (National Portrait Gallery, London)

supported the deposition of James II and rejected the 'divine right of kings' to do more or less as they wished (which included remaining kings). They would find particularly congenial a theology which emphasized both the need for social stability after the Revolution (as suggested by a comparably law-abiding behaviour in nature) *and* a continuing role for Providential intervention (for example, the 'miraculous' east wind which brought William and Mary safely to Plymouth). Both these ingredients were found plentifully in Newtonianism. And it is not hard to show that Newtonianism was embraced by many other churchmen who rejected the Toryism of the alternative High Church Party, which supported the whole apparatus of the establishment, including a divinely-privileged monarchy. The critical question is whether this is sufficient evidence for ascribing a socio-political purpose behind the effusions of natural theology in the eighteenth century. A plausible view is that, early in the century, this may have been *one* of a number of factors favouring the adoption and exposition of Newtonianism in England. To go beyond that, especially for the later years of the century, is to lose oneself in a labyrinth of circular arguments, ambiguous labels and unsubstantiated assertions.

In conclusion, then, Newtonianism was used in England by Christian apologists as a response to the challenge of atheism, and that is probably a sufficient explanation for phenomena like the Boyle Lectures. It also adds a further dimension to our understanding of how Newton's philosophy fared in the land of his birth.

10.2.5 Popularizing Newtonianism

Clearly one effect of the Boyle Lectures and the multitude of later books expounding natural theology was to bring Newtonian science to the notice of a wide public. Such popularization of Newtonianism was, of course, incidental to the main purpose of the writers. In other places promotion of Newtonianism was the avowed intention.

Most obviously, there were books. Newton's *Principia* (as has been mentioned) did not appear in English until after his death (in a translation by Andrew Motte, 1729). By then others had taken up cudgels on Newton's behalf, as Henry Pemberton, editor of the third (Latin) edition of the *Principia*, whose *A View of Sir Isaac Newton's Philosophy* appeared in 1728. Others as John Keill, Benjamin Martin and Samuel Clarke had expounded Newton's ideas in print, while several of the great man's friends, including Stukeley and Conduitt, later regaled their contemporaries with reminiscences of his life. General English dictionaries by 1700 could include scientific words, and Newton's name appears soon afterwards. A dictionary by Benjamin Martin, *Lingua Britannica reformata* (1749) unsurprisingly highlights Newtonianism and announces that in this book 'Sir Isaac Newton's Definitions and Doctrines have been solely regarded'. These topics featured prominently in the first scientific dictionaries to appear, John Harris' *Lexicon technicum* of 1704 and Ephraim Chambers's *Cyclopaedia: Or an Universal Dictionary of Arts and Sciences*, first appearing in 1728 (Layton, 1965). Even children were not neglected. In 1761 'Tom Telescope's' little book appeared, *The Newtonian System of Philosophy, Adapted to the Capacities of Young Gentlemen and Ladies*, published by John Newbery. Selling at least 25,000 copies in England, it also appeared in Ireland, Holland, Sweden, America and (in 1832) in Italy, a successful 'blend of moral instruction and rational entertainment' (Secord, 1985).

Figure 10.6 Popular science lecture in the eighteenth century: cartoon 'The Kentish Hop Merchant and the Lecturer on Optics'. (Royal Society of Chemistry, London)

Then there were lectures. For a tiny, elite minority Newtonianism was being taught at Cambridge by 1699 and at Oxford by 1704. But that was not the only provision. From the late seventeenth century, coffee houses in London had provided a platform for itinerant lecturers on a variety of subjects, not least natural philosophy. Many of these elected to expound Newtonianism, though often in a popular style for a non-numerate audience. A prospectus for 1725 proclaims that

> any one, though unskill'd in mathematical Sciences, may be able to understand all these Phaenomena of Nature which have been discovered by Geometrical principles, or accounted for by Experiments. *(J.-T. Desaguliers: cited in Stewart, 1986)*

Among the most popular lecturers were the London instrument-maker Benjamin Martin and the clerical inventor of the planetarium J.T. Desaguliers (both of whom published books on the subject). Nor were such lectures limited to London or the south. Several notable lecture courses are known to have been presented in Newcastle, Durham and Northumberland. It has been argued, particularly from these cases, that Newtonianism was accompanied by a strong commercial emphasis (Stewart, 1986). That is not to say that Newtonianism produced industrialism. It was rather that commercial entrepreneurs saw in science new ways of exploiting nature, and the lecturers whom they encouraged were shrewd enough to lace their cocktail of Newtonian philosophy with a strong dose of utilitarian technology.

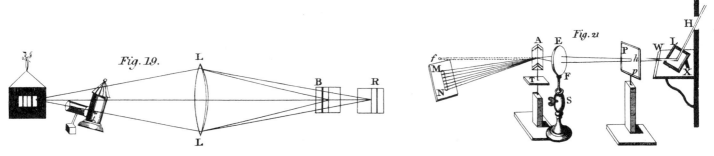

Figure 10.7 Optical experiments of Desaguliers. (Philosophical Transactions of the Royal Society, VI, 1713–23.)

10.2.6 Hindrances to Newtonianism

In England Newtonianism did not pass unchallenged. If the majority party of Church of England clerics, the Moderates, employed it in theological and even quasi-political argument, their High Church opponents within Anglicanism had recourse to another weapon. In so far as this was effective it helped to impede the universal acceptance of Newtonianism within England.

The alternative strategy centred round an ideology known as Hutchinsonianism, after its founder John Hutchinson (1674–1737). As steward to the Duke of Somerset he had been introduced to natural philosophy by his colleague John Woodward, amateur geologist and physician to the Duke. After prolonged study of the Scriptures, Hutchinson developed a world-view that differed from Newton's in several important respects. His theology was much more explicitly Trinitarian (with fire, light and air corresponding to the three Persons of the Trinity). He objected to Newton's use of non-mechanical forces and proposed instead an ethereal fluid that mechanically transmits the power it initially received from God. Unlike Newton, who conceived of God as the immaterial agent, and of 'intervening' from time to time within the universe, Hutchinson stressed his transcendence and 'otherness' from that universe which he had created.

It is easy to see how some of this commended itself to High Church Tories. Theologically speaking, Newtonianism had always struck certain questionable notes. If the universe were a vast interlocking mechanism, as seemed to be implied by most of Newtonian writing, where was the role for the Creator? If he were simply an absentee watchmaker, a 'mere' designer, then one is far from the caring God of Christian theology and into the heresy of deism. Several prominent Newtonians, as Clarke and Whiston, had been publicly discredited for tendencies in this direction associated with Arianism, another ancient heresy. On the other hand, if God were occasionally to tinker with his machine, as Newton seemed to suggest, then society, like nature, might not be so unchanging as a High Church Tory might wish: for example, the absolute right of kings might not be so immutable or 'divine', and might be blown away by a 'Protestant wind' delivering a new monarch to their shores. Which ever way Newtonianism were interpreted it could appear to be a dubious asset to those who saw themselves as guardians of the faith. The Hutchinsonians also felt that the emphasis on *natural* theology (a logical correlate to the watchmaker image) was at the expense of *revealed* theology (derived from the Bible).

This was not mere theological nit-picking, like counting angels on pinheads, but a strenuous attempt to correct what were seen at the time (and many have since recognized) as dangerous tendencies in the way the world was viewed, leading possibly to deism or worse. Nor was it simply an expression of High Church pique. Amongst those who viewed Hutchinsonianism favourably were the leaders of the Evangelical revival, such as John Wesley and some early Methodists. Many of their followers rejected a religion largely based on nature, whether in the name of Newton or anyone else, though they did not necessarily subscribe to Hutchinsonianism or belong to the High Church. Another group of people, progenitors of the Romantic movement, revolted against what they saw as the cold rationality of undiluted Newtonianism. As William Blake remarked, 'May God keep us from single vision and Newton's sleep!'. But then he also complained that 'Bacon's philosophy has ruined England', and his extreme anti-scientism was hardly typical of the age.

Thus while Newton was seen as a national hero by most educated Englishmen, the ideology that bears his name was by no means universally popular in his own country.

10.3 Newtonianism in the Low Countries

At the end of the seventeenth century Newton's science was generally viewed with cold admiration overseas. Something of his monumental achievement was recognized. The publication of his first letter on light (1672), his invention of the reflecting telescope, and the energetic promotion of his ideas by Henry Oldenburg (Secretary of the Royal Society until 1677) brought Newton's name to the fore in many European countries, but the details of his science were only vaguely understood. Partly this arose from its mathematical expression, a distinctively English phenomenon as it was to be seen.

Figure 10.8 Antony van Leeuwenhoek (1632–1723), Dutch microscopist, the first to observe protozoa in gut and bacteria in water. (Rijksmuseum-Stichting, Amsterdam)

First among the places where Newtonianism was received with any degree of warmth was that area of Europe generally known as the Low Countries, including the Netherlands and modern Belgium. The links with Britain had, of course, been recently strengthened with the accession of a Dutch prince, William, and his wife Mary to the British throne. There was now a strongly Protestant national culture on each side of the Channel, the Calvinist form of which had long encouraged in Holland the inquisitive study of nature. Despite its Protestant foundation (1574) the University of Leiden admitted all comers, and in the same tradition of tolerance all three competing cosmologies (Aristotelian, Tychonic and Copernican) had been taught in the previous century (see Chapter 3).

Two other features of Dutch life conspired to favour that empirical form of science which was represented by Newton's *Opticks* if not by his *Principia*. One was a growing number of instrument-makers whose products were invaluable for experimental science; they followed in the steps of A. van Leeuwenhoek (1632–1723) who made the microscopes that brought him fame as the discoverer of bacteria. The other was a deep admiration for the work of Francis Bacon, advocate of useful as well as inductive science (see Chapters 8 and 9). It has been pointed out (Schofield, 1978) that in the period from 1687 to 1727 at least nine editions of Bacon's works were published in Holland, and none at all in England. A culture so imbued with Baconianism would give a ready response to some of the ideas of Isaac Newton. One scholar has written:

> The Dutch physicists, whose writings were popular in England, too, when they were translated out of Latin, played a large part both in promoting Newtonian (and more generally English) empirical science on the Continent, and in transforming mechanics from the most forbidding of mathematical sciences to a pleasant series of demonstrations. They helped also, however, to spread the confused view that the *Principia* and the theory of gravitation were no less founded upon empiricism than was the *Opticks*. (Hall, 1963)

This is all the more remarkable when it is recalled that Descartes had lived in Holland for over twenty years and that Cartesianism, first taught at Utrecht, had won many followers in the Netherlands. But other and stronger influences were at work and Holland came to be the cradle of Newtonianism on the European mainland.

The lines of transmission from England to Holland are fairly clear. One of the earliest apostles of Newtonianism in Holland was W.J. 'sGravesande

Figure 10.9 University of Utrecht. (Universiteitsmuseum Utrecht)

Figure 10.10 W.J. 'sGravesande (1688–1722), leading Newtonian in Holland, working in optics and electricity. (From Michaud's Biographie universelle, *Paris, 1811.)*

(1688–1742). A graduate of Leiden and a geometer by inclination, he was sent as an embassy secretary with a delegation from the Netherlands to 'compliment' George I on his accession to the English throne in 1715. Here he saw scientific demonstrations by Desaguliers. Through the good offices of the redoubtable Bishop Burnet of Salisbury (with whose sons he was friendly) 'sGravesande was introduced at the Royal Society and made the acquaintance of Newton. It appears that on this visit he was thoroughly converted to Newtonianism, and became one of the first Europeans to embrace Newtonian physics with enthusiasm.

Returning home 'sGravesande was appointed professor of mathematics at his old university. His inaugural lecture justified the new alliance between astronomy and physics as proposed by Galileo and Newton. There is no question of 'sGravesande being a great discoverer, though he was committed to experiments, made some scientific instruments for them, and worked extensively in optics and electricity. He was less happy with speculation, though he used the same concept of force to account for phenomena in optics, cohesion and chemistry. It was on the exact nature of 'force', as applied to moving and colliding bodies, that he came to differ from Newton and on that topic embraced instead the ideas of Leibniz. Nevertheless, his supreme service to European science was as an advocate of Newtonianism. In 1720 he produced the first Newtonian textbook of general physics: *Physices elementa mathematica*, subtitled in its English translation by Desaguliers *An Introduction to Sir Isaac Newton's Philosophy*. Here he demonstrated a mastery not only over the physics of the *Principia* but also over the Englishman's treatment of short-range forces and the atomic structure of matter. Cartesian vortices, still popular in Holland, were banished from his book.

At Leiden 'sGravesande's teacher had been the notable physician, chemist and teacher Herman Boerhaave (1668–1738). Although no great discoverer, he justly enjoyed a towering European reputation in his own lifetime. He was one of science's great communicators, and men flocked to his lectures from all over the Continent. He held (sometimes simultaneously) chairs in medicine, chemistry and botany. As professor of physic (medicine) he established the practice of always having twelve beds available for clinical instruction. He had his imitators, of course, most notably in the medical faculty of the University of Edinburgh, whose curriculum was modelled on that at Leiden; during the eighteenth century no less than 40 occupants of Scottish academic chairs had studied at Leiden. In Boerhaave's 45 years there he taught 1,919 medical students, of whom 659 came from Britain or British colonies.

It is not certain where Boerhaave first encountered Newtonianism. As a student himself at Leiden he had, in 1692, attended lectures from the Scottish physician Archibald Pitcairne who had visited Newton that year in Cambridge and was later to consort with numerous Newtonian disciples. A greater stimulus probably came from his former pupil 'sGravesande.

Unsurprisingly, Boerhaave wrote a number of medical books. In one of these, *A Method of Studying Physick* (1719), he focussed on what we should call physics. Of this he says 'The prince and captain of all is Sir *Isaac Newton*, who knows as much as the rest of mankind put together'. Descartes did not get a mention. Elsewhere he wrote of 'the most exalted Newton', and of 'the happiest philosopher the world ever yet could boast, the great Sir *Isaac Newton*'. No empty rhetoric, these phrases display an immense enthusiasm for a man who had offered hope for a new understanding of those parts of natural philosophy for which experiment was crucial: light, electricity, magnetism, cohesion and (above all) chemistry. Boerhaave was no afficionado of Newton's celestial dynamics. It was the empirical basis of Newton's *Opticks* that attracted him.

Chiefly, however, Boerhaave was famous for his books on chemistry, which was closely integrated with medicine at that time. An unofficial version of his lectures was published in 1724 as *Institutiones et experimenta chemicae*, and put into English by Peter Shaw and Ephraim Chambers in 1727 as *A New Method of Chemistry*. These pirated versions were repudiated by Boerhaave who published his own *Elementa chemicae* in 1732; this appeared in English three years later, translated by Timothy Dallowe, as *Elements of Chemistry*, and in many other European languages also.

In both the 'official' and the pirated accounts of Boerhaave's chemistry there can be no doubt of his dependence on Newton. Apart from frequent adulatory references (which had not yet become a ritual necessity) the corpuscularian philosophy which runs through it is pure Newtonianism. One of its most striking notes is the great emphasis on fire. Not only is this the central agent of chemical change (a theme that was to dominate the century's chemistry). It is also an all-pervasive fluid, in continuous motion. There is something of Descartes in this, but Boerhaave saw fire as one element in a kind of Newtonian equilibrium, opposing the attractive forces between material particles which would, on their own, cause contraction to very small volume. 'Boerhaave clearly believed that the idea that there was Fire in everything could be given a Newtonian justification' (Love, 1972). Be that as it may the *Elementa chemicae*, in one or other of its translations, was an enormously influential textbook for the rest of the century and one of the major catalysts for the growth of Newtonianism in Europe.

One other figure deserves mention. Pieter van Musschenbroek (1692–1761) was a friend of 'sGravesande and, like him, had studied at Leiden. After teaching at Duisberg and Utrecht, he returned to Leiden as professor of mathematics in 1739. Musschenbroek was an indefatigable experimenter, often making his own apparatus. He worked at optics, regarding light as a

Figure 10.11 Herman Boerhaave (1668–1738), Dutch chemist, botanist and physician whose teaching courses in medicine at Leiden attracted students from all over Europe, and were of immense influence on the development of science and medicine in Scotland. (Rijksmuseum-Stichting, Amsterdam)

Figure 10.12 The Academy at Leiden. (Leiden University Library)

manifestation of fire. He attempted to subject magnetism to quantitative study, though was unable to detect 'Newtonian' regularities owing to the problems of short distances and variably shaped magnets. In chemistry he argued against the infinite divisibility of matter on the basis of experiment and adopted a Newtonian atomism. He is chiefly famed for making one of the most dangerous discoveries of the century: 'a new but terrible experiment which I on no account advise you personally to attempt', he wrote to Réaumur.

In 1746 he decided to try storing electricity, with the quaint idea of passing it into a bottle of water, and then sealing the bottle. He held the bottle in his hand, but on accidentally touching the gun-barrel used as a conductor he experienced 'a frightful shock'. As he said 'In a word I thought it was all up with me'. He had discovered the electric condenser and was fortunate to survive its involuntary discharge.

If his experiments brought him fame, his books extended his reputation, not least his posthumous textbook *Introductio ad philosophiam naturalem* of 1762. Hailed by one writer as a textbook truly representative of eighteenth-century physics (Heilbron, 1980), it is replete with lecture experiments, illustrations and clear exposition. Like his earlier writings it is shot through and through with Newtonianism. He was in good company with 'sGravesande. As van Swinden, author of Musschenbroek's entry in Michaud's *Biographie universelle* (1821), remarked of them both:

> It is in effect these two men, so eminently distinguished, to whom we owe the complete introduction of experimental physics and of Newtonianism into Holland. It is their lectures, their example, their works, which successively spread the light even well beyond their own country. They each worked separately, with the same zeal and equal success, but by different paths: 'sGravesande, a great mathematician gifted with real wisdom, in some way took for himself the mathematical part of physics ... Musschenbroek particularly applied himself to experimental physics in which he excelled and in which we owe to him a great number of discoveries.

Figure 10.13 Pieter van Musschenbroek (1692–1761), Dutch pioneer in the study of electricity. (Rijksmuseum-Stichting, Amsterdam)

Figure 10.14 The Leiden experiment. (From P. Benjamin, The Intellectual Rise in Electricity, 1895.)

10.4 Newtonianism in France

10.4.1 The Cartesian establishment

For the first quarter of the eighteenth century in France the ideas that reigned virtually unchallenged were those of Descartes. Certainly copies of the *Principia* had reached that country and the language, being Latin, would have been widely comprehensible. Nor were the French slow to perceive that, for all the dedication to mechanism manifested by Newton and Descartes, in almost all other ways they were poles apart (see Chapters 7 and 8).

Upon the death of a member of the Académie Royale des Sciences it was the responsibility of the Secretary to deliver an *Éloge* (eulogy). Many of these orations are important, not so much for their biographical information, as for the unwitting testimony they may offer to an 'official' view of contemporary science. Thus an *Éloge* by Condorcet on the mathematician G.P. de Roberval (1602–76) asserts that the 'sublime idea of universal gravitation had been presented by *physiciens* in a time before Newton', adding that 'it is uniquely in these exact and proved determinations [of planetary laws] that glory is due to Newton'. That and no more. Now Roberval's geometrical methods,

THE

ELOGIUM

OF

Sir *ISAAC NEWTON:*

BY

Monſieur *FONTENELLE,*

Perpetual Secretary of the Royal
Academy of Sciences at *Paris.*

LONDON:

Printed for J. Tonson in the *Strand,* and J. Osborn
and T. Longman in *Pater-noſter Row.*

MDCCXXVIII.

[3]

THE

ELOGIUM

OF

Sir *ISAAC NEWTON,*

BY

Monſieur *FONTENELLE.*

SIR Iſaac Newton, who was born at Woolſtrope in the county of Lincoln, on Chriſtmas day in the year 1642, deſcended from the elder branch of the family of Sir John Newton Baronet. The Manor of Woolſtrope had been in his Family near 200 years. The Newtons came thither from Weſtby in the ſame County, but originally from Newton in Lancaſhire. Sir Iſaac's Mother, whoſe maiden name was Hannah Aſcough, was likewiſe of an ancient family ; ſhe married again after his Father's death.

When her Son was twelve years old ſhe put him to the Free-ſchool at Grantham ; from whence ſhe removed

A 2

Figure 10.15 Title page and opening page of Fontenelle's Elegy for Newton *(1728).*

especially in relation to tangents, were independent of Descartes. Therein, according to his obituarist, lay his problem: 'If he had studied the geometry of Descartes, instead of opposing it, he would have been the first among his disciples'.[2] There was a price to be paid, even in death, for not conforming to a national cult.

This trend, of eulogizing Descartes even in homage to the memory of others, was begun in exaggerated form by Fontenelle, Secretary of the Académie Royale des Sciences from 1699 to 1741. Many years ago Herbert Butterfield reported how, 'on repeated occasions the identical fairy-story seems to take place', illustrating the tendency as follows:

> Of one person, Bernoulli, we are told that he saw geometrical figures by chance and immediately responded to their charms – proceeding later to a study of the philosophy of Descartes ... Régis was intended for the Church, and while he was getting tired of the length of time which he had to spend on an unimportant line of work he chanced upon the Cartesian philosophy and immediately was struck with it. Tournefort discovered the philosophy of Descartes in his father's library and recognised immediately that this was the thing his mind had been looking for. Louis Carré was the same – he was to have been a priest, but the prospect was disgusting to him; then he discovered the philosophy of Descartes which opened up a new universe for him. Malebranche was so transported on reading Descartes that he gave up everything else for the sake of the study of his philosophy. Varignon picked up a volume of Euclid by chance, and charmed to see the contrast between this and the sophistries and obscurities that he had been taught in the schools, he allowed himself to be led by geometry to the reading of Descartes who came to him as a new light. *(Butterfield, 1950, pp.146–7)*

Perhaps the most revealing *Éloge* delivered by Fontenelle was that for Newton himself. He spoke with restraint but revealed his deep antagonism to Newton's science with the remark that, though Descartes had apparently freed physics for ever from 'attraction and vacuum', they were now brought

Figure 10.16 Bernard le Bovier de Fontenelle (1657–1757), French philosopher and poet, author of Conversations on the Plurality of Worlds *(1686), and Secretary to the Académie Royale des Sciences. (Musées Nationaux, Paris)*

[2] *Éloges des Académiciens de l'Académie Royale des Sciences,* 1773.

back again by Newton. As late as 1752, when Fontenelle was 95, he was still defending Descartes against Newton, and advocating Cartesian vortices with enthusiasm.

Thus, to put it mildly, Newtonianism was not an immediate hit with the French. There are several possible reasons. The opposition of Fontenelle cannot be discounted, at least in his earlier, more influential period. His successful popularization of seventeenth-century science, *Conversations on the Plurality of Worlds* (1686), just predated the *Principia* by which it was however to be severely challenged. But there was more to it than that. A new movement in France was emerging, the *philosophes*. These were not merely 'philosophers' but writers on political, social and religious topics who concerned themselves with the formulation of a new world-view. Of this science was one, but only one, constituent. Partly in reaction against the obstructive conservatism of the Catholic clergy (as it seemed to them) the early *philosophes* cultivated a more secularized view of nature than Boyle, Kepler or (for that matter) Descartes had permitted themselves, and certainly in contradistinction to that of Newton himself.[3] And early in the eighteenth century the France of Louis XIV was becoming (in the words of that hostile aristocratic commentator, Saint-Simon) 'a long reign of the vile bourgeoisie', their new prosperity inevitably fostering a certain nationalistic pride. Nor must it be forgotten that until 1713 France was engaged with Britain in the War of the Spanish Succession. Without doubt nationalism offers one explanation for the French resistance to Newtonianism in the early eighteenth century. Finally, the sheer technicalities of his mathematics defeated even Newton's would-be supporters in France as elsewhere on the Continent. The calculus of Leibniz seemed to many the best way of solving problems of infinitesimals, but the Newtonian method of fluxions reigned supreme only in England.

10.4.2 *The Newtonian alternative*

It is broadly true that Cartesianism had little to fear from Newtonianism before 1720, though Fontenelle and his friends did not have it all their own way even in the early years of the century. The Abbé Pierre Varignon (1654–1722), a noted geometer, was a correspondent of both Leibniz and Newton, sought a reconciliation between them, and supported Newtonianism in France before 1720. His correspondence 'provided one of the happier interludes in Newton's old age' (Westfall, 1980, p.786). Both the *Principia* and the *Opticks* were available in France, and the subjects of review articles. The latter received a French translation in 1720.

Within the next few years the authority of Descartes began more visibly to yield. Varignon's successor in the Collège de France, J.P. de Molières (1677–1740), wrote *Leçons de mathématiques* (1726) and *Leçons de physique* (1733–9), reconciling Newtonianism with French, or Cartesian, philosophy. De Molières was a follower of Malebranche, as was his obituarist, the mathematician J.-J.D. de Mairan (1678–1771), who was another French supporter of Newtonianism, enjoying an international reputation.

Meanwhile scientific support for Newtonianism was to come from an unexpected quarter. The Académie des Sciences decided to address a problem of increasing concern to eighteenth-century astronomers: the shape of the Earth. Was it, or was it not, a perfect sphere? To find an answer they

Figure 10.17 Pierre Varignon (1654–1722), French mathematician. (Universitäts-Bibliothek Basel)

[3] There were, however, some who endeavoured to effect a synthesis between Catholicism and Cartesianism, most notably the philosopher N. Malebranche (1638–1715).

needed to measure accurately the lengths of several latitudinal arcs in different parts of the world. Accordingly, an expedition was despatched to Peru in 1735 to measure a 3° arc near the equator. One of the astronomers, Pierre Bouguer (1698–1758), had previously received a prize from the Académie for work on astronomical refraction which adopted a very non-Newtonian theory of light. Now, however, his Peruvian results convinced him on a different aspect of Newtonianism. His *La Figure de la terre* showed that the length of a meridian arc increases with increasing latitude. Hence Newton was right, at least in his prediction that the Earth was slightly flattened at the poles.[4] An identical conclusion was reached by his colleague P.-L.M. de Maupertuis (1698–1759) who was despatched on a similar mission to Lappland in the following year.[5] He had already espoused a Newtonian cause, as may be seen from his *Discours sur les différentes figures des astres*, published in 1732. Both men became prominent advocates of Newtonianism in the France of the 1740s. Their efforts were to receive powerful assistance of a bizarre kind from another unexpected source.

Figure 10.18 F.M.A. Voltaire (1694–1778), French philosopher, writer and proponent of Newtonianism. (Snark International, Paris)

No one could fairly accuse the French sceptic Voltaire of any degree of scientific prowess. Yet by a remarkable twist of events it was he who became the great popularizer of Newtonianism in France. The story is well known and can be briefly told.

Following involvement in a brawl, and subsequent commitment to the Bastille, Voltaire was exiled to England and there became impressed by the greater liberty of expression than in his own country. A converted Anglophile, he gained more than a nodding acquaintance with Newtonianism, in the first place, it seems, on account of its deistic overtones. For this resentful French exile Newton was to become a symbol of England, just as Descartes was a national symbol for France. And, in any contest between them, Newton had to win.

So on his return home Voltaire began a series of *Lettres philosophiques* with Newtonianism prominently displayed. He followed this in 1736 with his *Eléments de la philosophie de Newton* and another edition four years later. Together with his mistress, Madame du Châtelet, he established what amounted to a school of Newtonian studies. He corresponded with Maupertuis and other French mathematicians, and visited the Dutch Newtonians in Leiden. He encountered the writings (and the claims) of Leibniz, and flirted with the views of Christian Wolff (Chapter 12). Madame du Châtelet produced her *Institutions de physique* and a French translation of the *Principia*.[6] There is no doubt that, between them, they exerted a profound influence on the adoption of Newtonianism in France. Moreover, Britain was now at peace with France, and in the difficult later years of Louis XIV many Frenchmen looked across the Channel with some envy at the stability of government in Britain.

10.4.3 *Newtonian science in France*

In the field of mathematical physics French Newtonianism made its first conquests. On any account a crucial role was played by Maupertuis, a man convinced that attractive and repulsive forces are inherent in matter. He was

---

[4] An earlier member of the Paris Observatory, the Italian G.D. Cassini (1625–1712), had concluded from various observations that the Earth was actually flattened *at the equator*! This of course was totally incompatible with Newtonian dynamics.

[5] This was the expedition in which Celsius took part (see Chapter 12).

[6] Her *Institutions* also introduced Leibnizian ideas, apparently after her Newtonian phase.

one of the first to accumulate observational evidence in favour of Newtonianism, perceived its relevance to the shape-of-the-earth debate, and became a major influence on Voltaire. From 1732 he appears to have been a central figure in an informal group of Newtonians at the Académie.

Another was A.C. Clairaut (1713–65), colleague of Maupertuis on his Lappland expedition and author of yet another book on the shape of the Earth (1742). One of the most precocious mathematicians of all time (he read a mathematical paper to the Académie at the age of 13) he was a pioneer of three-dimensional analytical geometry. With the aid of Newton's inverse square law he wrestled with the problem of *three* interacting bodies and of motion within the solar system. He studied the disturbing effect of the Moon and planets on the Earth, demonstrating that this would cause an apparent 'wobble' in the motion of the Sun. In a similar way he used the disturbing effects of Jupiter and Saturn on the path of Halley's comet, to calculate within one month the date of its return in 1759. As one commentator remarked 'the return of this comet so near the predicted time was one of the most brilliant triumphs which the Newtonian theory had yet achieved' (Grant, 1852, p.104). His studies on the perturbation of the moon (lunar inequalities) led him at one stage to abandon Newtonian dynamics and to propose an alternative law:

$$F = Ar^{-2} + Br^{-4}$$

(where A and B are constants).

His colleague Buffon complained that physical laws must be simple, and not like that one. In due course Clairaut discovered that with correct calculations Newton's original law was sufficient (i.e. B = 0) and his faith in Newtonianism was restored. It was Clairaut who supervised the translation of Newton's *Principia* by Madame du Châtelet.

The mathematician Jean d'Alembert (1717–83) early achieved a great reputation in dynamical astronomy, being only six days behind Clairaut in submitting his work on lunar perturbations for the Académie prize in 1746. And, like Clairaut, d'Alembert eventually found his work a striking confirmation of Newtonian theory. He was able to give a sound mathematical basis to Newton's theory of precession of the equinoxes and several other

Figure 10.19 Alexis Claude Clairaut (1713–1765), French mathematician, famous for his work in dynamical astronomy. (Universitäts-Bibliothek Basel)

Figure 10.20 The Paris Observatory, built in 1667. (Ann Ronan Picture Library)

astronomical phenomena. In later life d'Alembert became perpetual secretary to the Académie and co-editor of the famous *Encylopédie*.

Further progress in applying and extending Newtonian ideas of dynamical astronomy was made by J.L. Lagrange (1736–1813) and P.S. de Laplace (1749–1827). Lagrange proposed a theory to explain why the Moon always faces the Earth in the same way, and propounded a dynamical theory of Jupiter's satellites. Although he did not live in France until 1787 (having worked in Turin and Berlin) he stimulated Laplace by presenting him with powerful general mathematical methods for astronomical application. The effort was not wasted for Laplace was able to show that variations in planetary orbits (inequalities) can indeed be accounted for in terms of Newton's inverse square law, and without recourse to special divine intervention. These investigations established the great durability of the solar system. Laplace's achievements, together with his own hypothesis for a nebular origin of the solar system, were expounded in his celebrated *Système du monde* of 1796 and his 5-volume *Traité de mécanique céleste* from 1799. This was both the ultimate vindication of Newtonian mechanics and the crowning glory of Newtonianism in France.

Before leaving the manifestation of Newtonianism in French dynamical astronomy it is pertinent to enquire *in what senses* the latter was truly Newtonian. Quite clearly this is a large question and only a few comments can be offered here. In terms of the six elements of the Newtonian creed (see Section 10.1) very few French 'Newtonians' were completely orthodox. Whether gravitation was action-at-a-distance was sometimes a moot point, though almost everyone agreed that Newton's basic mathematical analysis had been vindicated by the observed facts. However, as we have seen, even the inverse square law was temporarily abandoned in the face of daunting mathematical difficulties. Thus Clairaut was not a thorough-going Newtonian when he 'tinkered' with planetary laws, though like Newton himself he knew there were discrepancies between observed and predicted phenomena.

As for theology, Maupertuis inverted the deistic arguments of British Newtonians, arguing for God on the basis that gravitation was some kind of innate property of matter implanted as evidence for a Deity. For this reason he has been called 'a Leibnizian Newtonian', though that appellation has also been given to many of his colleagues, and even to Madame du Châtelet, largely on the grounds that (a) like most Continentals they used Leibniz's mathematical techniques, and (b) that they had Leibnizian friends (Schofield, 1978). Whether such 'guilt by association' may be sufficient warrant for regarding them as deviant Newtonians must be open to question. Newtonianism was mediated to the Continent not only by the *Principia*, but also by the *Opticks*. However, research into light was not a hallmark of French science until the early nineteenth century (when Fresnel took issue with Newton). Clearly the land of Descartes was more inclined to welcome the awesome mathematics of *Principia* than the empiricism of the *Opticks*.

The same tendency is evident, though not so completely, in the study of electricity. This flourished more obviously in eighteenth-century Germany and Holland than in France. Perhaps the first major French research in electricity tells us why. It was conducted by C.F. de C. Dufay (1698–1739), a chemist who became (1732) superintendent of the Jardin du Roi in Paris.

Dufay confirmed the results of Stephen Gray on electric conduction (see Section 10.2.2) and in particular his fundamental distinction between

Figure 10.21 Jean le Rond d'Alembert (1717–83), French mathematician and worker in acoustics and dynamics. Collaborator with Diderot in producing the great Encylopédie. *(Photographie Giraudon)*

Figure 10.22 Pierre Simon Laplace (1749–1827), French mathematician whose work disclosed the universality of Newtonian dynamics and addressed such problems as the stability of the solar system. (Mansell Collection)

Figure 10.23 C.F. de C. Dufay (1698–1739), Superintendent of the Jardin du Roi in Paris, Dufay is chiefly known for his electrical researches, recognizing two kinds of electricity. (Ann Ronan Picture Library)

conductors and non-conductors (insulators). In 1733/4 he published an important discovery of his own in static electricity. He found that a charged gold leaf was attracted by 'resinous' bodies (as amber, paper and silk) when rubbed, but it was repelled by rubbed 'vitreous' bodies (as wool, hair and glass). From this he concluded that there were two kinds of electricity, 'resinous' and 'vitreous', and that these were responsible for repulsion and attraction. Two bodies in the same class will repel each other, while bodies from opposite classes will attract each other. Here was a theory that had almost no analogy with Newtonian concepts of gravitation. Repulsion as well as attraction had to be considered, and electricity was to be thought of in particulate terms, not of forces acting at a distance.

The problem was compounded when other workers, as Robert Symmer (1759), explicitly articulated a two-fluid theory. Admittedly Newton himself had speculated whether electricity (but not gravitation) could be an 'effluvium', but that was one fluid, not two. When sparks were seen in electrical experiments or in thunderstorms the fluid or fluids became all too visible. Gravitation offered no parallel whatsoever. The inability to quantify its study led d'Alembert reluctantly to consign it to the dustbin of 'experimental physics' alongside magnetism, heat, optics and hydraulics; mathematization along Newtonian lines was manifestly inappropriate.

A student of Dufay was the Abbé J.A. Nollet (1700–70), an intrepid experimenter, exponent of a one-fluid theory of electricity, and inventor of an early form of 'electrometer' (the name is his). Although he has been hailed as a Newtonian on the grounds of his empirical prowess, he was also a showman, able to exploit the spectacular properties of the newly discovered Leiden jar (see Section 10.3). On one occasion the worthy Abbé discharged such a jar through a chain of 180 monks. Partly on account of his opposition to that dedicated American follower of Newton, Ben Franklin, Nollet has also been designated as an arch-anti-Newtonian (Cohen, 1956).

Much of our admittedly limited knowledge of French electrical work at this time does not really suggest a serious Newtonian agenda. Other foreign influences were strongly felt: the invention of electrical machines in England and Germany and the Leiden jar gave open invitations to spectacular demonstration. From the 1750s Benjamin Franklin appeared in France either in publication or in person. He won many admirers (Nollet notwithstanding), and may have persuaded some to a more Newtonian agenda towards the end of the century.

Franklin's work on atmospheric electricity inspired a few to emulate him. However, the overall impression is often more of a comedy bravado act than a scientific programme. Le Monnier got evidence of electrification (he was lucky to get nothing worse) by holding a 32-foot pole, with iron rods attached, into a thunderstorm. He later showed electrification occurred even in fine weather, an observation of great significance in the future. In 1752 Franklin's friend the botanist T.-F. Dalibard strayed from his usual field to set up an electrical experiment. A vertical and pointed iron rod, 40 feet high, was connected to apparatus in a shelter, and a thunderstorm awaited. Dalibard being absent at the crucial moment, observations were made by an assistant in the presence of a clergyman, who estimated the interval between sparks as the time taken by a *Paternoster* or an *Ave Maria*. For his pains the diligent cleric received a weal on the hand and smelt of sulphur. Similar results were reported elsewhere in France.

It was not until the 1760s that Newtonianism in the mathematical sense was clearly perceptible in European studies on electricity, by which time the

Figure 10.24 Jean Antoine Nollet (1700–1770), French Abbé noted for his intrepid experiments on electricity; on one occasion he is reported to have electrified an entire regiment at Madrid. (Ann Ronan Picture Library)

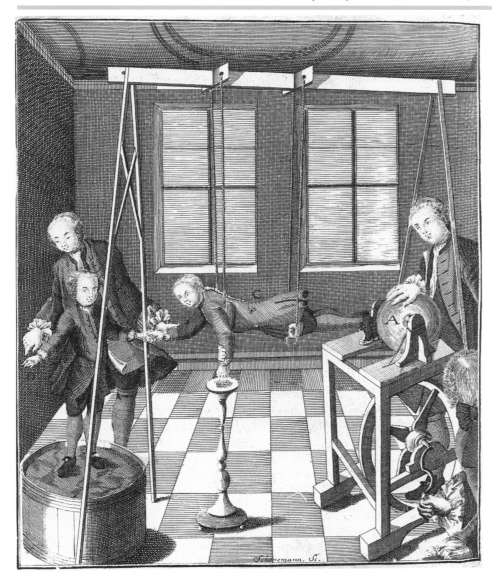

Schœnemann. Sc.

Figure 10.25 Abbé Nollet exhibiting Gray's experiment of the electrified boy. The youth is insulated from the surroundings by being supported from the ceiling with silk threads. He is charged with static electricity which cannot escape, so his hand attracts objects on the stool below him. (From P. Benjamin, The Intellectual Rise in Electricity, *1895.)*

subject was beginning to cry out for quantification. Perhaps Franklin was partly responsible for the movement in France. The French 'Newton of electricity' was an engineer, C.A. Coulomb (1736–1806), whose all-round interests have earned him the title of 'the first physicist'. His name is immortalized as the unit of electric charge (or flux). He measured electric charge on conductors and concluded that it lies on the surface only. With a torsion balance of his own devising he demonstrated an inverse square law for electrical repulsion (1784); using an oscillation method he demonstrated the same law for attraction (1785). The analogy between electricity and gravitation, hitherto so hard to believe, had been established beyond doubt. When he established the same inverse square law for magnetism the triumph of Newtonianism in French experimental physics was complete.

The pursuit of Newtonian chemistry in France followed a different course. There is much evidence to suggest that Newton's agenda for a quantified chemistry was well understood by the 1720s. His *Opticks* had appeared in French translation in 1720 and 1722. The work of his followers Keill and Freind had appeared in a pirated and anonymous *Nouveau cours de chymie suivant les principes de Newton et de Stahl* in 1723; this received complimentary reviews and was compared favourably with N. Lemery's *Cours de chymie* which was much older (from 1675), in the pharmaceutical tradition and thoroughly Cartesian in tone. Fontenelle went out of his way to criticize Newtonian chemistry, which implies that it constituted more than a minor nuisance.

The first apparent manifestation of chemical Newtonianism came in 1718 with a famous 'Table of different relationships observed in chemistry between different substances' by E.F. Geoffroy (1672–1731). Its language is not explicitly Newtonian, but its attempt to place substances in some kind of affinity order is exactly what Newton had suggested in the *Opticks*. Moreover, Geoffroy had visited England in 1698 and, as the Académie's official correspondent with the Royal Society, would have had much contact with English Newtonians. For some years his table was the only possible example of chemical Newtonianism in France. Then in the 1730s it was joined by several others.

Newtonian chemistry seemed to emerge from Newtonian physics. First came a discussion by Maupertuis in 1732 'on the laws of attraction', starting with the *Principia* but moving over to the operation of short-range attractive forces in chemistry. Keill and Freind are quoted and the attractive prospect is displayed of using such forces to explain phenomena like chemical precipitation. This was shortly followed by translations by G.L.L. Buffon (1707–88) of two unquestionably Newtonian English works: Hales's *Vegetable Staticks* (1735) and Newton's own *Treatise on Fluxions* (1740). Buffon was so committed to Newtonianism that he took issue with Clairaut over the latter's strategem to modify the inverse square law in astronomy and even insisted that the same simple relationship must apply in chemistry (1765). In 1753 Newtonian chemistry was being violently attacked in the *Encyclopédie* by G.F. Venel who, despite his assertion that 'it has never been adopted here', was clearly worried that it soon might be. Two of Buffon's protégés were about to ensure that it was.

In 1766 the chemist P.J. Macquer (1718–84) produced (anonymously) the first dictionary of chemistry. In his *Dictionnaire de chymie* he issued a clear manifesto for Newtonian chemistry, a position he had been unable (or unwilling) to take in his earlier and influential textbook *Elémens de chymie* (from 1749). Now, however, he called upon 'the zeal of persons skilled in mathematics and chemistry', for who else could probe the secrets of the atoms, or

> the reciprocal gravitation of these small particles to each other; which is variously modified, according to their size, their density, their figure, their extent, the intimacy of their contact, or the greater or less distance of their approach?

Adhesion and chemical affinities	
metal	force needed to lift disc (grains)
gold	446
silver	429
tin	418
lead	397
bismuth	372
zinc	204
copper	142
antimony	126
iron	115
cobalt	8

Macquer's colleague in this enterprise was Guyton de Morveau (1737–1816), a Dijon lawyer with great chemical skills. He advocated Newtonianism in an early paper on capillarity in 1773, and in an article on affinity in a supplement to the *Encyclopédie* three years later, suggested it would enable us 'to calculate affinities in the same way as the paths of the stars'. This intoxicating prospect was, he believed, fulfilled by his own experiments on measuring the force needed to lift identically-sized metal discs from a mercury surface, the order of adhesion being 'precisely the order of chemical affinities'.

As one author has remarked 'The aim of a quantified science of affinities was most keenly pursued in France', and 'it will be apparent that the late 1770s and the 1780s were the heroic days of Newtonian chemistry' (Thackray, 1970). However, for the greatest chemist of them all, Lavoisier, Newtonianism was but one component of his chemical thought, and a minor

Figure 10.26 P.J. Macquer (1718–84), French chemist, chiefly known for his textbooks which led the field for many years. (Ann Ronan Picture Library)

Figure 10.27 L.B. Guyton de Morveau (1737–1816), French lawyer of Dijon who developed an amateur interest in chemistry, was involved in the manufacture of soda, saltpetre and glass, and became a convert to the new chemistry of Lavoisier. (Ann Ronan Picture Library)

one at that. After his monumental work in the Chemical Revolution the Newtonian dream faded in France and reached its fulfilment in England.

Finally it may be added that even biology was not entirely immune from Newtonian influence. When Maupertuis sought to explain the reproductive process he did so in terms of living atoms acted upon by 'penetrating forces'. This clear echo of Newtonianism indicates it to be one of the 'penetrating forces' of eighteenth-century French scientific culture. However, the crude and unsatisfactory models of Descartes were warnings enough against attempting to mechanize life as Newton had mechanized the cosmos.

10.5 Conclusion: *webs of resistance*

If Newtonianism in one form or another flourished in England, France and Holland, what of the rest of Europe? In Scotland (of which more will be heard later; see Chapter 11) a strong Newtonian tradition developed, which is not surprising in view of its links to the other three countries. The Act of Union, the 'Auld Alliance' between Scotland and France and the fame of Boerhaave's Leiden ensured, in their different ways, a constant stream of two-way travel, both in people and ideas. However, on the Continent Newtonianism made heavy weather, with of course the exceptions of France and Holland. Spain and Italy were considerably less tolerant of Protestant ideas than their Catholic counterparts in Northern Europe, and it was impossible not to recognize the Protestant elements in the thinking of either Newton or many of his disciples. Nevertheless, some Catholics embraced Newtonianism with enthusiasm. Yet it is equally clear that the Protestant countries of Sweden (see Chapter 12), Switzerland and above all the German

Figure 10.28 Gottfried Wilhelm Leibnitz (1646–1716), German mathematician, philosopher and correspondent of Newton. (Roger-Viollet, Paris)

lands were at best lukewarm converts to Newtonian science. A partial explanation undoubtedly may be found in alternative cosmologies, competing with Cartesianism and Newtonianism alike. As Gillispie has put it 'the web of resistance to Newton was, in fact, complex' (Gillispie, 1958).

In the celebrated controversy with Leibniz may be found many of the critical issues, some of which Newtonianism had already encountered in its battles with the Cartesians. Gottfried Wilhelm Leibniz (1646–1716) was a mathematician, philosopher and man of letters who became librarian to the Duke of Brunswick in Hanover. At first his relations and correspondence with Newton were cordial, and apparently sincere admiration was expressed for the English philosopher. The famous dispute which arose between them came to a head with the publication in 1713 of the second edition of the *Principia*.

Newton and Leibniz differed from each other at a number of levels. Most superficially, perhaps, was the question of mathematical technique: fluxions (Newton) or differential calculus (Leibniz)? It was ironic that the problems posed by Newton were most conveniently solved by the method of his rival. Because it is a perennial problem in science to differentiate *at the time* between the substance and form of a discovery there is always plentiful opportunity for disputes about priority. Such was the case in point, with the added complication that the argument took on nationalistic overtones, with English and Germans each defending their own countryman.

To most physicists today the crucial difference between Leibniz and Newton lay in the understanding of dynamics. In the resolution of this difference, and in the clarification of the issues involved, lay a great advance for physical science. There was a bewildering variety of terms in use: force, motive force, quantity of motion, *vis viva*, etc. When bodies interact in an isolated system something is conserved; what exactly is it? To put it simply – too simply – Newton (and Descartes) asserted that it is *momentum* (Mv) that is conserved, while Leibniz said it was what we should call *kinetic energy* (Mv^2). That is a fairly important distinction.

Matters went even deeper than that, however. Consider the question of gravity. Is it or is it not a force of an unknown kind, or actually a mechanical cause? Newton was non-committal, but he thereby laid himself open to the charge of reintroducing occult causes into science. Leibniz asserted that this was reverting to Aristotle, with gravity an inherent but mysterious property of bodies. To this Newton replied that he had never envisaged an innateness of gravity within matter as Epicurean philosophy had supposed. He usually refused to speculate, claiming that he 'feigned no hypothesis'.[7] Or again, on the nature of space there were profound differences. While Descartes had thought of it in material and Leibniz in relational terms, for Newton space was ultimately an attribute of God. Indeed he was quite happy to think of space as a cosmic vacuum, just as he regarded matter as composed of atoms, for which conceits he was roundly admonished by Leibniz as early as 1692: 'I do not see the necessity which compels you to return to such extraordinary entities'.

That remark was made in private, as were most of their arguments until well into the new century. In 1715 and 1716, however, the pace quickened and a stern correspondence took place between Leibniz and Newton's spokesman Samuel Clarke, erupting into print in 1717. The disagreements thus laid open

---

[7] That this was untrue of his atomism is at present beside the point.

to public gaze were at the most fundamental level of all. They concerned the role of God within his universe. While Newton had invoked his intervention in planetary movements that would not (apparently) conform to Newton's own laws, Leibniz accused his rival of diminishing the Creator's skill. Is the universe so imperfectly made that God 'is obliged to clean it now and then ... and even to mend it as a clockmaker mends his work'? Clarke, on Newton's behalf, retorted that a universe built on Leibnizian lines, with no divine 'intervention', would have no role for God and therefore lead straight to atheism.

Whatever the philosophical and theological niceties of the case the knives were out and Newtonianism was now an issue of more than scientific importance. It has been well said that 'paradoxically his [Leibniz's] debates with Clarke did not sway opinion against Newton; rather they enhanced the importance of the *Principia* because Leibniz did not treat the book as a mere mathematical treatise' (Hall, 1963). At first that was certainly true, but people began to take sides. One author has argued that 'the Newtonian and Leibnizian groups of the 1720s developed a commitment to the mother scheme and took on the task of defending that system against the perceived threats of outside attacks' (Iltis, 1973). Ostensibly this was on the more obviously scientific issues. Some preferred the Leibnizian calculus, as we have seen. Others hesitantly sided with Newton over the question of force; the Newtonian doubts of 'sGravesande were precisely about this. As the century advanced it is remarkable how the German countries took the part of Leibniz, and it is difficult to exclude altogether considerations of national solidarity.

Leibniz and Descartes were not the only sources of anti-Newtonian ideas on the Continent, any more than Hutchinson had been in England. In chemistry G.E. Stahl (1660–1734) was another figure to be reckoned with. Professor of medicine at Halle and then (from 1714) Royal Physician in Berlin, he championed the independence of chemistry, particularly from the tyranny of the mechanical philosophy. Standing in the older tradition of Libavius (practical, analytical, useful) Stahl maintained a strongly anti-reductionist stance, declining to accept explanations in terms of hypothetical atoms. Reacting against the claims and the quasi-mysticism of the alchemists he emphasized utility and rationality. He became one of the founders of the phlogiston theory. It has been claimed that 'ultimately the approach to chemistry advocated by Stahl and his disciples became the rallying platform of the German chemical community' (Hufbauer, 1982). In those circumstances Newtonian atomism would have encountered a powerful alternative which, moreover, extended its hold over more than chemistry. Parts of experimental physics in Germany showed similar characteristics. In the 1740s, for instance, there was a rash of electrical machines for generating static electricity (Hausen, Bose, Winkler), where the utilitarianism was all too plain: the sensational production of ever larger electric sparks for the entertainment of the public.

Thus Newtonianism had to encounter challenges and resistance from the doctrines of Descartes, Leibniz and Stahl, to say nothing of ideologies of which we currently know much less. As with almost every other chapter in the history of ideas there is little to suggest that the passage of arms left either contestant unscathed, or that victory and defeat were ever absolute. In other words Newtonianism inevitably changed its character as the century progressed, and at the same time its disciples frequently embraced elements of rival cosmologies while keeping within the mainstream of Newtonian thought. It is for this reason that Schofield could write of a whole range of 'Newtonianisms' in the eighteenth century (Schofield, 1978). In the case of magnetism, for example, 'even the most advanced Newtonians continued to uphold orthodox Cartesian opinions throughout the first half of the eighteenth century' (Home, 1977). Many of the French Newtonianisms display aspects of Leibnizian thinking. Not all would go as far as Schofield in his designation of the mathematician Daniel Bernouilli. An enthusiastic propagandist for Leibniz's ideas generally ('Leibniz's bulldog'), he used the latter's mathematical methods to deal with Newton's laws of motion, and was at the same time reluctant to dispose of Descartes's vortices. So he has been termed 'a Cartesian Leibnizian Newtonian'! (Schofield, 1978)

Not all would favour such a fine semantic distinction, for in the making of it the very terms change their meaning: at what point is 'Newtonianism' so modified that it loses any recognizable qualities? Nevertheless behind such sophistries lie two facts from which few modern historians of science would dissent: that a dominant influence on eighteenth-century European thought was the package of ideas emanating from the work of Isaac Newton, and that the 'triumph of Newtonianism' was partial, incomplete and uneven.

Sources referred to in the text

Butterfield, H. (1950) *The Origins of Modern Science 1300–1800*, London, Bell.

Cohen, I.B. (ed.) (1958) *Isaac Newton's Papers and Letters on Natural Philosophy*, Harvard University Press.

Cohen, I.B. (1956) *Franklin and Newton*, Philadelphia, American Philosophical Society.

Gillispie, C.C. (1958) 'Fontenelle and Newton', in Cohen (*op. cit.*).

Grant, R. (1852) *History of Physical Astronomy*, London, Bohn.

Hall, A.R. (1963) *From Galileo to Newton 1630–1720*, London, Collins.

Heilbron, J.L. (1980) 'Experimental natural philosophy', in Rousseau and Porter (*op. cit.*).

Home, R.W. (1977) '"Newtonianism" and the theory of the magnet', *History of Science*, 15, pp.252–66.

Hufbauer, K. (1982) *The Formation of the German Chemical Community (1720–1795)*, University of California Press.

Iltis, C. (1973) 'The Leibnizian-Newtonian debates: natural philosophy and social psychology', *British Journal of the History of Science*, 6, pp.343–77.

Jacob, M.C. (1976) *The Newtonians and the English Revolution 1689–1720*, Hassocks, Harvester.

Layton, D. (1965) 'Diction and dictionaries in the diffusion of scientific knowledge: an aspect of the history of the popularization of science in Great Britain', *British Journal of the History of Science*, 2, pp.221–34.

Love, R. (1972) 'Some sources of Herman Boerhaave's concept of fire', *Ambix*, 19, pp.157–74.

Raven, C.E. (1953) *Natural Religion and Christian Theology*, Cambridge University Press.

Rousseau, G.S. and Porter, R. (1980) *The Ferment of Knowledge: Studies in the Historiography of Eighteenth Century Science*, Cambridge University Press.

Russell, C.A. (1983) *Science and Social Change, 1700–1900*, London, Macmillan.

Schofield, R. (1978) 'An evolutionary taxonomy of eighteenth-century Newtonianisms', *Studies in 18th Century Culture*, 7, pp.175–92.

Secord, J.A. (1985) 'Newton in the nursery: Tom Telescope and the philosophy of tops and balls, 1761–1838', *History of Science*, 23, pp.127–51.

Stewart, L. (1986) 'The selling of Newton: science and technology in early eighteenth-century England', *Journal of British Studies*, 25, pp.178–92.

Thackray, A. (1970) *Atoms and Powers: An Essay on Newtonian Matter-theory and the Development of Chemistry*, Harvard University Press.

Westfall, R.S. (1980) *Never at Rest: A Biography of Isaac Newton*, Cambridge University Press.

Willey (1972) *The Eighteenth Century Background*, Harmondsworth, Penguin.

Further reading

Boyer, C.B. (1968) *A History of Mathematics*, New York, Wiley.

Cantor, G.N. (1983) *Optics after Newton: Theories of Light in Britain and Ireland, 1704–1840*, Manchester University Press.

Cantor, G.N. and Hodge, M.J.S. (1981) *Conceptions of Ether: Studies in the History of Ether Theories 1740–1900*, Cambridge University Press.

Clow, A. (1950) 'Hermann Boerhaave and Scottish chemistry', in A. Kent, (ed.) *An Eighteenth Century Lectureship in Chemistry*, Glasgow, Jackson.

Elkana, Y. (1971) 'Newtonianism in the 18th century', *British Journal of the Philosophy of Science*, 22, pp.297–306.

Guerlac, H. (1979) 'Some areas for further Newtonian studies', *History of Science*, 17, pp.75–101.

Heimann, P.M. (1973) 'Newtonian natural philosophy and the scientific revolution', *History of Science*, 11, pp.1–7.

Heimann, P.M. (1978) 'Science and the Enlightenment', *History of Science*, 16, pp.143–51.

Kiernan, C. (1973) *The Enlightenment and Science in Eighteenth Century England*, Banbury, Voltaire Institute.

Meinel, C. (1984) '"... to make chemistry more applicable and generally beneficial" – The transition in scientific perspective in eighteenth century chemistry', *Angewandte Chemie, Int. Ed. Engl.*, 23, pp.339–47.

Park, B. (1895) *The Intellectual Rise in Electricity: A History*, London, Longmans.

Rousseau, G.S. and Porter, R. (1980) *The Ferment of Knowledge: Studies in the Historiography of Eighteenth Century Science*, Cambridge University Press.

Science in the Scottish Enlightenment

Chapter 11

by Michael Bartholomew and Peter Morris*

11.1 The Enlightenment in Scotland

11.1.1 Introduction

Inspired by the Scientific Revolution of the seventeenth century, the intellectuals of eighteenth-century Europe launched a dazzling programme for the extension of knowledge and for the promotion of human welfare. Their programme has become known as the 'Enlightenment', and their age is often called the 'Age of Enlightenment'.

The Enlightenment was a programme, rather than a set of completed achievements. Enlightenment thinkers produced few theories comparable with Copernicus's or Newton's in former centuries, or with Darwin's in the next. What makes them memorable is the vigour and confidence of their conviction that the universe – from the orbits of the planets to the workings of the human mind and of human society – is explicable, regular and lawlike, and will yield to the systematic application of rational, empirical, scientific procedures.

Enlightenment thinkers attempted to extend the realm of lawlike regularities beyond the physical sciences into biology, geology, medicine, psychology, politics, economics, history. Indeed, *wherever* knowledge was to be gained, it had to be scientific, empirical knowledge: it was the only sort that counted. Moreover, this knowledge, however abstract, should graduate into practical schemes for human welfare – into schemes for agricultural improvement, for industry, for better surgery and midwifery, for better laws.

There was to be no mystery. 'The unknown' signified only that which had not yet been understood: the Enlightenment recognized no category of 'the unknowable'. And the most potent source of light to dispel the darkness of ignorance, blind authority, and religion, was science.

The men (and one or two women) of the Enlightenment formed what one of the foremost historians of the movement has called a self-consciously cosmopolitan, European 'philosophic family' (Gay, 1973, vol.1, p.6). Inevitably, though, branches of the family took tinges of colour from the various national cultures within which they grew.

This chapter is concerned with science in Scotland, one of the most dynamic centres of Enlightenment thinking. Writers speak of the mid-eighteenth century as Scotland's 'Golden Age'. In order to get the flavour of this age, it

---

* Sections 11.1 and 11.2 were written by Michael Bartholomew; Section 11.3 was written by Peter Morris.

is necessary to take a very broad view of what we mean by 'science'. If we stay within the boundaries recognized by modern science faculties, we will miss most of what is distinctive about eighteenth-century Scotland. The interconnections and cross-fertilization between disciplines that we now regard as having little to do with each other is one of the remarkable features of the Scottish scene. Geologists associated with historians, economists with chemists, philosophers with surgeons, lawyers with farmers, church ministers with architects. Obviously, if we stretch the term 'science' too far, it disintegrates, but it is worth bearing in mind that the very term 'scientist' was not coined until the 1830s. Half a century earlier, a meeting of a learned society in Edinburgh, or Glasgow, or Aberdeen, would have brought together representatives of all the interests listed above, and they would all have recognized that they were engaged on a single project – namely, the pursuit of natural knowledge, by the light of observational, empirical methods, which in turn would lead to 'improvement' in the affairs of Scotland.

The Scottish conception of science and its purpose was neatly summed up in the programme of the Aberdeen Philosophical Society, or 'Wise Club' as it came to be known, founded in 1758: the Society aimed to investigate

> every Principle of Science which may be deduced by Just and Lawfull Induction from the Phaenomena either of the human Mind or of the material World; all Observations and Experiments that may furnish Materials for such Induction; the Examination of False Schemes of Philosophy and false Methods of Philosophising; the Subserviency of Philosophy to Arts, the Principles they borrow from it and the Means of carrying them out to their Perfection. *(Chitnis, 1976, p.200)*

The summary is useful too in showing how the meaning of key words has shifted since the eighteenth century. As the name indicates, the members of the Aberdeen Society were interested in philosophy, but they used the term to signify what today would be regarded as science. The word 'science' in their quite typical usage meant simply 'knowledge'. They were also interested in 'arts', by which they meant, not the fine arts, but skills or even trades: arts would have included activities like printing, or agriculture – it signified something close to the modern conception of technology. It is interesting to see that the practical Aberdeen Society stressed 'the subserviency of Philosophy to Arts', by which it meant that science provided a base for technology: science should ultimately, in their view, be in the service of technological application.

11.1.2 Origins of the Scottish Enlightenment

Before examining Scottish science in detail, we need a sketch of the particular Scottish historical background from which an astonishing cluster of intellectuals and ideas emerged. It needs to be said at the outset, however, that there is no scholarly consensus as to why a small, poor country in Northern Europe should have made such a disproportionately large contribution to the thought of the age.

The Act of Union, 1707

The event in Scottish history which tends to polarize opinion among scholars is the Act of Union with England, of 1707. The crowns of the two nations had been unified a century earlier, in 1603, when the Stuart James VI became king, not just of his native Scotland, but also of England, where he reigned as James I. But in 1707, Scotland gave up its parliament, and henceforth, the government of the country shifted from Edinburgh to Westminster. Some scholars have seen the Act of Union as precipitating a crisis in Scottish identity. Where, after 1707, might the intellectual energy of the nation be expressed?

The politically ambitious would speed to Westminster and join the scramble for office, shedding, in the process, their national loyalty. But what of those who remained in Scotland, yet who wished to contribute publicly to the nation's affairs? One route that might be predicted leads to the development and nourishing of a distinctive Scottish national culture, in protest against the loss of nationhood entailed by the Act of Union. After all, Scotland had its own languages – Gaelic in the Highlands, and Scots (a very markedly distinct form of English) in the Lowlands – and had its own unique culture and social system, especially in the Highlands. Perhaps we would predict the birth, after 1707, of a Scottish national, cultural movement.

This route was not taken. The leading lights of Scottish society came, almost wholly, from the Lowlands, and they directed their energies towards the establishment of an English-speaking, urban, civilized, commercial society that did not brandish Scottishness at every turn. Notably, they tended not to throw in their lot with the two Jacobite rebellions (of 1715 and 1745) which sought to restore the Stuart monarchy in Britain, and which embodied aspirations for Scottish national independence. The unwillingness of Scottish intellectuals to become identified with what they saw as a defeated, out-moded national culture is illustrated by one of the elegant deathbed utterances of perhaps the foremost intellectual of the age, the philosopher David Hume. He died, it was reported, 'confessing not his sins, but his Scotticisms': that is to say, he regretted not having succeeded in purging residual Scots phrases from his otherwise immaculate English prose.

For some scholars, then, the Act of Union had a 'traumatic effect'. It left the Scottish elite bereft of real political institutions, yet dissatisfied with the remnants of an ancient Scottish culture. They engaged, it is argued, in a search for a new 'cultural style' (Phillipson, 1973, 1981).

For other scholars, the origins of the Scottish Enlightenment are to be found not in a sudden trauma, but buried within long traditions in the Scottish economy and society. Scotland was certainly a poor, small country in the late seventeenth century, it is acknowledged, but a number of writers have looked hard at seemingly moribund institutions and found that commercial, scientific and philosophical life was stirring. For these writers, the Scottish Enlightenment was the flowering of Scotland's own indigenous traditions. Three areas of enquiry have been fruitful: the Church, the universities and the economy.

The Church

The Scottish Church seems an unlikely place to look for the stirrings of enlightenment. In 1690, the General Assembly of the Church of Scotland passed an act against 'the Atheistical Opinions of the Deists', and, in 1696, an eighteen-year-old Edinburgh University student was executed for denying some of the propositions of Christianity. The legacy of the Scottish, Calvinist Reformation, it seems, was one of conformism, intolerance and narrow-mindedness. But this is not the whole story. Another impulse from the Reformation itself was founded on the principle of critical scrutiny of Catholic tradition. This rational, critical impulse was felt by more liberal members of the Scottish Church, and was given typical expression by the Reverend William Wallace, a minister close to the pulse of Edinburgh

University life. He preached, in 1729, that there must be a

> hearkening to the voice of sound reason, the examining impartially both sides of
> the question, with a disposition always to adhere to the stronger side and to
> embrace the truth wherever it appears in spite of all prejudices, of all opposition
> and authority of men. This is what I can never censure or apprehend being
> capable of being carried to an extreme. *(quoted in Cameron, 1982, p.123)*

The tradition that Wallace represented grew steadily during the century, and
the 'Moderate Party' of the General Assembly, as it became known, was
receptive to – and in return made contributions to – Enlightenment thinking.

At a more general level, the intensely pious Calvinist tradition may have
flowed in unexpected, worldly directions. Calvinist zeal may have been one
of the ingredients in the development of Scottish industry and the economy
in the eighteenth century. Here is how a leading Scottish historian puts it:

> The singleminded drive that is seen so often in business, farming and trade in the
> eighteenth century, and which appeared in cultural matters in men as diverse as
> Adam Smith, James Watt and Sir Walter Scott, is strangely reminiscent of the
> energy of the seventeenth-century elders in the kirk when they set about imposing
> discipline on the congregation. Calvinism thus seems to be released as a
> psychological force for secular change just at the moment when it is losing its
> power as a religion. *(Smout, 1969, p.92)*

This is an attractive suggestion, but we should not underestimate the
problems inherent in transmuting a religious drive into a secular one.
Calvinism – indeed Christianity at large – teaches that human nature is
depraved. In 1717, in criticizing a Moderate minister, the Church Assembly
held that he had attributed 'too much to natural reason and corrupt nature'
(Cameron, 1982, p.119). Plainly, a number of radical intellectual moves had to
be made before human nature could be presented (as it was in the
Enlightenment) as notably *un*corrupt – as fundamentally social, and likely to
be virtuous, given a rationally organized society.

The universities

Turning to the universities, scholars have discovered that much more was
going on during the late seventeenth century than the unimaginative training
of young men for ministry in a dour church. Another legacy from the
Reformation in Scotland was a recognition of the need for education, and, by
the beginning of the eighteenth century, five universities, in four cities, were
well established. (England, a far larger country, had only two.) Research and
specialist teaching was held back by a system known as 'regenting', whereby
individual 'regents' taught every subject to undergraduates. Not until the
eighteenth century could lecturers break out of this generalist teaching of
often outdated material, and provide specialist courses. Even so, the
universities were not backwaters. The work of Shepherd, for example, has
shown that Newton's work was finding its way onto the syllabuses of
Scottish universities from the 1680s. She has also reconstructed syllabuses at
Edinburgh which show that the work of Copernicus, Galileo and Boyle was
being taught (Shepherd, 1982). And in a reconstruction of Hume's education
at Edinburgh University in the 1720s, Barfoot has found evidence that he was
alerted there to the latest developments in science (Barfoot, 1990).

Not all innovation came from beyond Scotland's borders, and that which did
was just as likely to have come from the Netherlands as from England,
especially in the field of medicine. There were powerful links between
medicine in Leiden and in Edinburgh. There were also entirely local
traditions in mathematics, chemistry and medicine.

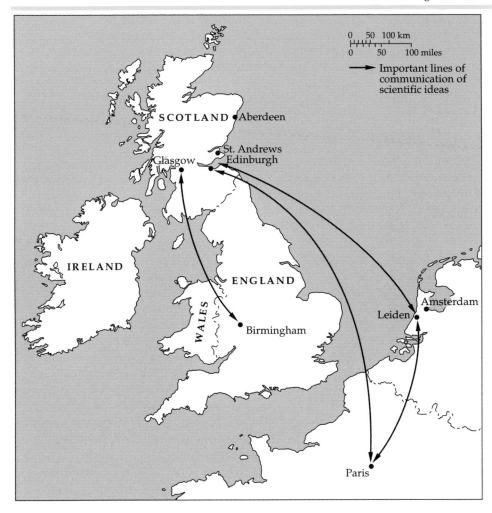

Map 11.1 The Scottish connection.

The economy

Turning lastly to the late seventeenth-century economy, a similar pattern of historical revision is revealed. Accounts stressing desperate poverty and backwardness have given way to accounts which indicate a more prosperous, vigorous state of affairs. In a survey of the Scottish merchant community, Devine has concluded that although the nation had not fully insulated itself against the calamity of bad harvests, its merchants were forward-looking and ready to innovate. They were not locked into conservative social hierarchies which inhibited commercial ventures. Sons of lairds became merchants; merchants bought land – it was an 'open' society. Here is Devine's conclusion:

> The business classes possessed the sophistication crucial to later advance. The merchant class made little intellectual contribution to the early Enlightenment; their function was more indirect, to help to provide, with the professional and landed classes, a social and material environment which was not resistant to change, whether in the cultural or economic spheres. *(Devine, 1982, p.37)*

It is from this background of mercantile openness that works like Adam Smith's *Wealth of Nations* (1776), the foundation text in the new social science of economics, came. From the same background, it is important to note, came the harsh industrial regimes of the early factories: enlightenment could sometimes be exploitation dressed up in new clothes.

No matter whether it is the supposed 'trauma' of the Act of Union, or longer, indigenous traditions which command historians' attention in their quest for the origins of the Scottish Enlightenment, there is no dispute about the general characteristics of the movement once it was underway.

11.1.3 The Enlightenment milieu

Clubs and societies

The milieu was urban. It was not a business of isolated individuals working in country estates, or of secluded academics, cloistered within unworldly universities. The scene was convivial, social. The focus was Edinburgh, although Glasgow and Aberdeen were active too. Cities were small. Even the capital was intimate enough for its intelligentsia to be able to meet regularly and casually. 'Here I stand, at what is called the Cross of Edinburgh', wrote an excited visitor, 'and within a few minutes take fifty men of genius by the hand' (quoted in Daiches, 1986, p.1). Perhaps the most characteristic expression of the conviviality and energy of the place was the club, or the society. Dozens of them were formed during the century, some short-lived dining and drinking clubs, some maturing into august scientific and medical bodies that still exist. Some, like the Poker Club (concerned with poking up sluggish intellectual fires, not card games), the Oyster Club or the Friday Club, at first sight seem frivolous – excuses, perhaps, for male claret-swilling – but behind the grandiloquence, serious issues were debated. The Oyster Club, for example, had among its founders the economist Adam Smith, the chemist Joseph Black and the geologist James Hutton – all pioneers in their fields and indebted to each other's criticism, help and stimulus.

Two societies can be singled out as being of fundamental importance in the discussion and dissemination of science. In 1731, the professors of medicine at Edinburgh founded the Medical Society of Edinburgh. The driving force was Alexander Monro, the first in a dynasty of three generations of Alexander Monros (known as *primus, secundus* and *tertius* – first, second and third) who dominated Edinburgh medicine. The Society published medical research and soon established for itself a reputation in European medicine. When Alexander Monro *primus* fell ill, Colin McLaurin, an Edinburgh University mathematician and Newtonian, broadened the Society's scope to include all 'philosophical' topics (in the eighteenth-century sense), and the name changed to the Philosophical Society. The membership is a rollcall of the Scottish Enlightenment: McLaurin himself, Joseph Black, James Hutton, Adam Smith, David Hume, the chemist and doctor William Cullen, and the philosopher Dugald Stewart. The Society flourished from 1737 until 1783. Within its boundaries, smaller, special-interest groups, like the Newtonian Club, operated. The Society as a whole achieved the highest possible status when it was given a royal charter in 1783, to emerge as the Royal Society of Edinburgh, the premier scientific society of the country.

Medicine did not fall by the wayside when the Philosophical Society broadened its scope. A student medical society, which met first in 1734, grew, within forty years, into the Royal Medical Society, which was chartered in 1778. And along the way, it developed the full infrastructure of a lively scientific academy – premises, a library, a museum, a laboratory, prizes, publications.

The historian Roger Emerson, who has made extensive studies of Scottish science, has assembled a useful identikit picture of a member of an Edinburgh Society. It brings out clearly the social background and the wide-ranging commercial and intellectual interests of the men who founded the clubs and societies. Emerson's picture is of a typical member of the Philosophical Society, in 1739: such a member

> was an active professional man from the landed gentry who was politically involved and who held a patronage post which enhanced an income not wholly derived from rents. Tied to Edinburgh and to Scotland by economic interests, various responsibilities, language, sentiment, and perhaps by his training in Scots law, he was a place seeker whose prospects outside Scotland were limited but within the kingdom reasonably good. Well educated and usually the beneficiary of foreign travels, he was aware of the backwardness and provincialism of his country, and patriotic enough to wish to remedy it. Relying on provincial institutions for his status and income, he sought to raise both through improvements which would modernize the country, and allow it and him to play greater roles in the world. His enlightenment, and the work of his academy, would be practical, non-literary, career-furthering and conservative of his position as a member of an economic, social, and intellectual elite dominating the kingdom's institutions. *(Emerson, 1979, p.173)*

Publishing

One of the strongest impulses in the Enlightenment was to codify knowledge and publish it widely. The most notable example of this impulse is the French *Encyclopédie*, 'a rational dictionary of the sciences, art and trades', published chiefly in Paris in the 1750s and 1760s, under the indomitable editorship of

Figure 11.1 An ideal of the Academy: the happy union of arts, science and technology. Note that the title does not signal the encyclopaedia's Scottish origin – a further indication of the movement's ambivalent attitude to nationhood. (From frontispiece to the Encyclopaedia Britannica, *3rd edn, Edinburgh, 1788. Reproduced by permission of the British Library Board.)*

285

Denis Diderot. The seventeen volumes of text and eleven volumes of plates were intended to summarize and clearly present everything that was worth knowing, from the construction of a water wheel or a glass manufactory to the latest theories in the psychology of perception. The impulse which drove Diderot was working in Edinburgh too. A number of encyclopaedias were started, but the venture which became the most famous was the *Encyclopaedia Britannica*, which started in the 1760s. *Britannica* was coaxed into life by the printer, William Smellie, a man who, though without formal academic qualifications, was a key figure in the dissemination of the work produced within Edinburgh. By the turn of the century, and with perhaps significantly less bashfulness about its origin, the *Edinburgh Review* was launched. This journal quickly achieved a British reputation and became one of the most influential reviews of science, politics, economics and the arts.

Architecture

Printing and publishing, then, had their connections with the Enlightenment programme. Architecture too was related. The Adam family of architects (the father and his two sons) moved in the Edinburgh circle of the intellectuals. The young Robert Adam, for example, attended both McLaurin's mathematics lectures and Monro's anatomy lectures at the university, and his home life was enlivened by regular visits from the leading lights of the city. As one contemporary described the household, in a rolling eighteenth-century sentence:

> The numerous family of Mr Adam, the uninterrupted cordiality in which they lived, their conciliatory manners and the various accomplishments in which they severally made proficience, formed a most attractive society and failed not to draw around them a set of men whose learning and genius have since done honour to that country which gave them birth. *(quoted in Fleming, 1962, p.5)*

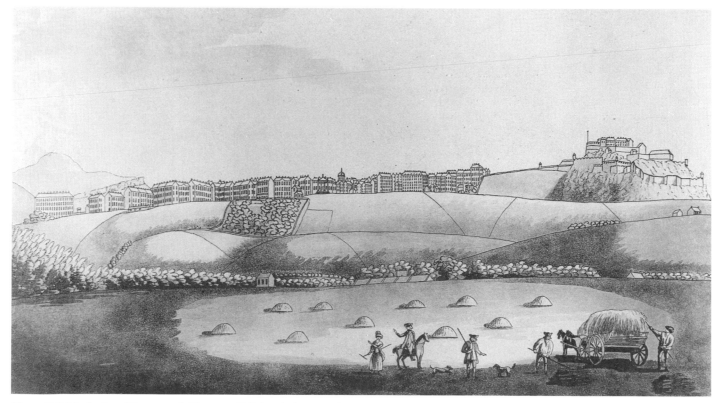

Figure 11.2 North view of the new and old towns of Edinburgh, from Inverleith, 1781. (By courtesy of the Edinburgh City Libraries.)

In the mid-century, Edinburgh was still an ancient city clustering around the castle and stretching down the hill to the neglected royal palace of Holyrood. But in 1752, the astute provost of the city, George Drummond, launched a plan to lay out a new town, beyond the North Loch, which would itself be drained. There were setbacks, but steadily there arose a rational grid of coolly elegant streets and squares, relieved by the occasional curve or gradient. As it arose, however, the New Town, as it became known, was failing quite to realize the grandeur implicit in the ground plan, and in 1791, Robert Adam, who was by then making his fortune in England, was called in to design a monumental square in order to demonstrate just what could be done with urban housing, if conceived on a grand scale. The result is Charlotte Square, in which rows of terraced houses, built for the prosperous bourgeoisie, are successfully subordinated to a conception of a single, palatial edifice.

It would be too slick to present the elegant, rational Edinburgh New Town simply and baldly as the embodiment of Scottish Enlightenment – especially as the leading lights of the movement preferred to stay over in the racier old town – but in tracing the networks of people and ideas that flourished in the city, the route that leads to architecture and town planning is not to be ignored.

The role of the Edinburgh Town Council

This route incidentally leads us to another important feature of the movement, namely, the role of the Edinburgh Town Council and its provosts. (The English equivalent would be a lord mayor.) Throughout the eighteenth century, the Town Council, with a policy of enlightened self-interest, promoted the city by sponsoring or patronizing its academic, medical and scientific life. The Council regarded the city's university, infirmary and medical school as institutions which, if given enough prestige, would not only stop the drift of Scottish students and their fees to foreign universities – especially to Leiden for medical training – but also reverse the flow and attract fee-paying students to Edinburgh from across Europe and America. Accordingly, it took an active role in the appointment of professors who would bring fame. As early as 1713, the Council minuted its reasons for appointing James Crawford to the chair of chemistry at the university: the appointment was made

> ... particularly considering that through the want of professors of physick and chymistry in this Kingdome the youth who have applyed themselves to study have been necessitat to travel and remain abroad a considerable time for their education to the great prejudice of the nation by the necessary charges occasioned thereby. *(quoted in Christie, 1974, pp.127–8)*

Another such appointment was that of Colin McLaurin, the mathematician and Newtonian, to the chair of mathematics in 1725. McLaurin had formerly been at Marischal College, Aberdeen, where he had taken a rather high-handed view of his teaching duties. Somewhat oddly, this did not count against him when he was recruited for Edinburgh. What counted *for* him was a growing European reputation: a rising star could be caught. The tempting modern analogy is with those town councils who invest in their cities' football teams. The perhaps more sober conclusion of the historian who has investigated this episode is that McLaurin's appointment guaranteed that 'the University of Edinburgh became an acknowledged centre for the diffusion of Newtonian mathematics, astronomy, and natural philosophy by the most gifted and accomplished British disciple of his generation' (Morrell, 1974, p.86). McLaurin also mended his lackadaisical attitude to lecturing, and taught courses which included surveying and gunnery: his classes were not just for aspiring young mathematicians; they were also to serve the practical needs of students who intended to become engineers and army officers (Christie, 1974, p.125). The architect Robert Adam, it will be recalled, also attended McLaurin's classes.

Regenting (the system of low-grade generalist teaching) came to an end in Scottish universities in the early decades of the eighteenth century, opening the way to the endowment of specialist professorships. In Edinburgh, for example, there were already chairs in natural philosophy, medicine and mathematics, surviving from the seventeenth century, but to these were added chairs in botany, anatomy, midwifery, chemistry, *materia medica* (the study of the materials, chiefly botanical, from which medicines were prepared), surgery, astronomy, agriculture. The patronage shown by the Town Council paid off: students did come, from home and abroad, and the number of graduates steadily rose.

The Town Council's investment in university teaching was shrewdly limited. Professors' salaries were not large. It was intended that the basic salary should be enhanced by a system that strikes terror into the heart of the twentieth-century academic: most of the income of eighteenth-century academics came from class fees paid by students. The stark and salutory implication was that poor lectures, attracting small numbers of students, would generate only a dismal income. Adam Smith, a successful professor at Glasgow University, and advocate of the market economy, recognized the compelling logic of the system:

> It is the interest of every man to live as much at his ease as he can; and if his emoluments are to be precisely the same, whether he does or does not perform some very laborious duty, it is certainly his interest ... either to neglect it altogether, or, if he is subject to some authority which will not suffer him to do this, to perform it in as careless and slovenly a manner as that authority will permit. *(quoted in Chitnis, 1976, p.140)*

Chitnis has compiled figures to show that class fees contributed much more to professors' salaries than did their basic salary. At the end of the century, for example, the professor of anatomy boosted a basic salary of fifty pounds to nearly a thousand (p.152).

In sum, then, the milieu of the Scottish Enlightenment was its university cities, where flourished groups of characteristically clubbable intellectuals, divided by no ideological rifts, all committed to the pursuit of natural knowledge, in the general context of a commitment to the improvement of Scotland's, and their own, fortunes. They were supported by civic authorities, by an enterprising commercial culture, by extensive international scholarly contact, and even by the moderate wing of the Church.

Within this milieu, a scientific and medical community had, by the middle of the century, reached maturity – a maturity which meant that it was independent of the accidental incidence of a handful of energetic individuals. By 1760 it had built itself an infrastructure of learned societies, journals, specialist university teaching and research, and last, but not least, connections with agriculture and industry. The scientific and medical community could reproduce itself: it wouldn't collapse at the death of one particular and influential member (Christie, 1974).

11.1.4 The leading figures of the Scottish Enlightenment

At this point, before we move on to look in greater detail at the work of a couple of characteristic and influential Scottish scientists, it will be useful to stand back and take a survey of the leading members of the scientific and medical community.

One of its most eminent members, Adam Smith, pioneered the discipline of economics, which is not customarily included within science today. But to exclude him from our survey would be to misrepresent the unfenced, boundary-free territory across which eighteenth-century intellectuals ranged. Smith was professor of moral philosophy at Glasgow University and associated regularly with the leading lights of the European philosophic community. He published the famous *Wealth of Nations* in 1776. Smith's concerns, however, were by no means purely economic. Along with less-well-remembered scholars, he was engaged in one of the fundamental enquiries of the Scottish movement as a whole, namely the enquiry into the nature of humankind and human society.

In the field of medicine, the Monro dynasty commands attention. Alexander Monro *primus*, trained at Leiden, was appointed by the Town Council in 1720 to be professor of anatomy. His grandson, Alexander Monro *tertius*, held the post in the 1840s, by which time Edinburgh medicine had developed the full range of institutions – university lectures, a teaching hospital, learned journals and societies. It should not be too readily assumed, however, that prestigious and well-supported medical institutions invariably led to improvements in patients' health. Historians of medicine have yet to resolve the question of whether eighteenth-century hospitals enhanced patient's chances of recovery or were, rather, 'gateways to death' caused chiefly by infections. The effectiveness of the most brilliant surgical skills – in amputating limbs, or removing urinary stones, for instance – was considerably diminished by shock and post-operative infections. Nor should it be assumed that medicine was solely a metropolitan affair, conducted by a handful of well-to-do physicians, surgeons and their students. Medical handbooks found their way into the households of citizens of moderate means. The most famous of these handbooks is William Buchan's *Domestic Medicine*, published in 1769 and running to 22 editions by 1822. Buchan was an Edinburgh-trained doctor, and his book embodied the rational, common-sense principles of the Enlightenment. In the absence of antibiotics, medicine was incapable of making spectacular breakthroughs in healing the sick, but books like Buchan's – with its sober calls for moderate living, for publicly-funded inoculation schemes, for an end to superstitious practices in child-birth and child-rearing (he recommended, for example, that fathers should play an active part in rearing their children and 'ought to assist in every thing that respects either the improvement of the body or the mind' (1769, p.7)) – did introduce the new medical thinking into the life of the community and led to modest improvements in its health.

Medicine was linked with the physical sciences, notably in the person of William Cullen, who lectured on medicine at Glasgow University before moving to Edinburgh in 1756. There, he combined research and teaching in both medicine and chemistry. He taught on the wards of the new Edinburgh Infirmary, was president of the Edinburgh College of Physicians, as well as holding the chair of chemistry at the university. He was a popular, pivotal figure in Scottish science and had a great influence on the young chemist Joseph Black (see Section 11.3).

11.2 James Hutton

11.2.1 Early career

James Hutton (1726–97) conforms fairly closely to Emerson's identikit picture of an intellectual of the Scottish Enlightenment. His chief scientific work was his *Theory of the Earth*, which was launched at meetings of the Royal Society of Edinburgh in 1785 and eventually expanded and published in two large volumes, ten years later, in 1795.

He was the son of a well-to-do Edinburgh merchant and was educated first at the city's university, where, like many students, he was particularly interested in chemistry. From Edinburgh University he took what was the natural route for young men who were keen to extend their studies in science: he went to Paris, and from there to the university which features again and again in the background to the Scottish Enlightenment – Leiden, in the Netherlands. The presiding spirit at Leiden was that of the doctor and chemist Hermann Boerhaave (1668–1738). Boerhaave's ideas influenced a generation of students, including those who returned to Scotland to establish the Edinburgh Medical School in 1726. Although Hutton graduated as a doctor at Leiden in 1749, he never practised regularly.

Instead, he returned to Edinburgh and set up a profitable chemical works which produced sal ammoniac (ammonium chloride) – a substance used as a flux in the metalworking trades and in the textile industry. Typically, Hutton was not averse to dirtying his hands, either with chemicals or with trade. Equally typically, he did not rest content as a successful chemicals manufacturer, but moved on into agriculture when he inherited two farms. He studied the latest agricultural techniques with a view to introducing them on his farms.

Farming, like the chemical industry, was unable to sustain his interest, and he moved on to geology. In making this move, though, he was able to take with him much of the knowledge he had derived from his earlier enterprises. Farming had prompted his interest in the structure of the earth's crust. Drainage schemes and quarrying opened sections through earth and rock which intrigued him, and in pursuit of his twin interests in agricultural improvement and the structure of the landscape, he travelled extensively around Scotland.

Eventually, in 1767, Hutton returned to Edinburgh, where he slotted comfortably into the Enlightenment milieu. He associated with Adam Smith, Joseph Black, the historian William Robertson, the anthropologist Lord Monboddo and the engineer James Watt. Through Watt, he met the members of the Lunar Society of Birmingham, a group of scientists, engineers and industrialists from the English Midlands. In short, Hutton was closely in touch with activities in a host of related and vigorous areas of enquiry.

11.2.2 Background to Theory of the Earth

The two volumes of *Theory of the Earth* embody a startlingly original conception of the processes which shape the earth's surface, and they contain some vivid observations, drawn from Hutton's travels. However, they are poorly organized, repetitive and sometimes obscure. In a most helpful survey of Hutton's work, from which I have drawn liberally, Jean Jones quotes from a wonderfully direct letter that a saddlesore Hutton wrote while on a field-

Figure 11.3 James Hutton (1726–97). (Scottish National Portrait Gallery)

trip in Wales: 'Lord pity the arse that's clagged to a head that will hunt stones' (Jones, 1986b, p.127). Such admirable conciseness is absent from the *Theory*, but the two volumes are a foundation text in the science of geology, and are well worth exploring.

In this brief account of his life, I have stressed the practical and commercial aspects of Hutton's life. However, another influence is at work in his geological theorizing: the book is very far from a handbook for coal prospectors. It is a grand attempt, as its title indicates, to establish the principles which govern the structure and shape of the earth's crust. Given the materials with which Hutton worked – rivers, rocks, volcanoes, oceans, fossils – it is plain that he could never formulate neat mathematical laws to account for landforms, but the drive of his theorizing is always to describe geological processes in terms of the interplay of two contending natural forces: elevation and erosion.

Also, it is equally plain that Hutton's work was inspired and regulated by his deistic religious beliefs. Deists put aside the Christian Revelation, with its scripture, miracles and incarnation, in favour of an unimpassioned belief in a Divine Architect whose sole purpose was to set the universe running. In so doing, deists who happened also to be geologists put aside the account in the book of Genesis of the formation and history of the world. Christians, on the other hand, were gripped by the powerful story of the seven days of Creation, of God's subsequent anger and the Flood. Not until the nineteenth century, and for some Christians not even then, did non-literal readings of the biblical Creation story start to make headway. Hutton's deism enabled him to sidestep all problems of harmonizing his theory with scripture. One of the remarkable features of the *Theory of the Earth* is the absence of references to the account of Creation which had possessed the European imagination for nigh on two thousand years: the Genesis story seems to have faded almost clean away in the blaze of the Enlightenment. Hutton made only oblique, but entirely civil, references to the biblical account. Here, for example, is how he handles the idea of the Flood:

> Philosophers observing an apparent disorder and confusion in the solid parts of this globe, have been led to conclude, that there formerly existed a more regular and uniform state, in the constitution of this earth; that there had happened some destructive change; and that the original structure of the earth had been broken and disturbed by some violent operation, whether natural, or from a supernatural cause.

He goes on to say that his own theory gives a perfectly satisfactory account of the phenomena supposedly resulting from a great cataclysm, and concludes:

> Therefore, there is no occasion for having recourse to any unnatural supposition of evil, to any destructive accident in nature, or to the agency of any preternatural [i.e. supernatural] cause, in explaining that which actually appears. *(Hutton, [1795] 1959, vol.1, pp.165–6)*

This is not to say that religious belief played no part in his theorizing. On the contrary, it was a powerful stimulus. Hutton's fundamental belief was that the earth has been formed for a purpose. That purpose is the support of life, and especially human life. Furthermore, in Hutton's view, the discovery of the way in which this purpose has been achieved leads enquirers to a noble conception of the Divine Architect.

Hutton's belief in a wise providential ordering of a world which, no matter how it changes, is always bountifully equipped to support life is not just a polite decoration to his work. It actively regulates his theorizing. This *teleological* view, stressing the purposeful drive towards an end, leads Hutton to assume, for example, that no matter how radically the face of the earth has been remodelled during geological time there has always been a harmonious relationship between land-mass and ocean: he could not conceive of the possibility of there ever having been a time when life on land was impossible. 'It is only required', he wrote, 'that at all times, there should be a just proportion of land and water upon the surface of the globe, for the purpose of a habitable world' (Hutton, [1795] 1959, vol.1, p.196).

'The purpose of a habitable world' is Hutton's answer to the teleological question 'What is the earth *for*?'. Moreover, in characteristically Enlightenment fashion, Hutton declares further that life is essentially happy:

> It is of importance to the happiness of man, to find consummate wisdom in the constitution of this earth, by which things are so contrived that nothing is wanting, in the bountiful provision of nature, for the pleasure and propagation of created beings; more particularly of those [i.e. humans] who live in order to know their happiness, and know their happiness on purpose to see the bountiful source from whence it flows. *(Hutton, [1795] 1959, vol.2, p.183)*

Such cheerful sentiments are a long way from the Christian tradition, strong in Scottish Calvinism, which asserted humanity's sinfulness.

11.2.3 Hutton's geology

'No vestige of a beginning – no prospect of an end'

Geologists are engaged on the business of reconstructing the earth's past and determining the agents of geological change. The only documentary evidence of the earth's origins and ancient past, and of the agents that had caused change, available to Hutton was the book of Genesis, and he had sceptically put it aside, along with miracles. But what if the processes that are *presently* observable were to be taken as the key to the past? How far might geological enquiry go with the assumption that what is now going on is all that has ever gone on – that the modern world presents an exhaustive catalogue of the processes that have shaped the world, and are continuing to shape it? Hutton's originality lies in his readiness to go all the way with this assumption. He produced a theory which pictured an earth in which 'the purpose of a habitable world' has perpetually been achieved by a set of perfectly balanced agents of natural destruction and renewal. Earth history has no direction: it is now, and always has and will be, in a steady-state. The challenge to the geologist is to show how the steady-state is maintained – to make a survey of the agencies of destruction and renewal at work in the landscape.

What were Hutton's agents of destruction and renewal? Briefly, he argued that rocks are formed at the bottom of the sea and are composed, first, of material eroded from the neighbouring landmasses. Continents are inexorably being eroded away, and their fragments are washed down rivers to the sea. Secondly, rocks are composed of the remains of sea-dwelling animals: calcareous rocks – limestones, chalk, marble – simply *are* the consolidated remains of countless populations of shellfish whose shells have sunk to the sea-bed. All this material, either from former continents or from former living things, consolidates on the sea-bed where, under pressure from the sea, it is baked by the subterranean heat of the globe (a heat which, in Hutton's view can be reliably inferred from the action of volcanoes). As ancient continents are relentlessly ground away, subterranean heat slowly upheaves sea-beds elsewhere and new continents are born. Nothing is permanent: all is in a flux of destruction and renewal.

In Hutton's account, geological time is directionless – it's not going anywhere: the earth has proceeded from no primeval state, and it will not culminate at some future final point. The steady-state of a habitable world can be projected backwards into the eternal vistas of the past, and can confidently be predicted, stretching into the equally endless vistas of the future. 'Time', he wrote, 'is to nature endless and as nothing' (Hutton, [1795] 1959, vol.1, p.15). And in one of the most memorable utterances in the history of geology – one in which Hutton exhibited an uncharacteristic eloquence – he concluded that his researches have shown that the present landscape is built from the materials of former landscapes, which in turn are built from yet earlier landscapes, which in turn stretch back in endless succession. Sounding the standard, eighteenth-century Newtonian note, Hutton wrote:

> For having, in the natural history of this earth, seen a succession of worlds, we may conclude that there is a system in nature; in like manner as, from seeing revolutions of the planets, it is concluded, that there is a system by which they are intended to continue those revolutions. But if the succession of worlds is established in the system of nature, it is in vain to look for anything higher in the origin of the earth. The result, therefore, of this physical enquiry is, that we find no vestige of a beginning, – no prospect of an end. *(Hutton, [1795] 1959, vol.1, p.200)*

How could Hutton be so confident that he could find 'no vestige of a beginning'? Other geologists had affirmed that rock strata could be sorted into a single sequence, stretching from 'primitive' rocks, formed when the world was young, up to modern rocks. Knowledge of the fossils (remains of living things) which characterize each rock formation was sketchy, but it seemed clear that there were rocks, low down in the sequence, which contained no fossils at all. It seemed reasonable, therefore, to say that the earth has developed uniquely, from a primitive, lifeless condition up to the present. Hutton challenged this by saying that there was, in effect, no such thing as a primitive rock. All rocks, no matter how low in the sequence, no matter how contorted, were formed, he argued, from the sorts of material that are still abundant in the world, and by the processes that are still observably at work in the landscape. If no fossils can be found in them, it is because they have been obliterated by the pressure and the heat which produced the strata.

Hutton's prosaic writing rarely does justice to the huge imaginative leap he made in grasping the explanatory potential of small, mundane modifications to the landscape – like the rolling of rocks downstream by rivers, or the accumulation of seashells on the sea-bed – when these modifications are given indefinite time in which to accumulate. It was remarkable to have been able to contemplate a mountainous country like Scotland, built seemingly of durable and stable rock, as, on one hand, having been built from strata laid down aeons ago beneath now vanished oceans, and, on the other, as potential raw material from which, in the immeasurably distant future, a new continent would be formed.

The Jedburgh unconformity

One concrete example from the *Theory of the Earth* will perhaps indicate the way in which Hutton could read features of the landscape as evidence of the action of forces acting over immeasurably long periods. He had been geologizing in the valley of Jed Water, near Jedburgh, in the Borders area between England and Scotland. From his observations in the neighbouring

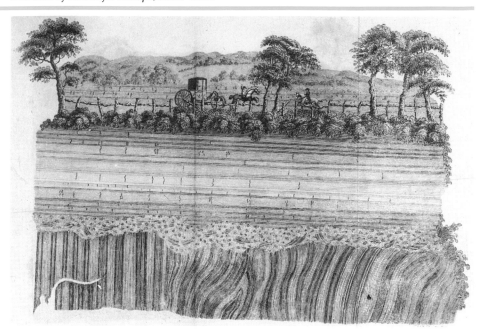

Figure 11.4 Unconformity near Jedburgh. (From J. Hutton, Theory of the Earth, *vol.1, Edinburgh, 1795, plate 3. Reproduced by permission of the British Library Board.)*

Teviot valley, he expected the Jed to be running over a bed of horizontally laid, soft strata which were sometimes exposed as sections alongside the river. However, in his own words:

> I was surprised with the appearance of *vertical* strata in the bed of the river, where I was certain that the banks were composed of horizontal strata. I was soon satisfied with regard to this phenomenon, and rejoiced at my good fortune in stumbling upon an object so interesting to the natural history of the earth, and which I had been long looking for in vain.
>
> ... above those vertical strata, are placed the horizontal beds, which extend along the whole country. *(Hutton, [1795] 1959, vol.1, p.432, my italics)*

What Hutton had found was what is now known as an 'unconformity': a junction between sets of rocks of quite different types, formed at quite different epochs. The Jedburgh unconformity was sketched by Hutton's travelling colleague, John Clerk, and appeared as a delightful engraving in the *Theory of the Earth* (see Figure 11.4).

How was the unconformity to be explained? Hutton proceeds, in the *Theory of the Earth*, by eliminating what he considers to be unsatisfactory explanations. For example, it is difficult to imagine that the upper, horizontal strata could have been laid down *before* the vertical strata beneath them: this would entail the subterranean building of vertical strata which somehow were 'cut off abruptly', in a straight edge, at the level where they met the overlying horizontal strata. Hutton rejects a number of other possibilities and then advances his own explanation. The strata which are now vertical were, like nearly all rocks, laid down horizontally, beneath the sea. As they were upheaved to form land, they were twisted into the vertical. Then

> by the effects of either rivers, winds, or tides, the surface of the vertical strata had been washed bare; and ... this surface had been afterwards sunk [beneath the sea] below the influence of these destructive operations, and thus placed in a situation proper for the opposite effect, the accumulation of matter prepared and put in motion by the destroying causes. *(Hutton, [1795] 1959, vol.1, p.435)*

That is to say, the upheaved vertical strata had been planed down by erosion, and had sunk again to the bottom of the ocean to become the bed upon which a new set of horizontal strata began to accumulate. Hutton fortifies this suggestion by pointing to the layer of boulders and stones that occur at the intersection of the two sets of strata: they are, he claims, fragments of the lower, vertical series, which became detached during the long period of erosion.

Now, this may all look a bit confusing to a reader unfamiliar with geology, or commonplace to a reader who knows the basics of the science, but it is worth spelling it out, in order to show the confidence with which Hutton could, with perfect equanimity, contemplate the building and erosion of huge landmasses. In the case of Jedburgh, he postulated the following sequence. There was once an ocean, where Jedburgh now stands, in which collected both the detritus of the neighbouring landmass and the detritus of tiny marine organisms. Horizontal beds of rock, composed of this detritus, were consolidated at the bottom of the ocean. Then, there was a period of upheaval which twisted and raised these beds vertically above the sea, where they were exposed to weathering and erosion for sufficient time for them to be planed down to a level. A period of subsidence followed, during which the rocks sank below the ocean again. A new sequence of horizontal sedimentary rocks consolidated on the base of the old, subsiding rocks. Lastly, the whole mass was upheaved yet again. Finally, the unconformity revealed itself to Hutton in the spectacular section cut by the humble river Jed. 'Finally', though is the wrong word to use, for Hutton said that there is 'no prospect of an end', the forces that wrought these titanic changes are still at work and will eventually drastically remodel the Borders landscape.

It took the best part of a century for Hutton's vision, transmitted through later geologists, to be sanctified, as it were, by the elite English culture and embodied famously in the verse of the English Poet Laureate Tennyson in the most widely read poem of the nineteenth century:

> There rolls the deep where grew the tree.
>   O earth, what changes hast thou seen!
>   There where the long street roars, hath been
> The stillness of the central sea.
>
> The hills are shadows, and they flow
>   From form to form, and nothing stands;
>   They melt like mist, the solid lands,
> Like clouds they shape themselves and go.
>
> (In Memoriam, *1850, section 123, lines 1–8*)

11.3 Joseph Black

Hutton can in many ways stand as a representative of the intellectuals of the Scottish Enlightenment. But they were not entirely homogeneous in their intellectual and religious outlooks. The chemist Joseph Black (1728–99) was a close friend of James Hutton (and Adam Smith), but the two men were quite different. Whereas Hutton was robust and disorganized, Black was pallid and precise. Hutton operated outside the universities, but Black was a lifelong academic. If Hutton gained his interest in geology from his industrial and farming activities, Black came to chemistry from his medical studies. Whereas Hutton was keen to speculate about the origins of the earth, even calling his book *Theory of the Earth*, Black insisted that it was only the facts that counted, and deplored all speculation and theorizing. Similarly, Hutton (like Black's colleague Cullen) made no secret of his deism, but Black's religious views remain an enigma even today and they played no part in his

Figure 11.5 James Hutton (1726–97) and Joseph Black (1729–99). (From J. Kay, A Descriptive Catalogue of Original Portraits, Edinburgh, 1836. Photo: National Library of Scotland, Edinburgh.)

scientific work. It would therefore be rash to assume that a case-study of a single figure, even one as illustrious as Hutton, can provide us with a complete picture of Scottish science in the eighteenth century. What light does Black's scientific activities shed on the Scottish Enlightenment and what were his major contributions to the development of European science?

Joseph Black was born in April 1728, not in Scotland but in France, the son of an Ulsterman, who was a wine merchant in Bordeaux, and his Scottish wife. After four years' education in Belfast, Black went to Glasgow University at the age of sixteen. Pressed by his father to choose a profession after he completed his arts course in 1748, Black decided to take up medicine. Black was not particularly interested in becoming a physician, but the medical course enabled him to continue the study of natural philosophy under the new lecturer in chemistry, William Cullen (1710–90). This was a crucial step in Black's career for Cullen was one of the first teachers of chemistry in British medical school to base his course on the general principles of chemistry, rather than *materia medica*.

11.3.1 Early research in Edinburgh

Magnesia alba

After four years with Cullen in Glasgow, Black transferred to Edinburgh to complete his medical studies. He then needed to select a topic for his MD dissertation, one which would involve chemistry, be of topical interest, and also touch upon a medical question. He decided to study the nature of causticity, the corrosive character of alkaline substances, such as quicklime (calcium oxide). He wrote to his father in December 1752 that he had chosen this topic because of a controversy between two Edinburgh medical professors, Robert Whytt (1714–66) and Charles Alston (1683–1760), stemming from their attempts to use limewater (a solution of calcium hydroxide in water) as a chemical means of dissolving excruciating urinary stones (Donovan, 1975, p.172).

Rather than become directly entangled in a dispute between two professors, Black chose another alkaline substance for his own investigations. This was magnesia alba (magnesium carbonate), which was of medical significance because it was taken (and is still widely used) for acidic indigestion and, to quote Black, 'it mildly loosens the bowels' (quoted by Donovan, 1975, p.193). This was important in a period when overeating of the wrong things and drinking often caused indigestion and constipation. His thesis, *De humore acido a cibis orto et magnesia alba* (Of the acid humour produced by food and of magnesia alba), was printed in June 1754. He did not achieve his original aim of producing a substitute for limewater by roasting magnesia and treating the product with water, because magnesium oxide, unlike quicklime, is totally insoluble in water. Nonetheless, Black carried out about thirty chemical experiments on magnesia and calcinated magnesia, which he called magnesia usta. The tentative and disappointing results of Black's thesis were transformed a year later in an essay he read to the Philosophical Society of Edinburgh entitled 'Experiments on magnesia alba, quicklime, and other alcaline substances', in which he extended his investigations to quicklime and potash.

Fixed air

It was well known that 'air' was given off by magnesia (or limestone) when treated with acids. Black sought to show that this 'air', which he called 'fixed air' (carbon dioxide), is also lost when magnesia is heated. Hampered by practical difficulties in his efforts to collect the fixed air liberated during the heating of magnesia, Black used a series of chemical reactions to prove his argument. He dissolved the magnesia usta in sulphuric acid to produce a solution of Epsom salt. This solution was treated with fixed alkali (potassium carbonate), which precipitated magnesia. This regenerated magnesia, after being washed and dried, had the weight and the properties of the original compound. As very little 'air' was given off during this sequence, the fixed air in the fixed alkali must have ended up in the magnesia. Black confirmed this by treating magnesia with sulphuric acid and then measuring the weight lost during this reaction, which was equal to the weight loss during calcination.

Black also noted that quicklime does not absorb ordinary air, but only the small quantity of fixed air contained in it. This implied that there were at least two chemically distinct 'airs', and Black knew that fixed air extinguished a candle. However, he was not interested in the chemical behaviour of gases, and although he carried out experiments which revealed that birds were unable to breathe in fixed air, he did not make any further contributions to the pneumatic chemistry he had so ably helped to found. (See Chapter 15 for the development of pneumatic chemistry in the second half of the eighteenth century.)

11.3.2 Heat research

Andrew Plummer (*c.* 1698–1756), the chemistry professor at Edinburgh, suffered a stroke in 1755, and the Town Council appointed Cullen as his conjoint professor, without consulting the stricken Plummer. Black, who had covered for Plummer until Cullen arrived, was appointed to Cullen's position at the University of Glasgow. This move also marked a change in the direction of Black's research. He now began to investigate the nature of heat, a central topic in eighteenth-century chemistry. It is important to realize that most chemists in this period regarded heat as a substance, if perhaps one without measurable weight, and the study of heat was therefore considered an appropriate field for chemists. Hermann Boerhaave devoted a long section to 'fire' in his famous *Elementa chemiæ* (1732). In his lectures, Cullen listed 'fire' as the second primary cause of chemical change, after the elective attraction (chemical affinity) – precisely the order of Black's research (Donovan, 1975, p.131). Black doubtlessly believed that some form of chemical combination took place between heated materials, such as water, and heat. At the same time, however, he was even more reluctant to hypothesize than Cullen. His work on latent and specific heats was not based on any theoretical foundation, except for a belief that substances possessed a capacity to take up heat. It is thus unwise to regard Black's research as constructing a *theory* of heat. Black simply sought to make clear the manner in which a given substance, most notably water, absorbed heat. This was in keeping with the Enlightenment philosophy that it was important to establish the causes of natural phenomena by examining the facts, without resorting to speculative assumptions or 'hypotheses'. As he later explained to his former assistant John Robison (1739–1805), he considered 'every hypothetical explanation as a mere waste of time and ingenuity' (in Robison, 1803, vol.1, p.vii).

Latent heat

The origins of Black's interest in the phenomenon of melting have been the subject of some debate. John Robison remarked, in his edition of Black's lectures, that Black had been struck by the simple fact that snow does not melt instantly on a sunny winter's day nor does a sharp night-time frost cause ponds to form thick layers of ice immediately (Robison, 1803, vol.1, pp.xxxvi–xxxvii). It is now generally agreed, however, that Black's interest in heat arose from his study of the temperature changes which take place when salts dissolve in water. Some salts give out heat, while others produce cold, and these differences forced him to think about the more general question of aggregation and heat. Several scholars, notably Henry Guerlac (1982, pp.15–16), regard Black's reflections on the observation of supercooling by Daniel Fahrenheit (1686–1736) as the crucial factor.[1] Arthur Donovan (1975, pp.224–5) argues that Black would have perceived a link between the fixing of 'air' by quicklime and the fixing of heat (so that it is no longer registered by the thermometer) by ice.

However he came to the question of why ice does not melt immediately the temperature rises above freezing, Black's experimental programme is clear. If the temperature – as measured by a thermometer – does not change while the ice is melting, can we be sure that the thermometer bears any relationship to heat at all, and if the temperature does not change, how can we measure the *quantity* of heat taken up by the ice? Black was able to confirm that a mercury thermometer was a reasonably accurate record of heat changes when no change of state occurred, by mixing equal volumes of hot and cold water and assuming that the temperature of the mixture was the average of the initial temperatures.

But how could the heat entering the melting ice be measured with the thermometer? Fortunately, Black recalled an experiment that a Scottish physician George Martine (1702–41) had published in 1740. He had put two

---

[1] Supercooling is the phenomenon whereby the temperature of undisturbed chilled water can fall below 32°F without freezing, but when the water is shaken, the thermometer rises to 32°F and remains there until all the water has frozen.

thin glasses, one containing water and the other mercury, in front of a fire; if the fire is a steady one, the quantity of heat entering each vessel should be the same. Black adapted the idea by measuring the rise in temperature of water in one glass, while ice was melting in another one.

He had to wait for the winter to arrive so he could obtain the necessary ice, and the key experiment was made in December 1761. One glass contained water that had been frozen using a snow and salt mixture and the other held water that had been chilled to 33°F; the room temperature was 47°F. After half an hour, the water temperature had risen to 40°F, but the ice took ten and a half hours to reach the same temperature. Black calculated that the extra heat required to melt the ice (its *latent heat*[2]) was equal to the heat required to raise the temperature of the water by 140°F. He then carried out a different experiment, which he later described as an 'obvious method' (Black, 1803, vol.1, p.122). He made a small block of ice, which was placed in hot water. Within a few seconds, the ice had melted and the temperature of the water had fallen from 190°F to 53°F. The ice, the mixture of melted ice and water, and the empty glass were all weighed. With this information, Black recalculated the latent heat of ice and the result this time was 143°F.[3]

Heat of vaporization

Black read a paper on these experiments to the Glasgow Literary Society in April 1762, and then turned to the investigation of vaporization. For reasons he himself found difficult to explain, Black was initially reluctant to accept that there was a similar heat of vaporization. This was in spite of the fact that he (and presumably many cooks) had observed that it took far longer to boil off water than it takes to raise water to boiling point. In October 1762, he devised a very simple experiment to measure the heat of vaporization. He took a flat-bottomed tinplated pan and heated small quantities of water in it, using a steady furnace. Knowing the initial temperature of the water (50°F), the time it took to reach boiling point (four minutes) and the extra time it took to boil off (twenty minutes), he could calculate the heat of vaporization. The quantity he obtained was 810°F.[4]

Almost exactly two years later, Black, and his student William Irvine, carried out the reverse experiment, namely the determination of the heat liberated when steam is condensed to water. Once again, Black displayed his penchant for the simplest apparatus. He used an ordinary laboratory still fitted with a condenser filled with water (at 52°F). The quantity of water condensed was measured and found to be at 132°F. The temperature of the water in the condenser was at 123°F. From this data, Black and Irvine calculated that the latent heat of steam was at least 774°F. This was obviously too low, but it was close enough to the 810°F Black obtained for the conversion of water to steam to show that the two processes were probably equal and opposite.

Black's work on the heat of vaporization provides us with an early example of the interaction between science and technology, because Black and Robison were close friends of James Watt (1736–1819), the pioneer of steam power. Watt was born in Greenock, but he trained as a scientific instrument-maker in London. On his return to Glasgow in 1757, he was appointed instrument-maker to the university, probably through his friendship with Professor

---

[2] Black devised this term, from the Latin *latet*, 'hidden' (Robison, 1803, vol.1, pp.xxxvii).

[3] The average was therefore 141.5°F, or 330 KJ/Kg in the modern SI system, close to the currently accepted value of 336 KJ/Kg.

[4] This is equivalent to 1890 KJ/Kg, rather less than the modern value of 2268 KJ/Kg.

Robert Dick, who may have introduced Watt to Black. Black and Watt entered a partnership with Alexander Wilson, later professor of astronomy, in November 1758.

Specific heats

Finally, we must consider Black's contribution to the discovery of specific heats, the fact that different substances take up heat at different rates. Two experiments on mercury and water had indicated the problem. Fahrenheit had found that mixing equal volumes of mercury and water produced a striking result. If the mercury was initially hotter than the water, the temperature of the mixture was less than the average, and the reverse was true if the water was originally hotter. Martine's experiment, which we have met in connection with the latent heat of ice, shed more light on this matter. When two glasses, one containing water and the other an equal volume of mercury, were placed in front of a steady fire, the temperature of the mercury rose twice as rapidly as the water. Black was able to solve these riddles. Mercury clearly had a lower capacity for heat than water, and hence it heated up (and cooled down) more rapidly. As he never published his conclusions, we know very little about his thinking on this question, but he may have arrived at this solution because he regarded the absorption of heat as a chemical process, and hence a function of chemical composition, rather than density, or bulk (as Boerhaave had suggested).

11.3.3 The Edinburgh professorship

Whytt, the Edinburgh professor of medicine, died in 1766 and Cullen was chosen to succeed him, largely with the aim of freeing the chemistry chair for Black. Black's transfer to Edinburgh was well received, and he fulfilled these expectations by being an excellent and popular lecturer. However, the Edinburgh chair also marked the end of his active research. One looks in vain for any sequel to his research on magnesia or his work on heat. With hindsight, foreshadowings of this change can be seen in Black's Glasgow period. He refused to publish his work on heat, and it was only made public when an unauthorized version, based on his lectures, appeared in 1770. Furthermore, in his last years in Glasgow, most of the research work was done by his assistants, William Irvine and John Robison.

However, the reasons are not too hard to find. He did not draw a salary as a professor, but had to rely on his lecture fees, and hence the number of students attracted to his course. Black's stock-in-trade was the elegant (rather than spectacular) lecture demonstration. With over 120 lectures to prepare and deliver between November and May, it doubtlessly reduced Black's scope for research, given his indifferent health. Black had become increasingly worried about his health – he had a bad chest – and probably felt that he did not have to prove his talents in chemistry now he had achieved his ambition of an Edinburgh chair. Furthermore, he was an active physician, and while his private practice was small, he was also a manager of the Royal Infirmary and eventually a 'Physician to the King in Scotland', in addition to his work on the sixth to eighth revisions of the *Edinburgh Pharmacopoeia*.

His medical work was overshadowed by his growing role as an adviser to industry. To quote Robert Anderson, a leading authority on Black:

> Black was consulted by a considerable number of industrialists on an extraordinary wide range of topics. In the surviving correspondence these include sugar refining, alkali production, bleaching, ceramic glazing, dyeing, brewing,

metal corrosion, salt extraction, glass making, mineral composition, water analysis and vinegar manufacture. In addition his opinion was sought on agricultural matters. *(Anderson, 1986, p.107)*

For instance, Black suggested that caustic potash (potassium hydroxide), prepared by the action of quicklime on potash, was a better bleach for linen than potash or sour milk. At first, the authorities were concerned that caustic potash would weaken the cloth, but the Irish Linen Board permitted its use in 1770.

Black never changed the structure of his lectures from his arrival in Edinburgh until his retirement 30 years later. While he updated individual items over the years, the unchanging structure became an obvious handicap in a period when chemistry was transformed. Clearly, the pressure on Black's time and his poor health partly explain the lack of any thorough revision, but it was also a reflection of Black's lack of interest in theoretical chemistry. He presented the phlogiston theory propagated by the English pneumatic chemists in his lectures without any great enthusiasm; their speculative conjectures were not to his taste. However, he was equally chary of the new chemistry from France, especially its systematic nomenclature.

The agent of change was Sir James Hall (1761–1832), a pupil of Black and James Hutton, who visited Paris in 1786. The earlier influence of an uncle and the heady experience of meeting Antoine Lavoisier (1743–94) converted Hall to the new chemistry. On his return to Scotland, he gave a paper to the Royal Society of Edinburgh on 'M. Lavoisier's new theory of chemistry' in the spring of 1788. Hutton defended the phlogiston theory in a later paper, but Black was characteristically silent. However, Lavoisier wrote to Black in September 1789 to inform him that he had been elected a foreign member of the French Academy of Sciences. It appears from a second letter from Lavoisier in July 1790 that Black had spoken guardedly in favour of Lavoisier's ideas. In a warm response to this second letter, Black declared his support for Lavoisier's chemistry, despite a few 'difficultys', and confirmed that he had begun to teach it in his lectures (text of letter in Donovan, 1979, p.245).

Although Black was now in his sixties, his eloquence and his dexterity with apparatus could still command the admiration of Henry Brougham (later Baron Brougham) in 1796. This was the last course Black delivered, and he handed his lecturing duties over to his former student Thomas Charles Hope (1766–1844), who had been converted to Lavoisier's teachings by Sir James Hall in 1788. Black's health now began to fail altogether, and he died suddenly in 1799.

Black had built up the reputation of the teaching of chemistry at the University of Edinburgh, but it did not continue to prosper after his death. Part of the blame must be laid at the feet of his successor, Hope, who has been described as 'dull, pompous and uninspiring' (Anderson, 1986, p.112). The Edinburgh tradition of teaching and lecture demonstrations to the exclusion of original research meant that it was unable to meet the challenge from the research-based German universities, most notably Giessen, in the 1840s.

Black's failure to prepare Edinburgh for the nineteenth century, and his personal failure to build on his initial achievements, can be traced to his indifferent health and his personality. Adam Smith once described his close friend as 'cool and steady' (Mossner and Ross, 1977, p.207). Black was a cautious and fastidious man, with a desire for precision, who was not given to enthusiasm and rash actions, amongst which he appears to have numbered scientific publications. It is significant that his only important publication, 'Experiments upon magnesia alba', was a direct consequence of his MD thesis. This unfortunate mixture of indolence and coolness limited Black's contribution to the Chemical Revolution.

Black's work on latent heat laid the foundations for Lavoisier's theory of heat as a weightless chemical element, caloric. But Black was more than an intellectual bridge between Newton and Lavoisier. By treating heat as a measurable quantity, which could be transferred from one body to another, Black paved the way for the development of thermodynamics, the science of heat, in the nineteenth century.

11.4 Conclusion

We have studied James Hutton and Joseph Black separately, but they can be properly understood only if they are considered as part of the close-knit community of philosophers and scientists which also included Adam Smith, David Hume, William Cullen and Dugald Stewart. For nearly seventy years of the eighteenth century, this group produced an intellectual ferment which placed Scotland at the forefront of the European Enlightenment.

By the end of the eighteenth century, Scotland had a mature scientific community, producing work which fed into both the wider European scientific and medical networks, and into Scotland's own developing industrial economy. The members of this community shared a common belief in the importance of reason, the goodness of humankind, and the serenity of nature. Equally, they shared a zeal for the commercial and agricultural improvement of Scotland's and their own fortunes. They were pioneers in several fields, particularly medicine, chemistry, geology, philosophy and economics. The advances they made underpinned the Industrial Revolution and the American Revolution.

What was, at the beginning of the eighteenth century, a small, poor, politically and culturally disorientated country, had, towards the end of that century, achieved a commanding status as one of the European centres of Enlightenment thought and practice.

Sources referred to in the text

Anderson, R.G.W. (1986) 'Joseph Black', in D. Daiches et al. (eds), *A Hotbed of Genius*, Edinburgh University Press.

Barfoot, M. (1990) 'Hume and the culture of science', in M.A. Stewart (ed.), *Studies in the Philosophy of the Scottish Enlightenment*, Oxford University Press.

Black, J. (1803) *Lectures on the Elements of Chemistry*, ed. John Robison, 2 vols, London, Longman Rees and William Creech.

Buchan, W. (1769) *Domestic Medicine*, Edinburgh, Barfour, Auld and Smellie.

Cameron, J.K. (1982) 'Theological controversy: a factor in the origins of the Scottish Enlightenment', in R.H. Campbell, and A.S. Skinner (eds), *The Origin and Nature of the Scottish Enlightenment*, Edinburgh, Donald.

Chitnis, A. (1976) *The Scottish Enlightenment*, London, Croom Helm.

Christie, J. (1974) 'The origins and development of the Scottish scientific community', *History of Science* 12, pp.122–41.

Daiches, D. (1986) 'The Scottish Enlightenment', in D. Daiches et al. (eds), *A Hotbed of Genius*, Edinburgh University Press.

Devine, T.M. (1982) 'The Scottish merchant community, 1680–1740', in R.H. Campbell and A.S. Skinner (eds), *The Origins and Nature of the Scottish Enlightenment*, Edinburgh, Donald.

Donovan, A.L. (1975) *Philosophical Chemistry in the Scottish Enlightenment*, Edinburgh University Press.

Donovan, A.L. (1979) 'Scottish responses to the new chemistry of Lavoisier', in R. Runte (ed.), *Studies in Eighteenth-Century Culture*, vol.9, University of Wisconsin Press.

Emerson, R. (1979) 'The Philosophical Society of Edinburgh, 1737–1747', *British Journal for the History of Science*, 12, pp.154–91.

Fleming, J. (1962) *Robert Adam and His Circle*, London, Murray.

Gay, P. (1973) *The Enlightenment: An Interpretation*, 2 vols, London, Wildwood House.

Guerlac, H. (1982) 'Joseph Black's work on heat', in A.D.C. Simpson (ed.), *Joseph Black, 1728–1799*, Edinburgh, Royal Scottish Museum.

Hutton, J. ([1795] 1959) *Theory of the Earth*, 2 vols, Codicote, Wheldon and Wesley.

Jones J. (1986) 'James Hutton', in D. Daiches et al. (eds), *A Hotbed of Genius*, Edinburgh University Press.

Morrell, J. (1974) 'Reflections on the history of Scottish Science', *History of Science*, 12, pp.81–94.

Mossner, E.C. and Ross, I.S. (1977) *The Correspondence of Adam Smith*, Oxford, Clarendon Press.

Phillipson, N. (1973) 'Towards a definition of the Scottish Enlightenment', in P. Fritz and D. Williams (eds), *City and Society in the Eighteenth Century*, Toronto, Hakkert.

Phillipson, N. (1981) 'The Scottish Enlightenment', in R. Porter and M. Teich (eds), *The Enlightenment in National Context*, Cambridge University Press.

Robison, J. (1803) Introduction, in J. Black, *Lectures on the Elements of Chemistry*, ed. J. Robison, 2 vols, London, Longman Rees and William Creech.

Shepherd, C. (1982) 'Newtonianism in the Scottish universities in the seventeenth century', in R.H. Campbell and A.S. Skinner (eds), *The Origin and Nature of the Scottish Enlightenment*, Edinburgh, Donald.

Smout, T.C. (1969) *A History of the Scottish People*, Glasgow, Collins.

Further reading

A most attractive collection of studies, to which this chapter is indebted has been edited by David Daiches et al. (1986). The beautifully illustrated volume contains a survey of the Scottish scene and studies of Hume, Adam Smith, Hutton and Black. It also contains suggestions for further reading.

Anderson, R.G.W. (1982) 'Joseph Black: an outline biography', in A.D.C. Simpson (ed.), *Joseph Black, 1728–1799*, Edinburgh, Royal Scottish Museum.

Daiches, D. et al. (eds) (1986) *A Hotbed of Genius*, Edinburgh University Press.

Doyle, W.P. (1982) 'Black, Hope and Lavoisier', in A.D.C. Simpson (ed.), *Joseph Black, 1728–1799*, Edinburgh, Royal Scottish Museum.

Guerlac, H. (1970) 'Black, Joseph', in the *Dictionary of Scientific Biography*, vol.2, New York, Scribner's.

Guerlac, H. (1957) 'Joseph Black and fixed air', *Isis*, 48, pp.124–51, 433–56.

Science on the Fringe of Europe: Eighteenth-Century Sweden

Chapter 12

by Colin A. Russell

12.1 On the margins of Europe

Most people know where Sweden is geographically: in area one of the largest countries in Europe, occupying today the eastern part of the Scandinavian peninsula, with a long seaboard on the Baltic. It extends northwards into the Arctic Circle and from the Middle Ages to 1809 included most of Finland as well. That said, it must be admitted that not many non-Scandinavians are aware of the part it played in European history and still less of its science. Even for some professional historians it is one of those 'kingdoms of Northern and Eastern Europe' that are conveniently ignored in mainstream studies as being rather on the fringe, politically and culturally as well as geographically. The wisdom of that strategy must be judged by specialists in political and cultural history. What is certain, however, is that to ignore Swedish *science* is not only to forget some of the most important advances in recent centuries but also to deprive oneself of the chance to examine the development of scientific ideas and institutions under unusual and unfamiliar conditions. The conclusions may well have an importance far greater than expected.

Science in Sweden, as elsewhere, was at first indissolubly linked to the Church. To take one example, the first (and for long the only) source of knowledge outside Sweden of that country's natural features was the *Historia de gentibus septentrionalibus* (History of the northern peoples) (1555) of Olaus Magnus (1490–1557), Archbishop of Sweden.[1] However, the history of Swedish science is inseparable from the growth and development of the Swedish nation itself. After many turbulent changes in the Middle Ages the Scandinavian peoples of Norway, Sweden and Denmark entered the fifteenth century as a united monarchy under the Danish queen. Swedish resentment at their subjugation culminated in victory over the Danes at the battle of Brunkeberg in 1471 and thereafter a new sense of Swedish nationhood. After a further brief spell of Danish rule in the next century the Swedes re-established the monarchy in the person of their victorious leader Gustavus Vasa (*r.* 1523–60). Under his dynasty, which lasted until 1720, Sweden became a great power and remained so through the seventeenth century.

It intervened in the Thirty Years' War, by mid-century had invaded Poland and Denmark, taken on both Pope and Emperor, and by 1699 was led by Charles XII into the great and eventually disastrous War of the North. At all times two considerations were paramount for Sweden: maintaining and extending its control of the lands bordering the Baltic, and the personal ambitions of its absolute monarchs, several of whom it must be said showed strong signs of mental instability. Attempting to invade Russia in 1708–9, Charles XII was not the only leader to experience defeat by the Russian

---

[1] This is said to contain the first printed reference to skis! As Sweden was now Lutheran, Olaus Magnus was Archbishop-in-exile, in Rome.

Map 12.1 Swedish lands in the eighteenth century

winter and army; by 1713 he had lost his German possessions, and in 1718 he was killed by a sniper while attacking Norway. In 1721 the Treaty of Nystadt saw not merely the secession of the Eastern Baltic countries to Russia but also the end of Sweden as a great European power.

Ironically, perhaps, it was the next half-century that was to witness the 'Age of Freedom' and at the same time the emergence of Sweden as a major scientific force in the Europe where its political humiliation had been so great.

The turbulent political and social changes of sixteenth- and seventeenth-century Europe were in part a legacy of the Protestant Reformation. Despite its comparative remoteness Sweden was far from unaffected by the convulsive upheavals within the Christian Church. Having shared in the prevailing Catholicism of the Middle Ages Sweden was set firmly in the

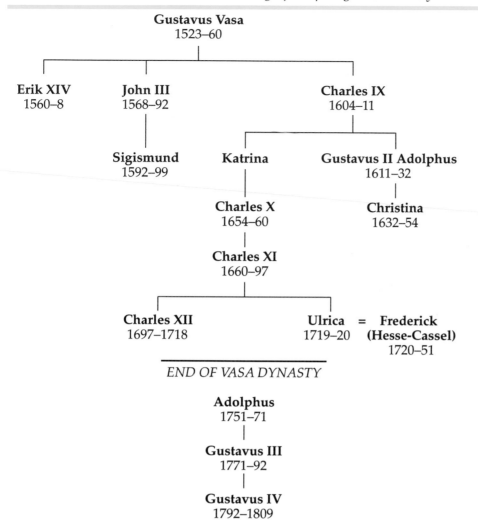

Gustavus Vasa
1523–60

Erik XIV
1560–8

John III
1568–92

Charles IX
1604–11

Sigismund
1592–99

Katrina

Gustavus II Adolphus
1611–32

Charles X
1654–60

Christina
1632–54

Charles XI
1660–97

Charles XII
1697–1718

Ulrica = **Frederick**
1719–20 **(Hesse-Cassel)**
1720–51

END OF VASA DYNASTY

Adolphus
1751–71

Gustavus III
1771–92

Gustavus IV
1792–1809

Figure 12.1 Swedish monarchs (with dates of reigns)

Lutheran camp by Gustavus Vasa. Of his three sons who succeeded him John tended towards the older branch of the Church, while active promotion of 'popish' institutions was the policy of his son Sigismund, who also managed to be King of Poland. Their Romanizing intentions were fiercely resisted by most of the clergy and eventually the absentee king was banished permanently to Poland and succeeded by his uncle Charles IX. The new king, and still more his successors, remained firmly committed to Lutheranism. On New Year's Day, 1593, the official Confessionist Church of Sweden was created within the Lutheran tradition (where it still remains). Thereafter unity in religion was seen as a prerequisite to national stability, a view that had profound consequences for science in Sweden, not least in the fortunes of the University of Uppsala.

12.2 The University of Uppsala

12.2.1 The emergence of a national institution

The fortunes of Swedish science are inextricably linked with those of the ancient university of Uppsala. The oldest university in Scandinavia, and for over a century the most northerly in the world, it was founded in 1477 as a manifestation of the rising national consciousness. The German universities were no longer to have a monopoly of providing education for Swedish students, any more than German academic immigrants were to be able to hold certain civic offices in Sweden itself. Uppsala, the ecclesiastical capital of Sweden, should be the site for a Swedish university. More precisely its foundation was a determinate effort to upstage the Danes who were known to be considering a similar move but who in fact were not able to establish their own university (at Copenhagen) until two years later.

On 27 February 1477, Jacob Ulvsson, Archbishop of Uppsala, received from the pope a charter to establish a university in his city. From the start the intention was to found a purely ecclesiastical establishment for the training of monks and priests for the Church. Science had the low profile of most mediaeval universities, although we are aware of the teaching of Euclid's geometry and Aristotle's physics. In astronomy the *Sphaera* of Sacrobosco was a prescribed text and by 1508 it was being taught by Petrus Astronomus, author of well-known astronomical *Tables* and maker of the famous clock in Uppsala Cathedral.

Probably because of its ecclesiastical character the University of Uppsala found it hard to survive the Reformation and subsequent tendencies to Germanize education. The increasing need for secular as well as religious education, combined with the remarkable success of the rival university at Copenhagen, led to royal efforts to inject life into the now somnolent institution at Uppsala. After a brief reawakening by Eric XIV in 1566 the university was promptly closed again by the joint efforts of his brother (the Catholic sympathizer John III) and bubonic plague. Despite opposition from John's similarly-minded son Sigismund the university was recreated on a new footing in 1593 by a synod of the newly-established Confessional Church and from now on became a bastion of Lutheran theology.

The university entered a new era of prosperity in the reign of King Gustavus II Adolphus, whose donations were on a magnificent scale. They included a very large gift of land, with over 300 farms on it. By 1620 Uppsala had become a major seat of learning, encouraged by successive monarchs, sometimes with such enthusiasm that chairs were created by royal *fiat* without the necessary means to finance them. Professors were imported from Germany and from the 1630s student numbers (though hard to estimate precisely) seem to have been unusually high. The Library, created by a donation from Gustavus II Adolphus in 1620, was augmented by collections captured from defeated Catholic enemies. Through piracy, if not through purchase, it was to become by mid-century one of the major libraries of Europe. Several splendid buildings which still adorn the University of Uppsala date from the seventeenth century. The Gustavianum, another gift from the king, was erected in 1622–5, its famous dome housing the Anatomy Lecture Theatre being added by Olof Rudbeck in 1662–3. He also laid out the botanical gardens later used by Linnaeus.

For a century, then, the University of Uppsala was a lively place with a growing European reputation and attracting increasing numbers of foreign students. There were the traditional facilities of theology, law, medicine and the liberal arts and sciences.

First impressions of the sciences are favourable. One yardstick of progress may be the attitudes to the new Copernican system. Some of Copernicus' own books found their way to Uppsala as trophies of war when the Frauenberg Library was captured. As early as 1593, when the university was reconstituted, his doctrines were taught by Laurentius Paulinus Gothus, professor of astronomy and eventually Lutheran Archbishop of Uppsala and

Figure 12.2 Gustavus Adolphus (1594–1632), as Gustavus II King of Sweden from 1611 to his death in the battle of Lützen. He was a generous benefactor of the University of Uppsala. (University of Uppsala)

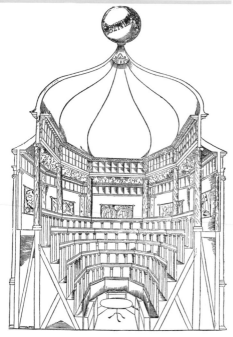

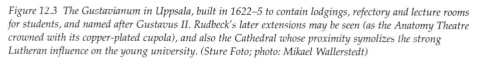

Figure 12.3 The Gustavianum in Uppsala, built in 1622–5 to contain lodgings, refectory and lecture rooms for students, and named after Gustavus II. Rudbeck's later extensions may be seen (as the Anatomy Theatre crowned with its copper-plated cupola), and also the Cathedral whose proximity symolizes the strong Lutheran influence on the young university. (Sture Foto; photo: Mikael Wallerstedt)

Figure 12.4 The famous Anatomical Theatre at Uppsala (1662–3), designed by Olof Rudbeck on lines similar to those in Leiden and Padua. The oval dissecting table was viewed by spectators standing in six tiered galleries, while important visitors were provided with seats at floor level. The theatre was in fact rarely used for its original purpose of public dissection since only bodies of executed criminals were available for that purpose. (From 'Atlantica' by Olaf Rudbeck, 1679)

Pro-Chancellor of the university. Copernicanism was stoutly defended by two other professors, Nils Celsius and Per Elvius. What is remarkable about these men, however, is that their defences were mounted in the last twenty years of the seventeenth century, a period when Copernican doubts had long been banished elsewhere.

The truth was that, for much of the seventeenth century, the University of Uppsala had been racked by dissension between two schools of thought, the Ramists and the Aristotelians, and for much of the time the latter were in the ascendant. An Aristotelian view of the universe was incompatible with any other that was not centred on an immobile earth, and that included Copernicanism. Ramus, or Pierre de la Ramée (1515–72),[2] was a French mathematician who sought to free astronomy from 'hypotheses', and for that reason preferred Copernicus to Aristotle. He found Swedish champions in several of the early professors and in the person of Johan Skytte, chancellor from 1622 to 1645, supported by his pro-chancellor the aged Gothus.[3] Yet, when it was waning elsewhere, Aristotelianism was actually on the increase in Sweden, and more and more of its supporters came to fill chairs vacated by the older Ramists – and this despite a prohibition of such philosophy by Skytte in 1640 and a peremptory order to the university bookshop to stock only Ramist books!

What had happened was the development of a new alliance between Aristotelianism and the Lutheran faith. This was a particularly Swedish phenomenon and stemmed partly from the perceived need for the

[2] Ramus was a supporter and correspondent of Rheticus (see Chapter 2).

[3] Originally the archbishop, by papal decree, was chancellor to the university, but in 1622 the king put his own man (Skytte) as chancellor, the archbishop becoming now pro-chancellor, a situation which lasted until 1934.

Figure 12.5 Olof Rudbeck (1630–1702), Swedish botanist and anatomist, professor of natural history and medicine at Uppsala and creator of the famous Anatomical Theatre. (University of Uppsala)

Figure 12.6 Illustration of Perdix Cinerea by Olof Rudbeck the Younger (1660–1740), showing the extent of naturalistic detail then becoming common. (University of Uppsala)

established Church to be a bastion of society and therefore rock-steady in its confessional attitudes. Just as in mediaeval Europe Aristotelianism had seemed an indispensable ally to Roman Catholicism, so now it appeared to offer a similar role to Lutheranism. There was, of course, nothing whatever in either the teachings of Luther, or of the New Testament from which they were derived, to justify such a *mésalliance*. The underlying rationale was not theological, but in the last analysis political, for the king was head of the Swedish Church, so stability of Church and State were closely interconnected. And thus it came about that a new scholasticism was to permeate the life and teaching of the faculties at Uppsala and for many decades significant scientific progress was inhibited.

The teachings of Descartes arrived rather late in Sweden, and with them a perception that Aristotelianism must be wrong and Copernicanism probably right. Descartes himself had been philosophical adviser to Queen Christina, and died in Sweden. But the defenders of the older cosmologies did not give up without a fight and the 1660s, 1670s and 1680s were marked by spectacular outbreaks of verbal violence at Uppsala. Sometimes students joined in the fun with windows, and occasionally bones, being broken. Not all the arguments were strictly about Aristotelianism or Copernicanism, of course, but these issues were seldom far below the surface. One of the most colourful figures in such disputes was Olof Rudbeck (1630–1702), a medical professor who overflowed with reformist zeal and became rector at Uppsala in 1661. He discovered the lymphatic system – the first major scientific discovery by a Swede (when he was twenty years old). He published a dissertation on Harvey's theory of the circulation of the blood (1652) and went on to study medicine at Leiden. He was a true polymath. Also a musician and composer, he wrote and directed a coronation piece for Charles XI; he executed biological paintings of considerable merit; he compiled a monumental work *Atlantica* to show that Sweden was the mother of the nations; he regarded himself as no mean architect and, as we have seen, left his mark permanently on the town of Uppsala. And Rudbeck would hurl himself into university disputes with explosive force. His career is an exaggerated indicator of the strengths and weaknesses of seventeenth-century Uppsala.

Rudbeck was a Cartesian. By the 1680s Swedish academics were prepared to join him (not that he seems to have been an ardent propagandist for Descartes). Only a few theologians remained in the Aristotelian camp, and with the advent of new scientific ideas from the south came the introduction of scientific instruments such as the air-pump, thermometer, and barometer. The time had come for Copernicus to be openly defended.

12.2.2 Uppsala in the Age of Freedom

With the death of Charles XII in 1718, and the humiliating termination of the wars in which Sweden had suffered economic ruin, the era of absolute monarchy also came to an end. The new democratic Age of Freedom (as the Swedes loved to call it) was now to be marked by mercantile rather than military expansion. It lasted until the accession of the King Gustavus III in 1772 and proved to be the half-century in which Swedish science first emerged as a national phenomenon. Much, though not all, of this development took place at Uppsala.

There are several paradoxes about the Age of Freedom. Despite the fact that Sweden had forfeited its position as a great European military power (or

Figure 12:7 Old Uppsala (University of Uppsala)

perhaps because of it?) the nation now entered a period of unprecedented prosperity and, as we shall see, scientific growth. Even more paradoxical are the fortunes of the University of Uppsala. Having been until 1718 the almost inalienable property of the Crown it might have anticipated a new-found liberation. Yet, for two reasons this was not to be the case.

The first inhibiting force was political. Two parties vied for government, the Hats and Caps. For most of the Age of Freedom the Hats were the ruling political party (from 1739 to 1765). Their commitment to a utilitarian philosophy and to mercantilism was deep but narrow. To reduce their trade deficit they sought by all means to promote manufacturing industry and agriculture. Their chief instrument was state subsidy, though they also recognized the importance of science in education. An Educational Commission of 1745–50 proposed that the youth of Sweden should be guided to economics and other practical 'sciences' and sought a radical change in the educational system to that end. Their proposals were too drastic for the University Senate, who threw them out. Inspection of their agenda reveals, however, a very restricted concept of science education, for the commission argued that all that was needed was teaching and that such scientific research as had been undertaken was no longer necessary! This attitude, combined with a scorn for all traditional learning, was hardly encouraging for a university. The party having, moreover, a low view of academic freedom it is not surprising that the University of Uppsala should soon have been the unwilling object of their attentions. They did not hesitate to meddle in university affairs and so Uppsala found itself the victim of popular-political instead of absolutist-royal manipulation.

The consequences were not entirely bad (as we shall see) but dependence upon government for senior appointments was, at best, vexatious, and, at worst, highly damaging. Moreover, by prohibiting the printing of Swedish works overseas they may have kept useful technical secrets to themselves, but they goaded to impotent fury an academic community that was beginning to understand something of the international character of science and learning to depend upon its contacts overseas. An infuriated Linnaeus (though himself a loyal Hat) was heard to vow that, were it not for his family, he would have accepted an appointment at Oxford where there was real academic freedom.

The second set of institutional inhibitions came from the Church. This was nothing new, of course, for Uppsala had long been an ecclesiastical

311

Figure 12.8 Samuel Klingenstierna (1698–1765), professor of mathematics and (later) experimental physics at Uppsala. (Royal Swedish Academy of Sciences)

ASTRONOMISKA OBSERVATORIUM

Figure 12.9 The first astronomical observatory at Uppsala, established by Celsius in Svartbäcksgatan and still standing (though rather changed) near the centre of the town's shopping area. (University of Uppsala)

Figure 12.10 Chemical laboratory of the University of Uppsala, in Västra Ågatan. It was completed by Wallerius in 1754, but nearly destroyed by fire in 1766, and is shown here just after its renovation in 1769–70 by Bergman. It may still be seen on its riverfront site. (University of Uppsala)

establishment. The need to preserve the state Lutheran Church from a Counter-Reformation or from the 'heretical' tendencies so visible in other Protestant countries led to an even narrower definition of orthodoxy than elsewhere, and to legal reinforcements of the protective measures inherited from the seventeenth century. It meant that the Church could still exercise much control on what was taught and written within the university. In practice this led to what amounted to almost a theological censorship of scientific matters. The scholastic attitudes of the mediaeval Catholic Church were reproduced in modified form, in the Lutheran Church of eighteenth-century Sweden.

Despite being squeezed between these political and theological millstones the University of Uppsala survived. It even managed to become the stage on which some of Sweden's first major scientific dramas were enacted (see Section 12.4). It became one of the few places where new and enlightening ideas from abroad could be discussed and where their influence could be felt (see Section 12.5). Also, in a curious way, the University of Uppsala was actually to *benefit* from political interference. For over twenty years the professor of astronomy Anders Celsius had been arguing for a chair in experimental physics (being more than willing to sacrifice the 'useless' professorship in Latin poetry for that purpose). Help was to come from an unexpected source.

Recognizing the need for well-trained government officials the Hats politicians had instituted 'civil service degrees', with a strong vocational element. For mining candidates these degrees were to include studies in physics and chemistry. In order to provide the necessary instruction they instituted, in February 1750, a professorship in experimental physics and another in chemistry; the former was filled by Samuel Klingenstierna (translated from his chair in mathematics) and the latter by Johan Gottschalk Wallerius. These were the first university appointments in science (other than astronomy, mathematics and medicine) in Sweden. Together with Celsius the two new professors were to join Linnaeus in making Uppsala a world leader in science. Already Linnaeus had restored the Botanical Gardens and Celsius had built a new observatory in Svartbäcksgatan; now Klingenstierna was to receive laboratory accommodation in the old hospital and Wallerius was to erect a purpose-built chemical laboratory at Västra Ågatan. Shortly afterwards a chair in practical economics (chiefly agriculture) was established, its occupant being referred to by less practically-minded colleagues as 'professor of manure'. Despite such philistinism towards

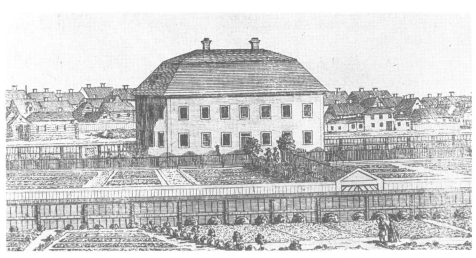

science and technology the University of Uppsala probably gained more than it lost from this unwelcome form of government interference.

It may seem paradoxical that the Age of Freedom did not witness an explosion of academic freedom in the University of Uppsala. Yet there was no inherent reason why political and academic freedoms should go together. To us this is surprising because we live on the other side of the 'Enlightenment'. There was in fact a rather selective osmosis of Enlightenment ideals and wider human values were not generally appreciated in Sweden until later than in Western Europe. This fact played a significant part in other aspects of Swedish science.

12.3 Other centres of Swedish science

Although the University of Uppsala was the chief centre of Swedish science in the eighteenth century it was not the only one. We can hardly do more than list a few of them.

12.3.1 Academies

The Royal Society of Science of Uppsala

In about 1710, when Uppsala was afflicted by the plague, and the university's lectures were suspended, certain professors responded to an idea of the industrialist and inventor Christopher Polhem and constituted themselves into a 'College of the Curious'. They met to discuss scientific matters and from the start were led by the university librarian (and later archbishop) Eric Benzelius. In 1728 it became the Royal Society of Letters and Science of Uppsala, the first such establishment in Sweden. With Celsius and Linnaeus among its secretaries it might have been expected to fulfil its early promise of lively discussions and printed *Transactions*. However events in 1739 diverted attention and interest from the Uppsala Academy, and its publishing activities ceased until much later in the century. It still exists with its journal *Nova acta*.

Figure 12.11 Eric Benzelius (1675–1743), second archbishop of Uppsala of that name and founder of what was to become the Royal Society of Science of Uppsala. (Royal Swedish Academy of Sciences)

Figure 12.12 The Palace of the Nobility, Stockholm, in about 1780. This large building on the right was the early home of the Royal Swedish Academy of Sciences. (Kungl. Biblioteket, Stockholm)

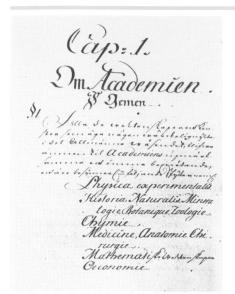

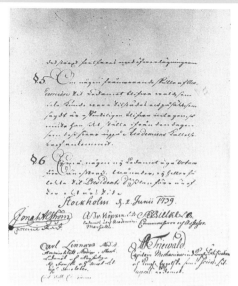

Figure 12.13 *The first and last pages of the Rules for the new Royal Swedish Academy of Sciences. Note the signatures of the founding members. (Royal Swedish Academy of Sciences)*

Figure 12.14 *Mårten Triewald (1691–1747), scientific lecturer and entrepreneur in technology. He introduced the Newcomen steam engine to Sweden (1726/7) and was a founder of the Royal Swedish Academy of Sciences. (Royal Swedish Academy of Sciences)*

The Royal Swedish Academy of Sciences

In 1739 Sweden joined the growing number of European countries with their own national academy: the Royal Swedish Academy of Sciences was established in Stockholm. Linnaeus was one of its five founders and its first president. Another founding member was Mårten Triewald (1691–1747), a prominent engineer who had been impressed by the example of the Royal Society during a visit to London. Before long its members included almost all the well-known names in Swedish science: Polhem, Linnaeus, Celsius, Wargentin, Klingenstierna, Swedenborg, Cronstedt, Bergman, Scheele, Berzelius and many others. It is now one of the world's leading scientific academies and (among many other things) awards and administers the prestigious Nobel Prizes in chemistry and physics.

From its beginning the Academy published a journal (the long-running *Vetenskapsakademiens handlingar*) and established a library (started by a donation by Linnaeus of his *Hortus Cliffortianus*). As first it had a strictly utilitarian purpose, being intended 'for the promotion of the useful sciences'. Election to its membership was a recognition of good service in the application of science. This was not necessarily limited to the scientific 'establishment'. Thus an early member was Eva Ekeblad, wife of the Hats' president, but in her own right notable for a short paper on the chemistry of potatoes (from which she discovered a distillation process to produce the highly alcoholic drink Brännvin). She seems never to have attended a meeting and was the only woman to be elected until 1945.

Figure 12.15 *Library plate of the Royal Swedish Academy of Sciences.*

The keynote of the early years of the Academy, for the rest of the eighteenth century in fact, was a practical kind of utilitarianism. This note was struck again and again by successive presidents. The industrialist Charles De Geer in 1744 stressed the economic advantages of entomology, for example, and in the following year the merchant Jonas Alströmer (1685–1761) spoke on 'Sweden's prosperity, if she wishes', stressing the development of manufacturing industry to begin to exploit the country's wealth of natural resources.

Figure 12.16 Jonas Alströmer (1685–1761), founding member of the Royal Swedish Academy of Sciences, experienced traveller and ardent advocate of Sweden's manufacturing industries. (The National Swedish Art Museums)

Figure 12.17 Pehr Wilhelm Wargentin (1717–83), assistant to Celsius at Uppsala, and from 1749 to 1783 secretary to the Royal Swedish Academy of Sciences. (Royal Swedish Academy of Sciences)

Both through its journal and its *Almanac* the Academy brought technical information about mining, agriculture etc. to a much wider audience than was (or is) normally the case with publications of a learned academy. By 1785 the *Almanac* had a circulation of nearly 300,000 copies. Only after 1818 did it lose its 'applied' image and become a more strictly scientific institution. In the previous century it nourished science by its journal, meetings and by plugging Sweden into the international network of scientific academies, including the Royal Society of London. Together with the University of Uppsala it became one of the two major centres of Swedish science for the rest of the century. There seems to have been little antagonism between them, a situation apparently unique in Europe.

A very influential person in the fortunes of the Academy was Pehr Wilhelm Wargentin (1717–83) who, from 1749 to 1783, was its longest-serving secretary. He had studied astronomy at Uppsala and was now to see the

Figure 12.18 A romanticized view of the astronomical observatory being constructed for the Royal Swedish Academy of Sciences (from its Proceedings *for 1749). (Royal Swedish Academy of Sciences)*

Figure 12.19 An astronomical motif from the Proceedings *of the Royal Swedish Academy of Sciences.*

Figure 12.20 Title page of volume 4 of the Proceedings *of the Royal Swedish Academy of Sciences. The recurring agricultural motif symbolized the Academy's utilitarian emphasis. (Royal Swedish Academy of Sciences)*

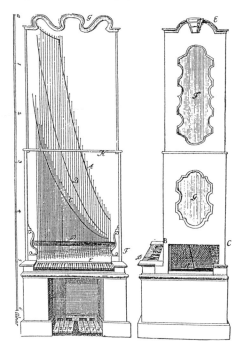

Figure 12.21 Upright clavichord by Nils Brelins: an engraving from the 1741 volume of the Proceedings *of the Royal Swedish Academy of Sciences, illustrating the interest in technology as well as science. (Royal Swedish Academy of Sciences)*

construction of the Academy's Observatory come to completion (1753). He became its first director. He was immensely hard-working in the Academy's service and under his leadership it prospered. However, after his death there was a slow decline in its fortunes. In large measure this seems to have been because his successor, Wilcke, though an excellent experimenter, did not have comparable skills in administration. Also many of the 'old guard' had died, some of the early expectations of economic progress had been dashed and there was in the country at large an air of disillusionment associated with the mysticism and later Romanticism that had had such a devastating effect on the fortunes of science in the University of Uppsala. But things were never as bad in the Academy and it entered the new century as the main centre of scientific excellence in Sweden.

12.3.2 Centres for industrial innovation

In accordance with the utilitarian spirit of the mid-century there was government encouragement for investigations likely to be relevant to industrial processes, most notably in the Bergskollegium (Board of Mines). This was one of the seven 'colleges' established in the early seventeenth century as government agencies for specific areas (law, commerce, finance etc.). They were collegiate, not in the sense of having students, but in the manner of their decision-making. At first the 'college' consisted of a *president* and several *assessors*, though later another grade of *counsellors* was added. One of the earliest of these was the Bergskollegium, or Board of Mines, formed in 1637 as a response to an extension of the great copper mine at Falun. It was the agent for government control in Sweden's mining industry. Its authority was almost absolute, and only in cases of homicide did it have to defer to the national criminal law. By 1700 it had enacted several hundred statutes dealing with a vast range of affairs from technology and marketing methods to the discipline of the workforce. Much remains to be discovered about the Bergskollegium and much confusion reigns over its changing role. However some recent work has thrown new light on the matter and revealed an important influence on the growth of at least some areas of Swedish science (Lindqvist, 1984).

Apart from a central administration in Stockholm, area boards and a section dealing with land and mine surveying, the Board had three divisions of obvious scientific importance:

Chamber of Assaying

Chemical Laboratory

Mechanical Laboratory.

The first of these, a Chamber of Assaying, was concerned with chemical analysis of metals so as to determine their purity. Several leading chemists were assayers here, including H.T. Scheffer and G. von Engeström. It consisted for much of the eighteenth century of one assayer whose lonely lot may have been relieved by the prospect (enjoyed by several of his predecessors) of promotion to the more populous Chemical Laboratory.

The Chemical Laboratory was established as long ago as 1639 though it had a chequered history for most of the next fifty years and, when it did function, appears to have been concerned with wider areas of chemistry than those directly connected to mining. In 1683 Urban Hjärne, a physician and chemical entrepreneur, received a royal appointment to the Laboratory as assessor and then director. Metallurgy then became the order of the day and some of the later directors were goldsmiths. Clearly some of its interests were now

overlapping with those of the Chamber of Assaying. Eventually some responsibilities of the Laboratory were transferred to the Master of the Mint.

One activity of the Chemical Laboratory was establishing a mineral collection. A rather similar function was performed by the Mechanical Laboratory, only here the collection was of working engineering models. It was established in 1700 in response to a suggestion by the inventor Christopher Polhem, just one of several such collections in Sweden. With Polhem its first director it served as a forum for both teaching and research. It started at Falun, where Polhem worked. In its first few years (1702–5) it witnessed Polhem's experiments on hydrodynamics, 'a very early example of science-based technology' (Lindqvist, 1984, p.71). Through recognition by the Bergskollegium Polhem was granted an annual salary, enabling him to provide the industry with tools, pumps and ore-raising equipment, to say nothing of larger civil engineering works for the country as a whole. He became known as 'the Archimedes of the North'. However his intense activity as an inventor led to neglect of the laboratory, and after his death it was replaced by the Royal Chamber of Models.

Figure 12.22 A painting of 1761 (by Sevenbom) showing (on the left) the building in Stockholm which housed the Board of Mines from 1674 to 1845. (Photo: Stockholm City Museum)

Figure 12.23 Christopher Polhem (1661–1751), inventor, technical director of the Falun copper mine and founder of the Mechanical Laboratory of the Board of Mines. (Royal Swedish Academy of Sciences)

Figure 12.24 Old Royal Palace in Stockholm, housing on its top floor the Royal Chamber of Models from 1757 to 1802. (Royal Library – National Library of Sweden, Maps and Prints Division)

Figure 12.25 Newcomen steam pumping engine at the Dannemora Mines, about 35 km north of Uppsala. The engraving (by Eric Beringius) accompanies Mårten Triewald's description of the engine, published in 1734. It displays some artistic licence in the interests of simplification and dramatic effect; thus the beams supporting the engine and other features have been turned through 90° and the chimney and flue through 180°. The human figures are all under-scale compared with the engine, and all have additional symbolic significance. (Engraving made by Eric Geringius in 1734 in Mårten Trewald's book Kort besrifning, om eld-och luft-machin wid Dannemora grufwor, *1734. Photo: Royal Swedish Academy of Sciences)*

The Royal Chamber of Models was founded in 1756 and brought together several separate collections, including those of the Bergskollegium, the Board of Commerce and the Royal Swedish Academy of Sciences. Located in the old Royal Palace in Stockholm, the Chamber became a place of pilgrimage for visitors from near and far, including France, England, Germany and even South America (cf. Chapter 14). Its function as a teaching institution was by example rather than formal lesson. It was particularly concerned with three problems: mechanization, fuel conservation and increased efficiency of machines. Pure science it did not support, but 'it provided tangible proof of the great importance attached by the State to technological development and education' (Lindqvist, 1984, p.29).

Not all initiatives for studying and teaching applied science came from the Bergskollegium, however. The Stockholm Mint, responsible for the country's coinage, eventually had its own laboratory with several distinguished chemists at its head. In 1747 the Swedish Ironmasters' Association was formed initially to supply information and financial advice to private iron-making firms. In response to concern about the products from the new blast furnaces it set up a bureau to supervise quality control, the five supervisors at the end of the century having all been trained in assaying at the Bergskollegium. Several individuals (as Rinman and Cronstedt) also set up their own chemical centres with small laboratories in the 1760s, the largest probably being that established at Falun by Johann Gottlieb Gahn, including a substantial library.

One effect of the establishment of mining laboratories was a quite spectacular growth in knowledge of mineralogy, making Sweden a very important European centre for that subject and others related to it, especially chemistry.

12.3.3 Other universities

It must be emphasized that Uppsala was not the only Swedish university in the eighteenth century. But for science it was by far the most important. In the reign of Gustavus II Adolphus universities were established in two of Sweden's Baltic provinces: at Tartu in Estonia and at Åbo (Finnish name Turku) in Finland.[4] In southern Sweden the University of Lund was founded in 1668. None of these reached scientific prominence in the eighteenth century, although Lund is famous as the place where Linnaeus commenced his studies (in 1727/8), and Åbo as the university of the very productive chemist Johan Gadolin (1760–1852) after whom is named the rare element gadolinium.

12.4 Some important Swedish achievements in eighteenth-century science

12.4.1 Linnaeus and his system

Carl Linnaeus (1707–78), who later was ennobled to von Linné, achieved world-wide fame in his own lifetime for successfully doing two of the things that many people find most boring in science: classifying and naming.[5] His lasting legacy to posterity is his Linnaean System of arranging and naming living creatures, whether plant or animal. Although its basis has long been known to be fundamentally flawed it still remains the starting-point for modern biological classification. Why was it so important?

For centuries plants (and to some extent animals) had been known by complicated names that were intended both to identify them and to convey something of their characteristics, place of origin or both. The rules of the game were hard to define and the result was a bewildering mixture of Latin names and folk-names. This was all very well with the relatively small

Figure 12.26 Carl von Linné (1707–78), otherwise known as Linnaeus, Sweden's greatest man of science in the eighteenth century, and father of modern biological taxonomy. (Royal Swedish Academy of Sciences)

---

[4] The Swedes also had Greifswald between 1631 and 1812; the university was not a Swedish foundation, though popular with Swedish students and staff.

[5] i.e. *taxonomy* and *nomenclature*, as they are technically known.

Some modern examples of Linnaean nomenclature

Cucumis sativus, cucumber

Rubus ideaus, raspberry

Cucumis melo, melon

Rubus fruticosus, blackberry

Rubus loganobaccus, loganberry

Cucumis and *Rubus* are genera and the other terms are species.

Monandria	1 stamen
Diandria	2 stamens
Triandria	3 stamens
Tetrandria	4 stamens
	etc.

(each class being divided into a definite number of orders)

numbers of organisms known in the Middle Ages. But with the Renaissance voyages of discovery multitudes of 'new' plants, and a good few hitherto unknown animals, now had to be adequately placed within the total scheme of all living things.[6] Frantic efforts to invent longer and longer names, each consisting of many words, merely added to the confusion for such names were often ambiguous and nearly always cumbrous. Linnaeus produced order out of chaos by devising a scheme in which each living thing was uniquely described by two words only – a *binomial* system. The first was its *genus*, the second was its *species*.

The terms are Latin (which made for universality in those days), but sometimes a stylized botanical 'Latin' of Linnaeus's own devising, thus avoiding the confusions encountered in Latin from the classical period or Middle Ages.

In order to classify the immense variety of plants then known Linnaeus had recourse to a theory that others had held before him but which he was to make all his own. It had been shown in 1694 by the German R.J. Camerarius that plants, like animals, have sexuality (reproducing through the sex organs which are stamens in the male and pistils in the female). Linnaeus encountered this discovery in reading *Sermo de sructure florum* (Speech concerning the structure of flowers) by Sébastien Vaillant (1717) to which he was introduced while still at school (the Gymnasium at Växjö) by his tutor Johan Rothman. Gradually he began to realize that plants may be effectively classified by the arrangement and number of these sexual organs.

> He went deeper and deeper into the study of biological reproduction and soon perceived what a superb aid to classification the reproductive organs offered. The sexual system of classification had been conceived. *(Lindroth, 1952, p.85)*

His system can be readily illustrated. Thus where stamens are of equal length and separate from the pistils plants are divided into the classes (see box).

The Linnaean system was conveyed to the world by many writings, above all by his *Genera plantarum* (1737), his *Species plantarum* (1753) and his massive *Systema naturae* which first appeared in 1735 and went through twelve editions in his own lifetime, in four volumes and 2,500 pages.

It must be said that Linnaeus did not find rigorous proof for his detailed theories about plant sexuality and, indeed, that he was not a great experimenter. What he did successfully accomplish was the provision of a systematic basis for all subsequent work in botany (and to some extent zoology). After a journey to Lapland in 1732 and a period in Holland from 1735 to 1738 (during which he visited France and England), his overseas travels were over. He was appointed to a medical chair at Uppsala in 1741, and his reputation rapidly became legendary. Students flocked to his lectures and many of them travelled far and wide to send back literally thousands of botanical specimens. These were all named and classified, and Linnaeus would often try to grow them in his own Botanical Garden. Naturalists from all over the world came to see him and Uppsala became for the last twenty years of his life the centre of European botany. When Linnaeus died all Sweden mourned, not least because his fabulous collection of specimens and books was to end in England, in the care of the institution later named in his memory, the Linnean Society.[7]

---

[6] This was true whether or not a 'great chain of being' was envisaged, stretching from the lowest of simple organisms up to the higher mammals and man. The point is that all naturalists had to incorporate their data within *some* overall scheme.

[7] The Linnean Society was founded in London in 1788 for the study of natural history. It continues to flourish.

Figure 12.27 Statue of Linnaeus (by Eldh) in the garden he restored at Svartbäcksgatan, Uppsala. (Sture Foto; photo: Mikael Wallerstedt)

Geniuses like Linnaeus cannot be classified into categories (like his own genera and species), nor can facile nationalistic explanations be given of their achievement. Nevertheless there seem to be several ways in which Linnaeus may be seen as particularly Swedish; he was, like all of us, a child of his own time and culture. More precisely, it is possible to identify at least three features of his work that characterized the science of his own nation in the eighteenth century.

First there was this Swedish sense of *utilitarianism*. His life's task was above all to prove to be useful (rather than, say, enlightening or provocative). He created a tool for which naturalists have ever since been grateful. So, we have reason to believe, was his intention. As a young man studying medicine his 'botanizing' expeditions were informed not only by the love of nature, but also by the knowledge of their practical value in the search for *materia medica*, for at that time most drugs came from plants. As one scholar has written:

> Linnaeus embraced the economic gospel of the times. The purpose of his journeys in the 1740s through the Swedish provinces, which were undertaken at the request of the Estates, was to make an inventory of all kinds of exploitable resources in the three realms of nature. (*Lindroth, 1976, p.129*)

In one other way a utilitarian note may be recognized in Linnaeus's approach to nature, and that is in the economy of terms needed for his nomenclature: only two per plant. It has been observed that he was a standardizer, using only a small number of words as a manufacturer may content himself with a small number of standardized parts – as was beginning to happen in armaments manufacture (Eriksson, in Broberg, 1979, pp.60–1). If this analogy means anything it points to a further assimilation of the utilitarian spirit of the age. Linnaeus's act of verbal frugality has been described by one scholar as 'an indexer's paper-saving device' (Stearn, 1957, vol.I, p.67).

The second characteristic of Linnaeus's work was that it was *ordered* or *systematic*. This could hardly be anything but an understatement. As was said in his own day 'God created, Linné organized'. For some reason this insistence on taxonomy was remarkably evident in Sweden. As far back as 1686 that country had gathered population data in an unusually systematic way (showing the remarkable fact that Sweden, alone of all European countries save Scotland, had an abnormally high female:male ratio, about 20:19). Then again the imagined world of Linnaeus's contemporary the Swedish mystic Swedenborg 'is so pathetically Swedish in its emphasis on order and the hierarchical system which reigns even between the immaterial spirits that, in short, one may call Swedenborg a Linnaeus in the world of spirits' (Lindroth, 1979, p.12). Even more obvious is the location in Sweden of the men who gave to mineralogy a thoroughly systematic basis, Cronstedt and Wallerius. Nor, by the nineteenth century, was chemistry exempt. J.J. Berzelius published in 1811 'An essay on chemical nomenclature' where he proposed a significant advance on the nomenclature of reforms of Lavoisier, giving each substance a systematic name, in Latin, and a comprehensive classification of all materials. Three years later he made this the basis also of 'A scientific system of mineralogy'. Berzelius the systematizer was to dominate European chemistry for the next two decades in a manner irresistibly reminiscent of Linnaeus. Berzelius's roots were back in the eighteenth century. Essentially empirical, this approach to nature is thoroughly Baconian (as it is in its utilitarianism), so it is hardly surprising that Linnaeus's system should fairly quickly have received a sympathetic reception in England.

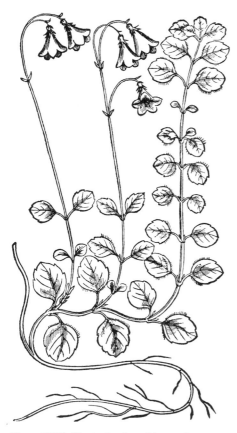

Figure 12.28 Illustration from Linnaeus's Species plantarum, *a facsimile of the first edition of 1753, vol.I. (Ray Society)*

Thirdly, Linnaeus's approach was fundamentally *static*, not dynamic. It was concerned, first of all, with a fixity of species. In that regard it was at first thoroughly Aristotelian (though without giving species an *eternal* duration). Later he was not so sure and permitted himself a certain degree of evolutionary speculation. But his approach was static in another sense, in that his operations could be reduced to a mere placing of objects in appropriate pigeon-holes. The operator's relationship with his objects was strictly passive, not interactional. Moreover, the basis of classification was morphological, not functional: how they appeared rather than what they did. Many of these features are characteristic of all taxonomy, but are particularly striking in the work of Linnaeus. When it became known in the salons of France and exposed to the spirit of the French Enlightenment it gained a very cool reception, and largely on account of its static nature. Buffon, who disagreed also about the fixity of species, opposed him bitterly. But there was surely another reason: 'To Buffon, and to everybody who reasoned like him, Linnaeus had killed living nature, whose real essence was dynamic force, change and everlasting movement in time' (Lindroth, 1979, p.15).

And if the Linnaean system was largely rejected in France its fate in Germany was less decisive. Although it appealed to the empirical stream of German science its static nature would be hard to reconcile with the dynamism of the new *Naturphilosophie* (a set of romantic attitudes to nature prominent in Germany at the end of the eighteenth century). But strong reactions had to wait until the new century. In Sweden itself only a few dared to argue along these lines, and they were followers of French thought. One was the industrialist and amateur naturalist Charles De Geer who had been educated abroad and whose massive seven-volume work on entomology displayed a contempt for catalogues and 'dead insects'. Another was the editor of the *Stockholmposten*, C.G. Leopold, a virulent critic of the University of Uppsala, of ancient tradition and especially of 'systems' whose sole function was 'to bore people'. There is no doubt whom he meant.

12.4.2 Chemistry

It is hard to exaggerate the importance of Swedish chemistry in the eighteenth century. For anyone familiar with the general history of chemistry a list of chemical members of the Swedish Academy of Sciences reads like a roll-call of honour. Equally, to enumerate the chief accomplishments of a relatively small number of men on the edge of Europe is to be reminded that science owes much more to Sweden than even the great work of Linnaeus. As late as 1886 no less than 40 per cent of the chemical elements found since the Middle Ages had been discovered in Sweden.

In addition an immense number of non-elementary substances was discovered in eighteenth-century Sweden, from formic acid (by dry distillation of red ants) to the pigment Rinman's Green. Many impure metals were described (including platinum) and the chemical nature of many minerals became clear. Rather than extending the list almost indefinitely we shall try to discover some underlying features of Swedish chemistry at this time. Certain generalizations do seem to be in order, including the following:

- Swedish chemistry stemmed largely from mineralogy;
- much of the chemical work was quantitative; and
- much experimental work centred on the use of the blowpipe.

Elements discovered by Swedish chemists

Element	Discoverer	Date
cobalt	Brandt	1730
nickel	Cronstedt	1751
oxygen	Scheele	1771
chlorine	Scheele	1773
manganese	Gahn	1774
molybdenum	Hjelm	1781
tantalum	Ekeberg	1797
selenium	Berzelius	1817
lithium	Arfwedson	1818
thorium	Berzelius	1828
vanadium	Sefström	1831
cerium	Mosander	1839
lanthanum	Mosander	1839
didymium	Mosander	1841
erbium	Mosander	1843
terbium	Mosander	1843
scandium	Nilsson	1879
holmium	Cleve	1879
thulium	Cleve	1879

Swedish chemistry stemmed largely from mineralogy

Apart from alchemy, the known ancestors of modern chemical science are medicine and mineralogy. The former included pharmacy and the latter included metallurgy. Whereas the University of Uppsala had a long tradition in medicine it is remarkable that most Swedish chemistry originated elsewhere (apart, that is, from the exceptional work of Scheele). The reasons are not hard to find.

Geologically speaking Sweden is a remarkably rich and diversified country. Its miners are therefore likely to encounter all manner of strange minerals, some of which hardly occur elsewhere. Given the insatiable collecting instinct of the eighteenth-century European, the Swedish propensity to organize and arrange and the unique wealth of material available it is not surprising that mineralogy should also find its Linnaeus. This happened with Wallerius (Section 12.2.2) whose *Mineralogia* of 1747 was 'the first properly arranged mineral system ever devised' (Lindroth, 1976, p.134). (Cf. Chapters 6 and 14.) Three years later he accepted the first chair of chemistry in Sweden. He was the first (in 1751) to introduce the now familiar distinction between pure and applied chemistry.

A detailed knowledge of minerals was promoted by the political atmosphere of utilitarian advance. Even before the Hats came to power the Bergskollegium in Stockholm had its own 'assessor' to evaluate the quality of materials, a public chemical laboratory for that purpose having been established by Charles XI in 1683. Chemists associated with quality control for mines included Hjärne (1641–1724), Swedenborg (1688–1772), Swab (1703–68) and Gahn (1745–1818).

The extraction of metals from minerals was important for export and home consumption. It involved much more than iron-making or copper-smelting (though at least one major chemist, J.G. Gahn, was connected with both). Precious metals used for coinage had to be prepared and then assayed, and a number of men got valuable chemical experience at the Mint in Stockholm, including Brandt (1694–1768), Engeström (1738–1813), and Hjelm (1746–1813). Such experience was not unique to Sweden but the ethos and some of the techniques were hardly found elsewhere.

Much of the chemical work was quantitative

It goes without saying that quality control on either minerals or precious metals must involve quantitative analysis. Europe had a long tradition of this in mine and Mint. It happened that in Sweden several important techniques were developed. Torbern Bergman, for example, stressed the need for pure reagents and of ensuring that products estimated (by weighing, for instance) were well-defined compounds. It is no accident that he, of all people, laid the foundation of quantitative chemistry by studying the 'affinities' of chemicals for each other (see also Chapter 15). He sought to discover how one acid (for instance) might displace another from a salt, and by measuring weights of reagents came near to an understanding of the laws of combining weights.[8] Bergman was probably the greatest analyst of the century.

---

[8] These in fact have nothing whatever to do with 'affinity' (an ill-defined concept in any case). Their study, called stoichiometry, led directly to the concept of equivalence and indirectly to the atomic theory of matter.

Much experimental work centred on the use of the blowpipe

The chemical blowpipe is a narrow tube through which a flame can be directed in a fine jet on to a hot substance. Under those conditions all kinds of reactions can be induced and subjected to close examination. The technique dates back at least to the seventeenth century. It was described in detail by Johann Kunckel who, though born in Holstein (1630), came to Sweden at the invitation of Charles XI as Minister of Mines and remained there until his death (1703). Thereafter it was occasionally used in Germany and elsewhere, but the locus of its real development was undoubtedly Sweden. It was used by Swab in 1738 and by Rinman in 1746 for the examination of antimony and of the ores of tin respectively. Cronstedt showed how minerals could be fused in its flame if mixed with substances like soda or borax, and this greatly extended its usefulness. So did the technique of placing the substances to be examined on a charcoal block, introduced by Engeström. Still further developments were introduced by Gahn. In 1779 Bergman published a comprehensive account of the technique of blowpipe analysis. Extensive European use of blowpipe analysis came only after the publication of Berzelius's book on the subject, published in 1820. Until then it was primarily a Swedish technique, and with its aid several new elements were discovered and much other progress was made in understanding the chemical nature of minerals. It was later and justifiably called 'the chemist's stethoscope'.

Despite the value of these generalizations about Swedish chemistry it is salutary to be reminded of an extremely important exception: a man of whom it is often said that he discovered more substances of fundamental importance than any other chemist. He was Carl Wilhelm Scheele (1742–86).

Scheele was an apothecary (pharmacist) in Stockholm and then Uppsala, at which place he made the acquaintance of Bergman and became his collaborator. From 1775 he lived in the small town of Köping. At Uppsala he discovered chlorine, a gas that would revolutionize the bleaching process in the future Industrial Revolution. He went on to isolate oxygen (independently of Priestley and Lavoisier). Other gaseous substances discovered included hydrogen fluoride, silicon tetrafluoride, hydrogen cyanide, hydrogen sulphide and arsine. It has been said that what Scheele did not breathe was hardly worth mentioning, and what he did breathe probably contributed to his early demise. Like Bergman he died in his forties. Chemistry was a hazardous subject in those days.

With the deaths of Scheele and Bergman in the 1780s Swedish chemistry fell into some decay until revived in the next century by Berzelius (and then not in Uppsala). For the University of Uppsala Bergman's passing meant the end of another era: the foreign scientists who had flocked to study, especially under Linnaeus and Bergman, were no longer to be a conspicuous feature of the student body, and Uppsala was to lose for 150 years its status as an international centre for science.

12.4.3 *The physical sciences*

The most notable figure in this field in eighteenth-century Sweden was Anders Celsius (1701–44). A member of an influential dynasty he became professor of astronomy at the University of Uppsala in 1730. Like many of his fellow-countrymen he travelled widely, in his case partly for astronomical observations. He studied the aurora borealis in Germany, and joined a French expedition to Lappland to measure the meridian arc (i.e. to discover whether

Figure 12.29 Anders Celsius (1701–44), early member of the Academy, professor of astronomy at Uppsala, and chiefly remembered today for the misuse of his name in connection with the centigrade thermometric scale which he used. (University of Uppsala)

the earth is slightly flattened at the poles – which it is). He established in 1737 his country's first astronomical observatory, conducted experiments on the relative brightness of stars, and examined terrestrial magnetism. In 1742 he described a mercury thermometer and described a scale in which the boiling and freezing temperatures of water are separated by 100° (and recognized that the former is pressure-dependent). This was more convenient than the 180° division of Fahrenheit (1714) or the 80° division of Reamur (1730). Within a few years of Celsius's paper a thermometric scale was in common use. Contrary to popular opinion, Celsius was not the first to invent the 'centigrade' scale, and the one in use today is not his! Celsius defined 0° as the *boiling* temperature of water and 100° as its *freezing* temperature.[9] But his fame is entirely deserved. One of his successors, Daniel Melanderhjelm, used Newtonian dynamics to calculate the equilibrium state of the universe.

Figure 12.30 Electrical machine of about 1740: static electricity was generated by placing a rubber against the rotating cylinder. This machine was probably the property of Samuel Klingenstierna. (University of Uppsala)

In physics Samuel Klingenstierna worked in geometrical optics, showing that chromatic aberration in glass lenses could be corrected (in contradiction to Newton). As a result the London instrument-maker John Dolland successfully constructed the first achromatic telescope. The European craze for electrical experiments was shared in Sweden and Klingenstierna obtained various 'electrical machines' (frictional generators) and other apparatus from England. Substantial work in this field was performed by Johann Carl Wilcke (1732–96) who from 1759 lectured at the Academy of Sciences in Stockholm and later became secretary of the Academy of Sciences. He examined the generation of electric charges by rubbing non-conducting materials together, and showed that two kinds of electricity are always produced. He inclined to a two-fluid theory of electricity, devised a primitive parallel plate capacitor, studied electrostatic induction, and worked on terrestrial magnetism. Most important of all were his discoveries of latent heat (1772) and specific heat (1781).

Figure 12.31 Johann Carl Wilcke (1732–96), Swedish pioneer in the study of electricity and of specific and latent heat. From 1784 he was secretary of the Royal Swedish Academy of Sciences. (University of Uppsala)

12.5 Science and religion

In the seventeenth century Sweden's established Lutheran Church controlled the University of Uppsala, and to that extent had a determinative role in most of the nation's science. With the extension of Lutheran scholasticism into the eighteenth century its influence remained considerable, although competing as time went on with pressures exerted by the politicians. In general terms its impact upon science grew less as scientific work was increasingly conducted outside any of the universities.

During the 1720s or 1730s visits of some Swedish academics to Germany led to an awareness in Sweden of the philosophical ideas of Christian Wolff (1679–1754). Born in Breslau, he became a student at Jena and then Leipzig, being specially interested in the application of mathematical methods to philosophical questions. He went to study with Newton's great continental rival, Leibniz, whom he later supported in his disputes with Newton and with whom he entered into a long correspondence. Wolff greatly extended Leibnizian ideas to form his own distinctive philosophy. The doctrines of Wolffianism involved a unified view of nature derived (as in mathematics) from a few fundamental axioms. (Ultimately these were the *principle of contradiction* and the *principle of sufficient reason*.) Method was everything. Human reason was supreme and, if only it were correctly applied, would

[9] The Celsius scale was probably inverted by the Academy's instrument-maker, Daniel Ekström.

lead inexorably to a world-view that included what we call science and theology. His attempts to be comprehensive soon got him into trouble when, in 1721, he spoke approvingly of the teachings of Confucius and he had to flee from Jena to Marburg.

Wolff's rationalistic creed was nevertheless conceived within a Christian framework and, indeed, regarded by its author as a legitimate tool for apologetics – i.e. for arguing the reasonableness of Christianity (as Locke would have put it). His method was that of *natural theology*, a characteristic theme in eighteenth-century Europe. He believed that a study of nature (in the natural sciences) could so display the evidence of intelligent design that only the most perverse would fail to discern the hand of God behind it all. Following Leibniz he considered this to be the best of all possible worlds, and thus supremely well constructed for the purpose of leading people to God. And, unlike other natural theologians of a deistic inclination, he paid more than lip service to the importance of *revelation*. In his view an important function of science was to disprove atheism, and also to provide a court to which all the struggling factions of the Christian Church could appeal.

These ideas, and the large corpus of writings in which they were expounded, gained wide circulation in Northern Europe as the century advanced, and they became particularly important in Sweden. They may have helped to maintain rigidity in the Lutheran Church. Here was a ready-made weapon for wielding against potential enemies. But it was also evident in the young students in Uppsala who cheerfully crossed and re-crossed what we should regard as the well-defined boundaries separating theology, philosophy, mathematics and the natural sciences. It is evident in the titles and contents of many dissertations from about 1730, but it acquired instant visibility with a public warning against new philosophical trends from the Chancellor, Cronhjelm, in 1732. His successor, Bonde, was less cautious and from the late 1730s until well after mid-century Wolffianism flourished at the University of Uppsala.

From science an early supporter of Wolffianism appeared in the person of Anders Celsius. In his own dissertation of 1728 he praises Wolff as 'the greatest philosopher of our time' and uses thoroughly Wolffian categories and arguments. Others followed, including several of Celsius's own students. Among them were the three brothers, Erik, Nils and Johan Gottschalk Wallerius.

Science reflected the Wolffian philosophy in a number of ways. Celsius was only the first prominent man of science to fall under its spell. In a thesis of 1732 (on the use of philosophy in promoting civil happiness) he argued for the priority of mathematics and physics amongst the sciences, and continually stressed the importance of *natural theology* (physicotheology).[10] For Celsius, Wolff was a stimulus both to a wide study of nature itself and to clear, logical thought in quantitative terms. Another student of the physical sciences was his student Nils Wallerius, a diligent experimenter who became a lecturer in physics at Uppsala in 1735. He was particularly interested in meteorology, and investigated the evaporation of water to such good effect that the prevailing theories of Leibniz on the subject were seriously challenged. Perhaps because he failed to achieve promotion he turned back to philosophy and theology and had become by mid-century one of the most aggressive upholders of the Wolffian orthodoxy in Church and university. It is probably going too far to suggest from the case of Wallerius that Wolffianism generally detracted from scientific endeavour by providing

---

[10] These terms will be familiar from earlier chapters, especially Sections 10.2.3 and 10.2.4.

alternative interests and occupations for disgruntled men of science. But since for several decades the University of Uppsala was riven with argument and debate over Wolffianism, chiefly among philosophers and theologians, it is not unlikely that in a few instances its diversionary effects were significant.

The question as to whether biblical and scientific doctrines were at variance loomed rather less frequently than we might have expected. Wolff himself denied vigorously that 'natural' and revealed religion were in conflict, and thought that the former should be used in the service of the latter. Similarly Linnaeus refused to believe that the 'two books' (of nature and Scripture) could ever be in contradiction. When eventually Wolffianism was attacked in Uppsala by Andreas Winbom (who occupied chairs in philosophy and theology) it was on the grounds of its methodology rather than its theological consequences.

To be sure there were areas of problematical interpretation, especially in the book of Genesis. Thus on the question of whether or not the six 'days' of creation were literal periods of 24 hours each Linnaeus simply said he was not sure. He accepted the story of the deluge but denied that it had been world-wide. As time went on he certainly modified his original view that the Ark contained all the species existing in his day, in accordance with his own work on cross-fertilization. But colourful stories of his 'timid' capitulation to ecclesiastical authority on this and other matters are often inventions of writers in the Victorian age.[11] As to the astonishing longevity of the Old Testament patriarchs (up to nearly 1,000 years) Linnaeus accepted it happily as an indicator of their healthier outdoor life-styles! On the other hand Swedenborg believed the astronomical 'year' in that epoch was shorter than ours. Where Linnaeus did clash with theologians it was often because he presumed to trespass on to their territory, if only to draw attention to the support given by natural theology (Lindroth, 1979, p.13). Such excursions would further undermine the social leadership of those embattled scholastics.

For Linnaeus, at least at the end of his life, the book of Genesis was to be interpreted in other than strictly scientific terms. The age of the earth, the extent of the Deluge, the fixity of species were all topics on which it was for science, not Genesis, to legislate. As early as 1730 he had claimed to set 'all prejudices' aside and become a 'sceptic', a word subsequently applied by conservative churchmen to Celsius, Klingenstierna and others holding similar attitudes towards Genesis. Yet the context of Linnaeus's remark suggests he meant something different from the 'Cartesian scepticism' attending the birth of much modern science; it implies that he 'simply wanted to return to Adam's original sound reason' (Broberg, 1979, p.33). And although some men of science may well have rejected Genesis *in toto* Linnaeus took it very seriously, even to the extent of relocating the Garden of Eden in a (probably) Swedish island, describing the Fall as the discovery of human sexuality and identifying the Tree of Knowledge as a banana tree whose leaves could cover human nakedness. Of his own divine vocation he felt increasingly certain (as Newton had done before him). His gratification at being called 'the second Adam' (because he, too, gave plants and animals their names) can well be imagined. Dr Broberg has written:

> Taking everything into consideration, Genesis was of enormous importance to him: it had inspired him to find paradise in nature, it had given his profession the status of *scientia divina*, and it gave his science its ultimate foundation. *(Broberg, 1979, p.39)*

[11] A classic (and notorious) case is the writing of A.D. White, especially pp.59–60 in the first volume of *A History of the Warfare of Science with Theology in Christendom* (1896).

If Swedish science in general was given a strong theological impetus in the eighteenth century, this happened especially through the promotion of natural theology, a characteristic feature of Wolffianism. Even here, however, other influences were at work, as may be demonstrated by the frequent references to the 'physicotheologies' of the English writers Ray and Derham, the latter being available in Swedish translation. There is indeed some similarity between Sweden and England in this respect. Wolff himself confessed special indebtedness to the English who, he said, had extended natural knowledge to the praise of God. These tendencies are especially evident in the work of the professor of philosophy at Uppsala from 1736, Petrus Ullén. He argued that God had created the universe with such care, with all of its component parts so interdependent, that it was not only the best possible, but also an eloquent testimony to his existence and goodness. Since he acted in a free and sovereign manner we must allow also for his right to introduce miracle, though normally the universe works through laws that he has also created. Man's duty, then, was to use natural theology not in a vain and speculative way but in a manner that emphasized the glory of God. The mandate for a careful and reverent study of nature was crystal clear.

That mandate was cheerfully accepted by many Swedish scientists who nevertheless steered clear of the convoluted arguments of their philosophical colleagues. A deep love of nature was often, if not always, combined with a simple piety that was more than mere convention. Linnaeus is an oft-quoted example, simply because he has been studied in such detail. How far his systematics owed anything to Wolff's approach is a moot point. Less uncertain is one origin of his natural theology, though here again caution is necessary.

> As a physicotheological author, Linnaeus was more advanced than most of the others ... He was not a Wolffian but may be regarded as a characteristic representative of the Wolffian spirit in a wider sense. All his interest in connecting natural science and religion and his endeavour to rebuke the atheists and deists with the aid of physicotheology were largely inspired by Wolff. *(Frängsmyr, 1972, p.233)*

The same spirit lingered on beyond the Age of Freedom and long after Wolffianism as such had disappeared from view. Linnaeus had bequeathed a nation-wide tradition of natural history and, in no small measure, natural theology. Towards the end of the century the young Berzelius offers an interesting example of the national trend; one of his teachers had been a student of Linnaeus and his step-father was a country clergyman with similar interests. On one occasion Berzelius had been reading a German classic, then popular in Sweden, Johann Arndt's *True Christianity*, which included a note of natural theology ('the creatures are the hands and messages of the Lord, and they shall lead us to Him'). His schoolboy reader confided to his diary a desire 'not only to become a better Christian, but also to change my way of living from now on', adding that 'On my way home I found some insects, and on coming home I rearranged my insect collection which is crowding the room and now includes 369 specimens, many of them rare ones'.

Such sentiments were not uncommon in eighteenth-century Europe, representing a combination of Christian piety, love of nature and Linnaean orderliness. Yet no understanding of Swedish science is complete that does not take into account these three intersecting themes.

12.6 The end of the century

From about the 1770s, Swedish science began generally to decline. Few major scientific advances were made and student numbers at the University of Uppsala were diminishing alarmingly. Some would add that science in Sweden has taken 150 years to recover. It is not at all clear why.

Towards the end of the century many of the great figures of Swedish science had passed from the scene, most notably of all Linnaeus. But unless one is committed to a particularly heroic view of science (where everything depends upon the great scientific genius) this offers no more than a very partial explanation. The geographical isolation of Sweden during the Napoleonic Wars can certainly account for some of its difficulties after about 1800, but hardly before.

One line of explanation has been to blame certain features of the Enlightenment. This complex movement (if that is what it was) presented some features that were positively beneficial to science, notably the early recognition of Lockean rationality and empiricism. Yet it is to certain aspects of the French Enlightenment that the decline of Swedish science has been attributed.

In 1753 Queen Ulrika created her own alternative 'academy', a kind of 'salon' for discussion of the distinctly French ideas to which she was greatly attracted. She was the sister of Frederick the Great of Prussia, and a correspondent of Voltaire and d'Alembert. In no sense was it 'scientific' but her academy's very existence underlines the general insularity of Sweden towards the French Enlightenment (which is why it was formed), and it may have played a small part in eventually injecting such ideas into Swedish culture. Chiefly affected by the advent of the new values was the University of Uppsala. They were well articulated by Kellgren (see Section 12.4.1) in his opposition to Linnaeus and other scientists in the University. His criticism amounted to this: 'About the only things that mattered – society and human nature – they taught nothing; for such teaching one had to go Voltaire, Hume, Rousseau and other heroes of the new age' (Lindroth, 1976, p.119).

Similar views were expressed by J.F. Neikter, professor of literature from 1785 and a disciple of Voltaire and Montesquieu. He became the spokesman of those for whom the human-centred enquiries diligently pursued by enlightened humanity were so much more interesting than the cold experiments of physical science. Manifestly it was impossible to hold back the tide that in other countries was running strongly. The chief bulwark against its advance had possibly been the Wolffianism that in Sweden had always been opposed to the more radical aspects of Locke's philosophy and the extrapolations from it made by the French followers of Diderot. Now Wolffianism had departed the floodgates were opened, though in fact a whole generation separated the effective demise of Wolffian ideas and the serious problems of Swedish science.

However, the extent to which the collapse of university scientific research was a direct consequence of the flooding in of French Enlightenment humanism must always be a matter of conjecture. What cannot be denied is the incompatibility of a secularized world-view with the natural theology that so stimulated science in the earlier part of the century. The almost total inversion of values that had for so long guided, directed and even inspired science in Uppsala may well have had a significant, if temporary, effect on scientific morale.

However, it is all too easy to exaggerate. Sweden was not Uppsala, and the misfortunes of the university there were not necessarily paralleled by the fate of science elsewhere. In fact they were not. The Royal Swedish Academy of Sciences, though not entirely unaffected by the ethos of the age, contrived nevertheless to prosper, and to become a kind of alternative focus for the nation's science. In so far as that is true it is surely because the conflicts between Wolff and Locke largely passed it by. Even though utilitarianism had in some senses shown itself to be a broken reed there was still enough success and therefore optimism to sustain the Academy through the closing years of the century. And the proliferation of centres of science-based technology left a possibility of real progress in mechanics, mineralogy and those sciences derived from them. Perhaps that is why the golden Linnaean age of Swedish science was followed by events that led to that country becoming a world-leader in one science for which it must always be remembered: chemistry.

Acknowledgements

I am glad to acknowledge many useful discussions with colleagues at the University of Uppsala, particularly Drs Broberg, Eriksson and Lundgren. With great patience they supplied much information, often as offprints of their papers. And I am specially grateful for much help and kindness from Professor T. Frängsmyr, not least for the loan of a pre-publication copy of his new history of the Royal Swedish Academy of Sciences and for his comments on an early draft of this chapter. None of them must, however, be held responsible for my own views on the development of science in their country.

Sources referred to in the text

Broberg, G. (ed.) (1979) *Svenska Linnésällskapets Årsskrift* [Yearbook of the Swedish Linnaeus Society], Commemorative Volume for 1978, Uppsala.

Frängsmyr, T. (1972) *Wolffianismens genombrott i Uppsala* [The emergence of Wolffianism at Uppsala], Stockholm, Almqvist and Wiksell.

Lindqvist, S. (1984) *Technology on Trial: the introduction of Steam Power Technology into Sweden, 1715–1736*, Uppsala Studies in the History of Science, 1, Stockholm, Almqvist and Wiksell.

Lindroth, S. (ed.) (1952) *Swedish Men of Science 1650–1950*, Stockholm, Almqvist and Wiksell.

Lindroth, S. (1976) *A History of Uppsala University, 1477–1977*, Stockholm, Almqvist and Wiksell.

Lindroth, S. (1979) in Broberg (*op. cit.*).

Stearn, W.T. (1957) *An Introduction to the* Species Plantarum *and Cognate Botanical Works of Carl Linnaeus*, London, Royal Society.

Weinstock, J. (ed.) (1985) *Contemporary Perspectives on Linnaeus*, Lanham, USA, University Press of America.

Further reading

Eriksson, G. (1979) in Broberg *(op. cit.)*.

Eriksson, G. (1986) 'La percée des sciences de la nature en Suède au XVII siècle. Sa signification et son impact', *Nouvelles de la Republique des lettres*, pp.29–43.

Frängsmyr, T. (ed.) (1983) *Linnaeus, the Man and His Work*, Berkeley, University of California Press.

Frängsmyr, T. (ed.) (1989) *Science in Sweden: The Royal Swedish Academy of Sciences 1739–1989*, Canton, USA, Science History Publishers.

Hildebrand, B. (1939) *Kungl. Svenska Vetenskapsakademien: Förhistoria, grundläggning och första organisation* [The prehistory, foundation and early organization of the Royal Swedish Academy of Sciences], Stockholm.

Science in Orthodox Europe   *Chapter 13*

by Colin Chant

13.1 Introduction

Five centuries ago, Eastern Europe was in a state of upheaval. The fall of
Constantinople in 1453 was a pivotal event in a 200-year campaign by the
Muslim Ottoman Turks against the Christian peoples of the Balkan Peninsula
of south-eastern Europe. By the middle of the sixteenth century, the Greeks,
Bulgarians and Serbians had been incorporated into the Ottoman Empire,
and Wallachia, Moldavia and Transylvania (approximating modern
Romania) reduced to vassal states. By contrast, in 1500 the Russian Slavs,
who had converted to Christianity in the tenth century, were poised to re-
enter the mainstream of European history. Following an era of domination by
nomadic Mongol Tartars from 1237 to 1480, the Russians proceeded to
accumulate a vast empire of their own. Under the first self-proclaimed
Caesar, or 'tsar', Ivan III (Ivan the Great, r. 1462–1505), the formerly
landlocked Muscovite principality reached north to the White Sea; and by the
seventeenth century's end, Russian possessions extended east all the way to
the Pacific Ocean. Finally, in the eighteenth century, footholds were
established on the Baltic and Black Sea coasts, military conquests heralding
Russia's arrival as a European power.

Russia's educated elite duly embraced the culture which nurtured the
Scientific Revolution. This account of science in Eastern Europe will in
consequence be heavily weighted towards eighteenth-century Russia. Even
so, no Russian scientist emerged who can be ranked unequivocally with, for
example, Linnaeus or Scheele in nearby Sweden. Why, then, consider Eastern
Europe at all – simply to present as broad a canvass of European science as
possible? A better reason is this: some consideration of European regions
resistant to the new science could well, by comparison, throw into relief those
features of West European culture which contributed most to its origins and
development.

The most that may be expected is illumination of one or two aspects of a
profound and complex problem. We saw in Chapter 1 that a range of features
of West European society may be *necessary* conditions of the Scientific
Revolution, but no single one is *sufficient* to explain it. Were any of these
features present in Eastern Europe? An obvious contrast exists between
Western Europe's numerous states and the two empires dominating the
majority of East Europeans; consider the contrast made in Chapter 1 between
the small city-states of classical Greece and the vast centralized monarchies of
the Near East. Underlying these broad political differences are notable social
and economic disparities. Although by 1800 the population of Russia was the
largest in Europe, it remained thinly spread, like that of the Ottoman Empire.
The centralized structure of both empires was manifest both in the size of the
two main cities (the populations of Moscow and Constantinople swelled to
some 175,000 and 500,000, respectively) and also in the dearth of other cities
with more than 30,000 inhabitants (in the Balkans, only Salonica, Sofia and

Adrianople, and in Russia, before the partitions of Poland added Vil'na, only the new capital of St Petersburg, founded in 1703). East European towns, moreover, lacked the semi-autonomous character of those in northern Italy and north-west Europe. They reflected economies in which finance, trade and manufacturing were considerably less developed, and in which the feudal system of serfdom in agriculture was actually being extended, at a time when

Map 13.1 The European possessions of the Ottoman and Russian empires at the end of the eighteenth century.

it was breaking up in the West. For most of the period (1500–1800), there was no trace in East European towns of the secular culture of Western Europe. Mediaeval Constantinople had a university, and the Ottomans tolerated a few Christian religious schools and academies apart from their own system of Islamic higher education; but it was only in eighteenth-century Russia that the first modern university and secular public school system was established. Printing reached Constantinople in 1494, and Moscow in 1553, but was often suppressed and generally confined to religious literature; again it was only in eighteenth-century Russia that privately owned presses were permitted to disseminate secular literature.

One feature associated with the efflorescence of scientific activity in the West was ever-present in Eastern Europe: the availability of classical Greek texts. If the rediscovery of these texts by West Europeans from the twelfth century on triggered off a scientific renaissance, why did the Greek scientific tradition fail to develop in Eastern Europe, in which the continuity of Greek learning was uninterrupted by barbarian incursions? It may well be that its very continuity precluded the special reverence for the ancients which typified the Western Renaissance (see Chapter 1). But a complete explanation of East European scholars' failure to add significantly to Greek science and mathematics, and indeed of Eastern Europe's subsequent resistance to the Scientific Revolution, must take account of the prevailing climate of ideas.

The development of ideas and beliefs cannot be fully understood without reference to social, economic and cultural change; nevertheless, as earlier chapters have shown, the particular *form* beliefs take may directly influence scientific activity. In any explanation of the relative neglect of science in Eastern Europe, the Orthodox religious beliefs of the great majority of the population have to be considered. The Orthodox Christian tradition stems from the division of the Roman Empire into two in the fourth century, and the subsequent fall of the western half in the fifth century. (The surviving eastern half is known from the seventh century as the Byzantine Empire, after the original Greek name, Byzantium, for the capital, Constantinople.) This cataclysmic change split the formerly bilingual Christian Church into a predominantly Latin-speaking western part and a predominantly Greek-speaking eastern part. Outright schism resulted from certain irreconcilable differences over liturgy and Church practice (notably the wording of the Nicene Creed and the marriage of priests) and an underlying political rivalry between the pope and the heads, or patriarchs, of the eastern national Churches; this process of schism (often treated as an event occurring in 1054) was certainly complete by the thirteenth century. The hallmark of eastern 'Orthodoxy', or 'right belief', in the face of Latin innovations and ambitions was rigid adherence to the writings of the first theologians, or Greek fathers, of the early Christian Church and the definitions of the seven general councils held between the fourth and the eighth centuries. The political defensiveness of the besieged Byzantine Empire was thus complemented by a reactionary religious ethos and a theology which particularly stressed human ignorance in the face of divine omniscience (Runciman, 1968, p.4). The Latin innovations which occasioned the schism nevertheless pale before the Aristotelian scholastic theology embraced by the Roman Church from the thirteenth century. Such a mixing of religion and pagan philosophy was anathema to the Byzantine Church, even though it generally tolerated commentaries on classical authors by lay philosophers within the bounds of Orthodox dogma.

Resistance to Western ideas among the Balkan Orthodox was further reinforced by the fall of Constantinople, after which the eastern patriarchs owed their allegiance to a Muslim sultan. Leadership of the independent Orthodox world passed to Russia, where theologians began to expound the doctrine of Moscow, the 'third Rome', with the tsar as leader and protector of the Orthodox Church. No fundamental change of beliefs was involved. Although the Russian Church was allowed its own patriarch in 1589, he recognized the primacy of the See of Constantinople. Orthodox Russians continued to regard Western ideas as heretical. Indeed, since there were no universities, schools or professions in sixteenth-century Russia, and since the language of the Russian Orthodox faith was Church Slavonic rather than Greek, even Byzantine learning made relatively little headway (Ryan, 1986, p.100). How then did science take root in Russia at all? The answer lies in its spectacular imperial growth, which Russia's rulers sustained by exchanging raw materials for the manufactures and technical expertise of Western Europe, despite the Church's implacable opposition to Western ideas and practices. Russian science consequently had something in common with science under the Ptolemies of Greek Egypt, the Chinese emperors and the Abbasid caliphs (see Chapter 1), and the monarchs of sixteenth-century Spain and Portugal (see Chapter 5), as well as those Byzantine emperors who encouraged the study of Greek science and mathematics (Vogel, 1967): its mainspring was imperial patronage.

13.2   The roots of Russian science

13.2.1   Early Western contacts

Peter the Great (*r*. 1682–1725), Russia's greatest imperial patron of science, has repeatedly been depicted as a visionary giant, singlehandedly dragging his benighted subjects into the dazzling glare of European culture. Indeed, with an ineffectual monarch in his place, Russia's cultural and scientific revolution could hardly have begun as and when it did. Nevertheless, Peter's Promethean mission to raise his country to the technical and scientific level of Western Europe was no mere autocratic idiosyncrasy; it is more roundly seen as a sharp acceleration of trends rooted in the sixteenth century, and impelled by the logic of territorial expansion.

Contacts with the West had resumed after the lifting of the Mongol yoke and increased during the reign of Ivan IV (Ivan the Terrible, *r*. 1533–84). In 1553, the English explorer Richard Chancellor, seeking a route to the Orient, came across Russia's White Sea coast; trade relations were established, leading to the formation in London of the Russia Company and the foundation of Archangel, Russia's first seaport. It was to Protestant Europe in general that Ivan looked for assistance in his drive towards the Caspian Sea in the south-east and the Baltic Sea in the north-west. Among the Western specialists to arrive in Russia were British and Dutch physicians and apothecaries, one of whom, James Frencham, founded the Moscow Court Pharmacy in 1581 (Appleby, 1983, pp.289–90).

Ivan the Terrible's dealings with the West were matters of expediency; like his subjects, he was zealously Orthodox and xenophobic. The first Muscovite ruler to be an enthusiastic Westernizer was Boris Godunov (*r*. 1598–1605), whose personal physician at one time was the Englishman Mark Ridley, a proponent of William Gilbert's theories of electricity and magnetism. Boris encouraged shaving, despite his compatriots' Orthodox abhorrence of this

Figure 13.1   Peter the Great (r. 1682–1725), painted by Kneller shortly before Peter's visit to London. (The Royal Collection. Reproduced by gracious permission of Her Majesty the Queen. Copyright reserved.)

Western practice as a defacement of God's image. With a view to opening Western-style schools in Moscow, he sent thirty young Russians to study in Germany, England, France and Austria. Only two chose to return, and Boris's attempt at modernization was in any case eclipsed by acute political instability. The so-called 'Time of Troubles' ended only with the election as tsar of Michael Romanov, who for much of his reign shared his authority with his father, one of the first patriarchs of the Russian Church.

The Romanov dynasty was to rule Russia from 1613 until 1917. Under Michael (*r.* 1613–45), we find, against a familiar backdrop of continuous warfare, renewed trade links with the West and further resort to foreign technical experts. One of these technicians, a Dutch merchant named André Vinius, set up the first Russian ironworks and ordnance factory at Tula in 1632. From this beginning, Russia grew to be the world's largest producer of pig iron by the second half of the eighteenth century. During Michael's reign, Western influences remained military, medical, commercial and industrial. Michael himself possessed a telescope and was interested in certain alchemical treatises written by his English physician Arthur Dee, son of the mathematician John Dee (see Chapter 8). But his dominating father, Patriarch Filaret, took it upon himself to protect the Orthodox faith from the taint of secular ideas through a rigid control of Moscow's seven printing presses (Appleby, 1983, p.293; Dukes, 1983, p.24).

Michael's son Alexis (*r.* 1645–76), though devoutly Orthodox, was obsessed with Western artefacts, curious about alchemy and astronomy, and may have flirted with the hermetic philosophy (Longworth, 1984, p.205). During his reign, Western influences deepened somewhat: not only technically (Dutch shipbuilders oversaw the germination of the Russian navy, in the form of a Volga river fleet, and foreign doctors staffed a short-lived medical school, founded in 1654), but also intellectually, as Latin theological influences penetrated a Russian Church increasingly internally divided over reforms of its own practices. The main source of Latin influences was Kiev in the Ukraine, until 1667 under Polish rule. Fedor Rtishchev, a member of the Moscow nobility, invited thirty Kievan monks to translate scholarly works, primarily of a religious nature. Amongst certain secular works they went on to translate were Vesalius's *De humani corporis fabrica* and the *Atlas novus* (1634) of the Dutch cosmographers Willem and Johan Blaeu. Although each translation remained in manuscript only, the Blaeus' work was 'the first writing available in Russian which alluded favorably to Copernicus and the heliocentric theories' (Vucinich, 1963, p.19).

The impact of these translations must have been minimal, but they are symptomatic of subterranean changes in Russia's political and cultural orientation, without which Peter the Great's reforms would have been inconceivable. Tension between Russia's Westward turn and its Greek Orthodox heritage had already surfaced in Alexis's reign, and the cultural indecision which ensued is nicely captured in the name given to a Moscow ecclesiastical school founded in 1687. The 'Slavonic–Greek–Latin' Academy, as it became known, was the only noteworthy Russian educational institution at the end of the seventeenth century (apart from an ecclesiastical college in recently annexed Kiev, officially raised to academy status in 1701). Its teachers were hostile to the new science; they nevertheless included Aristotelian physics in the curriculum. Direct acquaintance with Aristotle's works had been rare in Muscovite Russia; most Russian priests received their world-picture from the Greek fathers, notably John of Damascus and Cosmas Indicopleustes, whose sixth-century flat-earth *Christian Topography* graced every Russian monastery library (Vucinich, 1963, p.9; Ryan, 1986, pp.98–103).

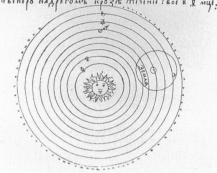

Figure 13.2 The Copernican system, from the seventeenth-century manuscript translation of the Blaeus' Atlas Novus. (From B.E. Raikov, Ocherki po istorii geliotsentricheskogo mirovozzreniya v Rossii, Moscow, Izdatel'stvo Akademii Nauk SSSR, 1947.)

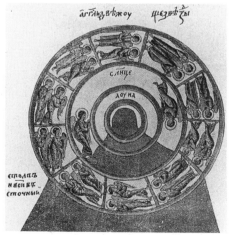

Figure 13.3 Russian version of the world-picture according to Cosmas Indicopleustes, with angels bearing the heavenly bodies. (From B.E. Raikov, Ocherki po istorii geliotsentricheskogo mirovozzreniya v Rossii, Moscow, Izdatel'stvo Akademii Nauk SSSR, 1947.)

More generally significant was a growing appreciation in the second half of the seventeenth century among Russia's educated elite that some kind of accommodation was required with Western learning and technique, the better to defend and extend Russia's religious and territorial interests. An outspoken proponent of this view was the Croatian Catholic missionary Yury Krizhanich (1617–83), most of whose years in Russia were spent in Siberian exile. An advocate of the reunion of the Roman Catholic and Russian Orthodox Churches under Russian leadership, he was nevertheless critical of the Orthodox Church's attitude to Western learning; he insisted that 'wisdom eliminates heresy, while the ignorance of illiterate people keeps heresy and superstition alive' (Letiche and Dmytryshyn, 1985, p.90). According to Krizhanich, if Russia were to compete with the West, the abilities of all social estates had to be mobilized, not least through a system of education based partly on the kind of practical science and technology with which he was familiar. Tsar Alexis was acquainted with Krizhanich; his ideas may therefore have influenced Alexis's son Peter and helped inspire the reforms which culminated in the St Petersburg Academy of Sciences.

13.2.2   Peter the Great and the founding of the St Petersburg Academy

It was soon apparent that Peter would follow his father's Westward path. The early years of co-rule with his elder step-brother Ivan V (1682–96) were marked by childish war games, followed by youthful revelry in Moscow's Foreign Quarter – this was an area to which foreigners had been restricted in the seventeenth century, to protect Russians from Western contamination. Here Peter's respect both for the Protestant religion and for West European navigational and shipbuilding know-how was nurtured. It was partly in search of such expertise that he undertook his Great Embassy of 1697–8, a fifteen-month tour including long stays in Holland and England, much of the time spent gaining his master's certificate in the dockyards of Saardam, Amsterdam and Deptford. Concerned above all with military and industrial technique, he nevertheless also tasted the scientific and intellectual life of those countries. In Holland, he visited Boerhaave and the anatomist Frederick Ruysch, witnessing surgical operations and anatomical dissections, at which he forced certain squeamish members of his suite to tear out the muscles of corpses with their teeth (Klyuchevsky, 1965, p.27); he was also introduced to the microscope by Leeuwenhoek. In England, he visited Oxford, the Royal Society, the Greenwich Observatory and, several times, the Royal Mint, of which Newton was warden. There is no conclusive proof that Peter met Newton, although Jacob Bruce, one of his retinue, did become known to Newton and Flamsteed, having stayed on to buy scientific instruments and receive mathematical instruction. It was largely through Bruce that Newton's ideas were introduced to Russia (Boss, 1972, p.15).

Russian life would never be the same again after Peter's tour. West European dress was made compulsory for the nobility, the Western calendar imposed, and a tax placed on all beards, excepting peasants and clergy. A tougher nut to crack was Orthodox aversion to the Western-style technical education which Peter saw as vital if Russia were to supply its own needs for trained personnel: officers, engineers, doctors, teachers and civil servants. But Peter's impatience for modernization was inflamed by humiliating defeat at the hands of Charles XII's greatly outnumbered Swedish army at Narva in 1700 and sustained during the ensuing 21-year Great Northern War.

At the close of the seventeenth century, Russia's educational system comprised the academies at Moscow and Kiev and a few church schools.

Numerous secular educational projects were spawned in the early years of Peter's personal rule. The most durable was the Mathematics and Navigation School, founded in Moscow in 1701. The first director was the afore-mentioned Jacob Bruce (1670–1735), a Russian-born engineer of Scottish descent, who soon handed over to one of three British recruits, Henry Farquharson (*c.* 1675–1739), a mathematics tutor from Marischal College, Aberdeen. The other recruits were young graduates of the Royal Mathematical School of Christ's Hospital, London, the first non-classical secular school in Europe and the model for the Russian institution. The British teachers were joined in 1702 by the first significant figure in Russian science, Leonty Magnitsky (1669–1739), by tradition a graduate of the Slavonic–Greek–Latin Academy. His *Arithmetic* (1703) was the first Russian mathematics textbook. To a large extent a compendium of translated West European sources, it included magnetic, astronomical and navigational data as well as pure mathematics; its very success as a textbook may even have contributed to the stagnation of Russian mathematics teaching over the half-century of its dominance (Okenfuss, 1973, p.334).

Figure 13.4   Jacob Bruce (1670–1735). (The Hermitage, Leningrad)

The Mathematics and Navigation School declined after part of it, including Farquharson, was moved in 1715 to St Petersburg, later to become its Naval Academy. Nevertheless, by 1716, some 1,200 had graduated from the pioneering Moscow institution (Black, 1979, p.24). Other schools of engineering, artillery, medicine and mining had quickly followed, but they generally struggled through a lack, not only of qualified teachers, but also of suitable or willing students. Peter's response to this problem was his decree of 1714 establishing elementary 'cipher' schools in all provincial capitals, with the aim of providing a mathematically oriented training, based on Magnitsky's textbook, for children of classes other than nobles and peasants. But these schools failed for identical reasons and were eventually absorbed by church schools and the military's garrison schools (Okenfuss, 1973, pp.338–40). If Peter's attempt to establish a domestic system of scientific and technical education was premature, he was at least more successful than his predecessors in having Russians educated abroad. Amongst his own entourage, the diplomat Peter Postnikov went to Padua and became the first Russian doctor of medicine, and his personal physician, Lavrenty Blumentrost, studied at Königsberg, Leiden and Halle. The tsar was especially keen to have Western secular books imported and translated, even if they offended the Church. One such was Huygens' *Cosmotheoros* (1698), written in Latin, but with a Greek title meaning 'spectator of the heavens'. This was a popular work including speculations about extra-terrestrial beings; the publication of Jacob Bruce's translation in 1717 marked the entry into the Russian public domain of Copernican-Newtonian cosmology (Boss, 1972, pp.50–67). Peter also founded the first public library, and the first museum, the Kunstkamera, opened in 1719 and initially built around the purchase of Frederick Ruysch's natural-historical and anatomical collection. But Peter's most novel and ambitious project was a national academy of arts and sciences to match those which existed in London, Paris and Berlin by the turn of the century. The academy proved to be his most enduring contribution to Russian science.

Peter's desire for a higher teaching and research institution, kindled perhaps by visits to the Royal Society, was reinforced by a correspondence with Leibniz, stretching from 1697 to Leibniz's death in 1716. Leibniz was a great enthusiast for Russia, both as a bridge between Europe and the East and as a 'treasure-house' for scientific research and exploration; he was especially curious about the possibility of a terrestrial link between Asia and America

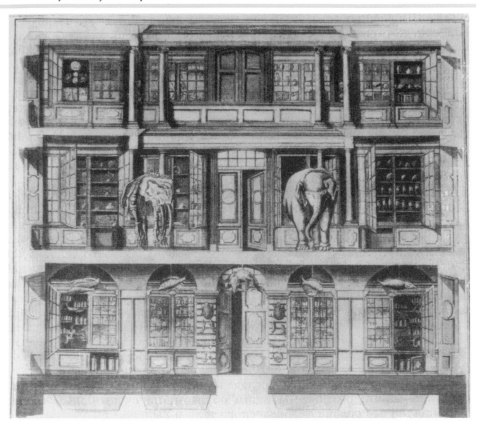

Figure 13.5 Zoological gallery of Russia's first museum, the Kunstkamera, 1741. (From N.A. Figurovsky (ed.) Istoriya estestvoznaniya v Rossii, vol.1, part 1, Moscow, Izdatel'stvo Akademii Nauk SSSR, 1957. Photo: British Library Board.)

and keen to promote the collection of astronomic and magnetic data, in the hope of solving the problem of determining longitude at sea (see Chapter 5). Much of Leibniz's advice was practical, concerning the procurement of books, equipment and scholars from abroad and the development of higher and lower education in Russia. He also advocated the creation of a special 'college' with sweeping powers over science, the arts, education and commerce. Likened by one Soviet scholar to Solomon's House in Bacon's scientific Utopia, *New Atlantis* (see Chapter 8), Leibniz's project involved 'a degree of centralization of all ... provinces of culture unknown at that time in any European state' (Kopelevich, 1977, p.34).

Despite his enthusiasm for Russia and his admission to its state service in 1712, Leibniz could not be enticed to Russia to help implement his grand scheme. Nevertheless, the tsar visited him three times between 1711 and 1716, and although much of Leibniz's advice was inappropriate to Russian conditions, these discussions played their part in bringing a national academy nearer the top of Peter's domestic agenda. A further stimulus was Peter's visit to Paris in 1717. Although its primary aim was to consolidate improving Franco-Russian relations, Peter seized the opportunity to acquire first-hand experience of French science and technique. Always fascinated by astronomy, he visited the Observatory three times, and an extraordinary session of the Académie Royale des Sciences was held in Peter's honour. By the end of the year, Peter's election to the Paris Academy had been confirmed, and it was soon after his French trip that Peter noted on an adviser's memorandum his aim 'to create an academy' (Kopelevich, 1977, p.46). In 1719, Lavrenty Blumentrost visited the German philosopher Christian Wolff (1679–1754), the foremost follower of Leibniz, both to get advice on the new project and to tempt him to Russia to supervise it. He failed in the latter objective; so too, on a similar trip in 1721, did Johann Schumacher, Blumentrost's assistant. In other respects Schumacher, whose journey also took in France, Holland and

England, was more successful, not least in recruiting a future academician, the French astronomer Joseph Delisle, and also in equipping the future Physics Cabinet with instruments and other materials.

Apart from its practical objectives, the strategic aim of Schumacher's journey was a further comparison of the academies at Paris, London and Berlin. But the establishment that emerged in St Petersburg on the banks of the Neva was no mere imitation of Western models. It was designed to meet Russian needs, sometimes against Western counsels. Wolff, who saw himself as Aristotle to Peter's Alexander the Great, agreed with Leibniz that Russia should give priority to the establishment of adequate schools. He also recommended that a university be set up before an academy, to provide proper training for future Russian academicians. Instead, the Project (a document drafted by Blumentrost and Schumacher in January 1724 and amended by Peter) proposed an institution incorporating all three of Leibniz's educational levels: an academy, a university and a 'gymnasium', or secondary school. As the document put it: 'In this way one building would with only small losses and also with great benefit achieve what is done by three separate bodies in other states' (quoted in Kopelevich, 1977, p.58). Peter and his advisers hoped that an initial influx of Western scholars would prime the pump of indigenous scientific talent: at the university, they would train the most able Russian students, who would in their turn teach the next generation in the gymnasium. The addition of teaching to the fundamental research duties of the academicians was at that time unique: they would be required to give public lectures at the university, and to prepare courses and textbooks to be translated into Russian.

In two respects, the projected institution owed much to Leibniz's Berlin Academy. First, a broad conception of science was adopted (*nauka*, the Russian word for 'science', is equivalent to the German *Wissenschaft*, meaning 'knowledge'). The professors were divided into three groups: a 'first class' of mathematical sciences (mathematics, astronomy, geography, navigation and mechanics); a 'second class' of physical sciences (theoretical and experimental physics, anatomy, chemistry and botany); and a 'third class' of humanities (rhetoric, ancient and modern history, jurisprudence, moral and political philosophy). However broad their conception of science appears from our standpoint, neither the Royal Society nor the Académie Royale des Sciences embraced the humanities. Second, the authors of the Project followed Leibniz in identifying the uses of science with national goals. Academicians were required both to augment and improve the technical basis of industry and commerce, and to carry out special tasks at the sovereign's behest. In that respect, the influence of the Paris Academy can also be detected.

One final way in which the projected St Petersburg Academy stood out was in the level of state funding and remuneration of staff. Its initial grant of some 25,000 roubles represented a quarter of all expenditure on education and medicine and surpassed public funding of the Paris and Berlin academies (the Royal Society being privately funded). Schumacher had heard about the inadequacy of salaries at the Paris Academy and the need for its members to supplement their income with outside work. The Project duly laid down that the salaries of academicians should be sufficient to support themselves and their families. This promise, despite its often laggardly fulfilment in the Academy's early history, leads Kopelevich to conclude that 'it was in Russia that there first arose the profession, broadly speaking, of the scientist, whose research work provided the means to support a family' (Kopelevich, 1977, pp.61–2). Kopelevich's thesis owes much to her very broad definition of 'profession'. Moreover, the St Petersburg Academy's novel features – the

Figure 13.6 Mid eighteenth-century view of the St Petersburg edifice housing the Kunstkamera and part of the Academy of Sciences. (From N.A. Figurovsky (ed.) Istoriya estestvoznaniya v Rossii, *vol.1, part 1, Moscow, Izdatel'stvo Akademii Nauk SSSR, 1957. Photo: British Library Board.)*

academicians' teaching duties and guarantee of adequate remuneration – can be seen as expedients, reflecting the main obstacles it would face during the eighteenth century: the lack of secular schools and difficulty in enticing distinguished Western scholars to Russia.

In his final attempt at modernization from above, Peter overrode the doubts not only of his main foreign adviser, Wolff, but also of his compatriot the engineer and historian Vasily Tatishchev (1686–1750), who ventured the opinion that Russia's educational institutions could hardly support such a learned society. Peter replied with a parable about an old man who built a watermill without a water course in order to stimulate his sons to dig a canal to it after his death. The tsar's motivation was complex and not wholly utilitarian; according to Schumacher, he expected that the Academy's work 'will bring us honour and respect in Europe. Foreigners will recognize that we have science, and will cease to regard us as barbarians, contemptuous of science' (quoted in Kopelevich, 1977, p.64). The character, and very survival, of the new institution owes much to Peter's personal initiative; so too does the place of science in Russian culture: 'he made science a "government science", a body of knowledge guided and guarded by the state' (Vucinich, 1963, p.73). But the need for such a science was ultimately propelled by Russia's snowballing imperial interests.

13.3 The rise of Russian science

13.3.1 The Academy under Peter's successors

The Imperial Academy of Sciences and Arts (to give its fullest eighteenth-century title) was a posthumous achievement. The Project was quickly ratified by Peter's Senate, on 22 January 1724, and work began in earnest on the recruitment of foreign scholars and, equally important, of students for the university and gymnasium. But little more than a year later, Peter was dead, with the first scholars yet to arrive in St Petersburg. It was Peter's widow,

Catherine I (*r.* 1725–7), born a Lithuanian peasant girl, whose decree of 20 November 1725 formally established the Academy, with Peter's former physician Blumentrost appointed its first president.

Lavrenty Blumentrost (1692–1755), himself the son of a German physician, was for some years the only Russian to hold a position of authority within the Academy. Of the sixteen scholars who held professorial status in the 1720s (a junior post of adjunct being introduced in 1727), all were German, apart from two French astronomers, Delisle and his half-brother Louis Delisle de la Croyère, and three Swiss mathematicians, Jakob Hermann and the brothers Nikolaus and Daniel Bernoulli. (The latter started as professor of physiology, switching to mathematics in 1730.) Other than Blumentrost, the main lay figure was his former assistant, the Alsatian Johann Schumacher (1690–1761), who from his position as librarian rapidly took over the Academy's administration. Apart from a brief spell of disgrace in 1742–3, when he was placed under house arrest pending charges of embezzlement and of favouring Germans over Russians, Schumacher dominated the internal affairs of the Academy until his retirement in 1759. Frequently at loggerheads with the academicians – foreigners as well as Russians, it must be said – he was called by Mikhail Lomonosov, the outstanding indigenous scientist of the period, *Flagellum professorum*, 'the scourge of the professors' (Black, 1986, p.20).

The arbitrary power exercised by Schumacher through his self-created Chancellery was only one of the contingencies which soured the first professors to academic life in St Petersburg, once the flush engendered by good facilities and good salaries had passed. During the reigns of Peter II (*r.* 1727–30), Anne (*r.* 1730–40) and Elizabeth (*r.* 1741–61), there was less commitment to science at court and a more grudging attitude to funding. It may be this change in the political climate, as well as the rigours of the north Russian winter, which partly accounts for the failure of some of the best of the foreign scholars to renew their initial contracts; by 1733, five had left Russia, including Daniel Bernoulli and Hermann. Three of the original professors died in the 1720s, including the 31-year-old Nikolaus Bernoulli. The international reputation of the Academy was sustained during the 1730s, partly through the efforts of the brilliant Swiss mathematician Leonhard Euler; but more foreign scholars, including Euler, left upon Elizabeth's accession in 1741. This turn of events, diminishing the quality of the Academy's research, is usually attributed to a pro-Russian backlash under Elizabeth, provoked by Anne's favouring of Germans at her court; such political swings could only exacerbate the 'vicious internal conflict between the so-called Russian and German factions' within the Academy (Vucinich, 1963, p.84). More recent historical accounts have played down these conflicts, whether at court (Dukes, 1982, pp.108–10) or at the Academy (Carver, 1980, pp.390–1). In Euler's case, his departure may simply have been due to an unrefusable offer from the Berlin Academy; he was never personally guilty of anti-Russian bias and maintained good relations with St Petersburg, continuing to have much of his work published there.

National sensitivities cannot, however, be written out of the early history of an institution which had first to rely on foreign expertise in order to compete with it. The foreign domination of the Academy had indeed increased under Anne, as Blumentrost was relieved of the presidency in 1733 and succeeded by a series of German court favourites. Schumacher's fall from the directorship of the Chancellery shortly after Elizabeth's accession allowed the Russian Andrei Nartov (1693–1756) to enjoy a temporary ascendancy. Nartov had been Peter the Great's personal lathe operator and from 1735 headed the

Figure 13.7 Leonhard Euler (1707–83). (Mansell Collection)

Academy workshops, which were responsible for instrument-making, printing and bookbinding. This was a rare position of eminence for a Russian at the time, and his later replacement of Schumacher represented some anti-German feeling; it was also a short-lived attempt to reorient the Academy away from its preoccupation with pure scientific research. He proved even more unpopular than Schumacher, who was soon restored. But in 1746 Elizabeth appointed a Russian president, the eighteen-year-old Count Kirill Razumovsky, her closest courtier's brother. Thereafter, Schumacher had to share his authority with a Russian, Grigory Teplov, Razumovsky's former tutor and, since 1742, an adjunct in botany at the Academy.

It was Teplov who collaborated with Schumacher on the preparation of the Academy's first charter of 1747. The arbitrariness of Schumacher's early administration is partly explained by Catherine's failure in 1725 to establish Peter's Project as the Academy's official regulations. The academicians played no part in the drafting of the document and unsurprisingly found themselves officially subject to the fines and regulations of the Chancellery; moreover, it was now clearly stated that their services were available to government departments on demand. Otherwise, the most significant provisions were, first, a much sharper distinction between academicians, who were primarily concerned with research, and professors, whose main task was to lecture at the university; and, second, the confinement of the Academy's research activities to the mathematical and natural sciences. Humanities, however, would be retained in the university curriculum. Although the new measure consolidated the position of foreigners established in the Academy, Elizabeth's reign was a period in which Russian scholars became more prominent, a trend which the charter was intended to reinforce. Vasily Adodurov (1709–80) had already in 1733 become the first Russian adjunct (in mathematics); an early product of the Academy's gymnasium and university, he went on to be Curator of Moscow University. Lomonosov and the poet Vasily Trediakovsky became the first Russian professors (of chemistry and rhetoric, respectively) in 1745.

The 'Russifying' tendencies of Elizabeth's reign were not inconsistent with greater investment in Western-style institutions: the budget of the Academy of Sciences was doubled in 1747; Moscow University was founded in 1755; and two years later, a separate Academy of Fine Arts was established. But it was unequivocally in keeping with the wishes of the next long-reigning tsar that Western ideas actually became fashionable amongst Russia's educated elite. The German-born Catherine II (Catherine the Great, *r.* 1762–96), was a devotee of the European Enlightenment and numbered Voltaire and Diderot amongst her correspondents. Her 'enlightened despotism' was always a balancing act: she introduced a system of secular elementary and secondary schools, encouraged debate over social and economic reform, and in 1783 allowed private individuals to operate 'free presses' (Marker, 1985, p.105); but she also strengthened the nobility's privileges and, in horrified response to the French Revolution, imprisoned leading liberal writers and subjected the recently established private printing houses to rigid censorship.

The opening years of Catherine's reign were encouraging for the Academy. In 1766, Euler was persuaded to return to St Petersburg on the promise of a reorganization of the Academy, allowing the scholars greater autonomy. Indeed, in the same year, the notorious Chancellery was abolished, and replaced by a commission of academicians. Although Razumovsky continued as a figurehead president until 1796, real authority was now vested in a crown-appointed director. Euler's remarkable output did much, again, to restore the Academy's reputation abroad, but internally, little good came of the 1766 reforms. The Academic Commission, itself abolished in 1783, had proved ineffective, largely because the post of director was generally filled by an overbearing and ill-educated court favourite. An exception to this rule was Princess Catherine Dashkova, director from 1783 to 1794: an admirer of the philosophers of the French Enlightenment, she did much in her early years to revive the flagging gymnasium and to encourage the dissemination of scientific ideas through the Academy's public lectures and publications.

The reaction against Western liberal ideas which characterized the final years of Catherine's reign was taken to extremes by her successor Paul (*r.* 1796–1801); it says much for the solidity of the Academy's place in Russian intellectual life by then that its output was undiminished in the last decade of the eighteenth century. By 1800, half of the sixteen academicians were Russian; indeed, after the initial domination of the Academy by foreigners, there had since the early 1750s been a rough parity between Russian and foreign scientific members, except for the decade following Euler's return (Schulze, 1985, p.325). Fluctuations in foreign membership partly reflected vicissitudes of domestic policy; but a gradual ascent of Russian membership from the 1750s to the 1790s attests to the success of the often ailing Academy university and gymnasium in training Russian scientists. With enrolments sometimes down to a handful of students, only the gymnasium survived to the beginning of the nineteenth century. By this time the existence of other schools and universities enabled the Academy, by its new charter of 1803, to confine its activities to research. But against all odds, the Academy's teaching arm performed a service to Russian science: of seventeen native adjuncts in science between the 1747 charter and the end of the century whose educational backgrounds are known, sixteen were taught at the Academy (Schulze, 1985, p.332).

The Academy has sometimes been portrayed as an island of foreign research of little or no relevance to Russian culture; the evident 'Russianization' of the Academy works against this interpretation, though it might still be argued that native scientists did little original work. The Academy's research will be discussed next, but no estimation of the Academy's place in Russian culture should overlook its translating and publishing work, whereby a Russian scientific vocabulary was fashioned and scientific ideas disseminated in popular works and textbooks. Indeed, the time spent by Russian members on these tasks must to some extent explain why their pure research output was lower than that of foreign scholars. Apart from its internationally respected series of scholarly journals (its *Commentarii*, 1728–41; *Novi Commentarii*, 1750–76; *Acta*, 1778–86; and *Nova Acta*, 1787–1806), the Academy Press also published a newspaper, the *St Petersburg Gazette*, the monthly *Notes* to which, from 1728 to 1743, constituted the first scientific journal in Russian (Schulze, 1985, p.312). The Press published over three-quarters of all secular books between 1727 and 1755 (Marker, 1985, p.45), and its *Monthly Compositions*, published in twenty volumes between 1755 and 1764, was the first journal to popularize science in Russia, albeit mainly among the educated elites of Moscow and St Petersburg. It was revived by Princess Dashkova in 1786, though by this time, many other private periodicals carried popular scientific articles to an increasingly literate urban public. The Translating Department of the Academy, known briefly in the 1730s as the Russian Council and including at that time Adodurov and Trediakovsky, has been called 'the first formal gathering of Russian intellectuals' (Vucinich, 1963, p.90). Its job of modernizing the Russian language passed to other agencies, most notably the Russian Academy of Letters, founded in 1783 and also directed by Princess Dashkova. The activities of the Academy of Science's Translating Department

Figure 13.8   Title page of the first Russian popular scientific and literary journal Monthly Compositions For Use and Entertainment, *January 1755. The motto beneath the imperial double-headed eagle reads 'For everyone'. (From B.E. Raikov,* Ocherki po istorii geliotsentricheskogo mirovozzreniya v Rossi, *Moscow, Izdatel'stvo Akademii Nauk SSSR, 1947. Photo: British Library Board.)*

nevertheless support the view that 'during the first three decades of its existence the Academy was the main instrument of the growing Westernization of Russian culture' (Vucinich, 1963, p.91).

13.3.2 Scientific research in Russia

The two main areas of research in which Russia won international recognition were at opposite ends of the scientific spectrum, though in their own ways distinctly Russian. Russia's most obvious appeal to the Western scientific community, articulated by Leibniz to Peter the Great, lay in the riches of its vast hinterland, as revealed by Russian-based natural historians and geographers. Less obviously grounded was its reputation in the mathematical sciences. In case this is too easily seen as an accident of foreign recruitment, it should be remembered that Peter the Great was concerned above all to establish a mathematically based educational system; it was apparently at the insistence of his adviser Jacob Bruce that six of the founding academicians were mathematicians (Boss, 1972, p.94).

The mathematical sciences

For the foundation of its mathematical research tradition, the Academy was indebted to members of the Basel school of Swiss mathematicians, initially to Jakob Hermann and two scions of the famous Bernoulli family, but above all to a relative of Hermann, Leonhard Euler (1707–83), the most prolific of eighteenth-century mathematicians. Euler was associated with the Academy throughout his adult life, not only during his two long spells in St Petersburg (1727–41 and 1766–83), but also as an honorary member with an annual pension, conscientiously earned, in the intervening years at the Berlin Academy. The son of a Calvinist minister, he was able, through his piously diligent and retiring character, to accommodate such notoriously difficult personalities as Schumacher and Lomonosov. His prolific research output was unaffected by the onset of blindness in his right eye through disease in 1738 and in the other shortly after his return to Russia; indeed, almost half his works were produced after 1765 (Youschkevitch, 1971, p.472). Euler was first and foremost a mathematician; his genius lay in the development of the calculus, and of mathematical analysis in general, in which he established the centrality of the concept of a function. He was, however, always concerned with practical applications and did notable work in optics, the physics of fluids and solid bodies, and celestial mechanics, especially the theory of lunar motion. His *Mechanica* of 1736 did much to establish the Academy's reputation and 'marked the first systematic use of mathematical analysis in a natural science' (Vucinich, 1963, p.94). Although this work was to some extent a mathematically updated treatment of Newton's *Principia*, Euler stood out against the Newtonian concepts of a vacuum and action-at-a-distance, a stance common among St Petersburg academicians before the mid-1750s.

The Academy boasted another accomplished mathematician in the Prussian-born Christian Goldbach (1690–1764), who contributed to the development of number theory in correspondence with Euler. Goldbach had two spells at the Academy between 1725 and 1742, largely as its conference secretary, but finally chose a career in the Ministry of Foreign Affairs. The most prominent physicists were Georg-Wilhelm Richmann (1711–53) and Franz U.T. Aepinus (1724–1802). Richmann, whose native Estonian town was incorporated into the Russian Empire by Peter's armies just before his birth, became professor of physics in 1741 and carried out original research on the measurement of

Figure 13.9 Georg-Wilhelm Richmann (1711–53). (From N.A. Figurovsky (ed.) Istoriya estestvoznaniya v Rossii, vol.1, part 1, Moscow, Izdatel'stvo Akademii Nauk SSSR, 1957. Photo: British Library Board.)

heat. Later, inspired by Franklin's experiments, he became the first, with Lomonosov, to investigate electrical phenomena in Russia, work tragically terminated by his electrocution during a St Petersburg thunderstorm. Richmann's chair was eventually filled, in 1757, by Aepinus, a north German who in 1755 had become professor of astronomy at the Berlin Academy under Euler. Aepinus had none of Euler's reservations about Newtonianism, and the considerable advance upon Franklin's explanations of electricity represented by his *Essay on the Theory of Electricity and Magnetism*, published by the St Petersburg Academy in 1759, was based partly on the idea of electrical and magnetic attraction-at-a-distance. In this respect, he was arguably more 'Newtonian' than Newton and his British followers (Home, 1977; see Chapter 10). Aepinus followed Goldbach in leaving the Academy for a career in the Ministry of Foreign Affairs; at one time a tutor to the future Catherine II, he went on to advise on the new system of public schools (Home, 1979).

Whilst at Berlin, Aepinus came to know Euler's Russian students Stepan Rumovsky (1734–1812) and Semen Kotel'nikov (1723–1806). Kotel'nikov, after some early papers in *Novi Commentarii*, was increasingly immersed in the teaching, translating, popularizing and administrative side of the Academy's work, becoming director of the library in 1797. Rumovsky was also heavily involved in teaching, text-book writing and translating – he translated Euler's popular work *Letters to a German Princess* (both the original French edition, 1768–72, and the Russian translation, 1768–74, were published by the Academy). He nevertheless won international recognition for his papers in mathematical analysis and, above all, for his astronomical work, for which he was elected to the Swedish Academy of Sciences. Following on the work of Joseph-Nicolas Delisle (1688–1768), professor of astronomy in St Petersburg from 1725 to 1747, Rumovsky compiled a catalogue of the astronomically determined geographical co-ordinates of 62 Russian localities, with a precision unsurpassed in his day (Kulikovsky, 1975, p.609).

The scientific exploration of Russia

Although pure mathematics and natural history were poles apart, the intervening areas of astronomy and cartography demanded both mathematics and the collection of data over wide areas. This explains why Rumovsky could head both the Academy's Observatory, from 1763 to 1803, and its Geographical Department, from 1766 to 1803, and why the Academy's work involved extensive journeys throughout the Russian Empire. Geographical exploration had been a necessary part of the process of territorial expansion since the reign of Ivan the Great, but it was under Peter that it first acquired a scientific dimension. Peter was especially anxious to improve the accuracy of Russian maps: cartography was one of the main activities of the press he set up under the direction of Jacob Bruce in 1705; and it was with a new map of the Caspian Sea that his envoys impressed the foreign academies with Russia's commitment to science. He also ordered several explorations of his Asian lands: most notably, a Siberian expedition (1720–7) led by the German physician Daniel Messerschmidt, whose natural-historical and mineralogical collections enriched the exhibits of the Kunstkamera; and, another of Peter's posthumous achievements, the first Kamchatka Expedition (1725–30). Led by the Dane Vitus Bering (1681–1741), part of its mission was to establish whether a land link existed between Russia and America, as Leibniz had earlier speculated.

Bering, who joined the Russian navy in 1703, was also entrusted with the overall planning of the second Kamchatka, or Great Northern, Expedition

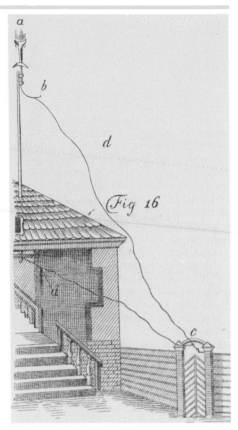

Figure 13.10 Apparatus for the experiments on atmospheric electricity leading to Richmann's death. (From N.A. Figurovsky (ed.) Istoriya estestvoznaniya v Rossii, vol.1, part 1, Moscow, Izdatel'stvo Akademii Nauk SSSR, 1957. Photo: British Library Board.)

Figure 13.11 Stepan Rumovsky (1734–1812). (From N.A. Figurovsky (ed.) Istoriya estestvoznaniya v Rossii, vol.1, part 1, Moscow, Izdatel'stvo Akademii Nauk SSSR, 1957. Photo: British Library Board.)

Figure 13.12 Representation of fur seal, sea lion and sea cow, first described by Steller. The sea cow inhabited the waters around Bering Island (shown underneath), but had been hunted to extinction by about 1765. (From F.A. Golder, Bering's Voyages: an Account of the Efforts of the Russians to Determine the Relation of Asia and America, *New York, 1922. Photo: British Museum.)*

(1733–43), a venture involving separate land and sea explorations. Several members of the Academy took part, at considerable risk; indeed, the harsh Siberian winter of 1741–2 claimed the lives of Louis Delisle de la Croyère and some of the Delisle brothers' astronomy students, all engaged in the establishment of geographical co-ordinates. Bering himself perished, on the island which now bears his name, during a sea voyage from Kamchatka to Alaska (1741–2). One of the few to survive was the German-born adjunct in botany, Georg Wilhelm Steller (1709–46). The study of marine animals which he undertook on the voyage made his name, though not before he contracted a fatal illness in a Siberian town. Before joining Bering, Steller travelled overland, catching up with another vital part of the expedition, headed by two other German academicians, the professor of chemistry and natural history, Johann Georg Gmelin (1709–55), and the professor of history, Gerhard Friedrich Müller (1705–83).

The result of this decade of exploration was a mass of historical, geographical, ethnographic, linguistic, astronomical, botanical and zoological data. Müller reckoned to have travelled 31,362 versts (about 24,000 miles), and his collected material enabled him to publish books and articles on Siberia for 35 years after his return (Black, 1986, pp.74–5). More than a thousand plants were described in Gmelin's *Flora sibirica* (1747–69), and even before the Academy began to publish the four volumes, Gmelin was credited by Linnaeus with discovering as many plants as all other botanists put together (Vucinich, 1963, p.100). A final product was the *Description of the Land of Kamchatka* (1756), a posthumous work by the Russian professor of natural history and botany, Stepan Krasheninnikov (1713–55), who participated in the expedition as a student of the Academy university. The book was of great interest in Europe, being translated into French, English, German and Dutch between 1764 and 1770 (Carver, 1980, p.400).

A second explosion of data resulted from the 'Academic expeditions' of 1768–74, the publications from which did as much as Euler's return to restore the Academy's international reputation. The scale and planning of the expeditions was unprecedented, though the hazards remained: in 1774, the professor of botany, Samuel Gmelin (nephew of J.G. Gmelin), died as a hostage in Dagestan (then part of Persia), and the professor of astronomy, Georg Löwitz, was killed in southern Russia during a major peasant rebellion. The dominating naturalist at this time was the German Peter Simon Pallas (1741–1811), a St Petersburg academician from 1767 to 1810; his *Travels Through Various Provinces of the Russian Empire* (1771–6) was read throughout Europe, and his theories on the formation of mountains and on variations in animal species influenced Cuvier and Darwin respectively. Regarded in his time as the equal of Linnaeus and Buffon, he did much to sustain the Academy's prestige beyond Euler's death. The outstanding Russian naturalist

Figure 13.13 Stepan Krasheninnikov (1713–55). (From N.A. Figurovsky (ed.) Istoriya estestvoznaniya v Rossii, *vol.1, part 1, Moscow, Izdatel'stvo Akademii Nauk SSSR, 1957. Photo: British Library Board.)*

Map 13.2 The scientific explorations of the Russian Empire. (facing page)

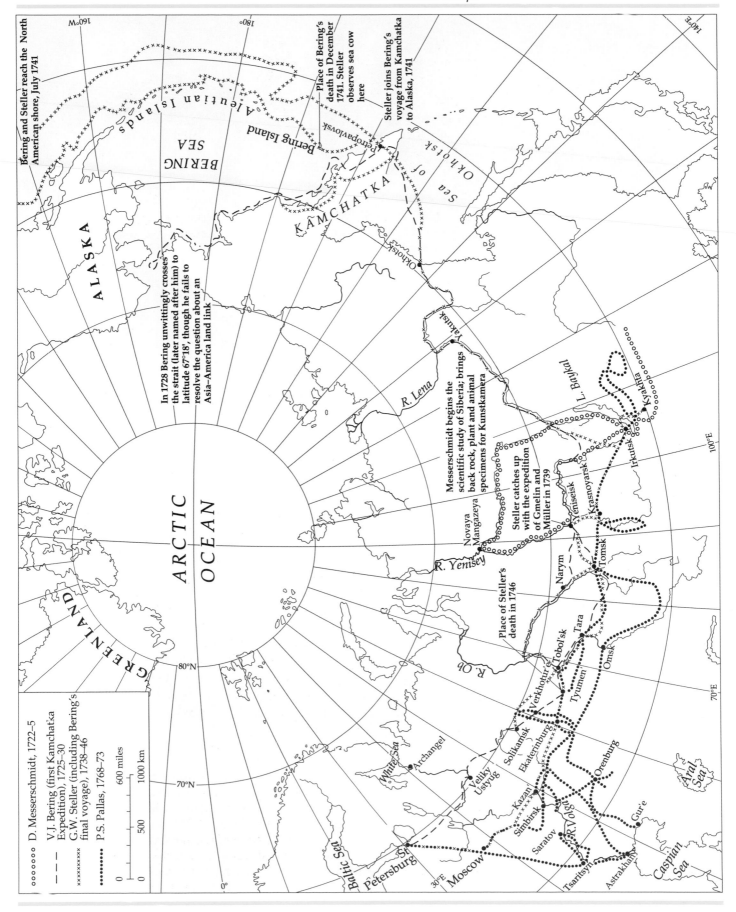

Bering and Steller reach the North American shore, July 1741

Place of Bering's death in December 1741. Steller observes sea cow here

Steller joins Bering's voyage from Kamchatka to Alaska, 1741

In 1728 Bering unwittingly crosses the strait (later named after him) to latitude 67°18', though he fails to resolve the question about an Asia–America land link

Messerschmidt begins the scientific study of Siberia; brings back rock, plant and animal specimens for Kunstkamera

Steller catches up with the expedition of Gmelin and Müller in 1739

Place of Steller's death in 1746

ALASKA

GREENLAND

ARCTIC OCEAN

BERING SEA

Bering Island

Aleutian Islands

KAMCHATKA

Sea of Okhotsk

Petropavlovsk

Okhotsk

Yakutsk

R. Lena

L. Baykal

Kyakhta

Irkutsk

R. Yenisey

Yeniseisk

Krasnoyarsk

Novaya Mangazeya

Narym

Tomsk

R. Ob

Tobol'sk

Tara

Omsk

Tyumen

Verkhotur'

Ekaterinburg

Solikamsk

Orenburg

Aral Sea

Veliky Ustyug

Archangel

White Sea

Baltic Sea

St Petersburg

Moscow

Kazan

Simbirsk

Saratov

R. Volga

Tsaritsyn

Astrakhan

Gure

Caspian Sea

160°W

180°

160°E

140°E

100°E

70°E

30°E

0°

80°N

70°N

D. Messerschmidt, 1722–5

V.J. Bering (first Kamchatka Expedition), 1725–30

G.W. Steller (including Bering's final voyage), 1738–46

P.S. Pallas, 1768–73

600 miles

1000 km

500

0

0

Figure 13.14 Peter Simon Pallas (1741–1811). (From N.A. Figurovsky (ed.) Istoriya estestvoznaniya v Rossii, vol.1, part 1, Moscow, Izdatel'stvo Akademii Nauk SSSR, 1957. Photo: British Library Board.)

Figure 13.15 'Fragrant poplar' from Pallas's Russian Flora. (From N.A. Figurovsky (ed.) Istoriya estestvoznaniya v Rossii, vol.1, part 1, Moscow, Izdatel'stvo Akademii Nauk SSSR, 1957. Photo: British Library Board.)

Figure 13.16 Ivan Lepekhin (1740–1802). (From N.A. Figurovsky (ed.) Istoriya estestvoznaniya v Rossii, vol.1, part 1, Moscow, Izdatel'stvo Akademii Nauk SSSR, 1957. Photo: British Library Board.)

was Ivan Lepekhin (1740–1802), another product of the Academy's gymnasium and university and a full academician from 1771. His European reputation was established by his four-volume *Diaries of a Journey Through Various Provinces of the Russian State* (1771–1805), much of which was translated into German and French. Like Rumovsky, he became involved in the linguistic work of the Russian Academy of Letters and was the chief translator of a ten-volume Russian edition of the French naturalist Buffon's celebrated *Natural History* (44 vols, 1749–1804).

Lomonosov

The foregoing sketch of science in eighteenth-century Russia shows that Peter's hopes of his distinctive national academy were by no means unfulfilled. Significant research had been conducted by native scientists, albeit still in the shadow of expatriates like Euler and Pallas. A place in this picture has yet to be assigned to the controversial Mikhail Vasil'evich Lomonosov (1711–65), who by one account in the tradition of Soviet hagiography 'was for Russia what Galileo was for Italy, Newton for England, Descartes for France, Leibniz for Germany, and Franklin for the USA' (Grigorian and Romanovskaya, 1986, p.53).

Lomonosov's rise to pre-eminence in the Academy is a tribute to the Petrine reforms. Born in the very north of Russia in Archangelsk province, the region where the English established trade relations in the sixteenth century, he was the son of a deacon's daughter and a literate, well-to-do owner of fishing and cargo ships. Although the area was untouched by the forced labour of serfdom, his family was officially of peasant status. The fourteen-year-old Lomonosov's scientific curiosity was awakened by Magnitsky's *Arithmetic* (see Section 13.2.2); in the pursuit of his studies, he had both to escape an antagonistic stepmother and, on reaching Moscow in 1731, to conceal his peasant status in order to gain admission to the Slavonic–Greek–Latin Academy. In 1736, he was sent to the Academy of Sciences' university and thence to the University of Marburg, where for three years he studied under Christian Wolff; he also studied mineralogy and metallurgy at Freiburg. In 1742, he became adjunct in physics at the St Petersburg Academy, though in 1743–4 he was placed under house arrest for insubordination, during which time he read Newton's *Principia* (Boss, 1972, p.171). From 1745 until his death in 1765, he was professor of chemistry. Thus Lomonosov's entire career at the Academy coincided with Euler's time in Berlin; but Euler disappointed Lomonosov's inveterate enemy Schumacher by reporting enthusiastically on some of the Russian's early papers, finding him 'endowed with the most fortunate ingenuity in explaining physical and chemical phenomena' (Leicester, 1970, p.8). Thereafter, Lomonosov began a lifelong correspondence with Euler.

Figure 13.17   Mikhail Vasil'evich Lomonosov (1711–65). (From B.A. Menshutkin, Mikhailo Vasil'evich Lomonosov: zhizneopisanie, *1911.)*

The breadth of Lomonosov's activity in his Academy years is remarkable, even by eighteenth-century standards: a breadth which inspired the poet Pushkin to describe him as 'our first university'. Apart from papers in physics and chemistry, he wrote on astronomy, meteorology, mining and metallurgy, devised improved Newtonian telescopes and navigational instruments, organized the Academy's chemical laboratory in 1748, headed its Geographical Department from 1757, opened a glass factory and designed glass mosaics; in addition to his scientific and technological endeavours, he wrote a Russian grammar and one of the first Russian histories, played a leading role in the foundation of Moscow University (1755), and is justly counted among Russia's major eighteenth-century poets.

Russian claims for his scientific greatness, which Western scholars generally find greatly exaggerated, rest mainly on his supposed anticipations of the law of conservation of matter and the kinetic theory of gases. Certainly, in a long letter to Euler in 1748 (much of its content repeated in a paper presented to the Academy in 1758) he argued, albeit in the most general terms, that:

> all the changes met in nature occur in such a way that if something is added to something, then this is taken away from something else ... Thus, as much material is added to some body as is lost to another ... *(quoted in Leicester, 1970, p.229)*

Moreover, in 1756, he anticipated Lavoisier's experiments by showing that there is no increase in weight when metal is heated in a sealed glass container; but since neither the experiment nor the paper was published, Lavoisier would have been ignorant of these findings. Lomonosov was not, however, as is frequently asserted, a general opponent of phlogiston theory (Leicester, 1970, p.21; see, e.g., Vucinich, 1963, p.109; on phlogiston, see Chapter 15).

Lomonosov's supposed anticipation of the kinetic theory of gases was itself predated by the more rigorous, mathematical account of an earlier St Petersburg academician, Daniel Bernoulli (1700–82). In general, Lomonosov's approach to physics and chemistry exemplified the 'mechanical philosophy' of his time. He believed, like others before him, that heat was no more than the motion of the ultimate constituents of matter, which he, following Boyle, divided into 'elements' and 'corpuscles'. He attempted similar mechanical explanations of light, electricity and gravitation, rejecting the Newtonians' gravitational attraction-at-a-distance as an 'occult quality' (Boss, 1972, p.175). No proper assessment of Lomonosov's scientific work can omit his indebtedness to Descartes, Boyle, Leibniz, Wolff and Stahl, or fail to recognize that much of the originality of his hypotheses lies in the speculative

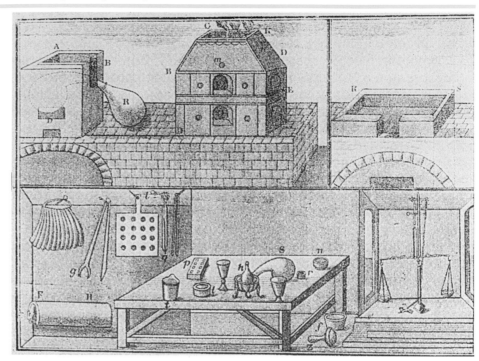

Figure 13.18 Lomonosov's chemical laboratory at the Academy. (From N.A. Figurovsky (ed.) Istoriya estestvoznaniya v Rossii, vol.1, part 1, Moscow, Izdatel'stvo Akademii Nauk SSSR, 1957. Photo: British Library Board.)

free rein allowed by the mechanical approach. There is as much rootedness in the past, as anticipation of the future, in Lomonosov's physical and chemical work, and it is exclusive attention to the latter which informs many Soviet judgements.

> [The other academicians] held their views with deep conviction, and, from the viewpoint of contemporary science, these views seemed more realistic to them than those put forward by Lomonosov, who lived not so much in the today, as in the tomorrow of science. Therefore he so often remained misunderstood by his contemporaries. *(Radovsky, 1961, p.178)*

Misunderstandings with colleagues characterized Lomonosov's time at the Academy. Always at loggerheads with Schumacher, he took patriotic exception to Müller's readings of Russian history, especially his thesis that the first Russian rulers were of Norman (Scandinavian) origin (Black, 1986, pp.115–16, 141), and had numerous other rows, notably with Aepinus. It was no simple matter of Russian against German: Rumovsky objected to the 'despotic power to which Mr. Lomonosov aspires' and criticized his lack of scientific depth (Black, 1986, p.128; Vucinich, 1963, p.113). Lomonosov was sufficiently esteemed in his day to be elected to the academies of Sweden (1760) and Bologna (1764), but he was rejected in Paris. Add to all this the fact that so many of his scientific writings remained unfinished and in manuscript, and the problems in assessing his work become oppressive. Nevertheless, the polymathic output of the most brilliant eighteenth-century Russian is surely better understood, less as heroic foresight, than as the scientific manifestation of the impulse behind Peter's modernizing drive: the desire, at all costs and on all fronts, to propel Russia into the first rank of European nations.

13.4   Science in the Balkans

Whether or not Lomonosov ranks as a scientist of world stature, there undoubtedly existed at the close of the eighteenth century a self-perpetuating Russian scientific community. No such community can be found among the

Orthodox peoples of south-east Europe, despite traditions of scholarship in the empires of the Byzantines, Serbs and Bulgarians before the Ottoman conquests. Only in the second half of the eighteenth century did the ideas of the Scientific Revolution percolate into the Ottoman Empire; and only during the nineteenth century, as Greece, Serbia, Bulgaria and Romania won independence, were the necessary secular schools, universities, scientific societies and academies instituted.

In Russia, the rise of science depended upon the ruling elite's energetic attempts to counteract their Orthodox subjects' hostility to Western learning. In the Balkans, however, this hostility was reinforced by the Turkish rulers' Muslim presumption of the superiority of their beliefs and techniques over those of the infidel. Resistance to Western science did not entail overt persecution of the Orthodox, and indeed, this did not happen. The Ottomans regarded the Orthodox, like all Christians and Jews, as 'people of the Book' worshipping the same God, and so tolerated their religious practices. All Orthodox were officially organized as one nation, or *millet*, within the Empire, with the Patriarch of Constantinople as secular head. In practice, the system resulted in slow demoralization. The Orthodox were subjected to heavy taxation, made to wear distinctive dress, forbidden to marry Muslim women, and prevented from opening new churches or schools, or from undertaking missionary work among Muslims. The sultans began by installing the most anti-Latin of the Orthodox leaders as Patriarch of Constantinople and then spread corruption down through the hierarchy by holding the office up to the highest bidder. The most obvious result was a remarkable turnover of patriarchs: there were 159 in nearly five centuries of Turkish rule, and only 21 died a natural death in office (Zernov, 1961, p.135). More fundamentally injurious was the *devshirme* ('gathering') system, under which, at five-yearly intervals, the strongest and brightest boys between the ages of 8 and 15 amongst the Balkan peoples were converted to Islam and trained as administrators, or more likely, drafted into the Janissaries, a crack infantry corps.

The *devshirme* system, however detrimental to the Orthodox Church, undeniably offered converts considerable advancement in Ottoman society; it was finally abolished by Mehmed IV (*r.* 1648–87), under whom the Empire began its slow decline after the failure to take Vienna in 1683. A new vigour in Western industry and military technique had become apparent, and some of the Turkish elite became more receptive to Western practices. At the court of Ahmed III (*r.* 1703–30), there was affectation of Western manners and dress, and in 1727 the printing of books in Turkish was allowed for the first time, until the suppression of the press in 1742. A military engineering school was founded in 1734 by a renegade French nobleman, though this, too, was shut down in 1750. Under Sultan Selim III (*r.* 1780–1807), the press was revived and a more earnest effort made to modernize the armed forces. European technicians were employed to supervise modern armaments factories and train Ottoman officers in new naval and land engineering schools, founded in 1784 and 1793 respectively. These Western influences should not be exaggerated; they reflected the traditional exemption of military matters from the general Muslim injunction on Western 'innovations' (Lewis, 1982, pp.224–5; Shaw, 1971, p.199).

Selim's brief modernization drive, ended by a conservative coup, was the ruling elite's reluctant response to mounting outside pressure from European powers, especially Austria, Russia and France. The educated Balkan Orthodox proved less resistant to Western ideas. The most outstanding example was Dimitrie Cantemir (1673–1723), Prince of Moldavia, whose

Figure 13.19   Sultan Selim III (r. 1780–1807). (Topkapi Saray Portrait Collection, Istanbul)

353

Figure 13.20 Prince Dimitrie Cantemir (1673–1723). (From C. Maciuca, Dimitrie Cantemir, *Editura Tineretului, n.d.)*

Figure 13.21 Peter the Great visiting Dimitrie Cantemir at Jassy, capital of Moldavia. (From C. Maciuca, Dimitrie Cantemir, *Editura Tineretului, n.d.)*

Descriptio Moldaviae (1701), apart from its geographical and economic information, popularized the Flemish natural philosopher J.B. van Helmont (1577–1644). Subsequently translated into English, French and German, it won him election to the Berlin Academy. Cantemir's entry into European intellectual life followed an alliance concluded with Peter the Great in 1711 against the Turks; the defeat of Peter's army led to Prince Dimitrie's flight to Russia, where he became a senator. His son Antiokh (1708–44) was educated at the Slavonic–Greek–Latin Academy and the St Petersburg Academy, and became a major Russian poet; he also translated the Cartesian Fontenelle's *Conversations on the Plurality of Worlds* (1686), though by the time the St Petersburg Academy published it (1740), Cantemir had become an ardent Newtonian (Boss, 1972, p.126).

The Cantemirs exemplify the greater cultural independence of the vassal states north of the Danube, where from the early seventeenth century, Roman Catholic influences had spread from Vienna. In the second half of the seventeenth century, ecclesiastical academies were founded at Jassy, the capital of Moldavia, and Bucharest, the capital of Wallachia. The academy at Jassy was based on the Kiev Academy and had a similar Aristotelian neo-scholastic curriculum. By the eighteenth century, both academies were teaching modern philosophy and science. Within the Ottoman Empire, the teaching of Aristotle had been revived at Constantinople's Patriarchal Academy during the 1620s by Theophilos Korydaleus (1570–1646), a Greek student of Cremonini at the University of Padua (see Chapter 4). Korydaleus began his teaching career at a new academy in Athens, and it was at such ecclesiastical schools that Greek monks, often educated in the West, restored the Byzantine tradition of classical learning within the bounds of Orthodox dogma. The founders of these schools were generally wealthy Greek merchants, whose trading in the Mediterranean and Black Sea areas brought them into contact with Western ideas and culture.

During the eighteenth century, graduates of the new Greek schools (in which the Turkish authorities usually took little interest) discovered in Enlightenment thought an ideological weapon against the Turkish yoke and its Greek Orthodox apologists. They were able to disseminate these ideas in Greek writings published outside the Ottoman Empire, above all in Vienna. These writings included translations from the best-known work of the French Enlightenment, the *Encyclopédie*, and a translation in 1794 of Fontenelle's *Plurality of Worlds*. Some were more directly scientific in content: a compendium called *The Way of Mathematics*, published in Venice in 1749, was compiled by Methodios Anthrakites (*c.* 1660–1748), who taught at a number of Greek schools and was condemned by the Church in the 1720s for preferring modern philosophy to the approved Aristotelianism (Henderson, 1971, p.17). The compendium introduced the Copernican system to the Greeks, though its translator preferred the geocentric system. A similar position was taken by another controversial teacher, Eugenios Voulgaris (1716–1806): his work *The System of the Universe* (1805) expounded Copernican-Newtonian cosmology, but on scriptural grounds favoured Tycho Brahe's system (Henderson, 1971, p.68; Pappas and Karas, 1987, p.240). Voulgaris is an appropriate figure with whom to end this section, partly because of his Russian connections: he left the Ottoman Empire because of the Greek Church's hostility towards modern philosophy, and eventually became librarian at the court of Catherine the Great and a member of the St Petersburg Academy. More importantly, his work raises the general issue of the relations between science and the Orthodox religion.

13.5 Orthodoxy and science

Voulgaris's preference for Tychonian cosmology was well judged, for in 1804 Beniamin Lesvios, a teacher at one of the 'most advanced' Greek ecclesiastical academies, was condemned by the patriarchate for teaching the heliocentric theory (Clogg, 1976, pp.264–5). Modern science was more welcome in Russia's theological academies and seminaries, which from Peter's time were supposed to include mathematics and physics in their curricula; from the 1780s, at any rate, the Kiev Academy began to teach modern, instead of Aristotelian, physics (Vucinich, 1963, p.163). The contribution of the church schools has often been overlooked; following the failure of Peter's secular schools, they carried the burden of education for most of the eighteenth century and 'played a profoundly important role in Russia's educational and professional growth' (Nichols, 1978, p.67). But the Russian Church was no champion of Western science. In the seventeenth century, when the Church monopolized printing, censorship and what educational institutions there were, Western scientific and philosophical ideas continued to be condemned as heresies, even after the introduction of Aristotelianism through the curricula of its ecclesiastical academies. Silvester Medvedev, a 'Latinizer' at the Slavonic–Greek–Latin Academy, was burnt at the stake in 1691 during a period of reaction, and the Academy's charter threatened a similar end to any 'magicians' caught disseminating secular natural-scientific knowledge (Vucinich, 1963, p.25). Even after control of the printing and censorship of secular books had passed to the Academy of Sciences, the Church's censure of heterodox theories could inhibit lay publishers: for example, articles by Joseph Delisle and Daniel Bernoulli attributing motion to the earth were not translated into Russian; Cantemir's translation of Fontenelle's *Plurality of Worlds* was blocked for ten years; and the Church imposed changes in a translation of Pope's *An Essay on Man* (1733–4), objectionable because of its Newtonian allusions and eventually published in 1757. It may also have been fear of theological censure which partly explains the relative inactivity and obscurity of Caspar Friedrich Wolff (1734–94), a German biologist whose important contributions to embryology were recognized only after his death, and who joined the St Petersburg Academy at Euler's initiative in 1767. The sensitive, and significant, aspect of Wolff's research was its repudiation of 'preformist' embryology: essentially the belief, consistent with the existence of fixed and immutable species, that development is simply the expansion of a microscopic, but fully formed, embryo (Vucinich, 1963, pp.87, 116, 183; Boss, 1972, pp.224–6).

It seems, then, that the Russian Orthodox Church was at best a reluctant sponsor of Petrine modernization. Its traditional stance was, as we saw in Section 13.1, backward-looking and hostile to Latin innovations. It had very little theology before the seventeenth century, and since preaching was in the vernacular, the clergy had no need for a classical education. In contrast with Protestants, the Orthodox valued the ritual aspect of Christianity above all else; hence the Old Believers, who broke with the Church over its mid seventeenth-century reforms, would rather burn themselves to death than accept any rewording of the liturgy or make the sign of the cross with three rather than two fingers. In general, the faith of Western Christians in the essentially rational nature of God's creation was absent in Russian Orthodoxy, which was

> dominated by a mystical acceptance of the universe as an entity ruled by a miracle-working divine caprice, which could not be comprehended by an inquiring rational mind – only felt, through the medium of awe-inspiring ritual.
> (*Vucinich, 1963, p.36*)

Science could hardly take root in such an antirationalist culture, unless the Church's intellectual monopoly was broken; steps to reduce ecclesiastical power were taken by Tsar Alexis, whose 1649 Code of Laws prohibited the Church from acquiring new land. This measure was fully enforced only in the reign of Peter the Great, who revolutionized Church–State relations in general. After the death of Patriarch Adrian in 1700, he kept the see vacant and finally in 1721 promulgated the lengthy *Spiritual Regulation*, through which the patriarchate was replaced by the Most Holy Synod, effectively a government department directed by a lay over-procurator. The Church's administration (though not its dogmas or liturgy) thereby came under State control. The document also laid down that mathematics and physics should be included in the curricula of ecclesiastical schools.

It would be tempting at this point to conclude that science was only established in Russia because of the subservience of the Orthodox patriarch to the imperial throne; the secular ambitions of the monarchy could thereby eventually prevail over the spiritual objections of the clergy. The relationship between science and Orthodoxy would accordingly be seen as one of 'open conflict' (Vucinich, 1963, p.183). The conflict metaphor is undeniably more applicable to Russian culture than to the scientific culture of Protestant Europe; there are, nevertheless, some qualifications to be made. First, antirationalism in the Church was by no means universal, especially after the seventeenth-century annexation of Kiev from Poland, when the intellectual life of the Church underwent a transfusion of rationalist, 'Latinizing' theological tendencies. Indeed, in the eighteenth century, the Church hierarchy fully endorsed the teaching at its schools of the curriculum of a Latin grammar school (Freeze, 1977, pp.90–5). Second, the champions of Western learning were often, as far as can be told, sincere Orthodox believers. Tsar Alexis, for example, the first Romanov to promote Western science actively, was 'the most devout of Russian Sovereigns' (Zernov, 1978, p.93).

The determination of Peter the Great's religious stance is more problematic. The Old Believers regarded him as Antichrist because of his Western-style innovations, and he was suspected of Protestant leanings by his enemies in the Church hierarchy. There is some evidence for the second charge: he mixed happily with Protestants in Moscow's Foreign Quarter, and some of them participated in Peter's 'Most Drunken Council', a debauched lampooning of the Russian Church, sometimes conducted before horrified foreign dignitaries. The reforming tsar was well versed in his religion and willing to discuss it with Protestants and Catholics in his European visits. He had several such discussions with Bishop Gilbert Burnet in England in 1698, but significantly, Burnet recalled that Peter 'hearkened to no part of what I told him more attentively than when I explained the authority that the Christian Emperours assumed in matters of religion and the supremacy of our Kings' (quoted in Cracraft, 1971, p.33). Although Peter's Church reform drew heavily on English and Swedish precedents (Muller, 1972, p.xxxvi), neither he nor his collaborators questioned the basic tenets of the Orthodox faith. His main supporter in the Orthodox Church was the Ukrainian Feofan Prokopovich (1681–1738), whom Peter made Archbishop of Pskov and who drafted the *Spiritual Regulation*. He was the first Orthodox ecclesiastic to be an enthusiastic Westernizer and to preach the compatibility of science and religion; initially a teacher at the Kiev Academy, he introduced mathematics and physics into its curriculum in 1707 and argued that theologians should interpret the scriptures 'not literally but allegorically' in order to accommodate Copernicus's heliocentric theory (Vucinich, 1963, p.57). In the *Spiritual Regulation*, a fifth of which was devoted to ecclesiastical schools, he

Figure 13.22 Archbishop Feofan Prokopovich (1681–1738). (From A. Sydorenko, The Kievan Academy in the Seventeenth Century, *Ottawa, University of Ottowa Press, 1977.)*

combatted the view that 'learning is responsible for heresy', like Krizhanich before him (see Section 13.2.1). He preached submission to the tsar's authority, equating the lapsed office of patriarch with 'papalism', and became the 'chief ideologist of the Petrine state' (Cracraft, 1971, pp.59–60, 263).

Belief in the compatibility of science and religion was predictably unquestioned at the Academy of Sciences, though its most popular articulation, in the form of Rumovsky's translation of Euler's *Letters to a German Princess*, comes clearly from the Swiss Calvinist tradition. Lomonosov described science and religion as 'two kindred sisters, the daughters of one supreme parent' and, invoking the authority of the Greek father Basil the Great, urged that 'reasonable and good persons must search for means to explain and avert all imagined conflict between them, following the aforementioned wise teacher of our Orthodox Church' (Lomonosov, 1950–83, vol.4, p.373). Like Galileo, he advocated a division of labour between natural philosophers and theologians:

> The mathematician reasons incorrectly, if he wishes to measure the divine will with a pair of compasses. So does the teacher of theology, if he thinks that one can learn astronomy or chemistry from the psalter. *(Lomonosov, 1950–83, vol.4, p.375)*

Although Lomonosov may well have imbibed his views on science and religion through his study with Christian Wolff at Marburg, and although Soviet scholars generally follow his best-known Russian biographer and detect deism in his writings (Menshutkin, 1952, p.83), he nevertheless remained overtly committed to the Orthodox faith. Other ostensibly Orthodox intellectuals could be cited, such as Vasily Tatishchev, who formed with Prokopovich and Antiokh Cantemir a 'Learned Guard', dedicated to the continuation of Peter's policies after his death. Both Lomonosov and Tatishchev, however, were critical of the Church, and anticlericalism, allied with scepticism about Orthodox dogma, came to prevail amongst lay intellectuals, especially during the reign of Catherine the Great. Catherine herself was a German Lutheran by birth and a deist by conviction (Madariaga, 1981, p.120). In 1764, she secularized all Church land and 'converted the Church into a salaried, bureaucratic organization' (Freeze, 1977, p.64); but she ensured support in the hierarchy for her Westernizing policies by advancing the ecclesiastical career of her protégé Platon Levshin (1737–1812). He became Metropolitan of Moscow in 1775 and did much to improve the education of the clergy.

Any compatibility mooted between science and the Orthodox religion needs to be carefully stated. Some historians have argued that the values and beliefs of Protestantism were conducive to the development of European science and, relatedly, that British Nonconformity spurred the Industrial Revolution. It would be difficult to assign similar roles to Orthodox beliefs. It is noteworthy that the Russian Nonconformists, the Old Believers, were the most hostile of all to Western science; and although they became, like British Nonconformists, disproportionately active in industry and commerce, what they shared was surely not religious motivation, but social and economic disabilities (they were debarred from public office and subjected by Peter to double taxation). The most that can be said is that the political subservience of the Orthodox Church permitted imperial patrons of science to surmount its traditional antirationalism, partly through their power to advance their supporters in the hierarchy. But the impetus for scientific development remained secular. In the Ottoman Empire, the sultans mostly rejected Western ideas, as did the Russian autocracy in the wake of the French Revolution and again after Napoleon's invasion of 1812. Under these circumstances, the Orthodox hierarchy revealed itself as a vastly more enthusiastic persecutor, than proselytizer, of Western science.

13.6   Conclusion

The highs and lows of science in Orthodox Europe before 1800 were a barometer of the relations of Church and State within distinctive imperial contexts. In the Ottoman Empire, complicated though these relations were by the ruling class's staunch Muslim traditionalism, their net effect was redoubled resistance to the West – and this despite the Ottomans' grudging recognition from the end of the seventeenth century of the growing superiority of Western military and industrial technology. Until the French Revolution, the Russian tsars had less difficulty than their subjects in allowing their imperial interests to override any Orthodox qualms about Western innovations. Their difficulty was not so much the objections of a subservient hierarchy, but the near absence in their feudal domain of the social, economic, educational and theological conditions for the emergence of self-perpetuating traditions of scientific research.

Russia's scientific revolution had to be a revolution from above. In a feudal society, the creation of institutions necessary to support a tradition of scientific research depended upon imperial patronage. How successful were the Romanov tsars? In the longer term, the institutional innovations of the eighteenth century led to the world-ranking achievements of Lobachevsky, Butlerov, Mendeleyev, Mechnikov and Pavlov in the nineteenth. With that perspective, the debate over Lomonosov's originality looks less crucial. In any case, no comprehensive account of native eighteenth-century research could pass over the work of Vasily Severgin (1765–1826), another product of the Academy's gymnasium and university, and its professor of mineralogy from 1793: his pioneering study of Russian minerals extended the tradition of the scientific exploration of Russia. In the physical sciences, Vasily Petrov (1761–1834) conducted original experiments in the 1790s within the framework of Lavoisier's theory of combustion, and in the early 1800s, joined the select band of early investigators of the electric battery. Petrov is especially significant as both his education and the best of his research took place outside the Academy of Sciences (he became a professor of the Medical-Surgical Academy) – evidence of some broadening of the Russian research base. It might still be doubted whether native scientists had realized Peter the Great's expectations by 1800; what cannot be denied is that a scientific tradition had become established that was peculiarly Russian.

Figure 13.23   Vasily Severgin (1765–1826). (From N.A. Figurovsky (ed.) Istoriya estestvoznaniya v Rossii, *vol.1, part 1, Moscow, Izdatel'stvo Akademii Nauk SSSR, 1957. Photo: British Library Board.)*

Sources referred to in the text

Appleby, J.H. (1983) 'Ivan the Terrible to Peter the Great: British formative influence on Russia's medico-apothecary system', *Medical History*, 27, pp.289–304.

Black, J.L. (1979) *Citizens for the Fatherland: Education, Educators and Pedagogical Ideals in Eighteenth Century Russia*, Boulder, Col., East European Quarterly.

Black, J.L. (1986) *G.-F. Müller and the Imperial Russian Academy*, Kingston and Montreal, McGill-Queen's University Press.

Boss, V. (1972) *Newton and Russia: the Early Influence, 1698–1796*, Cambridge, Mass., Harvard University Press.

Carver, J.S. (1980) 'A reconsideration of eighteenth-century Russia's contributions to modern science', *Canadian-American Slavic Studies*, 14, pp.389–405.

Clogg, R. (1976) 'Anticlericalism in pre-independence Greece c. 1750–1821', in Baker, D. (ed.) *The Orthodox Churches and the West*, Oxford, Ecclesiastical History Society/Basil Blackwell.

Cracraft, J. (1971) *The Church Reform of Peter the Great*, London, Macmillan.

Dukes, P. (1982) *The Making of Russian Absolutism 1613–1801*, London and New York, Longman.

Freeze, G.L. (1977) *The Russian Levites: Parish Clergy in the Eighteenth Century*, Cambridge, Mass., Harvard University Press.

Grigorian, A.T. and Romanovskaya, T.B. (1986) 'The scientific achievements of M.V. Lomonosov', *Archives Internationales d'Histoire des Sciences*, 36, pp.45–53.

Henderson, G.P. (1971) *The Revival of Greek Thought, 1620–1830*, Edinburgh and London, Scottish Academic Press.

Home, R.W. (1977) '"Newtonianism" and the theory of the magnet', *History of Science*, 15, pp.252–66.

Home, R.W. (1979) Introductory monograph to *Aepinus's Essay on the Theory of Electricity and Magnetism*, Princeton, NJ, Princeton University Press.

Klyuchevsky, V. (1965) *Peter the Great*, London, Macmillan.

Kopelevich, Yu. Kh. (1977) *Osnovanie Peterburgskoy Akademii Nauk*, Leningrad, Izdatel'stvo Nauka.

Kulikovsky, P.G. (1975) 'Stepan Yakovlevich Rumovsky', in Gillispie, C.C. (ed.) *Dictionary of Scientific Biography*, vol.11, New York, Charles Scribner's Sons.

Leicester, H.M. (ed.) (1970) *Mikhail Vasil'evich Lomonosov on the Corpuscular Theory*, Cambridge, Mass., Harvard University Press.

Letiche, J.M. and Dmytryshyn, B. (1985) *Russian Statecraft: the 'Politika' of Iurii Krizhanich*, Oxford, Basil Blackwell.

Lewis, B. (1982) *The Muslim Discovery of Europe*, London, Weidenfeld and Nicolson.

Lomonosov, M.V. (1950–83) *Polnoe Sobranie Sochinenii*, 11 vols, Moscow and Leningrad, Izdatel'stvo Akademii Nauk SSSR.

Longworth, P. (1984) *Alexis: Tsar of All the Russias*, London, Secker and Warburg.

Madariaga, I. de (1981) *Russia in the Age of Catherine the Great*, London, Weidenfeld and Nicolson.

Marker, G. (1985) *Publishing, Printing, and the Origins of Intellectual Life in Russia, 1700–1800*, Princeton, NJ, Princeton University Press.

Menshutkin, B.N. (1952) *Russia's Lomonosov*, Princeton, NJ, Princeton University Press.

Muller, A.V. (ed.) (1972) *The Spiritual Regulation of Peter the Great*, Seattle and London, University of Washington Press.

Nichols, R.L. (1978) 'Orthodoxy and Russia's Enlightenment, 1762–1825', in Nichols, R.L. and Stavrou, T.G. (eds) *Russian Orthodoxy Under the Old Regime*, Minneapolis, University of Minnesota Press.

Okenfuss, M.J. (1973) 'Technical training in Russia under Peter the Great', *History of Education Quarterly*, 13, pp.325–45.

Pappas, V. and Karas, I. (1987) 'The printed book of physics: the dissemination of scientific thought in Greece 1750–1821 before the Greek Revolution', *Annals of Science*, 44, pp.237–44.

Radovsky, M.I. (1961) *M.V. Lomonosov i Peterburgskaya Akademiya Nauk*, Moscow, Izdatel'stvo Akademii Nauk SSSR.

Runciman, S. (1968) *The Great Church in Captivity: a Study of the Patriarchate of Constantinople from the Eve of the Turkish Conquest to the Greek War of Independence*, Cambridge, Cambridge University Press.

Ryan, W.F. (1986) 'Aristotle and Pseudo-Aristotle in Kievan and Muscovite Russia', in Kraye, J., Ryan, W.F. and Schmitt, C.B. (eds) *Pseudo-Aristotle in the Middle Ages: the 'Theology' and Other Texts*, London, Warburg Institute.

Schulze, L. (1985) 'The Russification of the St. Petersburg Academy of Sciences and Arts in the Eighteenth Century', *British Journal for the History of Science*, 18, pp.305–35.

Shaw, S.J. (1971) *Between Old and New: the Ottoman Empire Under Selim III, 1789–1807*, Cambridge, Mass., Harvard University Press.

Vogel, K. (1967), 'Byzantine science', in Hussey, J.M. (ed.) *The Cambridge Medieval History*, vol.IV, part II, London and New York, Cambridge University Press.

Vucinich, A. (1963) *Science in Russian Culture: a History to 1860*, Stanford, Calif., Stanford University Press.

Youschkevitch, A.P. (1971) 'Leonhard Euler', in Gillespie, C.C. (ed.) *Dictionary of Scientific Biography*, vol.4, New York, Charles Scribner's Sons.

Zernov, N. (1961) *Eastern Christendom: a Study of the Origin and Development of the Eastern Orthodox Church*, London, Weidenfeld and Nicolson.

Zernov, N. (1978) *The Russians and their Church*, 3rd edn, London, SPCK.

Further reading

Black, J.L. (1986) *G.-F. Müller and the Imperial Russian Academy*, Kingston and Montreal, McGill-Queen's University Press.

Boss, V. (1972) *Newton and Russia: the Early Influence, 1698–1796*, Cambridge, Mass., Harvard University Press.

Lewis, B. (1982) *The Muslim Discovery of Europe*, London, Weidenfeld and Nicolson.

Vucinich, A. (1963) *Science in Russian Culture: a History to 1860*, Stanford, Calif., Stanford University Press.

Acknowledgements

The author is indebted to the Course Assessor and Course Team colleagues for their comments, and also to Professor J.L. Black, Professor R.W. Home, Dr Antony Lentin and Dr W.F. Ryan for their detailed advice and criticisms; the main divisions and emphases remain his responsibility.

Establishing Science in Eighteenth-Century Central Europe

Chapter 14

by Gerrylynn K. Roberts

14.1 Introduction

Following the Thirty Years' War (1618–48) the authority of the Habsburg emperors over the whole of the Holy Roman Empire continued to give way gradually to greater sovereignty for some three hundred independent states. Although these shared a common language, they participated in a bewildering variety of military and commercial alliances and were divided along confessional lines according to the preferences of their rulers.

Although pursued mostly by exceptional individuals in a rather fragmentary manner, the sciences – especially the more mystical aspects of alchemy which could be allied with the themes of the Counter-Reformation – remained influential in the courts of Central Europe in the seventeenth century, particularly in the Habsburg court of the Holy Roman emperors. At the same time, as a new political balance emerged between the authority of the emperor and that of the princes who ruled the states, the importance of a state role in technology was reasserted in order to facilitate economic recovery (Evans, 1979, part 2, ch.5, and part 3). Important also in the aftermath of the Thirty Years' War was the way, in institutional terms, that individual territories sought to cope with these changes. Seventeenth-century developments laid the ground for the subsequent establishment of science in state-sponsored educational institutions, and despite the emergence in Germany of certain scientific academies based on foreign models, especially the important Berlin Academy (see Chapter 9), it was the educational context that was to characterize the pursuit of science in eighteenth-century Central Europe.

14.2 The German universities

As early as 1609, when the chair of chemiatry was established at the University of Marburg (see Chapter 6), there were already 26 universities in the German-speaking lands. A further eleven were founded during the seventeenth century, the last of which (that created in Prussian Halle in 1694) initiated an important phase of reform in university education; a further four universities were founded in the eighteenth century (Paulsen, 1895, pp.239–40).

As Ornstein put it

> ... this remarkable increase was not the consequence of increased enthusiasm for learning. It was due in part to the establishment of centers of Lutheran and Calvinistic teaching, in part to the decentralized condition of Germany, to the fact that the ruler of even a small territory wanted his own university. *(Ornstein, [1913] 1975, p.226)*

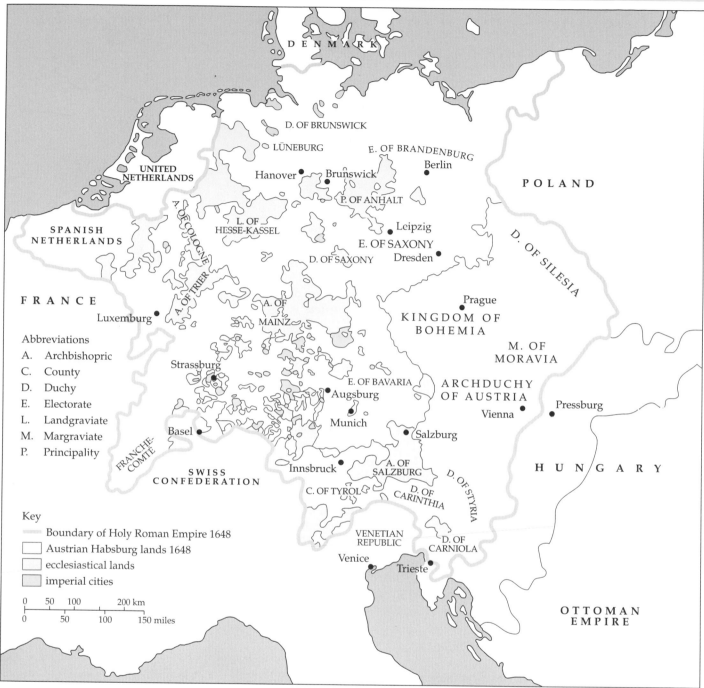

Map 14.1 The Holy Roman Empire in 1648. The Holy Roman Empire was divided into 234 distinct territorial units, 51 free cities and innumerable estates of imperial knights. This map shows only some of these divisions, but gives a hint of the political complexity of the area.

This trend increased in the years after the Thirty Years' War. Universities became the local agencies for the professional education of the growing number of secular and ecclesiastical officials needed for the exercise of territorial sovereignty. Formally, German universities were all state, archiepiscopal or municipal institutions. Before the Reformation, the procedure for founding universities was that a territorial ruler would call a new university into existence, authorize its funding and grant it corporate rights. Papal approval was then sought, as it was that which provided the authority for the holding of examinations and the granting of degrees. It became common also to seek the authority of the emperor. Along with other European universities in the mediaeval tradition of *studium generale*, they

Map 14.2 *The establishment of universities in Central Europe to the end of the eighteenth century. (Drawn from information in F. Paulsen, The German Universities: Their Character and Historical Development, tr. E.D. Penny, New York, Macmillan, pp.239–40.)*

recruited from a wide area and they had the power to grant a licence for teaching in any part of the world; the status of their degrees was widely recognized and approved. After the Reformation, it proved difficult for some Protestant foundations to secure papal and imperial approval (as was the case with Libavius's institution; see Chapter 6); this may have been a contributory factor in the increasing localization of education. Whereas at the time of Agricola and Paracelsus it was common practice to study at a number of universities in different places, in the seventeenth century, students were encouraged to study at their local universities. This of course had the effect of bolstering individual territorial coffers by discouraging the movement abroad of students' money, but it also helped to create a locally trained officialdom with the appropriate confessional orientation for a particular state. The seventeenth-century system was thus less open than previously, less universal and more marked by government efforts to control instruction (Paulsen, 1906, pp.36–7).

The universities were all organized on the mediaeval four-faculty basis – theology, law, medicine and philosophy – although several did not have all four. Theology was always the chief of the faculties in both Catholic and Protestant universities. Law was second in importance, particularly with the development of the states' sovereignty in the seventeenth century. This was followed by medicine, always the smallest of the three professional faculties until the nineteenth century. Numerically the most important faculty was arts, or philosophy as it became. It was the case from mediaeval times that study in this faculty was preparatory to entering any of the others, and until well into the nineteenth century it was not considered to provide a complete education. At the end of the seventeenth century, 23 of the universities were Protestant. Aristotle remained at the centre of the curriculum in all universities and Latin was the common language of tuition.

German-speaking universities which begin to teach chemistry in the seventeenth century

Marburg 1609
Jena 1630
Rostock 1640s
Leipzig 1668
Duisburg 1671
Erfurt 1673
Altdorf 1677
Strassburg 1685
Helmstedt 1688
Frankfurt-an-der-Oder 1690s
Tübingen 1693
Cologne 1698

As for the teaching of science in the seventeenth-century German universities, the Marburg chair of chemiatry, quite apart from being an early chemical chair, was an exception in being seen initially as a keystone of an ideal system of arts and sciences (see Chapter 6; Moran, 1985, pp.121–2). It soon, however, became merely an adjunct to the teaching of medicine which was a common characteristic of chemical teaching in the seventeenth century.

By 1720, six universities had salaried professors with formal responsibility for teaching chemistry (Hufbauer, 1982, p.33 and Appendix II), but the teaching at many of these institutions was minimal. The Protestant university of Altdorf was exceptional. It was founded by the city of Nürnberg in 1573 and achieved university status in 1622. Both modern physics and chemistry were taught there. One professor, who had founded a society for experimentation, repeated for his students experiments which he had seen at the Accademia del Cimento in Florence (see Chapter 9). The professor of astronomy had an observatory built and equipped by prosperous townspeople; and the university funded the building and equipping of a chemical laboratory. Botany was rather more generally favoured, and several universities set up botanical gardens in the seventeenth century. Although the medical faculties were not large, some did adopt modern Italian practice by building anatomical theatres for dissection – Freiburg in 1620, Jena in 1629, Altdorf in 1637 and Würzburg in 1675 (Ornstein, [1913] 1975, pp.230–1).

Despite this level of expansion and evolution in the seventeenth century, the German universities fell into disrepute.

> At the end of the seventeenth century the German universities had sunk to the lowest level which they ever reached in the public esteem and in their influence upon the intellectual life of the German people. The world of fashion, which centred at the princely courts, looked down upon them from the heights of its modern culture as the seat of an absolute and pedantic scholasticism. *(Paulsen, 1906, p.42)*

For the most part, the universities did not adapt their curricula; the professoriate was too powerful. Sterile theological faculties remained dominant as new orthodoxies replaced, in both Protestant and Catholic institutions, the intellectual stimulus of the sixteenth and early seventeenth centuries. Their independent corporate structure, which was rooted in the Middle Ages, allowed the universities to be isolated from the rest of society. This separateness served to marginalize them. Students gained a reputation for brutishness and depravity, and their numbers fell. The nobility, whose sons had previously attended the universities in large numbers, turned to what they saw as the more relevant education of the new *Ritterakademien*, or 'knightly' academies.

> These stressed the building of character over knowledge for its own sake, in keeping with the German version of the Renaissance idea of the gentleman. The academies offered worldly breadth, rather than scholastic depth; and a grand tour, or, occasionally, a short visit to a foreign university completed the ideal cosmopolitan gentlemanly education begun in the academies, especially for those noblemen bent on a diplomatic or civil service career. *(McClelland, 1980, p.33)*

Excluding the Austrian universities, just before the Thirty Years' War the then twenty German universities had a total enrolment of almost 8000 students, an average of 400 students each. By 1700, the now twenty-eight universities had a total enrolment of roughly 9000, but twenty of them had less than 300 students. Heidelberg, with only 80 students, was scarcely viable (McClelland, p.28). When Leibniz sought to promote his comprehensive reform of learning, it was not to the universities that he turned. Rather, he looked for royal patronage of an independent academy on the French model, resulting in the establishment in 1700 of what later became the Prussian Royal Academy of Sciences (see Chapter 9).

14.2.1 University reform in the early eighteenth century

There were, however, pockets of reform within this generally bleak picture – the universities of Halle, Göttingen and Erlangen founded in 1694, 1737 and 1743 respectively.

Halle

It was by accident, rather than by plan, that Halle attracted a group of active strong personalities and became a leader of university reform (McClelland, 1980, p.34). Halle was a vigorous, wealthy town enjoying an economic boom in the closing years of the seventeenth century. The new university's nucleus was a Ritterakademie which was transformed into a university by the Elector of Brandenburg-Prussia after the reformer Christian Thomasius (1655–1728) sought refuge there. A professor of law, who from 1679 taught contentiously in the vernacular German instead of Latin at the conservative University of Leipzig, Thomasius published a German-language periodical from 1688 which was critical of the universities. He was against pedantic scholasticism and residual Aristotelianism, as well as against narrow confessional religiosity. Although a Lutheran himself, his published stance on narrow orthodoxy risked imprisonment in Saxony and led to his expulsion from his Leipzig post (Ornstein, [1913] 1975, pp.233–4). Once re-established at the Halle Ritterakademie, Thomasius attracted so many students that that institution soon developed into a university. The theological faculty was inspired by August Hermann Francke (1663–1727), a Pietist who had already established a range of educational institutions at Halle primarily aimed at helping the poor through instruction in the vernacular in religious subjects as well as practical instruction in the manual trades. Devoted to the study of the Bible and practical Christianity, Pietism stressed active faith and the contemplative life rather than scholastic subtlety. It has been argued, therefore, that Halle's rather more worldly theological faculty was at the least not a brake on developments in the other faculties (McClelland, 1980, pp.34–5). Some scholars would argue even more positively that Francke and

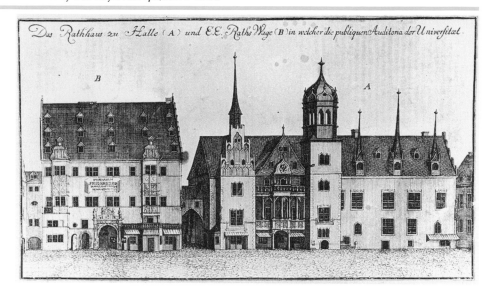

Figure 14.1 *(top) Halle University. The Brandenburg Elector Frederick III obliged the city of Halle to make available to the university several rooms of the Ratswaage (the building where marriages were celebrated). This building (B in the illustration) was joined to the Rathaus (Town Hall, A in the illustration) by a covered passage at the first floor. (bottom) The Great Auditorium in the Ratswaage served the university until 1834. (Martin-Luther University, Halle-Wittenberg)*

other late seventeenth-century Pietists viewed natural science favourably as a route to understanding Divine Revelation via an understanding of the divine laws of nature (Merton, 1984).

In 1707, Christian Wolff (1679–1754) came to Halle as professor of mathematics in the philosophical faculty. He soon added the teaching of physics to his brief. Wolff argued for the separation of theology and philosophy, basing the latter on the modern sciences of mathematics and physics (see Chapter 12). His philosophy, a development of Leibniz's teachings, became very prominent at German universities in the eighteenth century and spurred a reorientation towards a view of nature as controlled by thought and reason. He too wrote in German as well as Latin. His textbook of 1713 on pure and applied mathematics was a standard for some fifty years; his works on physics stressed the importance of experimentation to confirm ideas. As well as his rational theology, his teachings about physics spread rapidly within the Central European universities in the first half of the eighteenth century (Heilbron, 1982, pp.35–6, 132–4).

Rightly interpreting the attractions of the Ritterakademie, Thomasius organized a curriculum that included gentlemanly accomplishments such as fencing and foreign languages as well as the formal requirements for participation in the Prussian civil service, including his modern teaching of the law. This proved to be an attractive combination for the sons of the nobility and Halle became very successful, having more than 500 students by 1700 (McClelland, pp.35, 28). The particular group that Thomasius's curriculum succeeded in attracting was crucial to the university. Not only did he succeed in winning the nobility back to a university, but they were a particularly lucrative cohort to attract since they paid higher fees than commoners and brought prestige to the institution. Halle did not retain this intellectual lead. Wolff was expelled from Prussia in 1723 by King Frederick William I, when accusations of atheism were levelled against him (Heilbron, p.36). This made Wolff something of a hero of academic freedom, much sought after by other institutions. He was subsequently reinstated at Halle by Frederick the Great in 1740, but the university's leadership could not by then be restored.

Göttingen and Erlangen

By contrast with rather *ad hoc* developments at Halle, Göttingen was deliberately set up to be a reforming university. Rivalry between Hanover and Prussia was one of the motives, but so also was the need for a well-educated bureaucratic class. The main instigator was a noble civil servant, G.A. von Münchhausen, who had studied law at Halle, and that university was taken as a model. From the outset, Göttingen was made attractive to the nobility with a mixed curriculum much as at Halle, again for reasons of finance as well as prestige. Indeed, it was more a university of the nobility than a royal institution, in that the nobility provided the principal financial backing. Unattractive university characteristics, such as theological or sectarian controversy, were deliberately forbidden, which, it has been argued, stimulated subsequently a degree of academic freedom (McClelland, 1980, p.39). Indeed, Catholics were positively encouraged to attend this Protestant institution. Professional appointments were kept in government hands, rather than allowing the faculties their traditional guild-like

Figure 14.2 *An interior view of the first university library at Göttingen. (Städtisches Museum, Göttingen)*

privileges, in order to avoid the employment of contentious scholars. While perhaps worrying to the eye of a modern reader, this practice circumvented traditional problems of nepotism and favouritism and allowed the recruiting of staff with international reputations and prestigious publications. These recruits were paid well; the new university soon had a strong reputation and was successful in attracting students. A superb library was a major attraction and, unusually, a Society for Sciences was set up in parallel with the university in 1742 to further scientific activity along the most modern lines (see Chapter 9).

Although Halle and Göttingen, along with Erlangen, which was deliberately modelled on the latter, were successful, enrolments continued to decline. In the 1760s, total student numbers had dropped to roughly 7000; and with even more universities, average enrolments were down to about 220 per institution. Until quite late in the eighteenth century, Halle, Göttingen and Erlangen were the only reformed universities. And it is noteworthy that all three were new foundations; there was no impetus to reform in the long-established universities.

> Halle, Göttingen, Erlangen, and a few others were able to overcome such lethargy [of the traditional corporate universities] because the state virtually ran them and intended to align university life with the perceived needs of state and society. In other universities, however, habit and privilege often successfully beat back attacks by enlightened (and sometimes benighted) princely bureaucracies and ministers inspired by the second reform movement of the eighteenth century. Only then, when enlightened despotism reached its zenith and the unreformed universities (especially in Catholic states) had few defenders left, were the principles of the new universities put into practice on a large scale. *(McClelland, 1980, p.57)*

14.2.2 Science in the universities

Given this rather parlous state of the German universities in the eighteenth century, what then was the fate of science in them? At both Halle and Göttingen, one of the main locations of science teaching was in the medical faculty; and at both institutions, the medical faculty was weak. Yet it was during this period that science became embedded in the universities; by the 1780s, it becomes possible to discern the emergence of a self-conscious chemical community in Germany and most universities taught experimental physics. Although what we would now call physics was the most prominent of the sciences in eighteenth-century Central Europe as well as elsewhere, the case of chemistry's move into the universities has been more thoroughly researched, perhaps because of its significant role in German institutional development in the nineteenth century. In the eighteenth century, chemistry represented a new type of experimental science – practical, applied and goal-oriented. Institutionally, it linked the intellectual with the useful and this was the basis of its establishment in the universities.

Chemistry

That chemistry was drawn into the medical faculties during the seventeenth century was crucial for its development. Institutionalization in this manner helped to differentiate it as a subject separate from alchemy and also gave it a practical orientation of pharmaceutical utility for intending medical men which served to remove it from traditional philosophical speculations. However, there were also disadvantages in the German system. Medicine was a hierarchical subject with the most prestigious and lucrative chairs attached to subjects at the head of the hierarchy. Individual professors advanced by being promoted to the more prestigious chairs. As it was a low-ranking, auxiliary subject, individuals had every incentive to leave chemistry behind as they progressed, to hold their chemical position in combination with other responsibilities, or to delegate it to assistants (Meinel, 1988, pp.93–5). During the eighteenth century, however, the teaching of chemistry changed in two important ways. The subject gradually gained independent status within medical faculties and it began to be taught more widely for non-medical reasons.

The economic doctrine of cameralism, a German version of mercantilism, was influential at Halle and informed the curriculum there, including the teaching of chemistry (Walker, 1971, pp.146–8). The aim of this doctrine was to meet centrally the administrative and economic needs of the numerous small sovereign territories with a view to achieving social harmony as well as economic benefit. The role of the central State was less to control than to take an overview in order to help achieve a balance among different local interests (Walker, ch.5). The economic aims were self-sufficiency and increased revenue by limiting imports, encouraging trade, increasing the working population and exploiting territorial natural resources. Mining, of course, had long provided a precedent (see Chapter 6). Thus the Austrian Court Chamber (Hofkammer, from which the doctrine's name was derived) developed economic policies to be applied across the Habsburg territories and it managed a number of important industries (Evans, 1979, pp.162–4). There was clearly a role for chemistry here.

From the time of its foundation, chemistry was taught at Halle by the second professor of medicine, George Ernst Stahl (1660–1734). He had responsibility for theory of medicine, physiology, dietetics, *materia medica*, botany and anatomy as well as chemistry. His scientific ideas are discussed in Chapters 10 and 15.

Stahl drew on the work of J.J. Becher (1635–82) in developing his ideas on chemical theory and its uses. Becher was well known among his contemporaries for his alchemical and Paracelsian chemical ideas as well as his writings on a range of other subjects. He travelled widely in Europe, eventually settling at the Habsburg court in Vienna. In addition to scientific work, Becher made important contributions to the development of cameralism. Gaining the support of Emperor Leopold I (who was interested in alchemy), Becher established in 1676 in Vienna a House of Crafts and Trades (*Kunst- und Werkhaus*) to promote Austrian manufacturing, thus giving the government an important role in fostering industry. It was in fact a technical teaching and research institute which had a glassworks as well as chemical and metallurgical laboratories (Meinel, 1984, p.344).

In Stahl's view, chemistry and medicine needed to be kept completely separate, as chemistry could throw no light on vital functions. He was also opposed to contemporary French and English attempts at corpuscular explanations of chemical phenomena (see Chapters 7 and 8). Rather, in common with other Germans interested in chemistry, Libavius's categorization of chemistry as an independent science with its own methods and concepts was, for Stahl, the key.

> According to Stahl, 'true chemistry' was 'a rational, deliberate, and comprehensible investigation and processing' of natural substances that led to 'fundamental knowledge'. The true chemist, consequently, was inspired by a 'truly rational enthusiasm for research', a desire to find 'the true knowledge of the material composition of natural substances', and an eagerness to illuminate 'the

Figure 14.3 George Ernst Stahl (1660–1734). (Archiv Martin Luther University, Halle-Wittenberg)

truth of natural composition for its own sake'. As he made his enquiries, he could expect 'intellectual pleasure, clear knowledge, and moderate advantages and benefits'. *(Hufbauer, 1982, p.8)*

Stahl not only developed Becher's chemical theories (see Chapter 15), but also followed up the applied side of his work. Stahl's *Zymotechnia fundamentalis* of 1697 was an important study of fermentation in such basic processes as brewing and the preparation of other alcoholic substances as well as vinegar making. He also wrote treatises on metallurgy, assaying, dyeing and the preparation of saltpetre – all economically important chemically based subjects. Stahl had a great many pupils at Halle. Through them, as well as his own influential second career as royal physician in Berlin from 1714 until his death in 1734, Stahl's programme, outlined by the turn of the eighteenth century, was to be of major importance in liberating chemistry from its secondary position to medicine. It must be remembered that, at the same time, Stahl was developing a theoretical chemistry, so that the discipline also had claims to be more than a mere practical, empirical enterprise and could claim esteem as a university subject. What enabled the favourable acceptance of this particular new discipline was the spread of Enlightenment ideas (see Chapter 11) in the middle decades of the eighteenth century (Hufbauer, 1982, ch.2).

The teaching of chemistry in medical faculties was the immediate area of expansion for the discipline, and medical contacts were an important route for the transmission of the new views of chemistry. By 1780, the number of salaried professorships for teaching chemistry in the medical schools of both Protestant and Catholic institutions had grown to 28, more than a four-fold increase since 1720. Eight of those had access to institutional laboratories (Hufbauer, p.34). The main stimulus to this growth was a new emphasis on the pharmaceutical applications of chemistry. In the first place, physicians sought to enhance the status of their profession by requiring more theoretical knowledge, including knowledge of the preparation of drugs, although the actual task of compounding them remained one for the apothecary. At the same time, as the enlightened states saw it as in their interest to take a greater role in the improving of daily life, stricter examinations for apothecaries and regulations for inspecting their activities were brought into effect. This of course brought a new requirement for relevant knowledge on the part of those doing the examining and inspecting as well, hence the stimulus to chemical teaching. A tendency has been charted from roughly mid-century for the newer chemical positions to be seen as of equal status with the more traditional chairs, and therefore to be held throughout an individual's career, rather than used as a stepping stone up the medical hierarchy (Meinel, 1988, p.103).

Interestingly, the more conservative Catholic universities apparently funded such developments rather more generously than the more readily reform-oriented Protestant institutions. At the latter, if student demand were high enough, a professor could simply organize a new course and fund it, including the cost of demonstrations, from student fees. Thus, though subject to the vagaries of intellectual fashions, innovation was quite simply achieved and was not dependent on preliminary institutional expenditure. But of course fees could not be raised indefinitely. At the Catholic universities, however, curricular innovation was not a matter for individual professors, but rather strictly controlled by regulations. Furthermore, students could not be charged fees, so there was little incentive to innovate unless the authorities could be persuaded to provide funds. Once the Catholic authorities were persuaded (and arguments regarding competition with the Protestant foundations could be quite persuasive) their funding of chemical instruction tended to be quite generous.

This level of institutional penetration of chemistry had a number of effects which were significant for the eventual definition of the discipline. In the first place, chemistry became a much more obvious subject. Anyone attending a university would become aware that there was such a discipline and that it had some degree of formal backing. For the not inconsiderable number of apothecaries who were subject to periodic inspection, the subject was particularly visible. Furthermore, most chemical teaching, even when focused on pharmacy, considered some wider aspects of the subject, so the new posts helped to broaden the public image of chemistry. Perhaps related to this, several students who began their chemical studies with narrow pharmaceutical intentions, gradually developed careers as chemists, rather than as pharmacists.

Also, many of the new teachers began to publish their work. Formally, the function of the German universities in the eighteenth century was to teach, not to advance knowledge. Indeed at Göttingen, a parallel scientific society had been set up along with the university precisely to take on the research function that was argued to be inappropriate for the main institution, although it is unclear whether this intended division was ever more than a rhetorical justification for a financial expedient. The plan was that the Society of Sciences (Societät der Wissenschaften) would supplement professorial salaries so that selected members of the professoriate could be freed of part of their teaching loads to devote themselves to the Society's activities. The Society's aims were

Figure 14.4 Emblem of Göttingen Societät der Wissenschaften. (From Commentarii, *1751.)*

> ... to increase the realm of knowledge with new and important discoveries, encourage professors both to write solid works and to apply themselves to their lectures, and to spur students to praiseworthy zeal for science and good morals. *(Heilbron, 1982, p.122)*

However, given the problem of falling enrolments in the eighteenth century, professorial publication was often encouraged, not for the sake of advancing knowledge, but to increase the fame and therefore the competitiveness of their institutions (Hufbauer, 1982, pp.38–9; McClelland, 1980, p.43). While this is not to be confused with the nineteenth-century German research ideal stressing the prestige of the advancement of knowledge in specialisms as judged by peers, it is possible to discern the beginnings of the valuing of publication as a criterion of excellence (Turner, 1974 and 1981). None the less, the achievement of sufficient reputation in a field for an individual's publications to bring fame to an institution reinforced the trend of the chemical professors dedicating themselves to that subject rather than moving on in medical teaching.

From the middle of the eighteenth century, another trend in the creation of chemical teaching posts becomes discernible: the teaching of chemistry in connection with the academic development of cameralism in the universities (Meinel, 1988, p.97ff). Stahl's many pupils were particularly well placed to promote this development. Chemistry was promoted explicitly as serving the economic interests of the territorial states, through agriculture, commerce and technology. In this manner, it gained public and institutional support as an independent discipline, not associated with medicine and likely to be wealth-producing in its own right. Clearly, chemistry was presented very much as an Enlightenment science, the teaching of which in universities would enhance the whole of society. Sweden led the way in establishing new chemical chairs for this reason in the philosophical rather than the medical faculties: Uppsala (1750), Lund (1758) and Åbo (1761) (see Chapter 12). These chairs served not a scientific curriculum, but the economic or administrative curriculum.

In terms of giving an independent discipline of chemistry academic identity, the work of Johann Gottschalk Wallerius (1709–85) was important. He was the first person to hold the independent chair of chemistry at the University of Uppsala and he published a book in which he redefined the subject. Chemistry had long been bedevilled by a distinction between theory and practice, that is between theoretical science and the practical arts. The former was discredited because of its associations with alchemy, not to mention Aristotelianism; while the latter was disadvantaged as an academic subject because of its association with the workshop. At the same time, of course, from the seventeenth century, theoretical development was linked to practical work. Following precedents in mathematics, Wallerius redefined, or perhaps reworded, this distinction as pure and applied chemistry. The new terminology made it possible for all work in chemistry, whether intellectual or manual, to be valued for what it revealed. Wallerius defined several branches of applied chemistry: medical, stone, salt, fire, metal, glass, economic and arts or crafts. Each was to be interpreted as an independent science with both theoretical and practical aspects (Meinel, 1983, pp.126–7). Wallerius's definitions were soon taken up in the German textbooks. They fitted well with the Enlightenment emphasis on scientific rationalism, utility and improvement.

In Germany, cameralistic chairs of chemistry, outside of medical faculties, were established in Giessen, Mainz, Heidelberg, Dillingen, Bonn and Marburg. In this way, chemistry became part of the education of German administrators, giving it yet wider exposure, and providing expertise for the running of eighteenth-century states. For chemistry, this move had the effect of opening up new perspectives of professionalization and giving it a new strategy for institutionalization (Meinel, 1988, p.99). In fact most such chairs failed by the end of the century, as the promised direct linkage between the study of chemistry and commercial activity did not materialize. Meanwhile, a new trend was developing toward locating chemical chairs in philosophical faculties as Göttingen and Halle carried on being innovators among the universities and began to emphasize more the ideal of the academic as scholar, or even researcher, a concept which emerged not from the sciences but from historical, biblical and philological studies.

At Göttingen when the famous chemist Johann Friedrich Gmelin (1748–1804) was appointed to the chemical chair in the philosophical faculty in 1775, he was advised:

> ... for his own and the university's good that he [sh]ould limit his view and choose a narrower horizon where he can concentrate his powers. The general mistake of the Germans is that they undertake too many sciences, handling them in an elementary way, often just copying. The great utility of the universities is that one divides the sciences into small parts and gives each man a small and limited responsibility. *(quoted in Hufbauer, 1982, p.41)*

Gmelin *did* publish, on pharmacy and chemistry, but that did not enable him to earn an adequate living from the chemical chair. In the event, he had to relinquish the chemical post in favour of a medical chair which brought with it the lucrative income from the fees generated by examining medical students (Meinel, 1984, p.346). The fate of the Ingolstadt (1773) and Halle (1788) chairs was similar. Jena (1789) was the first university to create a permanent chemical chair in its philosophical faculty, but this was a residue from its declining cameralistic teaching rather than an innovation. It was not until the early nineteenth century that the consequences of reforming pharmaceutical training led to the incorporation of chemistry teaching systematically into philosophical faculties.

While the institutionalization of chemistry outside of the medical faculties was not particularly successful in eighteenth-century Germany, the various attempts were important both in bringing chemistry to a new public and in demonstrating that chemistry could have an identity and an institutional locus outside of the medical faculties. Furthermore, German chemists, even when located in medical faculties, began to have a national disciplinary identity separate from their medical affiliations. This was enhanced by the establishment by Lorenz Crell of the first successful chemical journal in 1778, the *Chemisches Journal*.

Crell trained in medicine at the University of Helmstedt in the Duchy of Brunswick, where he first studied chemistry. He then travelled to Paris, Edinburgh and London. Apparently, working with Cullen and Black in Edinburgh (see Chapter 11) was especially formative for him. In 1774, he took up a medical chair at Helmstedt where he had responsibility for *materia medica*. In establishing the journal, Crell was anxious to secure his own reputation as well as to promote the science. Arguing that books were not suitable media for publication of ongoing experimental work, Crell organized the journal to encourage the publishing of individual experiments in all areas of chemical interest. He offered to double-check experimental results himself, and the journal was also planned to include translations of the proceedings of foreign scientific societies to help German chemists keep up to date. Indeed, he wished expressly to further German chemistry and was successful in establishing a forum for their discussions. It was in his journal that German views on the exciting developments in French chemistry were aired (Hufbauer, ch.5). By the end of the eighteenth century then, there was a self-conscious chemical community in Germany.

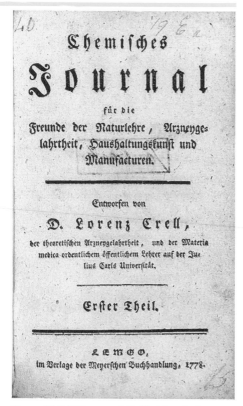

Figure 14.5 Title page of Crell's Chemisches Journal, *1778. The journal was for 'friends of the study of nature, medicine, economics and manufactures'. (Reproduced by permission of the British Library Board.)*

Physics

Physics too attained a new status in the German universities during the eighteenth century. Its pattern of establishment in the universities as a new experimental subject is rather more complex than the case of chemistry, and has been rather less thoroughly investigated. The establishment of chemistry was ultimately a process of separation, of its emancipation from the medical curriculum to be an independent subject. The case of physics was more a gradual merging of three quite distinct traditions: Aristotelian natural philosophy, natural magic and applied or 'mixed' mathematics (Hankins, 1985, ch.3). The traditional book-based subject, deriving from Aristotle and covering in a qualitative way natural bodies in general, gave way gradually to a quantitative, experimental study of certain specific topics. As it became more quantitative, its boundaries with applied mathematics were also defined. By the end of the eighteenth century, the standard coverage for physics included motion, gravity, elasticity, cohesion, hydrostatics, pneumatics, optics, heat, electricity, magnetism, elementary astronomy and geophysics – the biological heritage from antiquity had been effectively excluded (Heilbron, 1982, ch.1). The new experimental physics had come to mean '... the use of a quantitative and experimental method to discover the laws governing the inorganic world' (Hankins, 1985, p.3).

By 1700, almost every continental university had a post dedicated to physics, albeit of the traditional type. The principal exception in Germany at that time was the physics teaching at Altdorf, which was based from the late seventeenth century on experimental demonstrations. Wolff's teacher at Jena, an Altdorf graduate, was also an exception. Yet by the 1740s, the teaching of experimental physics was beginning to be widespread in Germany.

> This new activity, when expressed in alluring demonstrations, brought a wider audience than literary physics could command, and, by its requirements of space and equipment, made the professor a more expensive – and consequently a more valuable – member of the faculty. His prestige and value also rose outside the university, at least among those concerned to modernize instruction; for experimental physics, like economics, history, vernacular instruction and Cameralwissenschaften, breathed the spirit of the Enlightenment. *(Heilbron, 1982, p.126)*

In contrast to Britain and Holland, the stimulus in Germany was not Newtonian, but Leibnizian (see Chapter 10). Wolff's influence was important. His 2000-page textbook, *Generally Useful Researches for Attaining to a More Exact Knowledge of Nature and the Arts*, published in 1720–21, gave details of his own lectures and demonstrations and described quite precisely the construction of the apparatus and instruments required. By mid-century, Wolffians had chairs in the Protestant universities at Marburg (the exiled Wolff himself from 1723–40), Kiel, Leipzig, and Uppsala.

As with chemistry, it was the case that professors were not expected by their institutions to be researchers, but by the 1790s there was a clear trend toward preferment being given to those who had done research. It was a requirement of Tobias Mayer's appointment at Göttingen in 1750 that he undertake research, and he was later able to barter an invitation to move to the Berlin Academy for a promise of better pay and research facilities from the Göttingen authorities. But Göttingen was probably exceptional in this emphasis by the end of the century. Given the generally low student numbers, and the fact that physics was only a marginal part of the curriculum of the professional faculties – medicine in this case, but more commonly located in the less popular philosophy faculties – experimental physics had to rely on being almost a form of entertainment to attract a student body. (Some teachers complained that students came simply for the entertainment and not to do physics.)

As was the case in chemistry, the Catholic universities dominated by Jesuits were generally slower to change to the new curriculum. The fact that the movement in Germany was led by the work of the clearly non-Catholic Wolff contributed to this perhaps as much as conservatism; the Jesuits had long been, after all, important teachers of mathematics. More systematic reforms were undertaken in scientific instruction in Austria as part of the educational reform programme of the Habsburg empress Maria Theresa (1717–80, Holy Roman empress from 1740). Although ardently Catholic herself, she defended the rights of her crown against Rome and sought in the Habsburg territories to cope with the issue of territorial sovereignty by modernizing from the centre. A characteristic enlightened absolutist leader, her programme of financial and social reforms was strongly influenced by cameralistic ideas of the importance of an organized administration and an efficient economy. Introducing modern scientific curricula into the Jesuit-run schools was an important reform. Aristotle remained influential, even in the Austrian universities which did accommodate experimental physics, until 1752 when the empress instigated thoroughgoing changes, including the introduction of the vernacular and the elimination of 'useless' subjects such as metaphysics. It was important that the curriculum should be related to the common concerns of people and states. Even after their suppression in 1774, the Jesuits, as the main Catholic teachers, remained influential. However, the consequent financial changes did allow them to adopt modern textbooks.

Maria Theresa's reforms included the furnishing of the universities with collections of instruments and apparatus. As in the case of chemistry, the Catholic institutions, because of the way they were organized, actually fared rather better for resources. The Jesuits could not charge fees, and as members of a religious order, they had no personal wealth, so all equipment had to come from their institutions or the State. In the Protestant institutions, professors had to provide their own equipment out of fees. There was scope for abuse as wealthier colleagues could 'buy' physics chairs on the strength of the collections that they could afford. It was really only as the first generation of professors began to die off that Protestant institutions began to face the problem of how to replace their apparatus. In a number of cases, universities purchased the collections from their deceased professors' estates. There were concomitant problems of storage and technical support.

Again echoing the case of chemistry, by the end of the eighteenth century, there was a German-language periodical, the *Journal der Physik* which was started in 1790 with the aim of reporting experimental results. It had probably been stimulated by the success of the French-language *Journal de physique* founded in 1773. The aim was to get beyond the parochialism of the learned societies and allow experimental physicists to communicate quickly with peers having common interests. Thus by the end of the eighteenth century, physics too was fully incorporated in the institutional scientific life of Germany.

14.3 The mining academies

As well as favouring the enhancement of certain kinds of science teaching in the universities, the Enlightenment emphasis on economic improvement through the application of knowledge as a duty of the State led to the establishment of a new range of institutions where science and technology could be pursued, the mining academies. Dedicated primarily to training the officials who would guide the mining enterprises of the various states, these institutions were an exemplification of cameralism. Their focus was applied science, the improvement of technology and an eventual increase in national wealth and well-being. But Wallerius's pure science also flourished at some of them. The mining schools founded in the eighteenth century were established for the most part long after the respective state mining administrations; they were stimulated not only by changed views on education, but also by changing economic and political circumstances within the German states.

14.3.1 The Freiberg Mining Academy, Saxony

Technology and education in Saxony

The significance for science of the Swedish mining civil service was discussed in Chapter 12. There were elaborate organizations in some of the German mining regions as well. It was toward the end of Agricola's lifetime that a

Figure 14.6 Diagram of a section of a multi-mine drainage system with artificial lakes, channels and water-wheels. Built channels (K) link a series of lakes (T); in this case only one is shown, via underground streams (R) to a series of mines (S). The entrances and exits to the mines for the controlled water are marked (A). Taking advantage of the natural contours of the land, the water passes from higher to lower areas generally joining a river well down the system. In this particular example, after passing through two mines where it may have been used to drive underground water-wheels for pumping and/or ventilation, the water is used lower down the system to drive a water-wheel above ground for an ore-dressing station (W). In the area around Freiberg in Saxony, such artificial systems, the earliest sections of which date from the 1560s, were so skilfully engineered that they achieved a steady drop of a foot per thousand feet. The various Saxon systems covered an area 60 by 35 miles. (From O. Wagenbreth and E. Wächtler, Der Freiberger Bergbau: Technische Denkmale und Geschichte, Leipzig, VEB, Deutscher Verlag für Grundstoffindustrie, 1988, p.62.)

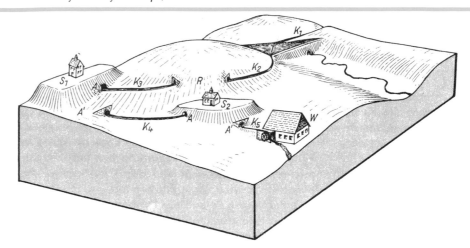

mining civil service was set up in Saxony by the Elector Maurice in 1542 (see Chapter 6). The Saxon silver mines had by then reached a high level of technological development and were tremendously prosperous. The role of the mining civil service was to organize and supervise all aspects of mining, including technological development.

Already from 1560, state positions for chemistry and mineralogy were created at the Museum for Mineralogy and Geology in Dresden, the seat of the Saxon court and government (Porter, 1981, p.549). It was here that the College of Finances met. Subject to the authority of the Elector, the College had to approve all projects and it had ultimate financial authority. The seat of the Higher Council of Mines (Oberbergamt), which had direct responsibility for running all of the Saxon mining districts, was in Freiberg, about 25 miles west of Dresden. Freiberg itself was one of the most important mining districts in the Erzgebirge (Ore Mountains) within the jurisdiction of the Saxon mining service and had its own District Mining Council (Bergamt). A high level of technical knowledge was required at all levels of the mining administration. An unofficial system of training by apprenticeship within the mining service for officials with technical responsibilities, such as mining surveyors and assayers, began during the seventeenth century in Freiberg. Towards the end of the century, students began to apply to the Oberbergamt on an *ad hoc* basis for funds to study with local officials and, apparently, these funds were fairly readily awarded. There is also evidence that local examinations were given in surveying and assaying (Baumgärtel, 1965, p.116).

This system was put on a more formal footing in 1702 when, on the initiative of the President of the Higher Council of Mines (Oberberghauptmann), an annual scholarship fund for intending mining officials was authorized by the Elector. Importantly, the fund was instituted to support Saxon students of mining and metallurgical sciences who, in return, had to make a commitment to work in the service of Saxony for a specified period. Foreigners, that is non-Saxons, who began to seek training in significant numbers from about this time, were still expected to pay. The scholarships could be used to finance local apprenticeship, or indeed university study, as well as study in other mining areas where technologies relating to different metals might be learnt. A knowledge of law was essential for the higher mining officials and this justified the expenditure on university courses. The Saxon universities of Leipzig and Wittenberg both had relevant curricula. The establishment of the fund was not linked to any immediate development of educational institutions in Freiberg. However, the regulations were such that the money went to those doing the teaching rather than to the students, so there was every incentive to teach. From 1702 until the founding of the Freiberg Mining Academy in 1765, some 128 scholarships were awarded, and five of the recipients subsequently became teachers in the Academy. The fund's historical importance then is that it marked the beginning of explicit state support for technical education within the context of training officials for state service and helped to create a climate in which such education would be valued and fostered (Baumgärtel, 1965, p.121).

There were not only developments within the official context of the mining civil service, but also private initiatives. In 1712 Johann Friedrich Henckel (1679–1744) came to Freiberg as town physician. He had studied medicine at Jena with the same professors who had influenced Stahl, and subsequently took his MD from Halle where he would have been part of Stahl's circle; he subsequently promoted the Stahlian programme. He took the opportunity of being in a place like Freiberg to pursue mineralogical and chemical researches, resulting in a number of important publications and an invitation to membership in the Prussian Academy of Sciences. His reputation attracted students and the attention of the Saxon mining service. In 1732, he was made physician to the mines and a year later a councillor in the Freiberg Oberbergamt. With these official positions, he was given state funds to expand his personal chemical laboratory and a responsibility for investigating Saxony's mineral wealth. Over the next decade, he attracted numerous students from all over Europe including some who were subsequently to become famous in mineralogy and chemistry: such as A.S. Marggraf, who became a mineralogical chemist and held important positions in the Berlin Academy; J.R. Spielmann, who held the first chemical chair at Strassburg; and M.V. Lomonosov (see Chapter 13). What distinguished his teaching from that available at universities at the time was the intimate linkage between the development of scientific ideas and a specific practical context (Baumgärtel, 1965, p.122).

Henckel's successor was Christlieb Ehregott Gellert (1713–95). Unlike most of the figures considered so far, Gellert's introduction to chemistry was not through medicine. He studied philosophical sciences at Leipzig and began his career as a teacher for the St Petersburg Academy's Gymnasium. He was also an adjunct member of the Academy's Physical Class and is likely to have studied chemistry there with Lomonosov. He returned to Germany in 1744 and, after some unsuccessful attempts to gain established posts, from 1749 he filled the void left by Henckel's death and became a private teacher of metallurgical chemistry in Freiberg. In 1753, he was appointed a councillor in the Oberbergamt and had responsibility for analysing minerals and testing mining machinery and smelting processes. From 1762, he was also the Chief Smelting Administrator. When the Mining Academy was founded in 1765, he became the foundation professor of metallurgical chemistry (Hufbauer, 1982, p.182; Baumgärtel, 1965, pp.123–5).

Gellert too attracted students from all over Europe. He translated from Latin into German an influential textbook on assaying by Johann Cramer, who had studied at Halle and would later be influential in the Habsburg mining service. He also wrote his own textbooks on metallurgical chemistry and analysis. His researches were largely practical, focusing on improvements in various metallurgical processes.

Lest the Saxon administration be seen as overly enlightened with regard to promoting technical training, it is important to note that a number of proposals for instituting training in relation to mining were unsuccessful. For example, in 1710, a technical university was proposed unsuccessfully by the town fathers for Freiberg (optimistically referred to as the August University after the Elector, Frederick August I). In that year, the administration of smelting and metallurgy had been brought into state hands along the lines of the long-established system of administration of mining in Freiberg (Baumgärtel, 1965, p.135). Later, in 1729, the administration refused a request from the University of Wittenberg to establish a chair of chemistry and mining on the spurious grounds that chemistry was already taught there (Hufbauer, 1982, p.43). And a proposal just after Henckel's death that a mining school be established in Freiberg with a chemical chair and laboratory as well as a scientific academy devoted to mining matters, particularly researches on geology and ore deposits, was also turned down (Baumgärtel, 1965, p.137–9).

Establishment of the Academy

In terms of the financial success of the industry, the issue of establishing a mining school may not have been seen as urgent, or perhaps even necessary, by the Elector. By 1720, the silver output of the Freiberg district had at last climbed back again to the level obtained at the outbreak of the Thirty Years' War. By 1750, output had grown even more to equal the peak achieved in the sixteenth century (Wagenbreth and Wächtler, 1988, p.84). What finally triggered approval for the idea of establishing a mining academy in Freiberg was the outcome of the Seven Years' War (1756–63). Effectively, Saxony was bankrupted by the war, a wide-ranging international conflict which had involved several European states and their colonies. Locally it was a conflict between the Habsburg and Prussian rulers over the territory of Saxony's neighbouring state, Silesia, but also over the larger issue of imperial versus territorial sovereignty. Saxony had been drawn in as a combatant on the Prussian side, and Saxon territory was fiercely contested. Its economy and administration were so disrupted that a Reconstruction Commission was set up in 1763 to institute reforms which would put Saxony back on its feet again. A key action of this commission was to call back Friedrich Anton v. Heynitz (1725–1802), who had studied in Freiberg with both Henckel and Gellert, from his posts as vice-president of the Higher Council of Mines of the Duchy of Brunswick and head of mining in the Harz mountains.

Figure 14.7 Portrait of Friedrich Anton v. Heynitz (1725–1802) in full regalia of office. His hat carries the Saxon coat of arms; and the buckle at his waist carries the crossed hammer and chisel, the symbol of mining; and in his right hand he carries the axe of office. The valuing of technology by the State thus depicted is typical of Central Europe. (Bergakademie Freiberg)

Figure 14.8 Picture of original premises of the Mining Academy. (Bergakademie Freiberg)

Von Heynitz took the position of Minister for Mining and investigated the existing administrative structure, including the working of the scholarship fund. He concluded that it was inadequate as a means of securing the necessary training for modern mining officials. The new Oberberghauptmann in Freiberg, Friedrich Wilhelm v. Oppel (1720–69) concurred, and resurrected the failed plan for a mining academy and an associated mining school which had been put forward in the 1740s. Von Heynitz presented it to the Regent for the under-aged Saxon Elector as a very economical way of regenerating the mines, and the establishment of the Freiberg Mining Academy was authorized within two days, on 13 November 1765. It was to be under the administration of the Oberbergamt, and the Oberberghauptmann was its director. Indeed, his official residence provided the original premises. The new institution opened in the spring of 1766. An associated school for the sons of lower mining officers was opened in 1777, and the children of mine workers were also encouraged to gain a degree of literacy (Baumgärtel, 1963).

The early years of the Academy

Saxon students were paid for by the State. The initial academic curriculum consisted of pure mathematics, mechanics (including hydrostatics and hydraulics), technical drawing, mineralogy, metallurgical chemistry and smelting technology, mining surveying, mining arts such as the location of ore deposits and the uses of machinery, assaying, and the construction of instruments, apparatus and models. In the early years, the Academy's curriculum had to be completed by attendance at a university to study mining law, but in 1785, a chair in that subject was added to the Freiberg curriculum. A number of the professors also had active roles within the mining administration (Baumgärtel, 1965, pp.143–4). As part of their work students accompanied these professors on mining inspection tours. From that experience, as well as from doing practical work in the Academy's own mine, intending mining officials gained a very full knowledge of the system that they would ultimately administer.

All of the initial professors were active workers in their fields, such as J.F.W. Charpentier, who was responsible for mathematics and the theoretical aspects of mining surveying; he made important innovations in geological mapping. Gellert's metallurgical work has already been mentioned. Gellert was succeeded by Wilhelm Lampadius (1772–1842) who had trained for pharmacy and studied at the University of Göttingen with Gmelin. He too held posts within the mining service and eventually became a member of the Higher Smelting Office. Lampadius undertook a number of applied projects outside of mining. For example, he sought sources of sugar from native vegetables and did extensive, though ultimately uncommercial, work on the possibilities of sugar beet. It was he who introduced gas lighting to Germany, installing it at a metallurgical works near Freiberg. The work of the professors reinforced each other as well. Lampadius wrote an important textbook on chemical mineralogy in support of the views of his colleague, Abraham Gottlob Werner.

Werner (1749–1817) had been an early student at the Academy. He enrolled in 1769 with the intention of following a career in metallurgical administration. As a good student, he was persuaded to aim for the Saxon administration, and so finished his education with the study of jurisprudence at the University of Leipzig. While there, he became interested in the study of languages and also pursued mineralogical interests, publishing in 1773 what became a very important classificatory work in mineralogy, *Von den aüsserlichen Kennzeichen der Fossilien* (On the external characteristics of

Figure 14.9 Title page of Bericht vom Bergbau, *1769, with an artist's impression of the room housing the Mining Academy's mineral collection. This illustration projects a very direct relationship between technology, represented by the symbol of mining, the crossed hammer and chisel, and science as represented by the mineralogical museum. Direct acquaintance with mineral specimens was considered to be an essential part of the mining students' education, and funding for the establishment of a museum was included in the Academy's foundation document. In addition to being custodian of the Academy's collection, Werner also established his own private collection for teaching purposes. This was later purchased by the Academy from his estate. (Bergakademie Freiberg)*

minerals). On the strength of this, he was offered a chair of mining and mineralogy together with the curatorship of the mineral collection back at the Freiberg Mining Academy.

Werner also became an inspector of mines in the Saxon mining service. He spent the whole of his career there, more than forty years, making the Mining Academy a world-famous centre for the study of mining and geology. Drawing on empirical knowledge gained from miners as well as his own field studies and wide reading of contemporary and past scientific literature in many languages, Werner developed a set of ideas about the formation of the earth's crust which, though not in themselves correct, helped to launch a new science of geology as distinct from mineralogy (Ospovat, 1976). He was also one of the few authors since Agricola to write about the origins of ore deposits. Werner was a charismatic teacher who published little, but none the less achieved great fame and influence by propagating his ideas through the classroom and through a voluminous international correspondence (Laudan, 1987, ch.5). Many of his students subsequently set up mining institutions modelled on Freiberg in their native lands. It should also be noted that the output of Saxon silver increased steadily until after the middle of the nineteenth century, and other metals such as lead began to be exploited in bulk. Thus the particular context of the work of the Freiberg Mining Academy was fruitful for the development of both science and technology.

14.3.2 *The Royal Hungarian Mining Academy, Schemnitz, Hungary*

Similar general considerations apply to a number of mining academies set up elsewhere in Central Europe in this period, at Schemnitz (1735, 1770), Prague, Berlin (1770) and Idria (late eighteenth century); as well as Ekaterinburg in Siberia (1763), St Petersburg in Russia (1773), and the École des Mines in Paris (1783). However, all of them had quite distinct local circumstances governing their foundations. As it was also located in a small mining town, rather than a national capital, the Academy at Schemnitz (Schemnitz is now Banská - tiavnica, Czechoslovakia) provides a useful comparison with that at Freiberg.

Technology and education in the Habsburg lands

In the early eighteenth century, the principal industrial income of the Habsburg territories was derived from mining enterprises, and anxious for revenue, the Austrian Habsburgs took a number of steps to enhance mining output. There were some mining areas in Austria, but the principal ones were in Upper and Lower Hungary and the Banat (see Map 6.2). Mining practice was very similar to that of the Erzgebirge. Silver accounted for about half the income from the mines, with copper and gold making up the rest. Unlike Saxony, where individuals could own mining enterprises which were administered by the State, in the Habsburg territories all mining and smelting enterprises were the property of the Crown. This meant that a place like Schemnitz, which was actually in Lower Hungary, was subject to the control of the Imperial Mining Chamber in Vienna and independent of the Hungarian Chancellory. Furthermore, the Crown did not necessarily take a long-term view of mining development, as for example when mercury from the rich deposit at Idria was allowed to be exported because of its high price, rather than husbanded for the domestic amalgamation process.

Individuals with scientific and technical interests had long been attached to various branches of the Habsburg court (see Chapter 6), and there was an elaborate imperial mining administration. Officials were trained by apprenticeship, and it was possible to nominate candidates as early as 1605. However, this proved to be demanding on the time of the officers, and posts were available only irregularly, so the system fell into disuse. New regulations in 1725 revived the system and also instituted a scholarship fund in Schemnitz. Unlike the Saxon system, the Habsburg scholarships went to the students rather than the teachers, so there was not the same incentive for developing instruction. By 1733, Samuel Mikoviny (*d.* 1750), a teacher noted for his work in mathematics and cartography, was located at Schemnitz and paid by the imperial authorities. A curriculum based on practical mechanics was formalized into a mining school in 1735. By contrast with the chemical and metallurgical focus of early teaching at Freiberg, mechanical engineering was seen to be the most pressing need because of the complexity of the machinery required to deal with the drainage problems of the very deep mines (Farrar, 1971, pp.2.2–2.3). Indeed, the Imperial Mining Chamber had already been active in introducing Newcomen engines as a means of pumping. J.K. Hell, a Schemnitz student of the 1740s, later a chief mining engineer, developed a hydraulically powered pumping engine for similar reasons. An assaying laboratory in the district was made available for demonstrations. Other mining subjects were taught by local officials.

When Mikoviny died, the reputation of Schemnitz declined rapidly. The famous French metallurgist, Gabriel Jars, toured Central European mining districts in 1758–59 and reported a very low standard at Schemnitz, much to the embarrassment of the authorities. At the end of the Seven Years' War, an imperial commission was set up to investigate the furthering of mining and metallurgy as part of the post-war reconstruction effort. An important member was Gerhard van Swieten, a former pupil of Boerhaave at Leiden, physician and adviser to Maria Theresa and professor of medicine at the University of Vienna, where he had introduced a chemical laboratory and a botanic garden. Van Swieten had been involved in a number of Maria Theresa's educational reforms, but it was only after the war that he began to promote chemical instruction for administrators. An extraordinary arrangement was devised: a practical mining school, which was meant to complement theoretical teaching at the University in Prague, was given Imperial funding at Schemnitz in 1763.

The chemist Nicholas Jacquin (1727–1817), who had also studied at Leiden, was recruited to introduce experimental classes. Students actually performed experiments linked to the lectures themselves. Jacquin was a follower of Joseph Black (see Chapter 11) and repeated for his students Black's experiments on fixed air. Jacquin was joined by the Jesuit mathematician Nickolaus Poda, who lectured on mining machinery. But the programme of instruction was not very successful. The fact that apprenticed students were dispersed around outlying mining towns and the division of instruction between Prague and Schemnitz made the whole system unsatisfactory, and the institution was reconstituted formally as the Mining Academy in 1770. The formal status of the Freiberg Academy, fairly recently gained, was probably influential.

The early years of the Academy

The three-year course at the Mining Academy at Schemnitz was organized from 1770 into three divisions: mathematics, chemistry and metallurgy, and mining sciences. The first was in the hands of Poda, while mining sciences was taught briefly by Christoph Traugott Delius (1728–79). He had studied at Wittenberg and Vienna as well as at Schemnitz. He came back to the

Academy as a teacher having had experience as director of mining operations in the Banat. While at Schemnitz from 1770 to 1772, he wrote an important and influential textbook on mining sciences, which became a standard work. He may well have known of a textbook on mining practice by the Saxon mineral inspector J.G. Kern (*d.* 1745) that had been produced at Freiberg in 1769. The two pioneering texts had a common approach and similar organization, though that of Delius was more up-to-date and had more examples than the Kern text, which was based on a manuscript prepared some thirty years previously. Delius left to answer a call to be court councillor in the Department of Mines and the Imperial Mint in Vienna. He was succeeded by Thaddeus Peithner, who had previously lectured at Prague, but then the chair was allowed to lapse.

The division of chemistry and metallurgy was dealt with by J.A. Scopoli (1723–88) after Jacquin had taken a post in Vienna. Scopoli had come to chemistry through the more traditional route of medical training followed by a post as a town physician in the mining area of Idria, where he had also lectured on metallurgy to mining officials. This is thought to have been prompted by van Swieten in an effort to establish a course similar to that at Prague (Farrar, 1971, p.3.6). At Schemnitz, he concentrated on mineralogy, and did research on the production of a better quality copper at the request of the Imperial Mining Chamber. He too took preferment at a Habsburg university after a decade at Schemnitz, moving on to a professorship of botany and chemistry at Pavia, where Boscovich and Volta were at work. For the following twelve years, chemical and metallurgical teaching was undertaken by Anton Ruprecht von Eggesberg, who had himself been a pupil of the Academy and then studied with Bergmann at Uppsala. He concentrated on bringing the laboratories up to modern standards and collaborated with Ignaz Born in developing the European amalgamation process.

There was a high turnover of teachers at the Schemnitz Academy, possibly because Schemnitz was just a small mining town in the midst of a dramatic mountainscape but remote from any centre of culture or other scientific activity. It was a three-day journey from Vienna and also distant from Prague. Freiberg, by contrast, was within a day's reach of two important and historic cities, Dresden and Leipzig. The former was the seat of Saxon government and the centre of cultural and intellectual activity of the State. The latter was an international centre of trade and learning, with a university and an enlightened bourgeois public. Freiberg itself had existed since the discovery of silver there in the twelfth century and even had a ducal residence as well as being the administrative centre of the Saxon mines. Mining was very much at the centre of Saxon life, but at the periphery of Habsburg life, despite the importance of its output (Baumgärtel, 1965, pp.131–2). Furthermore, although the Freiberg Academy was tied in to the Oberbergamt, that was a local tie and the relationship between the two bodies seems to have been one of positive reinforcement, rather than bureaucratic interference. Schemnitz was geographically, politically and administratively well removed from the Imperial Mining Chamber, which made relations difficult. The authorities' view of the Schemnitz Academy altered with imperial financial needs as well as with shifting political and fiscal arrangements between the Hungarian rulers of Slovakia and the imperial crown. The sudden calling of Delius from Schemnitz to Vienna well illustrates the problem of central versus local interests. So towards the end of the eighteenth century, the Academy at Schemnitz was not in the same thriving state as the Academy at Freiberg and neither was Slovakian silver production which declined from roughly the first decade of the century (Teich, 1975, p.310). Thus the two contemporary mining academies, though they had similar curricula by the 1770s, operated in very distinct contexts.

14.3.3 *The mining academies and the development of science*

It is clear that many improvements in mining and metallurgy were brought about by the applied research work at Freiberg and Schemnitz and also that an excellent training for the efficient running of mines was developed. The extent to which mineralogical and chemical theory *per se* contributed to such developments is perhaps more open to question, but Wallerius's views on the equivalence in status of pure and applied chemistry was certainly characteristic of the mining academies. It is equally clear that pure science was pursued as an integral part of the activities of the mining academy professoriate. For example, all aspects of Werner's work at Freiberg contributed to the development of his geological theories.

The Born amalgamation process

The development of the Born amalgamation process at Schemnitz indicates another aspect of this relationship. Ignaz Born (1742–91) was a typical Enlightenment figure with a wide range of roles – '... scientist, technologist, collector of minerals, civil servant, publisher, editor, organiser, Freemason and anti-clerical ideologist' (Teich, 1975, p.309).

Born, who had been a student at Schemnitz, was a councillor of the Imperial Mining Chamber and Mint working in the context of declining silver production in the closing decades of the eighteenth century. Through his official position, he became familiar with the literature on the amalgamation process that had been operating in Spanish America since the sixteenth century, which made use of that property of mercury whereby it could dissolve silver in the crushed ore to form an amalgam. The precious metal could then be separated by distillation from the amalgam. The traditional European process was smelting, followed by cupellation with lead, much as described by Agricola (see Chapter 6). The main advantage of amalgamation over smelting was that it required much less fuel, an important consideration in coal-poor, deforested areas. And it was considered to be more efficient for extracting precious metals from low-grade ores, such as were increasingly being resorted to in the Hungarian mines. As Austria possessed rich mercury deposits in Idria, Born saw great potential in developing a European version of the process.

He was convinced that his adaptation of the Spanish American process would work because he based it on what he thought was the sound chemical theory of phlogiston (see Chapter 15). He thus exemplified the Stahlian programme in full, even though he had to reject the idea of phlogiston being a constituent of silver, and therefore contributed to the undermining of that theory (Teich, p.314–19; 322–4). For Born, the guidance of theory and the proof of experimental evidence were critical.

> ... in these small operations, ... the nature, chemical properties and substantial parts of the stuff and mixture, must be exactly determined by experiment, without which the best theory is useless, and the result precarious and unsatisfactory though it is undeniable that, with good chemical principles and theory, the reasons of every operation and phenomenon are better understood, and difficulties more easily remedied, than with mere empirical practice and experiment. *(quoted in Teich, 1975, pp.318–19)*

Born petitioned the Habsburg emperor Joseph II to authorize an official public trial of the method in the presence of experts in return for a share of the profits that he confidently expected to result from its implementation. The trials in 1785 were sufficiently conclusive to induce the emperor to authorize the setting up of a full-scale process and to grant Born a share of the anticipated profits in return. Skleno, near Schemnitz, was chosen as the site and the professor of chemistry and metallurgy at Schemnitz was asked to work on scaling up the process for full industrial operation while Born remained in Vienna. Reports on the first full-scale trial were sufficiently encouraging to make some Viennese officials suggest that the process be introduced throughout the Habsburg mining regions immediately. But local officials at Schemnitz wanted more trials. The results of those trials were seemingly quite favourable, but the process did not receive assent for general use in Austria. Born died in debt. Although not at the trials himself, Gellert in his capacity as head of smelting operations in Freiberg took up the idea of amalgamation and revised it to achieve even greater fuel savings for use in Saxon circumstances. An amalgamation works using Gellert's adaptation was set up near Freiberg by Charpentier, and the process was successfully worked there for roughly a century.

The Skleno conference

The trials which were to decide the fate of the process were set up in Skleno in 1786 in the presence of an international gathering of eminent metallurgists from Spain, Sweden, England, Norway and, of course, Saxony. In effect, it was a scientific congress to investigate all aspects of a specific set of phenomena and it lasted three months. Each expert filed a report comparing amalgamation with smelting, and the local district physician commented on medical aspects. Published together, they amounted to a form of congress 'proceedings'. Some of the experts were motivated to do further work on the chemistry of the process themselves. Perhaps most significant, a proposal emerged from this meeting for the setting up of an international Society of Mining Sciences (Societät der Bergbaukunde). The group sent out a statement to potential members indicating that its main purpose was to act as a forum for communication to circumvent prevailing secrecy and to encourage further developments in mining science in order to benefit practice. Peer judgment, such as took place at the Skleno gathering, would be the real test for developments. In 1789, the first volume of its journal, *Bergbaukunde*, was published. The Society failed with Born's death in 1791, but its Enlightenment values of a sharing of rational knowledge for mutual improvement were in any case likely to have come into conflict with industrial values of competitiveness and secrecy. On the one hand, the Skleno gathering and the Society which it spawned can be understood as typical of eighteenth-century organizations devoted to utilizing science to achieve 'improvement'. On the other hand, they can also be seen as marking the beginning of a shift toward specialization which was to be characteristic of the nineteenth century.

14.3.4 The technological context of science

If Born's work on the amalgamation process produced evidence which contradicted the dominant eighteenth-century German phlogiston theory of chemistry, the mining academies and the tradition of mining sciences also helped to create a particular outlook about the nature of substances which underpinned the Chemical Revolution of the end of the century (see Chapter 15). It was in the course of the elaboration of a chemical approach to mineralogy in the mining-academy context that the assayer's operational conclusion that the end point of an analysis was a simple substance (one which could be separated no further [see Chapter 6]), was turned around so that the idea of what was a simple substance was defined by analysis rather than by theoretical preconceptions (Porter, 1981, p.544). The chemical mineralogist's aim was to classify minerals according to their chemical composition, the better to aid practice. This required elaborate and careful analytical procedures. The impressive record of the Swedish chemists in discovering new elements (see Chapter 12) is further evidence of this approach. In the context of the mining academies, the transition from the aim of aiding practice to more abstract scientific aims was readily made as was the transition from mineralogy to chemistry.

As a group, the mining academies founded in the second half of the eighteenth century were important in providing an additional locus for scientific posts and activity, albeit state-funded for particular applied purposes. The Mining Academy at Freiberg, and that at Schemnitz, may be exceptional examples. However, it is important not to overlook the contribution of such institutions to scientific developments in the particular way in which they brought together theory and practice. Freiberg in particular, played an important role in the advancement of various sciences, both in terms of their intellectual development and the establishment of their social identity.

Sources referred to in the text

Baumgärtel, H. (1963) *Bergbau und Absolutismus: Der sächsische Bergbau in der zweiten Hälfte des 18. Jahrhunderts und MaBnahmen zu seiner Verbesswerung nach dem Siebenjährigen Kriege.* in Freiberger Forschungshefte no. 44, Kultur und Technik, Leipzig, VEB Deutscher Verlag für Grundstoffindustrie.

Baumgärtel, H. (1965) *Von Bergbüchlein zur Bergakademie: Zur Entstehung der Bergbauwissenchaften zwischen 1500 und 1765/1770,* in Freiberger Forschungshefte no. 50, Geschichte des Bergbaus und Hüttenwesens, Leipzig, VEB Deutscher Verlag für Grundstoffindustrie.

Evans, R.J.W. (1979) *The Making of the Habsburg Monarchy, 1550–1700: An Interpretation*, Oxford, Clarendon Press.

Farrar, D.M. (1971) The Royal Hungarian Mining Academy, Schemnitz: some aspects of technical education in the eighteenth century, University of Manchester, MSc.

Hankins, T.L. (1985) *Science and the Enlightenment* (Cambridge History of Science Series), Cambridge University Press.

Heilbron, J.L. (1982) *Elements of Early Modern Physics*, Berkeley, University of California Press.

Hufbauer, K. (1982) *The Formation of the German Chemical Community (1720–1795)*, Berkeley, University of California Press.

Laudan, R. (1987) *From Mineralogy to Geology: The Foundations of a Science, 1650–1830*, University of Chicago Press.

McClelland, C.E. (1980) *State, Society and University in Germany, 1700–1914*, Cambridge University Press.

Meinel, C. (1983) 'Theory or practice? The eighteenth-century debate on the scientific status of chemistry', *Ambix*, 30, pp.122–32.

Meinel, C. (1984) '... to make Chemistry more applicable and generally beneficial – the transition in scientific perspective in eighteenth century chemistry', *Angewandte Chemie Int. Ed. Engl.*, 23, pp.339–47.

Meinel, C. (1988) '*Artibus Academicis Inserenda*: Chemistry's place in eighteenth and early nineteenth century universities', *History of Universities*, 7, pp.89–115.

Merton, R.K. (1984) 'The fallacy of the latest word: The case of "Pietism and science"' , *American Journal of Sociology*, 89, pp.1091–121.

Moran, B.T. (1985) 'Privilege, communication and chemistry: the hermetic alchemical circle of Moritz of Hessen-Kassel', *Ambix*, 32, pp.110–26.

Ornstein, M. ([1913] 1975) *The Role of Scientific Societies in the Seventeenth Century*, New York, Arno Press Reprint (based on her 1913 Columbia University Dissertation).

Ospovat, A. (1976) 'Abraham Gottlob Werner', in C.C. Gillespie (ed.), *Dictionary of Scientific Biography*, vol.14, New York, Charles Scribner's Sons.

Paulsen, F. (1895) *The German Universities: Their Character and Historical Development*, tr. E.D. Penny, New York, Macmillan.

Paulsen, F. (1906) *The German Universities and University Study*, tr. F. Thilley and W. Elwang, London, Longmans, Green & Co.

Porter, T.M. (1981) 'The promotion of mining and the advancement of science: the chemical revolution of mineralogy', *Annals of Science*, 38, pp.543–70.

Teich, M. (1975) 'Born's amalgamation process and the international metallurgic gathering at Skleno in 1786', *Annals of Science*, 32, pp.305–40.

Turner, R.S. (1974) 'University reformers and professorial scholarship in Germany, 1706–1806', in L. Stone (ed.), *The University in Society*, vol.2, Princeton University Press.

Turner, R.S. (1981) 'The Prussian professoriate and the research imperative, 1790–1840', in H.N. Jahnke and M. Otte (eds), *Epistemological and Social Problems of the Sciences in the Early Nineteenth Century*, Dordrecht, Reidel.

Wagenbreth, O. and Wächtler, E. (1988) *Der Freiberger Bergbau: Technische Denkmale und Geschichte*, Leipzig, VEB Deutscher Verlag für Grundstoffindustrie.

Walker, M. (1971) *German Home Towns: Community, State and General State, 1648–1871*, Ithaca, NY, Cornell University Press.

Further reading

Heilbron, J.L. (1982) *Elements of Early Modern Physics*, Berkeley, University of California Press.

Hufbauer, K. (1982) *The Formation of the German Chemical Community (1720–1795)*, Berkeley, University of California Press.

The Chemical Revolution

by Noel Coley

15.1 Early eighteenth-century chemistry in industry and medicine

The growing importance of industry and various developments in medicine focused attention on chemistry, the science of matter, in the eighteenth century. New gases were discovered and fresh ideas about combustion, calcination and respiration led to Lavoisier's oxygen theory and other changes which amounted to a conceptual revolution. With the proliferation of scientific journals from the late eighteenth century onwards, chemists in all European countries were ultimately obliged to revise their ideas; consequently, chemistry became the first truly European science by the end of the century. In this chapter we shall consider some of the most important developments leading up to this situation.

The crafts and industries of eighteenth-century Europe relied heavily on traditional techniques, although the potential of science for improving them had been recognized already in the seventeenth century and the new scientific societies in England and France had tried to put the sciences to work, yet the accounts of crafts and trades in the French *Encyclopédie*, for example, were still based on traditional methods. The same was true of other European countries in the early eighteenth century. German mining still owed much to the practices described in Agricola's *De re metallica* (1556) and the extraction of metals like copper, lead, tin and zinc from their ores depended on techniques which had been practised for centuries. Small iron works were found all over Europe, particularly in heavily wooded areas of Sweden, Germany and England where plentiful supplies of charcoal were available for smelting. Much of the wrought iron produced by these small ironworks was used by local blacksmiths to make horse-shoes, farm implements and domestic items. Other processes such as brewing ale, making soap and glass, tanning leather, or washing, bleaching and dyeing cloth were also carried out by age-old methods in small individual works close to urban and rural settlements scattered across Europe.

In eighteenth-century England, however, these crafts, together with others like spinning and weaving, were forced to develop in response to the demands of a rapidly growing population and thriving young industries. Most of the new industries which grew up in England during the Industrial Revolution of the second half of the century depended in some way on the use of chemicals. Thus the growing demand for cloth resulted in an increased call for soap, which in turn created a demand for alkalis. Traditional methods of bleaching had depended on exposing the cloth for long periods to air and sunlight while moistening it with sour milk. This was only slowly replaced by a dilute solution of sulphuric acid, but towards the end of the century chlorine made from common salt was used for bleaching. Dyes, traditionally extracted from plants like madder, saffron and indigo, were also needed. The craft industries all used rule-of-thumb methods differing in detail from one area to another – a motley collection of skills and procedures often depending on chemical changes which worked, but were not understood.

In medicine, physicians relied mainly on plant extracts thought to possess healing properties, though a few mineral substances introduced in the sixteenth century by Paracelsus were also used. Natural mineral waters were prized for their healing properties. Some were hot, others effervesced as 'fixed air' (carbon dioxide) escaped from them. Their therapeutic value was recognized throughout Europe, but though they were widely used, their chemical composition remained a mystery. Ideas about the healing powers of medicines were enshrined in ancient lore, but the precise nature of their action on the body was not understood. The apothecary was considered to be a dispenser of medicines – merely an assistant to the physician. Yet the apothecary shared the physician's responsibility for treating the patient, since many of the substances among the *materia medica* were highly poisonous and great care was needed in preparing medicines from them. Like the assayer of precious metals, the apothecary's methods required manual skill and accuracy in weighing and measuring which were to make an important contribution to the development of chemical analysis and other laboratory techniques in the eighteenth century.

While the applications of chemistry in industry and medicine were vitally important, there had long been another quite different approach to chemistry: that of alchemists, who used occult methods to discover the hidden secrets of nature. Their observations were recorded in secret signs and symbols to keep them out of the hands of the uninitiated. Esoteric alchemy was mainly concerned with changing base metals into gold and discovering the secret of eternal youth. The aims were futile, but the work of the alchemists made some useful contributions to chemistry. They discovered new substances, including several elements, and introduced some useful practical techniques, such as sublimation, crystallization and distillation – techniques adopted and developed by eighteenth-century chemists and applied to experimental work in the laboratory and working methods in the chemical industries.

By the mid-eighteenth century crude chemical methods were beginning to appear in some industries where the need for economy encouraged care in the use of raw materials. The quantities used were controlled to ensure maximum yield of desired products with the minimum of waste. Simple

Figure 15.1 An eighteenth-century chemical laboratory for metallurgy and assaying. Note the many furnaces and the precise balances in the window on the right. (From William Lewis, Commercium philosophico-technicum *[The philosophical commerce of the arts], London, 1765. Photo: Reproduced by permission of the British Library Board.)*

ideas about chemical reactions developed to satisfy the needs of industry, but processes were carried out with little chance of understanding how they worked. Most chemists before Lavoisier accepted traditional ideas about the four elements (earth, air, fire and water) or three Paracelsian principles (sulphur, mercury and salt), which allowed only qualitative accounts of chemical changes to be given. Observations often lent credence to the four-element theory, whilst many still believed that metals were formed in the earth by combination between sulphur and mercury, an idea in line with Paracelsus's *tria prima*.

15.2 Chemistry in Germany: the phlogiston theory

In the last quarter of the seventeenth century the German chemist J.J. Becher proposed a new form of the three Paracelsian principles, in which he described variations of the earthy principle (salt), including an 'oily earth' found in all combustible substances and thought to be released during combustion. These ideas were taken up and extended by Becher's contemporary, the German chemist George Ernst Stahl (1660–1734), who renamed the oily earth 'phlogiston'. Thus, according to the phlogiston theory the principle of combustion is contained in the fuel rather than the air. In Germany, where the smelting of metal ores with charcoal to extract the metals was widely practised, this theory seemed quite reasonable and, moreover, it was a German idea. Charcoal was known to burn leaving very little ash and it was therefore easy to assume it to be rich in phlogiston given up to the metal ore during the smelting process. During ordinary burning and respiration phlogiston was thought to escape into the air until the air became saturated and could absorb no more, at which point combustion ceased. These ideas seemed to be supported by innumerable observations of ordinary occurrences both in industrial processes and in the common events of everyday life, for example, the smelting of lead or the burning of a candle.

Stahl's authority as one of the leading chemists in late seventeenth-century Germany was considerable, but the concept of phlogiston was not widely adopted outside Germany before the middle of the eighteenth century. In France, for example, although the foundations of chemistry had been built on German sources, the best known textbooks of chemistry in the early eighteenth century were those of Nicolas Lemery (see Chapter 7) and Boerhaave, neither of which mentioned phlogiston. But from the middle of the eighteenth century there was an increase in French industry including metal extraction, and this created a demand for German chemical texts, several of which were translated into French in the 1750s. These works introduced the concept of phlogiston to French chemists, and in 1756 a version of Lemery's *Cours de chimie* appeared with a commentary in phlogistic terms. In another development at the Jardin du Roi, G.F. Rouelle had revived French chemistry from about 1742, basing his lectures largely on German sources. He adopted a modified phlogiston theory, which was elaborated during the following three decades into a universal principle by which a great variety of chemical observations could be explained. Phlogiston was, however, ill-defined. It was often equated with 'matter of fire', but at other times it was said to be an 'immaterial spirit', and there were so many versions of the theory that it could be changed to suit all circumstances.

Sometimes phlogiston obscured the properties of substances rather than clarified them. For example, the known gases were all thought to be forms of common air with different degrees of phlogistication. Hydrogen, methane (fire damp) and carbon monoxide, all considered rich in phlogiston, were generally regarded as 'inflammable air' without distinction. They would all only burn in common air. In this process the common air, by absorbing phlogiston from the combustible material, was said to become phlogisticated and would no longer support combustion. Consequently, any gas which would not support combustion was called 'phlogisticated air'. This included

Map 15.1 Centres of chemical activity in eighteenth-century Europe.

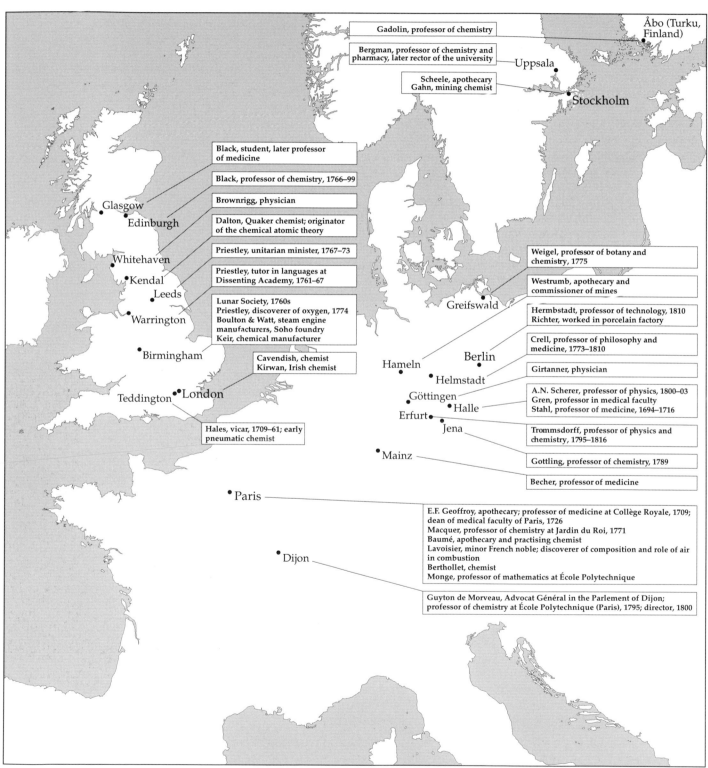

carbon dioxide, sulphur dioxide, nitrogen and some other gases. Another unfortunate consequence of the phlogiston theory was that metals such as iron, lead, tin, copper, mercury, silver and gold, and non-metals like sulphur, carbon and phosphorus (which are all chemical elements), were thought to contain phlogiston and therefore be *more complex* than their combustion products, an observation which appeared to be supported by experimental evidence.

This inversion of the chemical nature of the elements and their oxides was to become a serious problem for the phlogiston theory. Stahl was aware that when a metal was calcinated, despite the supposed loss of phlogiston, there was always a *gain* in weight, but relying on his German contemporaries, he ignored the work of Boyle, Hooke and Mayow, who had suggested that the increase in weight of metals on calcination was due to the fixation of igneous or 'nitro-aerial' particles from the air. The English chemists were moving in the right general direction when they suggested that during combustion and related processes something was taken up from the air, but the phlogiston theory turned this idea on its head. Attempts to explain the gain in weight were often based on confusion between weight and density since metal calces often seemed to be less bulky and denser than the metals themselves.

Yet, despite these and other difficulties, some chemical changes were satisfactorily explained by phlogiston and useful chemical discoveries were made using phlogistic principles. The chemical individuality of many gases was slowly realized during the eighteenth century, though a clearer idea of the part they play in the composition of solids and the chemical changes involved in their evolution was needed before progress towards a more productive chemical theory could be made.

> *The phlogiston theory*
> All combustible substances release phlogiston during combustion. Metals release phlogiston as they are calcinated (in fact, the metals are oxidized by oxygen in the air, i.e. metal calx equals metal oxide).
>
> Metal calx + phlogiston = metal
>
> If lead oxide is heated with charcoal (rich in phlogiston), metallic lead is obtained. The metal is therefore considered *more complex* than its calx. Phlogiston is also evolved during respiration.

15.3 Chemistry in England

Following the work of Hooke, Boyle and Mayow, the nature of gases became a focus of interest in eighteenth-century England. Boyle and Mayow had shown how gases could be collected in bladders, and Mayow showed how they could be manipulated by transferring them from one vessel to another. Several different 'damps' found in coal-mines were well known, especially in mining communities. Some of these burned or exploded when mixed with air; others extinguished flames and were clearly quite different. The methods of collecting and manipulating gases were improved by the work of English amateurs like Stephen Hales and Joseph Priestley, while physicians like Joseph Black in Edinburgh (see Chapter 11) extended knowledge about their chemical properties. Thus the heavy, non-inflammable gas evolved when acids reacted with chalk was called by Black 'fixed air' and it was known that inflammable air was given off when the same acids reacted with iron. Though these 'factitious airs' were still thought to be modifications of common air, their discovery contributed to a better understanding of chemical composition and reactions.

15.3.1 Stephen Hales

Stephen Hales (1677–1761), vicar of Teddington, distilled a great variety of substances in order to observe the products. He was not particularly original and his experiments were generally crude, but he took care to record the

Figure 15.2 Stephen Hales (1677–1761), perpetual curate of Teddington and amateur chemist. (From Stephen Hales, Vegetable Staticks, *London, 1727. Photo: National Portrait Gallery, London.)*

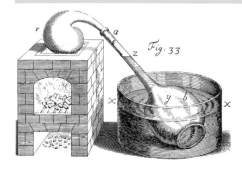

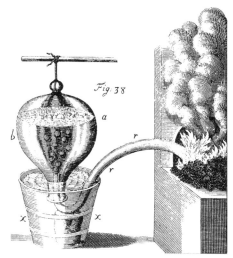

Figure 15.3 Hales's apparatus. Hales was the first to show how 'airs' could be expelled from solid matter by strong heating and collected in a vessel over water. His apparatus for collecting the airs evolved in destructive distillation was crude. It consisted of (bottom) a bent iron gun-barrel containing the material to be heated, placed in a strong fire with its open end under water in the neck of a large glass bell-jar, full of water and suspended over a bucket of water. In another method (top) he used a glass or iron retort luted to a long-necked flask pierced at the bottom to admit a siphon for drawing off the air evolved. Hales thus separated the generator from the collector, an important step in the manipulation of gases. In these flasks he distilled various natural products including blood, tallow, hartshorn, wood, beeswax, sugar, sea-salt, nitre, olive oil, bladder-stones, gall-stones and many other natural substances. His methods were later developed by Priestley, Cavendish, Lavoisier and others. (From Stephen Hales, Vegetable Staticks, *London, 1727. Photo: Ann Ronan Picture Library.)*

results and his methods were based on Newton's hint that various kinds of air could be obtained by heating dense bodies, or by fermentation. In his best known book, *Vegetable Staticks* (1727), Hales applied Newtonian mechanical principles to the study of plants. He set up experiments to measure the pressure of sap rising in the stems of plants and determined the proportions of water, air and earthy matter or ash in the various parts of plants including leaves, roots and stems. Having collected the products, however, he was content to measure their weights or volumes without investigating their chemical properties.

Hales also collected the airs expelled by solids during strong heating. His conclusions were vague and his greatest weakness was that he did not recognize the qualitative differences between the various gaseous products. He regarded air as an element existing in inelastic form in many substances from which it could be released. On combustion and respiration, Hales followed Mayow, though he did not accept Mayow's suggestion of a 'vivifying spirit' as a constituent of the air and argued that the loss of ability by the air to support combustion was due to absorption of the air as a whole with consequent loss of elasticity. He also suggested that bad air contained 'gross vapours' and thinking that this could be overcome in confined spaces by ventilation, Hales devised ventilators for prisons, hospitals and ships. These were introduced in a few isolated cases, but they were never widely adopted.

15.3.2 Chemistry and medicine

Among the upper classes in eighteenth-century England, gout and kidney-stones, or bladder-stones were common. These diseases resisted medical treatment and the search for a remedy for bladder-stone in particular provided a useful aim for experimental chemistry which was to have important theoretical consequences. As the stone was ultimately fatal and only surgery gave hope of a cure, it was highly desirable to find a less drastic form of treatment. The best chance seemed to lie in dissolving the stones *in vivo* and a number of more or less harmful chemical medicines were tried, including solutions of caustic akalis. There were many false claims of which Mrs Joanna Stephens's secret remedy for the stone was the most notorious example in England. Some chemists were interested in these problems, and advances in theoretical chemistry stemmed from their efforts to find an effective medical treatment. Joseph Black in Edinburgh investigated the differences between mild and caustic alkalis, especially magnesia alba (magnesium carbonate), in his search for a more effective solvent for the stone. His work led to the discovery of fixed air as a constituent of the mild alkalis, which could be released from them by the action of acids or transferred from one to another (Hankins, 1985, pp.89–91).

A different approach to the search for methods of treating these and other common diseases involved the use of natural mineral waters. In several European countries attempts were made to discover the chemical composition of these natural medicines. Since many mineral waters contained dissolved gases, efforts to improve the methods of analysing them encouraged the study of pneumatic chemistry. In Sweden Bergman worked on mineral water analysis, while in France mineral springs were regarded as national assets requiring careful control and regulation. Since they originated in the earth, the composition of mineral waters was often linked with geological studies. In England William Brownrigg, a physician at Whitehaven in Cumberland, published two papers in *Philosophical Transactions* describing

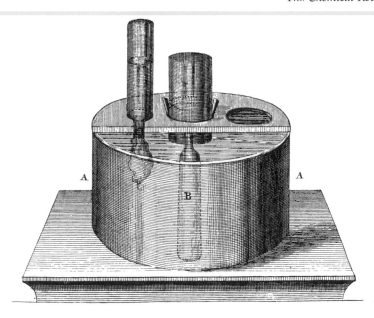

Figure 15.4 Brownrigg's apparatus for collecting the fixed air (carbon dioxide) expelled from Spa water. (From Philosophical Transactions of the Royal Society, *vol.55, 1765, p.224.)*

a 'mineral elastic spirit' in the mineral water of Spa in Belgium which he found to be similar to the choke-damp (fixed air) of the mines. Brownrigg observed that when the dissolved air was driven out of the mineral water by heat an earthy precipitate was formed, though he was unable to explain this observation.

15.3.3 Joseph Priestley

Joseph Priestley (1733–1804), Unitarian minister and amateur chemist, began to make experiments on gases in about 1770, when he was living in Leeds near to a brewery. He discovered a ready source of fixed air (carbon dioxide) above the fermenting liquors in the brewer's vats and, having read Brownrigg's paper of 1756, he tried the effect of dissolving fixed air in water to make an artificial mineral water. Priestley explained his method in a pamphlet in 1772; William Falconer, a Bath physician, recommended Priestley's 'soda water' as a cure for the stone in 1776, and the drink achieved popularity among the upper classes.

Priestley's first important paper on gases was read at a meeting of the Royal Society in March 1772. In it he gave an account of apparatus for preparing and manipulating gases in which he acknowledged the work of Hales, Brownrigg and Cavendish. Priestley's apparatus included a pneumatic trough, with a shelf for collecting gases over water or mercury (see Figure 15.6). After a brief discussion of his early experiments with fixed air and the preparation of soda water, Priestley turned to the air in which burning matter had been extinguished. He remarked that after burning every combustible he had tried, except brimstone, the volume of air was reduced by one-fifteenth or one-sixteenth. The residual air precipitated limewater and was heavier than common air. He concluded that flame causes common air to deposit the fixed air it contained. Later observations showed that an animal could still live in air which had extinguished a candle, and in about 1771 he noticed that this vitiated air was restored by growing plants. As a phlogistonist, Priestley thought the improvement of the air was due to the growing plants imbibing phlogistic matter from the vitiated air as part of their nourishment. Later, in 1779 he realized that the green matter in plants (chlorophyll) was responsible for the changes which occurred only in the presence of light.

Figure 15.5 Joseph Priestley (1733–1804), Unitarian minister and amateur chemist; he discovered oxygen in 1774. (Mary Evans Picture Library)

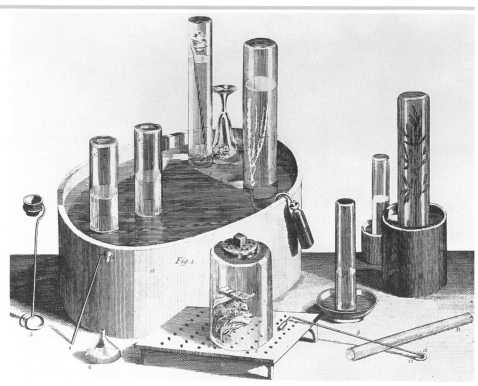

Figure 15.6 Priestley's apparatus for the manipulation of gases. Note the familiar pneumatic trough and gas jars. (Science Museum, London)

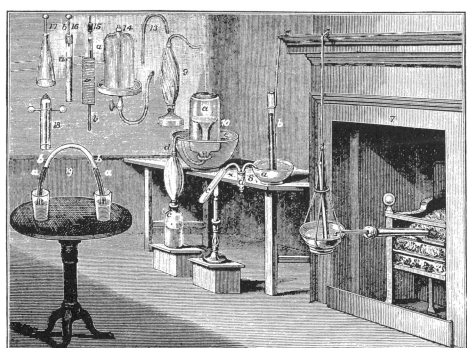

Figure 15.7 A corner of Priestley's laboratory. (From J.R. Partington, A History of Chemistry, *vol.3, Macmillan, London, 1962.)*

Priestley also prepared 'nitrous air' (nitric oxide) by heating nitric acid with metals such as brass, iron, copper or tin. He found that when mixed with common air, nitrous air gave red fumes and when the experiment was carried out over water about one-fifth of the common air and as much of the nitrous air as necessary to cause the reaction disappeared. This prompted him to devise a chemical method for testing the 'goodness' of air samples. He mixed equal volumes of atmospheric air and nitrous air in a jar over water and after contraction had ceased, transferred the remaining air to a narrow graduated tube. Its volume was called by Priestley the 'standard' of the air.

He could find no difference between the standard of the air in towns and in the country, but the air in a crowded room had a standard of 1.31 compared with fresh air (1.25) and air in which a candle had burnt out (1.43). On the other hand the air expelled from boiling water was better than common air.

Of all Priestley's discoveries in 1772, 'dephlogisticated air' (oxygen) was undoubtedly the most important. He obtained it by heating the red calx of mercury by concentrating the heat of the sun with a large burning lens. He observed that a candle burned vigorously in this air and a glowing splint of wood sparkled in it like paper dipped in a solution of nitre. A mouse confined in the new air remained active for twice as long as one confined in common air, and after being removed from the receiver it revived and appeared unharmed – the remaining air was still much better than common air. For some time Priestley confused the new air with nitrous oxide, overlooking the fact that unlike nitrous oxide it was insoluble in water and that a candle burned in it with a much brighter flame than in nitrous oxide. His nitrous air test showed that the new gas was about five times as good as common air.

Priestley observed that during respiration the air became charged with phlogiston and he suggested that this must have escaped from the blood into the air in the lungs. As to the nature of phlogiston itself, Priestley concluded that it was the same as light, but in a different state, a view also adopted by the French chemist P.J. Macquer (see Section 15.5.2). Yet, when metal calces were heated in inflammable air (hydrogen) confined over water or mercury, Priestley found that the inflammable air was absorbed and the calx changed into the metal. This seemed to suggest that phlogiston might be identical with inflammable air and that it was combined in metals just as fixed air was combined in chalk. Priestley also thought the metal should weigh more than its calx, but when this was never found to be the case he decided that some of the original calx must have sublimed.

15.3.4 Henry Cavendish

Priestley, a radical theologian who supported the ideals of the French Revolution and advocated them from the pulpit, was a prolific writer of sermons and religious tracts and studied science as a hobby. His contemporary Henry Cavendish (1731–1810), by contrast, was retiring in the extreme. Cavendish, who belonged to a noble family, published very little, but devoted his whole life and much of his fortune to experimental work. Educated in mathematics and natural philosophy, Cavendish made some important discoveries in both physics and chemistry. His house on Clapham Common in London was fitted out with a laboratory, forge and observatory; he also had a large library in his town house which he made available to his friends, even employing a librarian to manage the loan and return of the books. In the contemporary fashion for pneumatic chemistry, he studied the properties of gases, especially those produced by chemical methods, the so-called 'factitious airs'. His first published chemical memoir appeared in *Philosophical Transactions* in 1766 and dealt with inflammable air (hydrogen), fixed air (carbon dioxide) and an air produced during fermentation and putrefaction (mainly carbon dioxide).

In 1775 Priestley observed that dephlogisticated air exploded with inflammable air very vigorously. Later he found that Volta had used an electric spark to explode a mixture of inflammable air (hydrogen) and common air (later shown to be a mixture of oxygen and nitrogen) in a

Figure 15.8 The Hon. Henry Cavendish (1731–1810). A rich man, Cavendish was extremely shy. This is almost the only portrait of him. (Ann Ronan Picture Library)

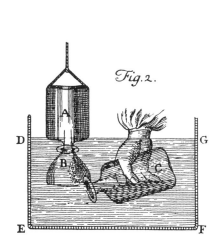

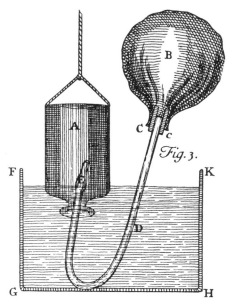

Figure 15.9 *Cavendish's apparatus for the manipulation of gases. (From J.R. Partington, A History of Chemistry, vol.3, Macmillan, London, 1962.)*

graduated tube so that the proportion of inflammable air decomposed by the common air could be measured, and in 1781 Priestley repeated some experiments of this kind in glass globes. He observed that moisture was deposited on the glass, but this did not strike him as important, and the discovery that the moisture was water was made by Cavendish, who also showed that the weight of water formed equalled the combined weights of the gases used up. Cavendish also found that the water was acidic and later showed that this was due to the formation of a small amount of nitric acid by combination during the explosion between phlogisticated air (nitrogen) and dephlogisticated air (oxygen). In the 1780s there was a long controversy between Priestley, Cavendish and James Watt about the composition of water. The problem was not resolved until Lavoisier's oxygen theory was introduced, when the chemical composition of water proved a crucial test for the oxygen theory of combustion.

Figure 15.10 *Cavendish's apparatus for finding the weight of gas expelled from a solid, such as chalk, by a dilute acid. (From J.R. Partington, A History of Chemistry, vol.3, Macmillan, London, 1962.)*

Cavendish on the nature of phlogiston
Cavendish found that when a given weight of a metal, like zinc, iron or tin, was dissolved in an acid, the same volume of inflammable air was released regardless of whether hydrochloric or sulphuric acid was used. This observation seemed to him to be conclusive proof that the inflammable air came from the metal, not the acid, and it convinced him that the phlogiston theory was correct.

metal (calx + phlogiston) + acid = salt (calx + acid) + inflammable air (phlogiston)

15.3.5 James Keir

While living in Birmingham, Priestley became associated with the Lunar Society, established in the 1760s by a group of Midland entrepreneurs, natural philosophers and savants, including the chemical manufacturer, James Keir (1735–1820). Trained in medicine at Edinburgh, where he became a friend of fellow-student Erasmus Darwin, Keir established himself in the Midlands from about 1770. He managed a glass factory at Stourbridge, worked with Boulton and Watt in their Soho foundry and in 1780 entered into a partnership in a chemical works at Tipton. Here he manufactured nitric and hydrochloric acids, litharge, red and white lead, an alloy of brass and iron (Keir's metal) and most importantly, alkali (sodium carbonate) for use in glass-making. However, Keir soon began to use his alkali to manufacture soap, and it was as a soap works that the Tipton plant became well known, attracting European visitors just as Boulton and Watt's Soho works did. In his industrial activities Keir used an empirical knowledge of chemical reactions to improve the processes he worked, especially the alkali process, but he was also attracted by a more intellectual approach to the subject. In 1771 he made an English translation of P.J. Macquer's *Dictionnaire de chimie* (1766) to which he added footnotes and comments derived from his own knowledge and practical experience. He remained a phlogistonist like Priestley, but toyed uncertainly with Lavoisier's new ideas. Like many practising chemists he took refuge in the empiricism which served him so well in his industrial operations.

15.4 Chemistry in Sweden

In Sweden there was a lively interest in science from the beginning of the eighteenth century, as it appeared that improvements in mining, metallurgy, agriculture and other practical arts could be secured by scientific means. The value of chemistry, especially chemical analysis, was recognized and methods of identifying metals and their compounds were devised (see Chapter 12), including in particular the use of the blowpipe, which was introduced in about 1738. In addition to this interest in mineral substances, pharmacy, *materia medica* and the applications of chemistry in medicine also attracted considerable attention among Swedish chemists.

15.4.1 Torbern Bergman

Torbern Bergman (1735–84), the leading eighteenth-century Swedish chemist, gained an international reputation for his skill as a chemical analyst and his work on elective attractions. In 1779 he published several general treatises in which the results of his analyses of minerals, mineral waters and salts were summarized. Bergman's efforts to improve methods of mineral water analysis were among the earliest. From about 1770, his own health declined and he was advised to drink the mineral waters of Seydschutz in Bohemia, Spa in Belgium and Pyrmont and Selz in Germany. Large quantities of these and other foreign mineral waters were imported annually into Sweden, but they were expensive and often arrived in poor condition after long journeys over land. Bergman identified the salts and gases in these mineral waters, and their proportions, so that mixtures with the same compositions could be prepared. His analytical methods depended on the formation of precipitates which he filtered off, dried and weighed. To obtain the best results it was necessary to know the chemical composition of the compounds precipitated, and Bergman listed the weights of precipitates obtained from 100 parts of each metal. His results were not very accurate, but the principle, which was sound, led to the German chemist J.B. Richter's law of equivalent proportions in the 1790s. The causes of chemical affinity also interested Bergman, who thought that it was akin to gravitational attraction, so that when one element was displaced from a compound by another, the change must be due to the difference between such attractive forces. Bergman organized his work on chemical affinity on the basis of this idea and in 1775 published his important *Dissertation on elective attractions*, containing a table similar to that of Geoffroy (see Section 15.5.1) showing the relative chemical affinities of various

Figure 15.11 The Swedish chemist Torbern Bergman (1735–84). (From a painting by Laurent Pusch, 1778. Courtesy of the Curator of the Art Collection of Uppsala University.)

substances. Baumé in Paris had observed that when mixtures were heated dry in a crucible the reactions were often different from those heated in solution, and Bergman, recognizing this, divided his table into two sections, with 36 columns of reactions in dry conditions and 50 in solution. The table was later increased to 43 and 59 columns respectively.

Bergman's work on elective attractions was an important step towards understanding chemical combination, but some fundamental re-thinking was needed on the nature of the chemical element and the composition of compound bodies. Bergman, Lavoisier and others began the process of quantifying the results of chemical experiments by their systematic use of the balance, but as more chemical compounds were discovered, a more rational method of naming them was needed and in 1787 a group of French chemists proposed a new system of chemical nomenclature based on the oxygen theory and linked to the composition and properties of substances.

15.4.2 C.W. Scheele

In 1770 C.W. Scheele (1742–86), an apothecary's assistant who had recently moved to Uppsala from Stockholm, was introduced to Bergman who, impressed by Scheele's extensive knowledge of chemistry, suggested that he might investigate the properties of pyrolusite (manganese dioxide), a recently discovered ore. The most important outcome of this work was Scheele's discovery of chlorine, announced in 1773. He found that pyrolusite dissolved in cold hydrochloric acid to form a brown solution which on heating gave off a greenish choking gas with the smell like aqua regia. The gas attacked animal and vegetable matter and all metals, including even gold. In line with the phlogiston theory, which Scheele accepted, he thought the new gas must be dephlogisticated acid of salt (hydrochloric acid *minus* hydrogen) for, like Cavendish, Scheele took phlogiston and hydrogen to be identical. In the following year he was elected a member of the Royal Swedish Academy of Sciences, an unprecedented honour for an apothecary's assistant. At about this time, too, Bergman mentioned that Scheele had made soda from common salt and had also observed that soda is formed on a paste of quicklime and brine. These observations were of interest to the chemical industry in Sweden, as both would be possible new ways of making soda for glass or soap manufacture.

Figure 15.12 A posthumous portrait of the Swedish apothecary Carl Wilhelm Scheele (1742–86). Discoverer of many new chemical compounds and elements, including oxygen in 1766. There is some doubt about the authenticity of this portrait. (Chemical Society Library)

Scheele on the reactions of 'fire air'

While working at Uppsala between 1770 and 1773 Scheele discovered 'fire air' (oxygen). It seems that he first obtained the gas by heating pyrolusite with concentrated sulphuric acid, but soon afterwards he prepared it by heating other compounds, including mercury oxide, silver carbonate, magnesium nitrate and saltpetre. Describing the new air as colourless, odourless and tasteless, Scheele found that a candle burned much brighter in it than in common air. His discovery of oxygen and its properties were discussed by Scheele in 1777 in a little book, *On Air and Fire*. Bergman said in an introduction that he had repeated the experiments and could confirm Scheele's results; he also stated that Scheele's discovery preceded Priestley's. This is probably true, but Scheele, like Priestley, failed to recognize the true role of oxygen in combustion and calcination. Instead he continued to explain these processes in terms of the phlogiston theory.

metal (calx + phlogiston) + oxygen (fire air) =
metal oxide (calx) + heat (phlogiston + fire air)

Neither Bergman nor Scheele lived to see the development of Lavoisier's theory, though both were aware of his work. Bergman's great prestige as a chemist, coupled with the fact that he supported the concept of phlogiston while suggesting many successful applications of chemistry to problems in mining, mineralogy and assaying, which were so important to Swedish industry and commerce, may account at least in part for the fact that Swedish chemistry remained so long in the phlogistic tradition.

15.5 French chemistry before Lavoisier

French chemistry had traditionally been based on German sources, and in 1723 a book, which may have been compiled from the lectures of E.F. Geoffroy and G.F. Boulduc at the Jardin du Roi (see Chapter 7), was published anonymously in Paris. It offered a new approach to chemistry based on Newton's theory of matter and Stahl's concept of phlogiston. This book was the first to introduce phlogiston into French chemistry, but it was little known and did not displace Lemery's famous *Cours de chimie*. The phlogiston theory remained relatively unknown in France until the middle of the century.

15.5.1 Étienne François Geoffroy

Étienne François Geoffroy (1672–1731), the son of a wealthy apothecary, studied botany, chemistry and anatomy in Paris, then pharmacy at Montpellier and later took up medicine. He visited London in 1698 as physician to Count Tallard, the French Ambassador Extraordinary, and was elected a Fellow of the Royal Society. He then travelled in Holland and Italy and in about 1700 took over his father's business in Paris. After this he studied medicine further, became MD in 1704 and three years later began to teach chemistry at the Jardin du Roi. In 1709 he became professor of medicine

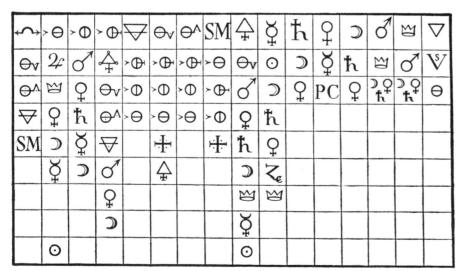

Figure 15.13 Geoffroy's affinity table. He recognized different degrees of rapport between substances which combine; those with greater rapport for each other combined more readily to form stable compounds, at the same time releasing components with which they had weaker rapport. This table, the first of a long line of affinity tables, was based on Geoffroy's own experiments and those of others. (From J.R. Partington, A History of Chemistry, vol.3, Macmillan, London, 1962.)

in 1726. Thus, although remembered chiefly as a chemist, Geoffroy was firmly established as a teacher of medicine. His career illustrates the close ties which existed between chemistry and medicine and clearly shows, if proof were needed, that medicine was regarded as the nobler discipline. Geoffroy was among the first French chemists to adopt the concept of phlogiston, which he called *principe huileux ou Soufre Principe*.

As befitted a physician Geoffroy regarded the calces of metals as dead bodies from which the soul had been displaced as the metal combined with fire, while its weight simultaneously increased. This idea of the displacement of one component by another led him to construct a table of affinities, or 'rapport', published in 1718. Geoffroy's table (Figure 15.13) was far from perfect, but the principle aroused interest among chemists and attracted critical attention in France and elsewhere throughout the eighteenth century. Bergman's table of elective attractions was a later version of the same idea (Crosland, 1980, p.397).

15.5.2 Pierre Joseph Macquer

Pierre Joseph Macquer (1718–84), the leading French chemist of the eighteenth century before Lavoisier, made some important discoveries and in particular contributed to the systematization of chemistry. His text-book, *Élémens de chimie théorique* (1749), replaced Lemery's *Cours de chimie* (see Chapter 7), which had become old-fashioned. His *Dictionnaire de chimie*, published in two volumes in 1766, was the first of its kind. It was translated into English for the benefit of English chemists by James Keir in 1771. In the *Dictionnaire* Macquer began the difficult process of improving chemical nomenclature. He was a staunch supporter of the phlogiston theory and regarded fire as made up of infinitely small material particles agitated in the most rapid motion, and so essentially a fluid similar to phlogiston. Macquer considered the phlogiston theory to be a reliable guide to the interpretation of chemical observations and insisted that experimental results, far from destroying the theory, constantly provided new evidence in favour of it. He never adopted Lavoisier's ideas, although he gave a full account of them in the second edition of his *Dictionnaire* (1778).

For Macquer the calces of all metals were made of the same earth from which phlogiston, the element of fire, had been expelled more or less completely. Phlogiston, he thought, was present in an almost pure state in the exhalation from glowing charcoal (carbon monoxide). Macquer was not much concerned with quantitative experiments, but he thought that the addition of phlogiston to the calx of a metal should result in an increase in weight. Finding that the opposite was the case for lead, he was unable to explain away this result. After the introduction of Lavoisier's theory Macquer suggested that phlogiston was 'matter of light', released when 'pure air' (oxygen) was fixed during combustion and the calcination of metals. While light was the material form of fire, heat was not a substance but a particular state of existence, related to the degree of agitation of its parts. The fact that oils, fats, resins, spirits, ethers and so on were derived from vegetable sources and were rich in phlogiston, seemed to confirm the view that it was matter of light, absorbed and fixed from sunlight by plants. Lavoisier later ignored Macquer's distinction between heat and light, but attached the same status to heat as a material substance without weight as Macquer had given to his matter of light.

15.5.3 Antoine Baumé

Antoine Baumé (1728–1804), an apothecary who had been a pupil of Geoffroy and was lecture demonstrator to Macquer for 25 years, owned a large manufacturing laboratory where he prepared various chemical compounds including sal ammoniac, red precipitate (mercury II oxide), tin and mercury chlorides and sugar of lead (lead acetate). He also invented processes for gilding metals, dyeing in two colours, bleaching silk, purifying saltpetre and some other techniques like the distilling spirit of wine. As a manufacturing chemist Baumé was an empiricist; he clung to the phlogiston theory, attacked Lavoisier's ideas and avoided mention of him wherever he could. Phlogiston, Baumé thought, was a compound of a vitrifiable earth and fire, a view he never relinquished, nor did he accept the discoveries of pneumatic chemistry and the idea of gases with distinct chemical identities taking part in chemical reactions. In his everyday applications of chemistry to working practices Baumé saw no need for such theories; he could prepare his compounds and conduct his processes satisfactorily without them.

15.6 The Chemical Revolution begins in France

The Chemical Revolution resulted from developments in theoretical chemistry which took place towards the end of the eighteenth century in several European countries. There were at least six principal changes: including the displacement of phlogiston by oxygen; the improvement of accuracy in quantitative methods; an empirical definition of the chemical element; the development of theories of chemical affinity; a new system of chemical nomenclature; and improved analytical methods, especially for the analysis of organic compounds.

15.6.1 Lavoisier and the oxygen theory

Important discoveries in most of these areas were made in France by Antoine Laurent Lavoisier (1743–94), who showed that common air is a mixture of oxygen with 'mephitic air', which he at first called *mofette*, later *azote*. (Chaptal introduced the name 'nitrogen' in 1790.) Lavoisier showed that oxygen is absorbed from the air during combustion, calcination and respiration, and this, together with the discovery of the composition of water, formed the basis for his new oxygen theory. But the oxygen theory necessarily developed from phlogistic chemistry, and there are strong similarities between the first formulation of the new theory and its predecessor. Phlogiston, the principle of fire, or matter of heat, was thought to be combined in all combustible bodies; similarly oxygen gas was also thought to consist of the oxygen principle and 'caloric', Lavoisier's new name for the matter of heat. By itself the displacement of phlogiston by oxygen was no more than the very beginning of the new chemistry and in the eighteenth century there were still some chemical changes which were explained more satisfactorily by phlogiston than by its rival, oxygen.

In the autumn of 1774 Priestley travelled to Paris with his patron Lord Shelbourne for whom he was acting as secretary, and at a dinner at which Lavoisier was present Priestley mentioned his discovery of dephlogisticated air by heating mercuric oxide. Lavoisier was aware that mercuric oxide gave off an air on heating, but he did not then know that this air supported

Figure 15.14 Antoine Laurent Lavoisier (1743–94), from a sketch made during his imprisonment before his execution. (Bibliotheque Nationale, Paris)

combustion so much better than common air, and in the following spring he repeated Priestley's experiments. Lavoisier then worked on this problem for the next two years and in 1777 presented no fewer than nine memoirs to the Academy in Paris, all on topics related to combustion and calcination. Among these was a description of an ideal experiment which showed that the air is made up of at least two components, only one of which takes part in the calcination of mercury. He outlined a new theory of combustion and calcination in which he said that only in one kind of air, pure air (called by Priestley 'dephlogisticated air'), was involved. During combustion and calcination, heat and sometimes light are disengaged and the pure air is destroyed. At the same time the calcinated metal increases in weight by an amount equal to the quantity of pure air absorbed.

Lavoisier also observed the formation of (a) sulphuric acid from burning sulphur, (b) fixed air, the 'chalky acid', from the combustion of carbon-containing compounds and (c) phosphoric acid from phosphorus. This led him to conclude, wrongly, that combustion always resulted in the formation of an acid, and consequently he proposed the name 'oxygen' (acid former) for pure air. (Guyton de Morveau, the Dijon chemist, also thought that 'vital air' was the true universal acid.) Lavoisier's new theory explained the production of heat in combustion by locating caloric, his term for the matter of fire, in the oxygen gas, whereas Stahl had located phlogiston in the combustible material. Lavoisier did not claim in 1777 that his new theory was a rigorously tested alternative to phlogiston; he merely offered it as a new hypothesis to account for the production of heat and light in combustion, suggesting that it seemed more probable and contained fewer contradictions than the old phlogiston theory. There was still a long way to go before the new theory would displace the old.

15.6.2 *Lavoisier and the composition of water*

Seven years later, in 1784, the controversy still unresolved, Lavoisier presented to the Paris Academy a memoir on the composition of water, still commonly regarded as one of the elements. Following Cavendish's experiments (see Section 15.3.4) Lavoisier suggested that water was a compound between inflammable air (hydrogen) and oxygen. This was a crucial step in the arguments supporting the oxygen theory, for until the composition of water had been determined, the reactions between metals and

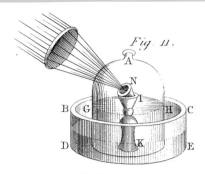

Figure 15.15 Lavoisier's apparatus for the calcination of mercury. The sun's rays are focused by the lens onto the mercury suspended inside a bell-jar full of air confined over mercury. As the red mercury calx is formed part of the air is used up. (From Oeuvres de Lavoisier. *Photo reproduced by permission of the British Library Board.)*

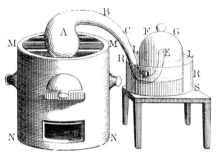

Figure 15.16 Lavoisier's apparatus for the calcination of mercury by heating the mercury for prolonged periods over a charcoal furnace. The surface of the heated mercury becomes covered with red mercury calx as part of the air is used up. (From Oeuvres de Lavoisier. *Photo reproduced by permission of the British Library Board.)*

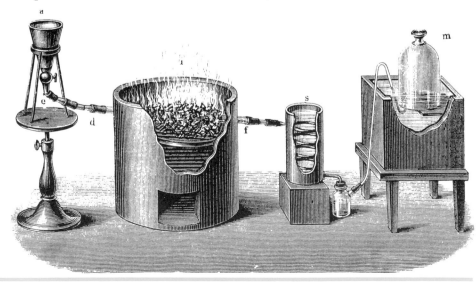

Figure 15.17 Lavoisier and Meusnier's apparatus for obtaining hydrogen from steam and red-hot iron. Steam is passed from the flask on the left through the red-hot gun-barrel. Iron oxide is formed inside the barrel and hydrogen collects in the bell-jar on the right. This experiment showed the composition of water. (From J.R. Partington, A History of Chemistry, *vol.3, Macmillan, London, 1962.)*

acids in solution could be explained more simply by phlogiston than by the oxygen theory. Lavoisier began by recalling an experiment he had made with Bucquet in 1777 in which they had shown that the product of burning inflammable air in oxygen was water. This seemed to show that water was a compound between the two gases, and Lavoisier, at the suggestion of Laplace, who had assisted him in the experiments, now offered an alternative explanation for the evolution of hydrogen when a metal like zinc was dissolved in an aqueous solution of an acid. He argued that the metal first decomposed water to form the metal oxide and hydrogen, which was released, after which the metal oxide dissolved in the acid to form a salt.

If this were the case, water must be a compound. Lavoisier set out to test the assumption that it could be decomposed, assisted by J.B. Meusnier, an army officer who was concerned with improving the reliability of military balloons filled with hydrogen. Meusnier wished to find an economical method of producing inflammable air in large quantities, and in 1783–84 Lavoisier and he decomposed water by passing it slowly through a red-hot iron gun-barrel. Lavoisier later (1789) described a similar experiment to decompose steam by passing it over red-hot iron in a glass tube. After repeating these experiments many times he concluded that the iron became oxidized as inflammable air (hydrogen) was liberated from the water or steam. In 1786 the results of some experiments by Monge, who synthesized water by exploding oxygen and hydrogen in a glass globe with an electric spark, were published. A large quantity of water was formed, and although the results were not very accurate they encouraged Lavoisier to repeat the experiment in 1787; in the following year he stated that water was oxygenated hydrogen.

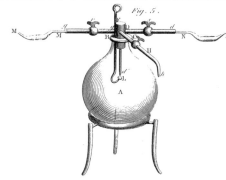

Figure 15.18 *A mixture of two volumes of hydrogen and one of oxygen in the large flask is exploded by means of electric sparks. The only product is water. Lavoisier determined the composition of water by synthesis as well as analysis. (From* Oeuvres de Lavoisier. *Photo reproduced by permission of the British Library Board.)*

Phlogiston and oxygen theories compared

Experiments dissolving metals, such as zinc, in acids were crucial for the new oxygen theory; it was not until the chemical composition of water had been discovered that the oxygen theory was able to explain the changes occurring when a metal dissolved in an acid as satisfactorily as the phlogiston theory. Thus, before 1789 the oxygen theory had no explanation for the prod-uction of inflammable air when a metal dissolves in an acid, whereas this posed no problem for the phlogiston theory.

1 Phlogiston theory

metal (calx + phlogiston) + acid = salt (calx + acid) + inflammable air (phlogiston)

2 Oxygen theory before 1789

metal + acid = salt (calx? + acid) + inflammable air (?)

The oxygen theory could not explain the composition of the salt or the origin of the inflammable air. The phlogiston theory gave a better explanation of this reaction before the composition of water was discovered in 1789. As soon as this was understood, the oxygen theory became viable.

3 Oxygen theory after 1789

metal + water (hydrogen oxide) + acid = salt (metal oxide + acid) + hydrogen

Thus when a metal dissolves in a dilute acid, water (hydrogen oxide) is decom-posed to form the metal oxide, which then combines with the acid to form the salt, and hydrogen, which is released.

When Lavoisier published his *Traité élémentaire de chimie* (1789), in which he summarized all the new ideas, the main planks of the Chemical Revolution were in place. Yet the transition to the new theories of chemistry was by no means complete in 1789; in fact it was only just beginning and there would be a long period of controversy before the new theory was generally accepted.

15.6.3 Lavoisier's contributions to organic analysis

Turning to the chemistry of natural substances Lavoisier again began from the observation that plants decompose water by taking up the base of inflammable air and setting free vital air (oxygen). He thought this was the mechanism by which vegetable oils were formed. Guyton de Morveau proposed the name 'acidifiable base' for a substance which unites with oxygen to form an acid. He introduced the term 'radical' to describe these acidifiable bases, and Lavoisier acknowledged this by accepting the idea of compound radicals in natural products. The latter had always been 'analysed' by destructive distillation, which yielded a motley collection of gases, oils, watery mixtures and earthy residues, all of which were complex and variable in composition and appearance, according to the precise conditions of the analysis. Yet they were considered the fundamental constituents of the organic material.

Oxidation analysis made quantitative determinations of the proportions of carbon, hydrogen and oxygen in organic substances possible – the first step towards fixing the composition of organic compounds in terms of their elements. The new analytical method was based on the oxygen theory and Lavoisier's definition of the chemical element as the final stage of analysis. Although crude and inaccurate, the method represented an enormous advance on distillation analysis. Lavoisier also emphasized the value of the chemical balance for quantitative work, insisting on the conservation of matter, the balance-sheet of the elements, in all chemical reactions. He began the systematization of chemistry and showed, by his analyses of organic matter and his work on natural processes such as fermentation and respiration, that organic matter and the functions of living organisms could be studied by the same chemical and physical methods as those which applied to mineral substances. Indeed, Lavoisier contributed to more areas of modern science than almost anyone else in the eighteenth century (Holmes, 1985, pp.xv ff).

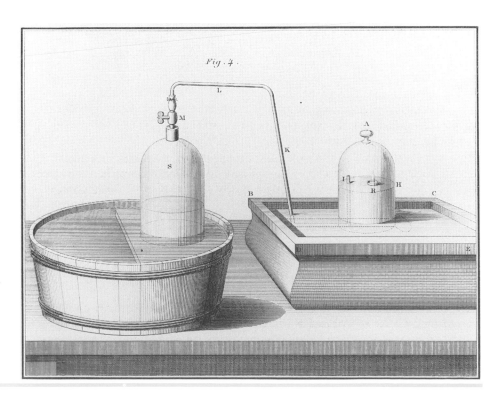

Figure 15.19 Lavoisier's apparatus for burning spirit of wine with oxygen to form water and carbon dioxide. BDEC is the marble basin containing mercury. Bell-jar A, on right, immersed in mercury, is the combustion chamber, filled initially with ordinary air. R is the spirit of wine lamp, floating on mercury. On left, jar S, containing vital air, is immersed in water. Vital air is supplied intermittently to replenish that consumed in A, by opening valve M. (From Oeuvres de Lavoisier. *Photo reproduced by permission of the British Library Board.)*

But for all his important discoveries, Lavoisier made several errors which remained to be corrected in the nineteenth century. In his work on respiration and animal heat, for example, he observed that the air undergoes chemical changes in the lungs similar to those occurring when carbon burns. This led him to suggest that caloric contained in the oxygen is released in the lungs, where it mixes with the blood and is distributed throughout the body. If this were the case all the heat of the body would be released in the lungs and none in the other organs or tissues. The antiphlogistic theory retained the concept of caloric, an imponderable matter of heat, combined with oxygen. The heat and light evolved in combustion therefore came from the caloric contained in oxygen gas and in this respect the new theory offered little advance. Moreover, from his combustion experiments Lavoisier concluded that all acids must contain oxygen, and that combustion could occur only in oxygen, both of which were mistaken. It must, however, be said that the antiphlogistic theory had the great advantage that it allowed oxidizable substances like carbon, sulphur, phosphorus and the metals to be considered as simple elements rather than compounds of the calx or acid with matter of heat.

Figure 15.20 Late in his career Lavoisier began to make quantitative experiments on physiological functions. In this experiment he measures the products of human respiration. The drawing was made by Madame Lavoisier who is seated at the table on the right. The illustration shows Lavoisier measuring the oxygen consumed and the fixed air (carbon dioxide) given off in human respiration. (Mansell Collection, London)

15.7 The reception of antiphlogistic chemistry

The reception given to Lavoisier's theory varied from one country to another depending on a variety of factors. In France, as might be expected, it was welcomed and adopted quite rapidly. In Edinburgh, too, where there were strong ties with France, Joseph Black saw the advantages of the oxygen theory and adopted it almost as soon as the French chemists. On the other hand, in Germany, the homeland of chemistry and birthplace of the phlogiston theory, most chemists dismissed the oxygen theory as a foreign fad from a country which could not be expected to produce significant chemical advances. German chemists ridiculed, ignored or doggedly resisted the new theory well into the 1790s. In England, too, Priestley, Cavendish and others eschewed the new French chemistry. Phlogiston had provided working explanations of most experimental observations and, since the oxygen theory did not always improve on, or even match, phlogiston, there was no incentive to change. In Sweden, isolated from the mainstream of scientific advance, a similar situation prevailed and the new theory did not receive much serious attention before the end of the century.

15.7.1 France

In the decade up to 1776 many converts to the oxygen theory were made among French chemists. C.L. Berthollet accepted it by 1785, as is shown by his explanations of the action of sunlight on a solution of 'oxymuriatic acid' (chlorine), the oxidation of alcohol and ether and on the combination of vital air with oils, all of which were based on the oxygen theory. A.F. Fourcroy observed that the discovery that water is a compound of hydrogen and oxygen removed doubts and uncertainties for most chemists and that from 1787 the new theory was widely accepted in France. Guyton de Morveau adopted the oxygen theory in 1787 soon after coming to Paris to instigate the reform of chemical nomenclature which was based on the oxygen theory. Lavoisier and his supporters used the new nomenclature in all their publications after 1787, and chemists who wished to understand their work had to learn how to think in antiphlogistic terms. Fourcroy began to teach the oxygen theory in 1786/87, generally calling it *la chimie pneumatique*, although in the second edition of his *Élémens d'histoire naturelle et de la chimie* (1786), Fourcroy gave Lavoisier full credit for the new theory, even though in his *Traité élémentaire de chimie* (1789) Lavoisier had said that the new ideas had been so much discussed among French chemists that it was difficult to identify the author of each point. The *Traité* used the new nomenclature and was based on the oxygen theory; it set the seal on the antiphlogistic revolution in France, and in 1792 Lavoisier staked a personal claim to the oxygen theory.

15.7.2 Britain

Outside France one of the first to accept Lavoisier's views seems to have been Joseph Black in Edinburgh. Richard Lubbock, a pupil of Black who adopted Lavoisier's theory in his MD dissertation in 1784, says that Black had shown little confidence in the phlogiston theory for many years and in 1784 he had abandoned it in favour of the oxygen theory. The majority of chemists at that time were less convinced about the absolute superiority of the antiphlogistic theory and preferred to weigh the relative merits of the two rival theories. William Nicholson (1753–1815), a schoolmaster, civil engineer and author, chose to explain both theories, for while he thought that the existence of phlogiston was far from certain, he felt there were difficulties in accounting for some facts without it. He preferred the antiphlogistic theory, but thought that both theories were equally probable.

In 1784 the Irish chemist, Richard Kirwan, published his *Essay on Phlogiston* in which he accepted Cavendish's view that phlogiston and inflammable air (hydrogen) were identical. He supported his argument with experimental observations. For example, it was known that metals which dissolve in acids with the evolution of inflammable air displace other metals from solutions of their salts without releasing any gas; consequently it seemed likely that inflammable air was a constituent of all these metals. This was borne out by the observation that most metal calces are reduced to their metals by heating in inflammable air, which they absorb. Inflammable air was also known to be evolved when metals were heated with steam. Moreover, Kirwan noted that the same inflammable air was evolved by the action of alkalis on certain metals and he thought it must be free phlogiston made fluid by combination with elementary fire. He was also interested in chemical affinity and many of his arguments depended on apparent contradictions between reactions postulated by the oxygen theory and the regular order of affinities.

Kirwan's arguments were refuted by Guyton de Morveau, Lavoisier, Laplace, Monge and Fourcroy in a French translation of his book published in 1788. By then, Kirwan, aware that the number of chemists who opposed phlogiston in England was continually increasing, had himself conceded Lavoisier's reversal of the old hypothesis. It was Kirwan who coined the term 'antiphlogistic' as an appropriate name for the new doctrine. William Higgins, who claimed to have accepted the new theory from 1785, published *A Comparative View of the Phlogistic and Antiphlogistic Theories* in 1789, in which he showed that the antiphlogistic theory was capable of explaining all Kirwan's previous objections. There were, of course, men like Priestley, Watt, Keir and others who never abandoned phlogiston, though Cavendish is said to have done so in later life, after he had ceased to publish on chemistry. Kirwan's initial support for phlogiston was important too, because it influenced those German chemists who resisted the antiphlogistic theory, but even in Germany, there were some early converts to Lavoisier's system. Klaproth, Bucholz, Girtanner, Hermbstädt, Trommsdorff and A.N. Scherer, adopted it by about 1785, but reception of the new theory was slower in Germany than elsewhere. Fourcroy distinguished those German chemists who accepted the new theory completely from those like Crell, Westrumb, Richter, Göttling and Gren, who adopted some combination of the old and new theories.

Among those who opposed Lavoisier's theory in France, perhaps the most important was De la Metherie who was the editor of *Observations sur la physique* (later renamed *Journal de physique*) from 1785. It was in this journal that Lavoisier had published many of his early papers. Influenced by Priestley's experiments, De la Metherie thought that all combustible bodies, including the metals and diamond, contained inflammable air which he identified with phlogiston, a form of matter which could be transferred from one substance to another. The water formed by burning hydrogen in oxygen he regarded as being separated from the two gases with which it was previously combined. As the editor of the *Journal de physique* was opposed to the oxygen theory, papers by supporters of the theory were not accepted and in 1789, at the suggestion of Adet, Lavoisier and others founded the *Annales de chimie*, a new journal in which papers based on the oxygen theory could be published.

15.7.3 Sweden

Little historical research has so far been done on the reception of antiphlogistic chemistry in Sweden. It seems that the new ideas were not entirely unknown, though those conversant with them belonged neither to the industrial chemists nor to the universities (Lundgren, in Donovan, 1988, p.160). Peter Nicolas von Gadda, a civil servant and an amateur chemist elected to the Royal Swedish Academy of Sciences, was aware of the new ideas by the end of the 1780s, but he did nothing to promote them. Carl Axel Arrhenius, an army officer, heard about Lavoisier's new chemistry while in Paris in 1788, but being involved in military matters he too did not pass on what he had heard. The first published account of Lavoisier's new chemistry in Sweden came in an article on the current state of chemistry in 1795 by Anders Gustaf Eckberg, assistant professor of chemistry at Uppsala, but it was Gadolin, who had been a student of Bergman at Uppsala in 1778, who first published a full text-book version of the oxygen theory in Swedish. Gadolin, best known for his discovery of yttria, a rare earth in gadolinite, a mineral discovered at Ytterby near Stockholm in 1787, began to teach chemistry at Åbo, in Finland (Åbo was then under Swedish administration; it is now called Turku) in 1785. He adopted the oxygen theory from 1789 and after travelling in Germany, Holland, England and Ireland, returned to Åbo as professor of chemistry in 1797. His text-book, published in 1798, was the first in Swedish to explain the new system. Gadolin later published an important dissertation on affinity and a work on chemical mineralogy, showing the same interest in the practical applications of chemistry as had inspired Bergman and Scheele.

15.7.4 Germany

In Germany most chemists continued to think in terms of phlogiston, and by 1789 the Chemical Revolution had not even begun there; not a single advocate of Lavoisier's oxygen theory had yet appeared (Huffbauer, 1971, p.127). For the practical uses of chemistry in mining and metal extraction, so important in Germany, the phlogiston theory was quite adequate. There was little other chemical industry during the eighteenth century, and educational courses in the subject were not available until the 1790s in German universities. Gren taught chemistry at Halle after 1788; Hermbstädt was at Berlin from 1791; and Bucholz was professor of chemistry at Erfurt from 1799 where Trommsdorff was professor of physics. But it was not until the nineteenth century that German chemical education flourished. Yet, German chemists in the eighteenth century generally thought of their country as the homeland of chemistry. This outlook was fostered by Lorenz Crell, first as editor of the *Chemische Journal* 1778–81 and later as editor of the *Chemische Annalen*.

The two commonest notions about phlogiston in late eighteenth-century Germany were either that it was identical with inflammable air, as Cavendish and Kirwan had thought, or that it was a form of matter with negative weight, as Gren had suggested. As fresh discoveries rapidly followed each other from Sweden, Britain and France, German chemists began to fear that they were losing their predominance, and strenuous efforts were made to bolster the phlogiston theory. In 1787, for example, Crell wrote in the *Annalen*, 'My dearest wish was ... to work to the best of my ability to maintain the chemistry of our Fatherland ...' (in Schneider, 1989, p.16) and three years later in 1790 Gren said, 'Up till now I have tried to support the absolute weightlessness of phlogiston with several proofs ... I think that through this I have saved Stahl's German Phlogiston Theory' (in Schneider, p.16). Lavoisier's *Traité* was not translated into German until 1794; its introduction into Germany was then heralded by G.C. Lichtenberg in Göttingen with a scornful critique of French science: 'France is not the country from which we Germans are accustomed to expect lasting scientific principles. Short-lived marvels are the norm, which past experience has taught us to expect' (in Schneider, p.14).

Looking back in 1799, A.N. Scherer blamed nationalistic prejudice and indolence for the persistence of phlogiston, as there had been plenty of opportunities to study the new theory. By 1787 all of Lavoisier's important articles had been published in German, either in Crell's *Chemische Annalen* or elsewhere, but German chemists had regarded Lavoisier's new system as little more than a French fad. Some blamed Lavoisier's own slowness to publish a complete account of his system for the German failure to recognize that he had developed a coherent and powerful alternative to phlogiston. Between 1783 and 1785 C.E. Weigel, professor of chemistry at Greifswald, translated and published Lavoisier's minor papers in German, but just as Lavoisier began his systematic attack on phlogiston in 1785, Weigel ceased publishing his papers and for seven years until 1792 the new developments in French chemistry were not brought to the general notice in Germany. From the mid-1780s, however, Crell published letters from correspondents like De la Metherie and Kirwan urging German chemists to investigate the oxygen theory and give a judgement on it. At least three German chemists mentioned Lavoisier's denial of phlogiston by 1787. Gren and Westrumb recognized his importance as a chemist in 1786 and Ingenhjousz mentioned Lavoisier's challenge to Stahl's theory in his *Expérience sur les végétaux* (1787).

Although German chemists became aware of the French theory, it was not seen as a serious threat until the beginning of 1789. In that year, however, there appeared an account of the new chemical nomenclature in Crell's *Annalen* and a summary of the oxygen theory in Göttling's *Almanach*. Göttling was one of the first German chemists to take Lavoisier's work seriously, but after 1789 other German chemists became concerned about the new theory and two opposing camps began to form with phlogistonists still in the majority and a smaller number who were prepared to accept the antiphlogistic view. It required courage to abandon Stahl's doctrine and the antiphlogistonists were criticized for supporting French ideas. There was also an element of conservatism based on a more empirical approach among German chemists; older chemists especially felt more comfortable with the traditional theory and were afraid of displaying too much partiality for innovation. Crell also remarked that younger chemists brought up on the new theory would still have to learn the phlogistic system to understand the work of older chemists. Thus the influence of the phlogiston theory was maintained, especially in Protestant areas of Germany where there was a strong nationalistic feeling of community among chemists and a desire to conform to the traditional views of the phlogistonist majority.

Until the summer of 1792 Crell, Westrumb and Gren continued to defend the doctrine of phlogiston in the pages of Crell's *Annalen* and Gren's *Journal der Physik* to which they had ready access. Antiphlogistonists failed to argue their case in the pages of these journals thus leaving the most effective channels of communication among German chemists in the hands of the traditionalists. The main protagonists of the oxygen theory, J.A. Scherer, Girtanner and Hermbstädt chose to spread information about Lavoisier's new theory in monographs, new texts and translations. Hermbstädt persuaded F. Wolff to translate F.L. Schurer's defence of the oxygen theory in 1790, the French antiphlogistonist critique of Kirwan's *Essay on Phlogiston* in 1791 and J.A. Chaptal's *Élémens de chimie* in 1791/92. Meanwhile he himself published a new text presenting both systems in 1791 and a translation of

Lavoisier's *Traité* in 1792. Thus by the summer of that year there was plenty of opportunity for German chemists to become aware of the oxygen theory, and a controversy developed between Hermbstädt on the one hand and Gren, Westrumb and Trommsdorff on the other. Hermbstädt marshalled arguments and evidence for antiphlogistic chemistry against the critiques of Gren and Westrumb, but there was little discussion of the wider issues.

Instead the controversy centred on the crucial question of whether the red calx of mercury contained the basis of vital air (oxygen) as Lavoisier had stated. It was thought that reduction of the red calx would reveal its composition, and both sides agreed that the outcome of such experiments would determine the fate of the antiphlogistic system. The main problem for the phlogistonists was to explain how the red calx of mercury could change into metallic mercury in a closed vessel without access for phlogiston. There were two methods of preparing the mercury calx, which were thought to result in different products. On the one hand it could be obtained with some difficulty by heating metallic mercury for long periods in air. This yielded *mercurius calcinatus per se*. On the other hand, it was easier to dissolve mercury in nitric acid and obtain the red calx by heating the nitrate. The product of this method was usually contaminated with some nitrate, which was also known to yield oxygen on heating, so any oxygen evolved by heating this form of the red calx might have come from the nitrate rather than the calx.

In 1791 Westrumb and Crell heated *mercurius calcinatus per se* and, like Gren, failed to obtain any oxygen, though some water vapour was produced. Gren carried out a single reduction experiment in 1791; using red hot calx he obtained some water vapour. Wiegleb gave support to Gren's claims about *mercurius calcinatus per se* and thought that while the outcome of the reduction experiment had not yet been settled, Gren's results could well be correct. It seems almost incredible that such experienced chemists should have failed to obtain oxygen by these simple methods, whereas the same experiments carried out in England, France or Sweden were always successful.

Discussing the objections of German chemists to the oxygen theory in an article published in *Annales de chimie*, Berthollet charged Gren with harbouring nationalistic bias and sectarian prejudice and told young chemists wishing to test the theory for themselves to prepare oxide of mercury using nitric acid and reduce it at high temperature. In the same year Westrumb wrote to Gren agreeing with the view that water contained in calces was the source of gases evolved in reduction and claiming that the reduction of mercury calx made with nitric acid proved nothing. Gren published Westrumb's letter in his *Journal der Physik* along with a letter from the Belgian antiphlogistonist van Mons who had repeated the experiment more than ten times using dry mercury calx and obtained vital air (oxygen) every time.

The argument dragged on until in 1793 Gren finally abandoned the phlogiston theory, though not phlogiston itself. Instead he adopted a compromise proposed by J.B. Richter in which all the facts supporting the oxygen theory and most of its doctrines were accepted, but the concept of phlogiston was retained as matter of light, assumed to be present in every substance which reacted with oxygen. In the same year Westrumb also admitted to having changed his mind about the reduction of mercury calx, but he was unwilling to follow Gren's lead and reverted instead to his first idea that every calx is a compound of an earth and water. After this, while he remained a phlogistonist he took no further part in discussions about theories but turned his attention to the practical applications of chemistry in pharmacy, mineral water analysis and various crafts. Trommsdorff also withdrew from the argument and declared himself neutral. Thus by the autumn of 1793 three of the most active protagonists of phlogiston in Germany ceased their active opposition to the French chemistry: Gren adopted a compromise which rested on Lavoisier's system, but retained phlogiston in a marginal role; Westrumb turned away from theoretical debate; and Trommsdorff took up a positivist position. Resistance to the antiphlogistic revolution in Germany was over.

15.8 The legacy of the Chemical Revolution

The introduction of the oxygen theory together with rationalization of other chemical ideas established the fundamental principles of modern chemistry, laid the foundations for the development of chemical theory and facilitated the rapid advances which marked the early years of the nineteenth century. Lavoisier's definition of the chemical element was an essential foundation for Dalton's atomic theory, which in turn led to the laws of constant composition and multiple proportions. Chemical analysis also depended on the new definition of the chemical element. Moreover, the oxygen theory was a unifying force in chemistry. Unlike phlogiston, oxygen was a chemical entity, itself one of the elements which could be identified and isolated. Its role in combustion was both experimentally demonstrable and quantifiable. Thus, from being a subject influenced by tradition or national prejudice, chemistry gradually became a unified study based on fundamental principles which could be verified by experiment.

Chemists working in different European countries could now communicate on a common basis, publishing their results in the same international journals without risk of misunderstanding due to differences of definition or interpretation. This brought in a more significant change than almost any other aspect of the Chemical Revolution, and by 1800 chemistry had become an international science throughout Europe. Yet had this been all, chemistry would still have remained basically similar to its eighteenth-century progenitor. More radical changes yet, arising from further discoveries which the new antiphlogistic chemistry was able to assimilate, were still to come crowding rapidly upon each other at about the turn of the century and in the first decade of the nineteenth.

In 1800 Alessandro Volta, investigating Galvani's concept of animal electricity, showed that when two different metals such as silver and zinc are placed in contact with a dilute solution of an acid an electric current is generated. Volta's discovery of current electricity not only changed ideas about the nature of electricity, but also suggested a new approach to the theory of chemical combination. The influence of the Chemical Revolution becomes evident when it is considered that had Volta's discovery been made a quarter of a century earlier, it is unlikely that its connection with chemical combination and reactions would have become apparent, but within a year of Volta's announcement in *Philosophical Transactions*, William Nicholson and Anthony Carlisle in England used the electric current to split water into its component elements, hydrogen and oxygen, by the process of electrolysis.

In 1801 Humphry Davy, who in the eighteenth-century tradition of British chemistry had been engaged in research on the therapeutic properties of gases at Thomas Beddoes's Pneumatic Institution in Bristol, came to the

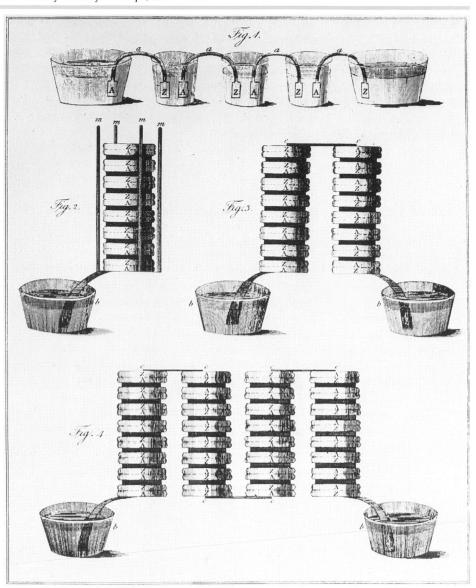

Figure 15.21 Volta's pile. With the discovery of current electricity from the simple cell a whole new era of chemical research began. New metals were discovered, new explanations for chemical combination were devised and the oxygen theory showed its superiority over older chemical theories. (Ann Ronan Picture Collection)

Royal Institution in London. Having learned of Volta's work and the discovery of electrolysis Davy demanded a large electric battery for research on the nature and properties of electricity. His experiments soon led to the invention of the electric arc-light, but in about 1807 he discovered several new metals, including sodium, potassium, strontium and barium. These discoveries vindicated Lavoisier's definition of the chemical element as the last point of analysis, for with the new analytical tool Davy showed that the alkalis and earths formerly regarded as elements (because there were no known means of analysing them further) could indeed be split up into simpler substances, new elements.

At about the same time John Dalton, the Quaker schoolmaster of Kendal, showed how the atomic theory, which in the seventeenth century had proved so valuable in physics, could also be applied usefully to chemistry. By thinking in atomic terms, chemical reactions need no longer be thought of as taking place between tangible quantities of material, but could be brought right down to interactions between the individual, invisible atoms themselves. It followed that each atom must exert a force of attraction or repulsion on other atoms and must also possess a weight. Dalton published his first atomic-weight tables in 1807, showing the relative weights of his atoms. Soon afterwards the French chemist J.L. Gay-Lussac showed that Dalton's indivisible atoms were incapable of explaining the reactions between volumes of gases. A single volume of a gaseous reactant yielded often two volumes of product and it seemed either that the observations were mistaken or the atoms of gaseous elements must after all be divisible. A solution to this problem was proposed in 1811 by the Italian physicist Amedeo Avogadro with his idea of diatomic molecules. There were however, too many doubts about this arising from other contemporary theories, not least Berzelius's idea that atoms carry electrical charges. All the atoms of a given element must have the same charge and as it was known that like charges repel, Avogadro's diatomic molecules seemed most unlikely. Avogadro could offer no experimental proof to support his hypothesis and it was to remain unaccepted by chemists for the next half-century.

Coupled with the idea that chemical reactions occur at the atomic level was the problem of how and why atoms combine. J.J. Berzelius in Sweden proposed a theory of chemical combination between atoms and radicals (groups of atoms). Supposing that each atom or radical carried either positive or negative electric charge, Berzelius was able to account for the combination of positively charged metals with negative oxygen and other non-metals or acid radicals. Furthermore, differences between the intensities of the electric charges would account for the relative stability of compounds and the displacement of one element or radical by another in chemical changes. Berzelius's electrochemical theory was first applied to inorganic and mineral chemistry, but it was later extended to the reactions of organic compounds and formed an important step towards the unification of chemistry in the nineteenth century.

Sources referred to in the text

Crosland, M.P. (1980) 'Chemistry and the Chemical Revolution', in G.S. Rousseau and R. Porter (eds), *The Ferment of Knowledge*, Cambridge University Press.

Donovan, A. (ed.) (1988) *Osiris*, 2nd series, vol.4. *The Chemical Revolution: Essays in Reinterpretation*, History of Science Society, University of Pennsylvania.

Hales, S. (1727) *Vegetable Staticks*, London.

Hankins, T.L. (1985) *Science and the Enlightenment*, Cambridge University Press.

Holmes, F.L. (1985) *Lavoisier and the Chemistry of Life*, Madison, University of Wisconsin Press.

Huffbauer, K. (1971) The Formation of the German Chemical Community (1700–1795), University of California, Berkeley, PhD thesis, Ann Arbor, Michigan, University Microfilms, ref. no. 71-789. Later published as *The Formation of the German Chemical Community (1720–1795)*, Berkeley, University of California Press, 1982.

Lavoisier, A.L. ([1790] 1963) *Elements of Chemistry*, tr. R.T Kerr, Introduction by Douglas McKie, Edinburgh, Dover.

Partington, J.R. (1962) *A History of Chemistry*, vol.3, London, Macmillan.

Schneider, H.G. (1989) 'The "fatherland of chemistry": early nationalistic currents in late eighteenth century German chemistry', *Ambix*, 36, pp.14–21.

Further reading

Coley, N.G. (1985) 'Chemistry to 1800', in C.A. Russell (ed.), *Recent Developments in the History of Chemistry*, London, Royal Society of Chemistry.

Crosland, M.P. (1980) 'Chemistry and the Chemical Revolution', in G.S. Rousseau and R. Porter (eds), *The Ferment of Knowledge*, Cambridge University Press.

Donovan, A. (ed.) (1988), *Osiris*, 2nd series, vol.4. *The Chemical Revolution: Essays in Reinterpretation*, History of Science Society, University of Pennsylvania.

Fox, R. (1971) *The Caloric Theory of Gases: From Lavoisier to Regnault*, Oxford University Press.

Gillispie, C.C. (ed.) (1970–76), *A Dictionary of Scientific Biography*, 16 vols, New York, Charles Scribner's Sons. On Bergman, see vol.2, pp.4–8. On Cavendish, see vol.3, pp.155–9. On Geoffroy, see vol.5, pp.352–4. On Lavoisier, see vol.8, pp.66–91. On Scheele, see vol.12, pp.143–50.

Guerlac, H. (1975) *Antoine-Laurent Lavoisier, Chemist and Revolutionary*, New York, Charles Scribner's Sons.

Hankins, T.L. (1985) *Science and the Enlightenment*, Cambridge University Press, ch.4.

Siegfried, R. (1989) 'Lavoisier and the phlogistic connection', *Ambix*, 36, pp.31–40.

Conclusion

Chapter 16

by David Goodman

In the period 1500–1800, Europe, or rather parts of Western Europe, became the world leader in science. Seville, Padua, Florence, Paris, London, Amsterdam, Uppsala and Edinburgh were the new sparkling centres of scientific activity. China and the lands of Islam, once in the van of scientific development, were now a long way behind Europe, their impressive mediaeval phases of scientific activity now extinguished. Europe alone had experienced the Scientific Revolution, and it was from there that science in its modern form would gradually spread to other parts of the globe. What had made Europe such a fertile nursery for modern science? Two aspects of European civilization were of special importance: a distinctive political structure and the distinctive mentality of its citizens.

16.1 European towns and science

In sharp contrast to China and the Muslim empires, where vast land masses were ruled from a single political centre, Europe was split up into several competing kingdoms, the consequence of the disintegration of the Western Roman Empire. Within Spain, France, England and, most of all, in Italy and the German-speaking lands there were numerous towns. Some of these towns had a continuous existence since the time of the Romans (Seville, Pisa, Milan, London); over one thousand others were founded in the Middle Ages (Amsterdam, Groningen, Marburg, Berlin, Prague, Danzig, Cracow).

It was not merely the quantity of towns which distinguished European society, but their quality. From the eleventh century on, Europe's towns had acquired degrees of self-government (for example many towns in Northern Italy; London in the twelfth century) which gave them notable privileges, such as exemption from direct royal taxation, control of local commerce and legal concessions. All of this meant that the towns of Western Europe were places of greater freedom than surrounding rural areas, that there was a division between two societies: urban and rural, sharply defined by the limits of town walls. Nowhere else in the world were these autonomous towns to be found; they were most numerous in the west of Europe, quickly becoming fewer and fewer east of the river Elbe. Their freedoms fell far short of democracy; these towns were oligarchies, ruled often by rich merchants, like the Medici of Florence who rose from merchants and bankers to become autocratic dukes. The distinctive environment of Europe's towns was conducive to the rise of modern science.

Sociologists used to argue that cities were bound to stimulate intellectual activity because of their concentrated populations. Cities were supposed to contain a higher proportion of individuals with genius and inventiveness than rural hamlets and villages, and to enable 'a more intensive exchange of

415

experience which is likely to result in a more rapid accumulation of knowledge and mental progress' (Sorokin, 1928, pp.388–9).[1] But the size of urban populations was not sufficient by itself to generate scientific activity in the towns of early modern Europe. Constantinople shows that. This was the largest city in Europe, perhaps 700,000 inhabitants at the end of the sixteenth century; yet it was intellectually drab and not a centre of scientific creativity. There was a flourish of literary and artistic activity in the final century of Christian rule, under the Palaeologi emperors, but no original scientific work was produced. And the same absence of scientific creativity continued in the centuries after Constantinople had fallen to the Turks. Constantinople was an 'urban monster' whose population devoured the products of the vast Ottoman Empire and the luxury goods of the West; 'in return the city gave nothing' (Braudel, 1972, vol.1, pp.347–51). It was the seat of imperial administration; its huge population enjoyed none of the benefits of autonomous government and displayed little of the intellectual originality of much smaller towns in Western Europe.

The unique complexion of Western Europe – multiple centres of political authority and a multitude of autonomous towns – generated fierce competition and the capitalist system. Trade, long-distance as well as local, was at the very heart of the life of many of Europe's towns. The most successful merchants became dominant in urban government. Some have seen this rise of the merchant class as the key to the explanation of Europe's Scientific Revolution. While China despised merchants,[2] putting them at the bottom of the social ladder, below farmers and artisans, and Islamic societies gave them decreasing prestige, Europe alone witnessed their social ascent (Needham, 1969, pp.184–8). The alleged importance of merchants for scientific development is that they are the people who 'put up the money for scientific discovery', financing research 'in order to develop new forms of production' (Needham, 1970, p.82). And even more intimately connected with the fabric of science itself, merchants 'bring together manual and mental work' (Needham, 1969, p.186), seen as the two essential components of modern science: the manual skills associated with setting up a scientific experiment, and the complementary intellectual processes of theory construction and quantification. Merchants certainly used numbers and simple calculations as much as anyone in their everyday business, and some of them were involved with manufacturing processes in which craftsmen worked with apparatus. Nevertheless it remains conjectural whether these aspects of Western capitalism really provided the rudiments of modern science, and difficult to see how merchant activity could lead to the generation of scientific theories.

But in other ways trade was intimately connected with Europe's rising science, as several chapters in this book have indicated. It was a wealthy merchant of the city of London, Sir Thomas Gresham (*d.* 1579), who founded and endowed the main centre of scientific activity in Elizabethan England. Capital from his estates, managed by the corporation of London and the Mercers' Company, provided the salaries of lecturers appointed to teach scientific subjects. But the lasting importance of Gresham College was that it was one of the seeds of the Royal Society. The early years of the Royal Society also show a marked interest in manufacturing processes, with the adoption of

---

[1] A survey of some of these theories is given in Sorokin (1928, pp.409–12).

[2] This is now denied. It is alleged that a merchant class arose in China in the eighth and ninth centuries, and acquired such influence that, by the time of the Sung dynasty, some historians believe the conditions were then 'ripe for the emergence of a modern capitalist society' (Twitchett, 1979, pp.26–31).

the Baconian programme of a 'history of trades', in the expectation that an investigation of the techniques of dyeing and other crafts would lead to scientific advances. Other chapters have pointed to notable scientific developments stimulated by capitalistic mining enterprises. The Fuggers of Augsburg, amongst the richest of European merchants, had invested their wealth from the late fifteenth century in the silver mines of Carinthia and the Tyrol. It was as an apprentice in the Fugger mines of Schwaz, and a student at the mining school set up by the Fuggers at Villach, that Paracelsus formed the strong chemical and mineralogical interests which would be so conspicuous in his influential alchemical philosophy. And in seventeenth-century Sweden commercial exploitation of iron and other valuable mineral resources caused the spectacular success of Swedish chemistry, leading to the discovery of new elements and the development of quantitative analysis.

The discussion in several of our chapters leads to the conclusion that of all forms of capitalism it was long-distance, maritime trade that most stimulated the rise of modern science in Europe. The voyages of discovery, at least in part, were motivated by the search for new sources of wealth. And those oceanic voyages came increasingly to depend on the art of navigation, which had the closest of connections with science. This development is most clearly seen in sixteenth-century Seville. The merchants of that city had been granted a monopoly of trade with the New World. To supervise the loading of their cargoes, the provisioning of ships for the Atlantic crossing, and the issuing of nautical charts and reliable observational instruments, were the duties of a special crown institution, the House of Trade. Map-making, scientific-instrument manufacture and astronomy applied to finding position at sea were all developed. From the sixteenth century French, English and then Dutch ships strove to enter the waters of the Iberian Empire in the Atlantic and Pacific, and break the monopoly which Portugal and Spain declared in the lucrative trade with their colonies in the Indies. All of these competing maritime powers fostered sciences associated with navigation. In sixteenth-century England William Gilbert's research on magnetism, one of the impressive experimental studies of the Scientific Revolution, was inspired by his interest in navigation. Part of the teaching at Gresham College was devoted to the dissemination of nautical science, and English Copernicanism was closely associated with strong navigational interests. And in the seventeenth century the Royal Observatory at Greenwich was established to provide astronomical data for perfecting navigation. In France the Académie Royale des Sciences was created partly for the same reason, and in the hope that its members would solve the intractable problem of determining longitude at sea.

All of these hectic navigational studies may have generated a heightened awareness of the need for accuracy in the scientists of the time. But a still more profound influence on science may well have come from the results of the voyages of exploration. The discovery of the New World with its remarkable flora and fauna shook the confidence of intellectuals who had put their trust in the ancients. The record shows that there were individual Portuguese and Spaniards who felt a keen sense of achievement in voyaging to regions unknown to the ancient Greeks and Romans, and making direct observations which contradicted classical authorities or exposed the limits of their knowledge. The Scientific Revolution was very much to do with the independent investigation of nature, freed from uncritical acceptance of the teachings of Aristotle, Ptolemy and Galen, and it seems reasonable to attribute some of this intellectual independence to the recent voyages of discovery.

At the same time that European merchants and bankers were financing the oceanic voyages, considerable capital was also being invested in the new printing houses which sprang up in Mainz, Strassburg, Seville, London, Paris, Wittenberg, Venice and numerous other towns. It was another powerful way in which science was fostered by Europe's peculiar urban network. By the seventeenth century the first printed scientific journals greatly facilitated the communication of experimental results and tabular observational data, presenting them as never before to a wide community of scientists for quicker comment and criticism. And those journals, the *Philosophical Transactions* and the *Journal des savants*, were the publications of the scientific academies which spread through Europe from the seventeenth century, again within the dynamic urban environments of Florence, London, Paris, and later in Berlin, Uppsala and Stockholm. By the end of the eighteenth century there were some thirty of them in the provincial towns of France alone. From the start these academies had been dedicated to promoting the revolutionary new science. Their officials included men like Robert Hooke, specially appointed to demonstrate and verify experiments before a gathering of members. Nothing quite as organized had existed before, and this mode of concentrating scientific investigation was peculiar to Europe.

The towns were also the home of Europe's mediaeval universities. They had not been the creations of municipal corporations but of princes and prelates. But municipal involvement is clear in some of the newer universities founded from the fifteenth century, notably Louvain (1425) and Basel (1459). In Valencia (1502) the university was funded and run by the town's ruling merchant oligarchy. Louvain and Valencia were both important for the new anatomy of Vesalius. And Chapter 11 has shown how in eighteenth-century Edinburgh the town council was responsible for the success of the medical school. Irrespective of how they were created, the universities were very much part of town life.

The science taught in the universities continued for the most part to be based on the mediaeval curriculum of revered classical doctrines of Aristotle, Ptolemy and Galen. Occasionally glimmers of the new science lit university teaching in the sixteenth century: Copernicanism at Altdorf and Rostock; Vesalian anatomy and Paracelsian philosophy at Valencia. Although the universities were frequently criticized as sterile centres of bookish learning, this was where many of the leaders of the Scientific Revolution were educated. The tension between the old science and the new, between university and scientific academy, was stimulating and productive. The universities had preserved continuity of learning as well as fostering reason in debate. Their fossilized curriculum was attacked by some of their graduates, steeped in classical learning and armed with an acute logic.

These tensions occurred only in Europe, almost entirely in Western Europe. The further east of the river Elbe one travelled, the fewer the towns. This was the backward rural side of Europe: villages, peasants and serfs dominated by an exploiting nobility. There were few towns, nothing of the freedom of the autonomous towns further west, and no university east of Cracow. That helps to explain the scientific backwardness of the eastern half of Europe. It also goes some way to explaining the absence of a scientific revolution in the empires of Mediaeval China and Islam. Those vast regions also lacked the dynamism of autonomous towns, competitive merchant capitalism, universities and scientific academies of the Western type. There scientific achievement had depended instead on fragile patronage by individual emperors and caliphs.

None of the West's leading scientific innovators worked in isolation; one way or another they all profited from Europe's lively urban culture. Copernicus for much of his life lived east of the Elbe, in Frauenberg, a very small town founded in the late thirteenth century. He described his home as a remote part of Europe, and the town could have provided him with little scientific stimulus. But he did not suffer from that academic isolation which may have been the most important check on the progress of ancient Greek science. Copernicus had as a young man left the backwater of his home to study at the universities of Bologna and Padua. There, in urban Italy, he had entered the mainstream of European intellectual life, becoming familiar with classical cosmological theories and influenced by Renaissance neo-Platonism. Later, back in Frauenberg, came the fruitful meeting with Rheticus, who had travelled from Wittenberg, a university town and the birthplace of Luther's Reformation. The importance of this urban culture for Copernicus is also apparent in the link with Nuremberg: it was to that lively German town of merchants, artisans and printers that he sent his manuscript *De revolutionibus* for publication, so effectively communicating with scholars throughout Europe.

16.2 Mentalities

To refer to prevailing beliefs within a particular society about the world and the general outlook, French historians introduced the term *mentalités*. And the translation 'mentalities' will fit the very broad set of attitudes and beliefs discussed in this section.

Because of the multiplicity of centres of political authority, there was a keen competitive spirit amongst early modern Western Europeans, which was not apparent in the civilizations of China and Islam. This competitive attitude was shown in capitalistic trade, colonization of new lands and the rivalry between Italy's city-states. Did this competitive spirit have any importance for the rise of modern science? It mattered in that scientific advances resulted from voyages of exploration, and that princely patrons vied to secure the services of the most talented men of science. And in Renaissance Italy individual scholars and artists seem to have been motivated by the need to excel in competition with others. Perhaps something of that can be seen in Galileo's career: his early ambition to achieve enough renown to secure an appointment at the prestigious University of Padua; his successful construction of a telescope, when news was circulating of its invention, and the way he used it to attract the attention of his future employer, the Grand Duke of Tuscany. He also competed with numerous other men of science for the prize offered by the king of Spain for a solution to the problem of finding longitude at sea. There was rivalry too between the new scientific academies in London and Paris. And bitter feelings when individual scientists believed their discoveries had been stolen and announced by others. The Jesuit astronomer Christopher Scheiner was infuriated that his publication on sunspots was followed by Galileo's *Letters on Sunspots* with a preface claiming priority of discovery – the preface was due to the Lincean Academy and Galileo himself is said not to have liked it. And in the same century the dispute between Leibniz and Newton over the invention of the calculus was long lasting.

But collaboration, the opposite of competition, was just as important for modern science. Bacon's ideal scientific institution, the Solomon's House described in his *New Atlantis*, and to some degree imitated by the Royal Society, portrayed scholars and craftsmen bringing their talents and skills together in a joint effort to penetrate nature's secrets all the more effectively. National sentiment amongst Germans in the late-eighteenth century does seem to have played a part in defending 'the phlogiston theory of the Fatherland' from the oxygen theory of Lavoisier (see Chapter 15). But it would be a mistake to present early modern science in terms of national rivalry. The 'republic of letters', a European intellectual community, would be a more accurate image for the period. As other chapters have indicated, the circle of natural philosophers in London around Oldenburg interacted with a similar French circle around Montmor. And the correspondence between natural philosophers organized by Mersenne covered a wide international network. There were no obstacles from narrow nationalism when Nicolas Le Fèvre left the Jardin du Roi to take up a royal appointment in London as professor of chemistry to Charles II; nor when later in the same century another Frenchman, Denis Papin, became successively curator of experiments at the Royal Society, and professor of mathematics at the University of Marburg. And when in 1736 Maupertuis went on the scientific expedition to Lapland organized by the Académie Royale des Sciences, some of the scientific instruments came from England, and Celsius, a Swede, was a member of the expedition. Soon after his return Maupertuis, a Frenchman, was appointed president of the Prussian Royal Academy of Sciences. In 1735 French and Spanish scientists collaborated in an expedition to Peru to make geodetic measurements. And in 1798–99 the world's first international scientific conference assembled in Paris to consider the introduction of the metric system. The government of revolutionary France had sent official invitations to various countries calling for expert assistance; in the end the standard metre and kilogram were fixed by the collaborative efforts of eight Frenchmen and eleven foreigners from Denmark, the Netherlands, Switzerland, Italy and Spain (Crosland, 1969).

The sharp difference in levels of scientific achievement in Western and Balkan Europe is explicable by a difference of mentality. In Western Europe from the twelfth to the fifteenth centuries the educated had felt an acute awareness of a lost classical heritage, and many were strongly motivated to try to recover it. That resulted in an intellectual drive which was not experienced in Balkan Europe, where Greek learning had survived and no loss felt. The neo-Platonism of the Renaissance was one of the fruits which stimulated modern science in the West.

There were also striking differences in mentality amongst Europeans, and between them and the peoples of the Near and Far East, attributable to the powerful conditioning of different religious beliefs. A Muslim historian of science is convinced that Christianity and Islam had totally different influences on the cultivation of natural science because of the varying value the two religions gave to reason:

> Christianity is essentially a way of love and its mysteries remain forever veiled from the believer, at least in the ordinary interpretation of the religion. Islam, on the contrary, is gnostic, and its final aim is to guide the believer to a 'vision' of the spiritual realities. That is why there has always been in Christianity the question of preserving the domain of faith from intrusion by reason, whereas in Islam the problem has been to overcome the obstacles placed before the intelligence by the passions in order to enable the believer to reach the very heart of the faith which is the Unity of the Divine Essence. *(Nasr, 1978, p.7)*

From this the author concludes that the Muslim faith's encouragement of reason explains the intense study of the mathematical sciences in the mediaeval Islamic world, while in Christendom any 'thirst' for rational explanations and causality had to be satisfied 'outside of the faith'. This misrepresents the relations between science and Christianity. It is true that in the mediaeval Latin West a tension between faith and reason was produced by the recovery of the pagan philosophy of the ancient Greeks. Some clergy had reacted by issuing bans on the study of science and philosophy. But other clergy looked to Aristotelian logic to elucidate Christian mysteries, and by the end of the Middle Ages this was the view which prevailed, a harmonious synthesis of Christian belief and Aristotelian philosophy. Far from being cast out from the camp of the faithful, reason acquired an exalted status in the Latin West, as is clear from the extraordinary importance attached to logic in the universities, where there was a strong, and sometimes controlling, ecclesiastical presence. The search for rational explanations of natural phenomena by Europeans, already developing in the later Middle Ages, was accepted as entirely compatible with Christian belief, even leading to a deeper appreciation of the greatness of God through the discovery of the natural world he had planned and created. That Christian mentality would continue in an unbroken tradition through the early modern period; the positive stimulus Christian beliefs gave to the sciences is conspicuous in many chapters of this book.

But in Eastern Europe, in the Balkans and the Russian Empire, a different version of Christianity, Orthodoxy, prevailed; and there the effect was to discourage the cultivation of natural science. Orthodoxy emphasized miracles and mysteries; the idea of the rationality of God's creation was absent in Russian Orthodoxy, and the use of reason discouraged by the stress on human ignorance (see Chapter 13). In the Orthodox Church of the Middle Ages, unlike the Catholic West, there had been nothing of the harmonious blend with Aristotelian philosophy. Right up until the eighteenth century the Russian Church remained hostile to the 'heretical' modern science of Western Europe and used its powers of censorship to prevent it entering the Russian Empire. That goes some way to explaining the slow scientific development of Russia. And similarly for Balkan Europe, where religious hostility to Western science was intensified after the Ottoman conquests. The Ottoman Turks, the dominant Islamic power throughout the early modern period, in sharp contrast to the mediaeval caliphs, gave no patronage to the sciences. Instead they minimized contacts with the West, preferring even to leave essential commerce to the Jewish inhabitants of the Ottoman Empire, so avoiding pollution through contact with the inferior infidel. And until the eighteenth century, apart from their employment of Jews migrating from the West, as court physicians in Istanbul, and the imitation of Western artillery, the Ottoman Turks turned their backs on Western science and technology as valueless compared with the glory of the Koran.

Further east, the peoples of China for over two millenia had lived in a society uninfluenced by belief in a Supreme Creator. Chinese philosophers over a considerable time span thought of the natural world as eternal, uncreated and inscrutable. The Taoists, empirical natural philosophers, are said by Needham to have distrusted reason. The most serious limitations in Chinese science, the failure to develop the idea of laws of nature, can plausibly be attributed either to the lack of belief in a planned creation or to Chinese antipathy to rational legal codes.

Belief in a monotheistic, rational Creator was not indispensable for conceiving nature in terms of laws – the ancient Greeks had shown that. But the importance of that belief in the Scientific Revolution of the sixteenth and seventeenth centuries is inescapable. The entire scientific thought of Kepler and Newton was inspired by the belief in a rational creation whose planning is discoverable by the human intellect. Their momentous discoveries of fundamental physical laws were the fruit of a religious view of nature. And the same applies to Robert Boyle, Mersenne and numerous other natural philosophers of the period.

But Merton's thesis that a particular type of Christian belief, Puritanism, was the motor behind English scientific advance in the seventeenth century remains controversial, partly because of the impossibility of defining 'Puritanism' with the requisite precision. The alleged Puritan values – hard work; independence of authority; the empirical search for useful knowledge to alleviate human suffering; the pursuit of natural science to reveal the goodness, wisdom and power of the Creator – all had a wide currency in seventeenth-century England and do not seem to have been the preserve of a single group. The Royal Society was not a Puritan club.

Nor is it easy to sustain the corollary to Merton's thesis that Roman Catholicism was an impediment in the centuries of the Scientific Revolution. In France and Italy Catholic natural philosophers, notably Descartes and Galileo, played major roles in bringing about the revolution in science. Strict censorship of scientific work and a reluctance to part with Aristotelianism were not peculiar to Roman Catholicism; both are apparent in the Lutheran Church of eighteenth-century Sweden (see Chapter 12). But with the stiffening of Roman Catholicism in its Counter-Reformation phase, especially after the trial in Rome of Galileo, it is difficult to avoid the conclusion that the attitudes of Catholicism in Italy, Spain and Portugal were a hindrance to independent scientific research.

The new science of the sixteenth and seventeenth centuries had regenerated the tensions between faith and reason, temporarily resolved by the mediaeval synthesis of Aristotelianism with Christian belief. This characteristically Western tension between the secular and the ecclesiastical, entirely absent in Chinese society and much less developed in Islamic society, could have a stimulating effect on science. An example of that is the influential cosmological theory of Tycho Brahe, a world picture invented to incorporate some of the features of the new astronomy of Copernicus with the traditional belief, scriptural and Aristotelian, in a stationary central Earth. From the tension came the idea of two books, the Book of Nature and the Bible, two paths to truths which could not be in conflict, because they both derived from God. But the success of this doctrine depended on how much freedom was given by the prevailing ecclesiastical authorities in the interpretation of the Scriptures. Galileo was allowed none and paid the penalty.

The whole question of why modern science was generated only in Europe leads to a consideration of the distinctive structure of European society and its distinctive intellectual traditions. And that was very much a creation of the Middle Ages: the establishment of autonomous towns with their universities and merchant capitalism; the intellectual life stimulated by the determination to recover a lost classical heritage; the Christian tradition powerfully affected by the tensions between faith and reason. All of these forces were still influential in the period of this book. On this view the conditions which favoured the rise of modern science in Europe were seeded centuries before in the Middle Ages. And the spark which triggered off the Scientific Revolution, causing it to occur when it did in the sixteenth and seventeenth centuries, may well have been the oceanic voyages of discovery.

16.3 Consequences

The view of the world in 1800 looked very different from that in 1500. The Earth was no longer the immobile centre of a finite universe, but a planet hurtling through celestial space, in orbit around the Sun, and subject to the universal laws of mechanics. But for the great majority of European scientists God remained important; the universe may have been a machine, but was invented and maintained by the Creator. Knowledge of the earth had grown enormously: new continents, unknown seas, accurate coastlines, precise latitudes, and finally accurate longitudes transformed the appearance of maps produced over these centuries. There was also a new understanding of the formation of the earth based on the study of its strata. And through the development of chemical techniques, the earth's material was found to be made up of many more elements than had ever been imagined. The variety of life on earth also seemed immeasurably richer after the discovery and colonization of the New World. The discovery of a multitude of marvellous exotic plant and animal species surpassed the mediaeval imaginings of mermaids, griffins and unicorns. And at the same time there was a revolution in the understanding of what had seemed to be the most familiar of all forms of life: humans. The anatomical parts of humans were investigated and more accurately portrayed than ever before. And from the mid-seventeenth century scientists, some much more easily than others, came to accept the image of a circulation of the blood.

By 1800 the educated recognized that the revolution in science had been due to the development of a scientific method: the combination of theory, often expressed in formal mathematical terms, and experiment. That in turn implied a willingness to approach the natural world in a fresh way, no longer relying on what the Ancients had said, but observing and experimenting with independence. It was a boldness which brought the reward of exhilaration in discovery. Confidence grew; and if in 1800 that led some to believe that there was little left in the world for Newtonian science to discover, the next two centuries would show how that confidence was entirely misplaced.

Sources referred to in the text

Braudel, F. (1972) *The Mediterranean and the Mediterranean World in the Age of Philip II*, tr. S. Reynolds, 2 vols, London, Collins.

Crosland, M. (1969) 'The Congress on Definite Metric Standards 1798–1799: the first international scientific conference?', *Isis*, 60, pp.226–31.

Nasr, S. (1978) *An Introduction to Islamic Cosmological Doctrines*, rev. edn, London, Thames and Hudson.

Needham, J. (1969) *The Grand Titration: Science and Society in East and West*, London, Allen and Unwin.

Needham, J. (1970) *Clerks and Craftsmen in China and the West*, Cambridge University Press.

Sorokin, P. (1928) *Contemporary Sociological Theories Through the First Quarter of the Twentieth Century*, New York, Harper and Row.

Twitchett, D. (ed.) (1979) *The Cambridge History of China*, vol. 3, Cambridge University Press.

Index